Cutting Edge Robotics 2010

Edited by

Vedran Kordic

Cutting Edge Robotics 2010
Edited by Vedran Kordic

Republished by InTech
Janeza Trdine 9, 51000 Rijeka, Croatia

Additional hard copies can be obtained from orders@intechopen.com

Cutting Edge Robotics 2010,
Edited by Vedran Kordic
p. cm.
ISBN 978-953-307-062-9

For EU product safety concerns:
IN TECH d.o.o., Prolaz Marije Krucifikse Kozulić 3, 51000 Rijeka, Croatia,
info@intechopen.com or visit our website at intechopen.com.

Meet the editor

Vedran Kordic is the founder and Board member of IntechOpen. After obtaining his Master's degree in Mechanical Engineering he continued his education at the Vienna University of Technology where he obtained his PhD degree in 2004. He worked as a researcher at the Automation and Control Institute, Faculty of Electrical Engineering, Vienna University of Technology until 2008. His studies in robotics lead him not only to a PhD degree but also inspired him to co-found and build the International Journal of Advanced Robotic Systems - world's first Open Access journal in the field of robotics. Vedran's background and experience remain crucial for the company's development as he is continuously engaged with overseeing the IT department.

Preface

Robotics research, especially mobile robotics is a young field. Its roots include many engineering and scientific disciplines from mechanical, electrical and electronics engineering to computer, cognitive and social sciences. Each of this parent fields is exciting in its own way and has its share in different books.

This book is a result of inspirations and contributions from many researchers worldwide. It presents a collection of a wide range of research results in robotics scientific community. Various aspects of current research in robotics area are explored and discussed. We have tried to investigate the most important research areas of a really wide scope of robotic science.

We hope you will enjoy reading the book as much as we have enjoyed bringing it together for you. The book presents efforts by a number of people. We would like to thank all the researchers and especially the chapter authors who entrusted us with their best work and it is their work that enabled us to collect the material for this book. Of course, great acknowledgments go to the people who invested their time to review all manuscripts and choose only the best ones.

Contents

1

Motion Control of Robots Based on Sensings of Human Forces and Movements

Tao Liu, Chunguang Li, Kyoko Shibata and Yoshio Inoue
Kochi University of Technology
Japan

1. Introduction

1.1 Requirement for arm therapy and clinical background

The percentage of aged persons is continuously increasing in many countries, which is becoming a social problem demanding concern from different fields including social science, medical science and engineering. This trend is particularly rigorous in Japan where the aged (over 65 years old) accounted for 20.8% of the total population up to 2006 (Statistics Bureau, 2007). In the elderly, the prevalence of physical deterioration is sharply high, and their physical deterioration generally leads to degeneration of some motor functions. Besides, Hemiplegic limb impairment after a stroke which is a common disease among the aged, is becoming a global issue. Both motor function deterioration and disability have an indirect influence on brain degeneration. Thereby, strength enhancement and function recovery are necessary in the aging society. Moreover, physical therapy resources are quite limited, and the rehabilitation therapy places a large economic burden on patients. Under these conditions, considerable interest has been stimulated in the development of upper limb rehabilitation robots which can act as a therapeutic aid for therapists in rehabilitation training.

1.2 Related rehabilitation robots and problems

Among the numerous robots designed to deliver arm therapy, MIT-MANUS (Hogan & Krebs, 2004; Krebs et al., 2000), ARM-GUIDE (Reinkensmeyer et al., 1999 & 2000), and MIME (Burgar et al., 2000; Lum et al., 2002 & 2004) are three representative devices that have been tested extensively on hemiplegic patients. MIT-MANUS can support patients in executing reaching movements in a horizontal plane; ARM-GUIDE and MIME can give training in a three-dimensional workspace. ARM-GUIDE allows the subject to exercise against the gravity and can be used as a diagnostic tool and a treatment tool for addressing the arm impairment in hemiparetics. With MIME the limb's position can be inferred from the robot's position based on measurements of the interaction forces. It was verified that the subjects who received MIME therapy had statistically higher gains in arm motor function by having the both upper limbs execute movements that mirror one another. ARMin (Nef et al., 2007) is another representative robotic device which can deliver patient-cooperative arm therapy. However, these robotic arms are heavy in weight and must be fixed on walls and poles, so the motion space is limited and patients are easily to feel excess fatigue. Otherwise, these robots are too complex to set up by patients themselves, thus, they are not suitable for carrying out rehabilitation training at home (Zheng et al., 2006).

A home environment makes it possible to increase duration that patients spend in rehabilitation activities, thus it can ensure a high level of intervention with adequate intensity and frequency that can improve the motor recovery (van Exel et al., 2005). In addition, the home-based rehabilitation can reduce economic burden to a certain extent. Therefore, the development of wearable robots which can be easily used in patients' home is a new tendency recently. For example, a new human motion tracking system using two body-mounted inertial sensors to measure upper limb motion (Zhou et al., 2008) was developed for home-based therapy. In this system, motion tracking is implemented with a pure position control and a visual feedback but without a sensible force feedback, thus operators can not be well informed about the exact status of the impaired limb. Since interaction conditions between the robot and the patient can vary considerably depending on the patient's kinetic capabilities and unpredictable reactions to therapeutic stimuli (Reinkensmeyer et al., 2000), the security and reliability of the system can not be ensured.

A force assistant master-slave tele-rehabilitation robotic system (Li & Song, 2008), which realizes impedance transfer by means of force transducers, enables therapists to experience the interaction force between the robot and the impaired limb, and thus increases the adaptability of the system. The systems introduced by Song & Guo (2006) and Peng et al. (2005) are also capable of force feedback. However, the force feedback control in these robotic systems is realized with force sensors, which has the drawbacks of introducing control complexity (both the force control and position control are needed), high system cost, and mounting difficulty. Otherwise, the operators of the above robots are the therapists rather than the patients themselves. That is, the patient is trained passively. Even though the therapist can optimize the therapy scheme according to the feedback force, but the degree of comfort of the patients can not be sensed. This is unfavourable to acquire good recovery effect.

RoboWear (Jeong et al., 2001), a wearable robotic arm with high force-reflection capability, can be operated by the patient himself/herself, but it needs two pressure sensors to realize force-reflection. The system introduced by Gang & Shuxiang (2006) realized self-assisted rehabilitation, but the training program is based on a virtual reality environment and the system is only suitable for training mildly affected limbs.

1.3 Research aims

Working from the above realization, a master-slave control scheme utilizing the healthy limbs of hemiplegic patients is presented for home-based wearable rehabilitation robots. With this control system, the patient can directly feel the interaction force between the robot and the impaired limb without force sensors, and can make a timely and proper adjustment to input force of the healthy limb according to the reflected force as well as the degree of comfort of the impaired limb. Besides, the movement trajectory is controlled by the patient himself/herself, this can increase the patient's motivation and activity, and can further enhance the recover progress (Hogan et al., 2006; Jack et al., 2001). Moreover, the energy generated by the master site is transmitted to the slave site, which can realize a kind of energy recycling.

2. Methods and materials

2.1 Theoretical analysis: force sensing

The master-slave control system contains two identical DC motors with the master motor being operated by the healthy limb and the slave one behaving in the same manner as the master. The master motor as a generator powers the slave motor which works as an actuator, to rotate and support the impaired limb in rehabilitation activities. The equivalent closed-loop circuit of the master-slave control system is given in Fig. 1, in which M_1 and M_2 represent the master motor and slave motor respectively.

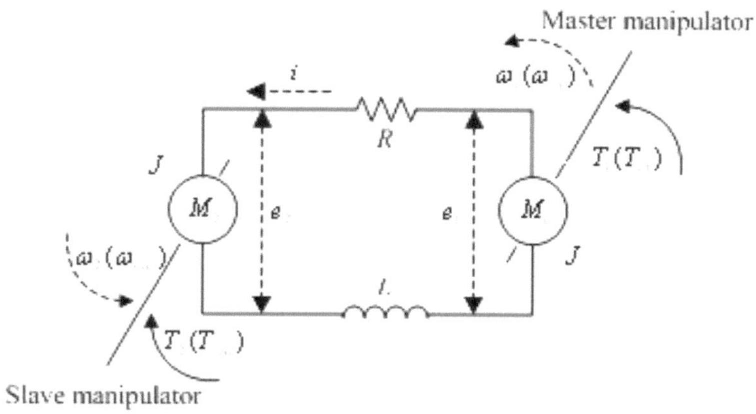

Fig. 1. Equivalent circuit of the master-slave control system

Based on the dynamics mechanism, the motion equation is written as

$$\begin{cases} T_1 = T_{M1} + T_0 \\ T_{M2} = T_2 + T_0 \\ T_{M1} = T_{M2} = C_T i \end{cases} \quad (1)$$

where T_1 and T_2 are the mechanical torques in the master and slave motor shafts, also represent the input torque obtained from the operator and the output torque used to drive workload (T_{in} and T_{out}); T_{M1} and T_{M2} are the electromagnetic torques, which equals to the multiplication of the motor torque constant C_T and the closed-loop current i, and has the same magnitude; T_0 is the unload torque caused by unload losses including mechanical energy loss, magnetic core loss, and added loss. According to (1), the relationship between the input and output torque can be re-expressed as

$$T_1 = T_2 + 2T_0 \quad (2)$$

This suggests that in order to drive the workload exerting the torque T_2 on the slave motor shaft, the input torque provided for the master motor should be no less than the summation of T_2 and $2T_0$. If the input torque is less than this summation, the operator can feel the difficulty and increase input force accordingly. This means that the system is capable of realizing force sensing without a force sensor. Thus, both the hardware and software design can be simplified to a great extent. In fact, since the two motors possess the same current, the master electromagnetic torque equals to the slave electromagnetic torque and varies following load variation. Then the input force can be adjusted according to the variation of the master electromagnetic torque. As well, unload torque, T_0, which mainly depends on the rotational speed, also has reflection in the master site. That is, the operator can regulate the input force according to the variation of workload and the perception of velocity. The force and velocity sensing characteristic enables the operator to control the movement trajectory of the two limbs by himself/herself without a trajectory definition program, so this simplifies the software design greatly.

2.2 Theoretical analysis: energy recycling

Based on the electrical mechanism, the dynamic voltage balance equation of the master-slave circuit can be written as

$$Ri + L\frac{di}{dt} = e_1 - e_2 = C_T\omega_1 - C_T\omega_2 \tag{3}$$

where R and L denote the armature resistance summation and inductance summation, respectively; ω_1 and ω_2 are the rotor speeds of the master and slave motors, also are the input and output speeds (ω_{in} and ω_{out}) here; and e_1 and e_2 are the armature voltages, which depend on the motor torque constant and rotor speed. e_2 is called as reverse EMF (electromotive force) since it has an opposite-directional current. As can be seen from (3), the energy generated by the master motor is transmitted to the slave motor except the energy loss in the resistance and inductance, thus the system can realize a kind of energy recycling. Furthermore, the smaller the current, the smaller the energy loss in the closed-loop (energy recycling) circuit will be, leading to ω_2 with a nearer approach to ω_1. Therefore, a small current is helpful for achieving an accurate motion tracking.

2.3 Master-slave control system design

During rehabilitation operation, in order to drive the impaired limb to imitate the motion of healthy limb correctly, a high motion tracking performance is necessary. As the analysis above, a small closed-loop current is preferable. However, it is difficult to achieve a small current to drive the impaired limb (a relatively large workload) with a DC motor directly. Otherwise, it is almost impossible to find motors with sufficient torques to support the impaired limb directly. Even if it is possible, the hemiplegic patients will be unable to wear the rehabilitation robot because high-power motors tend to be heavy. Thus, the gearbox mechanism is adopted here to reduce the closed-loop current for enhancing the motion tracking performance, and to increase the driving power of the system with small DC motors. On the other hand, even though the inside energy loss can be reduced by the gearbox mechanism, the yet existent energy loss makes it impossible to realize an acceptable motion tracking performance. Hence, the appropriate amount of energy is compensated for the closed-loop circuit to offset the inside energy loss, further achieving a high motion tracking property. The corresponding introduction is given below:

1) Gearbox mechanism: In order to acquire a symmetric mechanism, two identical gearboxes are employed in the master and slave sites. An equivalent circuit is shown in Fig. 2. The gear transmission relationship can be expressed as

$$\begin{cases} \omega_1 = N\omega_{in}, \omega_2 = N\omega_{out} \\ T_{in} = NT_1, T_{out} = NT_2 \end{cases} \tag{4}$$

where N is the gear ratio of the two gearboxes; ω_1, ω_2 and T_1, T_2 are still used to indicate the rotor speeds and the mechanical torques in the two motor shafts; ω_{in} and ω_{out} stand for the input and output speeds; T_{in} and T_{out} represent the input and output (load) torques of the system. The gear ratio can be selected by dividing the motor rated torque into the load torque, which can be estimated with the radius of the force arm in the slave site and the mass of the unhealthy limb.

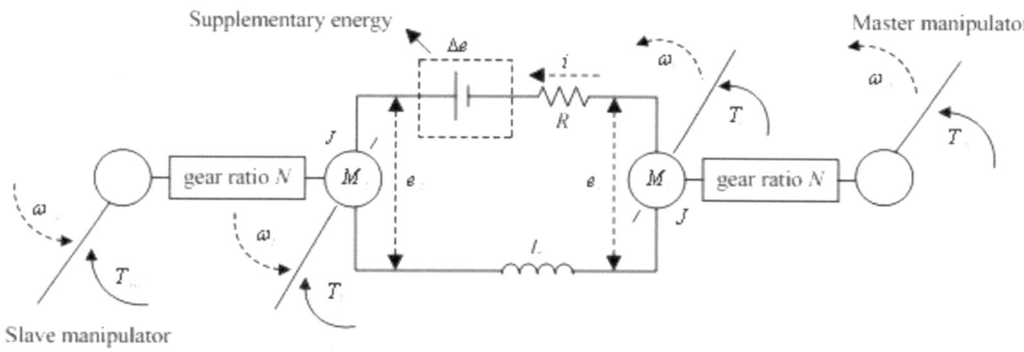

Fig. 2. Equivalent circuit of the system with gearbox mechanism

As can be seen from (4), the torques in motor shafts are minified N times compared to the input/output torques, and the rotor speeds are magnified N times compared to the input/output speeds. This leads the armature voltage generated by the master motor to be N-times magnified, and the current to be N-times minified (refer to (1) and (3)), while the electric power to be kept nearly constant. Thus, the energy loss in the energy transfer circuit (closed-loop circuit) can be reduced. This can increase the energy recycling efficiency in the electronic circuit, and is advantageous to realize motion tracking. However, the unload loss will be increased slightly due to the magnified speeds in motor shafts. Additionally, there is energy loss in the gearboxes. That is, in order to enhance the energy recycling efficiency in the closed-loop circuit, the operator should deliver a larger input power to drive the system. Besides, the gearbox mechanism can also increase the load-bearing capability, making the system have enough driving power to motivate the impaired limb without high-power motors, and being advantageous to reduce the weight as well as the volume of the system.
By combining (1) and (4), the relationship of the input electric power and output torque can be re-expressed as

$$T_{in} = T_{out} + 2NT_0 \tag{5}$$

The difference between the input and output torques is magnified N times, thus a larger input torque is required for driving the same load. But this does not impact the force sensing capability towards load variation so long as $2NT_0$ is not much larger than T_{out}. However, the variation of the unload torque, which relies on rotational speed variation, has a greater reflection in the master site. Based on the analysis above, we can conclude that the gearboxes with appropriate gear ratio have small impact on force sensing ability, but have a relatively large influence on the motion sensing ability. The gearboxes should be decided based on test experiments.

2) Energy supplement: The gearboxes can enhance energy recycling efficiency, whereas the reduced energy loss makes it impossible to realize an acceptable motion tracking performance yet. Therefore, the appropriate amount of energy is compensated for the closed-loop circuit to offset the inside energy loss, further achieving a high motion tracking property. The supplementary energy is added to the system as shown in the dashed line frame of the Fig. 2. The supplementary energy is regulated by adjusting the duty cycle of a pulse-width modulated (PWM) signal that is fed to an H-bridge driver and enables the driver to provide moderate energy for the closed-loop circuit. The compensated voltage, $e_{\sup}$, can be calculated as

$$e_{\sup} = \alpha U_s \tag{6}$$

where U_s is the supply voltage of the H-bridge driver, and α is the duty cycle of the PWM signal.

During operation, it is hoped that the input and output sites have the same movement behaviour. Here, based on the position difference and speed difference between the two motors, a motion tracking control (position-speed control) is carried out to calculate the required supplementary energy for the circuit. This controller can regulate the direction of compensated energy in accordance with the amplitude and direction of the rotational speeds, thus can assure the two motors to possess the same motion trajectory no matter in which side the master motor is located.

The power transmission flowchart of the master-slave control system is shown in Fig. 3, in which master/gear unit and slave/gear unit represent the side exerted with an active force and the side attached with a resistant force, P_{in} and P_{out} denote the input power and output power of the system; P_{1_in}, P_{1_M}, and P_{1_out} represent the input mechanical power, the electromagnetic power, and the output electric power of the master motor, respectively; P_{2_in}, P_{2_M}, and P_{2_out} are the input electric power, the electromagnetic power, and the output mechanical power of the slave motor, respectively; and $p_{\sup}$ is the compensated energy power for the inside closed-loop circuit. The various energy losses in the system are listed in Table 1. Mechanical loss, magnetic core loss and excitation loss are mainly caused by the mechanical fraction and alternative magnetic field towards armature core, and they are called as unload loss in general and primarily related to the rotational speed. Resistance loss and contact loss called as load loss are losses caused by the current in the armature circuit, and they change mainly following the current variation.

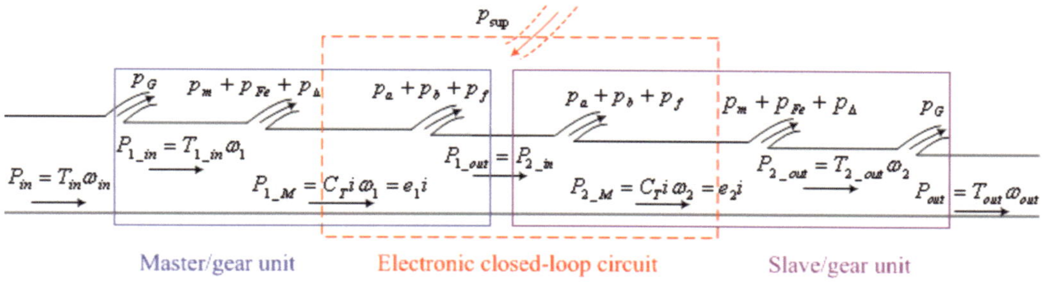

Fig. 3. The power transmission flowchart of the master-slave system

gear loss	mechanical loss	core loss	added loss	resistance loss	contact loss	excitation loss
p_G	p_m	p_{Fe}	p_Δ	p_a	p_b	p_f

Table 1. Various losses in the control system

As shown in Fig. 3, the relationship between the input power from the operator and the output electric power of the master motor is given by:

$$P_{1\_out} = P_{in} - p_G - (p_m + p_{Fe} + p_\Delta) - (p_a + p_b + p_f) \tag{7}$$

In the slave site, there is a reverse energy transmission flow, with the relationship between the input and output power expressed as

$$P_{out} = P_{1\_out} - (p_a + p_b + p_f) - (p_m + p_{Fe} + p_\Delta) - p_G \tag{8}$$

The supplementary energy, p_{sup}, is used to offset energy losses p_a, p_b, and p_f in the two motors. When the system achieves motion tracking accurately, ω_{out} equals to ω_{in}. According to (4), we can obtain

$$P_{1\_M} = P_{2\_M} \tag{9}$$

Thus the supplementary energy can be expressed as

$$p_{sup} = 2(p_a + p_b + p_f) \tag{10}$$

where p_a is the main part of the energy losses in the closed-loop circuit. When the energy loss in the closed-loop circuit is fully compensated, overall system efficiency depends upon the gearbox efficiency and the energy losses including p_m, p_{Fe}, and p_Δ.

2.4 Working modes

The force sensing capability and the symmetric mechanism make the system has no limitation in the orientation of the master and slave manipulators. That is, each side can act as the master manipulator; accordingly, the other site will act as the salve manipulator. This suggests that the system can work well no matter in which side the limb is unhealthy. With this master-slave control system, a patient's impaired limb can be exercised in different modes by coordinating the two limbs. The different training modes are explained as follows:

1) Passive mode: one motor is operated by the healthy limb and deemed as the master motor to drive the other motor which is connected with the impaired limb and deemed as the slave motor. When the healthy limb rotates the master motor, the slave motor imitates the trajectory of the master and supports the impaired limb to exercise. The movement trajectory and velocity are controlled by the healthy limb, and are subject to the acceptable motor capacity of the impaired limb. During the operation, the patient feels resistant force from the impaired limb, and adjusts input force of the healthy limb properly to achieve an expected movement trajectory and velocity within the range of motor capacity of the impaired limb. This control mode can be adopted when the motor capacity of the unhealthy limb is extremely weak.

2) Active-assistive mode: the impaired limb tries to rotate the slave motor; on the other side, the healthy limb feels the reflected force (amplitude and direction) in the master site and provides an auxiliary force to help the impaired limb complete the movement. In this mode, both the motors work in the generating state. The electric power generated by the master as auxiliary input electric energy is provided for the slave motor, to reduce the input force requirement of the affected arm, and to accomplish the movement even if the motor capacity of the affected arm is not strong enough. That is, when the impaired limb has insufficient ability to move, the auxiliary force can offer positive power to assist the impaired limb to carry on movement with an expected trajectory and speed. The movement trajectory and speed is dependent on the impaired limb's motor capacity as well as the auxiliary force from the healthy limb. This control mode can be used when the impaired limb has a mild motor capacity.

3) Active-resistive mode: the impaired limb tries to rotate the slave motor, and the healthy limb provides a reverse force to resist this movement. In this mode, the two motors still work in the generating state, but the master motor provides negative electric power for the slave to resist movement of the affected limb. Similarly, the movement trajectory and speed relies on the forces from the unhealthy and healthy limbs. This control mode can be used when the unhealthy limb has a certain recover in motor function and should be trained with an opposite acting force.

With the above working modes, the valid limb can provide a varying force to the movement of the affected one, ranging from full assistance, where the affected limb only can behave passively, to resistance, if the impaired limb has sufficient voluntary control ability.

3. Experimental study

3.1 Experimental platform

In order to verify the viability of the above approach, a preliminary test platform, as shown in Fig. 4, was built for experiments. Fig. 5 is the corresponding schematic diagram. The platform is composed of two identical motors (A-max 32 motor, combined with Planetary Gearhead GP 32 A, N=4.8 and Encoder HEDL 5540, maxon co. Switzerland), an H-bridge driver (LMD18200, National Semiconductor co. America), two torque transducers (TP-20KCE, Kyowa co. Japan), a torque signal amplifier, and a dSPACE control platform (CLP1104, dSPACE Inc, Germany). Here, the torque transducers and the torque signal amplifier are adopted to measure the input and output torques for verifing the force sensing ability of the system in our test experiments.

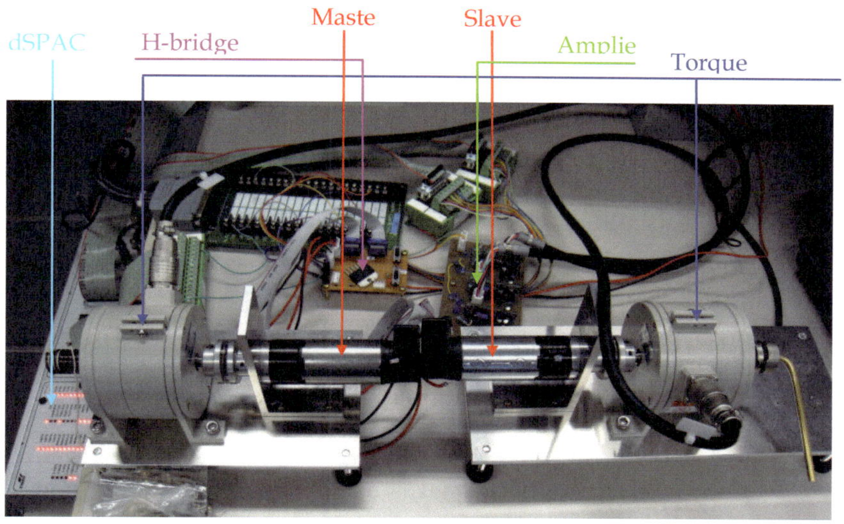

Fig. 4. The experimental platform of the master-slave control system

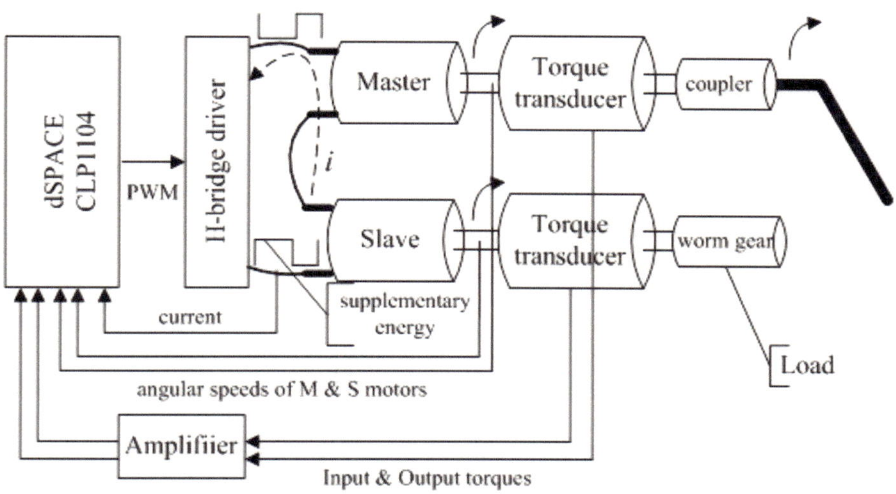

Fig. 5. Schematic diagram of the master-slave motor control system

3.2 Control flow

In the experiment system, the operator used the left hand to exert a force in the slave site and sensed this force reflected in the master site, then used the right hand to provide a moderate force (a driving force to overcome the resistance of the left hand or an auxiliary force to assist the left hand) to rotate the left hand. The speed and position information was detected with the increamental encoders, and the torque information was detected with the torque transducers. The CLP1104 collected the speed and possition information though the incremental encoder interface, and worked out the control quantity α of the PWM siganl with the motion control strategy, then though the PWM generation module in the slave DSP of the CLP1104 to offer PWM signal for the H-bridge driver, enabling the driver to supply the compensatary energy for the master-slave motor circuit. Simultaneously, the CLP1104 collected the datas of the current in the electronic closed-loop circuit and the torques in two sites through the AD module in the slave DSP system.

In each control period (one millisecond), the regulated duty cycle of the PWM siganl, the rotational speed, and the corresponding current were used to calculate the energy loss in the electronic circuit, the supplementary energy provided by the H-bridge driver, and the electromagnetic power of the slave motor (equals to master electromagnetic power in balance state). The corresponding calculation formula is

$$\begin{cases} p_{\sup} = \alpha U_s i \\ p_a = i^2 R \\ P_{2\_M} = e_2 i = C_T \omega_2 i \end{cases} \tag{11}$$

To ensure the global stability, a PID (proportional-integral-differential) control module was built in the DSP system for the position-speed feedback (motion tracking) control. Since the input speed (controlled by the operator) is not a constant, if the differential operation is applied to the speed difference between the input and output sites, the variation of the input speed may lead to overshoot and fluctuation of the whole system. Therefore, the differential operation is applied only to the output speed of the slave motor.

4. Evaluation experiment

4.1 Force sensing test

In this experiment, a variable resistant force was imposed on the slave motor site with the left hand; and in the master site, the operator felt this force and used the right hand to provide a reaction force to overcome the resistance and to rotate the left hand. The corresponding results are given in Fig. 7.

The curves of the input torque and the load torque have the same trend (see Fig. 7 (a)), which verifies that the system has force sensing capability. The operator sensed the load torque variation reflected in the master site, and adjusted the input force accordingly to maintain the speed with a certain variation regulation. The large difference between the input and output torque was caused by the unload torque and the torque amplification function of the gearboxes (refer to (5)). And the difference was almost constant when there was a small variation in the rotational speeds, because the unload torque mainly varies following the changes of the speeds. When the input force in the master site is considered as the driving force and the force attached in the slave site is deemed as resistance, we can say that the system worked in the passive mode; while when the force in the slave site is considered as the input force and the force added in the master site is deemed as the resistant force, we can say that the system worked in the active-resistance mode.

From Fig. 7 (b) and (c), we noted that the system possessed a good motion tracking performance. The maximum steady-state errors were 2.2950 rad/s in speeds and 0.0538 rad in position for the two DC motors. While the maximum steady-state errors in the input and output sides were minified 4.8-times (gear ratio) by the gearboxes. That is, the maximum steady-state errors of the system were 0.4781 rad/s in speeds and 0.0112 rad in position. Figure 7 (d) manifests that the supplementary energy was approximately coincident with the resistance loss. However, the former was slightly larger than the latter because the contact loss and excitation loss also occurred in the energy recycling circuit (refer to (10)). We noted that the energy loss was much larger than the electromagnetic power. This situation will be changed when we use gearboxes with a larger gear ratio which can further reduce the closed-loop current and increase the speeds of the motors.

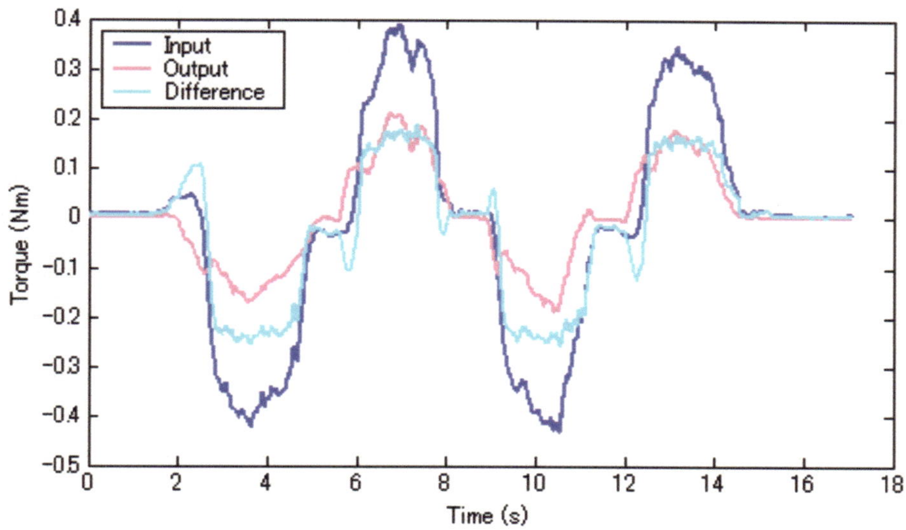

(a) A representative results of the input torque, the output torque and the difference between the input and output torques

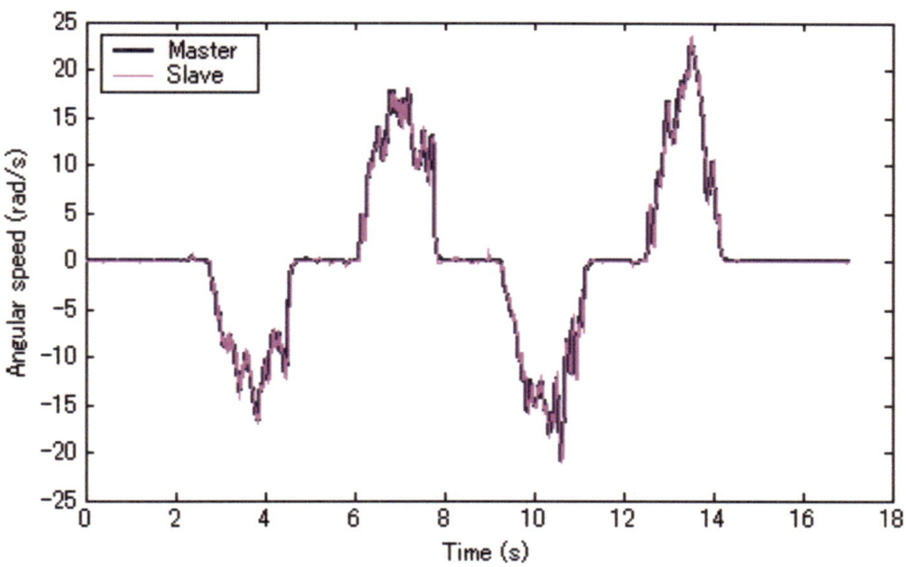

(b) Speed tracking curves of the motors

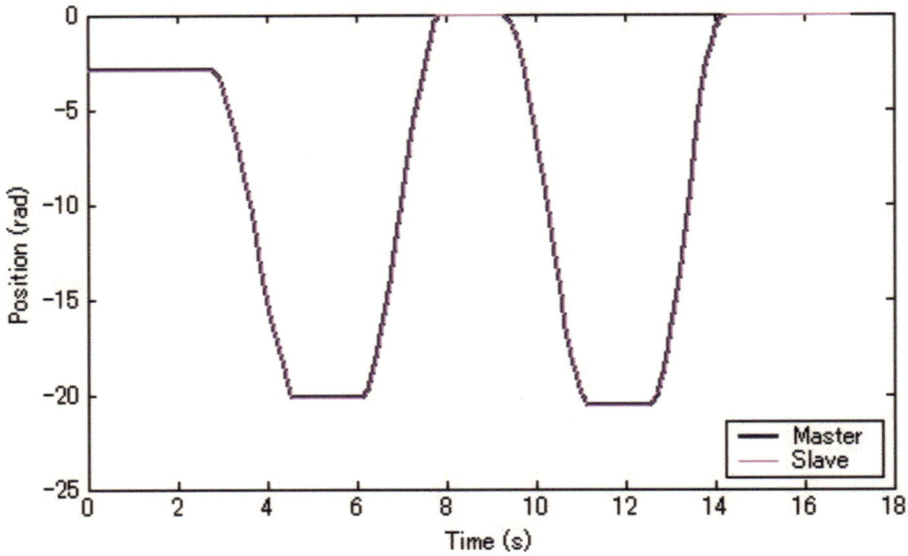

(c) Position tracking curves of the motors

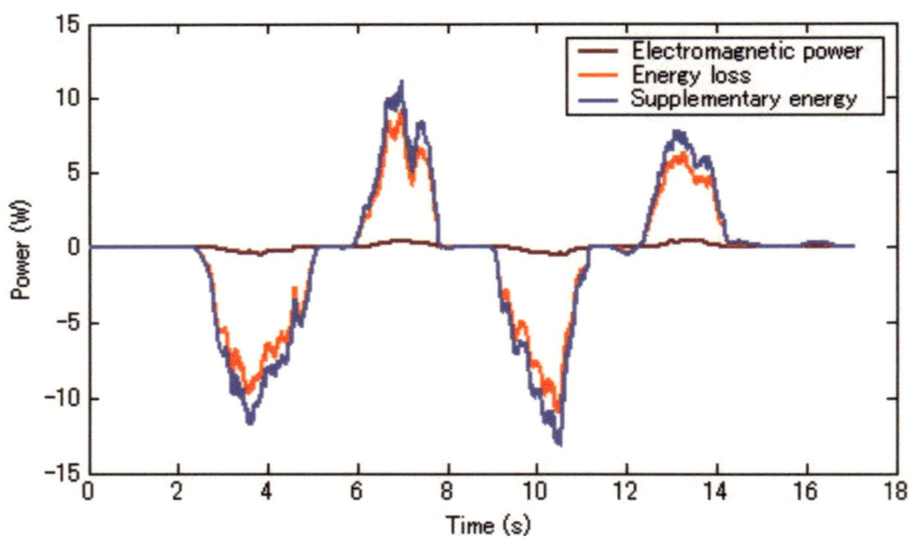

(d) The relation curve of resistance loss, compensated energy and electromagnetic power
Fig. 6. The results of the force sensing test

4.2 Active-assistance exercise test

In this experiment, the left hand exerted a force in the slave motor site; and in the master site, the right hand sensed this force, and provided an auxiliary force to assist the left hand to move. The corresponding results are given in Fig. 8.

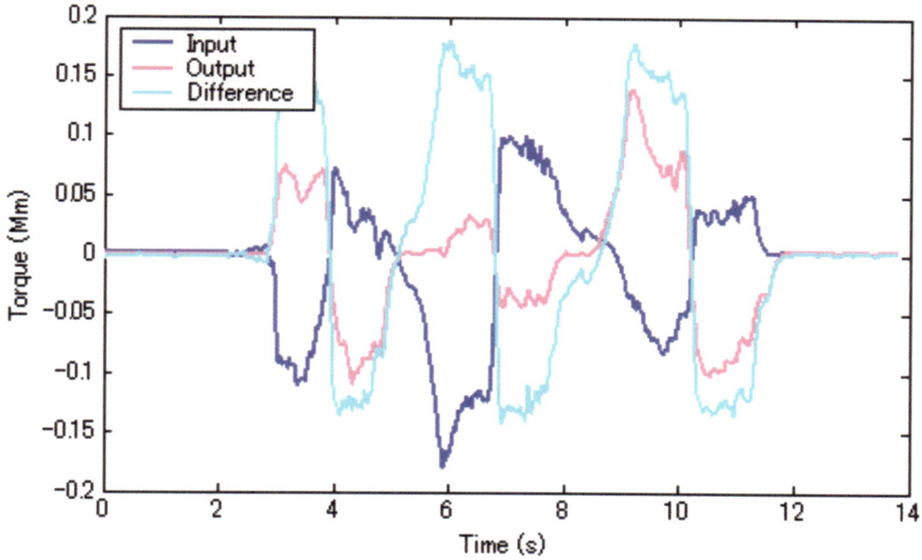

(a) A representative results of the input torque, the output torque and the difference between the input and output torques

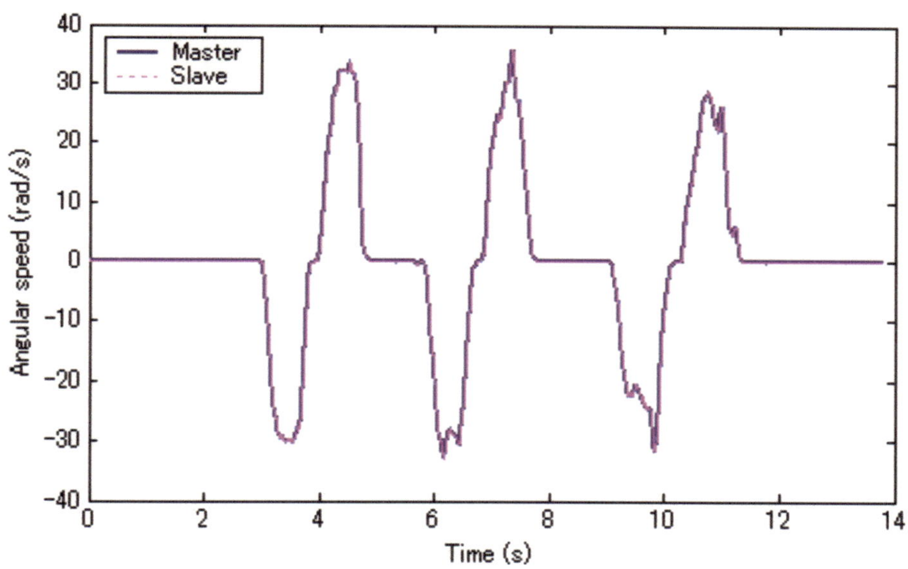

(b) Speed tracking curves of the motors

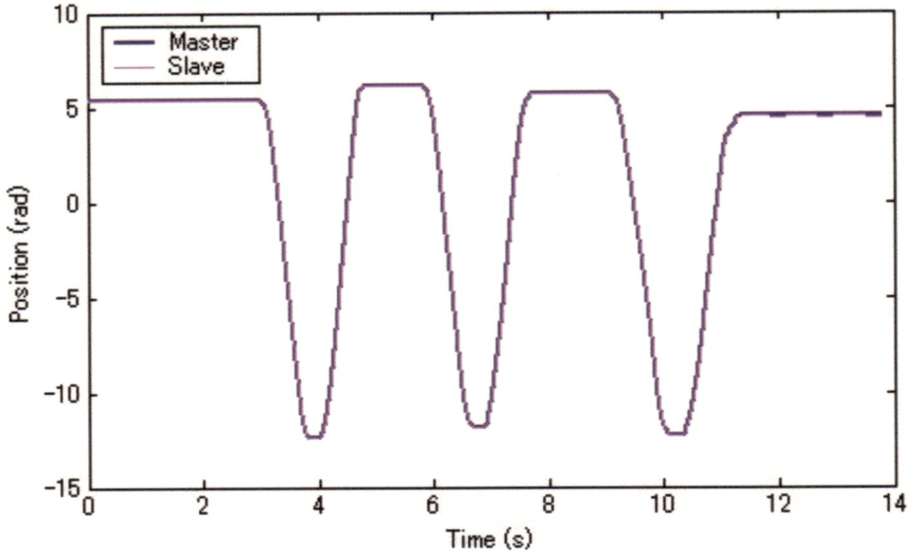

(c) Position tracking curves of the motors

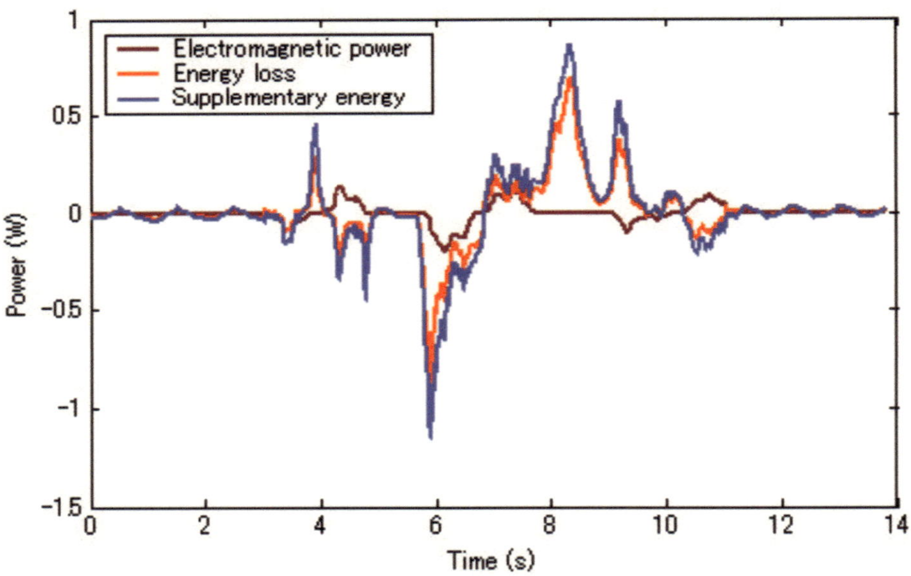

(d) The relation curve of resistance loss, compensated energy and electromagnetic power
Fig. 7. The results of the active-assistive working mode test

It can be seen from Fig. 8 (a), in order to keep the speed with almost the same variation trend, the input force provided for the master part was increased or decreased following the reduction or increase of the force exerted in the slave part (In order to test force sensing capability in the precious experiment, the forces in both sites with the opposite direction were defined to have the same sign symbol. Thus the assistive force here possessed an opposite sign symbol compared to the active force in the slave site). However, the difference between the two forces, which represents the summation of the two forces actually, had the same variation trend and was coincident with the speed variation. This verifies that the operator can regulate the input force moderately based on expected rotational speed and the reflected force of the slave site. Therefore, the system is capable of carrying out the active-assistive training. In this working mode, the system also possessed a good motion tracking performance. The maximum steady-state errors of the system were 0.2657 rad/s in speeds and 0.0067 rad in position. Meanwhile, figure 7 (d) manifests that the supplementary energy was used to offset the energy loss in the electronic circuit, in which the resistance loss accounts for the mainly part. As well, it is obvious that the supplementary energy had no relation to the electromagnetic power. This verified that except the portion used to compensate the resistance loss, the supplementary energy was used to compensate the contact loss and excitation loss in the circuit, rather than to provide power for the two motors.

5. Discussion

The experimental results verify the theory introduced above, and confirm the feasibility of the proposed control method. However, there are some drawbacks as follows:

1) Even though the driving energy is saved somewhat, the demand for energy supplement is rather high. In order to reduce the demand for supplementary energy, we can use motors with larger torque constant and smaller resistance to reduce energy loss (refer to (1) and (3)). Alternatively, we can increase the gear ratio of the gearbox in the master site. With N_1 and N_2 ($N_1 > N_2$) denoting the gear ratios of the gearboxes in the master and slave sites, the electromagnetic power of the master and slave motors can be written as

$$\begin{cases} P_{1\_M} = C_T i N_1 \omega_{in} \\ P_{2\_M} = C_T i N_2 \omega_{out} \end{cases} \tag{12}$$

which suggests that the master motor generates more electromagnetic power than the power required in the slave site when ω_{out} equals to ω_{in}. If the two gear ratios are matched appropriately, the following relationship can be achieved

$$P_{1\_M} - P_{2\_M} \approx 2(p_a + p_b + p_f) \tag{13}$$

Thus the demand for energy supplement can be reduced greatly. The viability of this method has been verified by practical tests employing geared DC motors of 1271 series (McLennan co. UK) with $N_1 = 43$ and $N_2 = 21$ in our experiments.

However, an asymmetric mechanism makes the system is limited to the collocation of master and slave parts, especially when the system works in passive mode. If the great amount of supplementary energy is needed to achieve motion tracking.

The healthy limb to drive the impaired limb. Moreover, the gearbox with a larger gear ratio in the slave site is easily to be destroyed, because it works in the back drivable state in our experiments. In conclusion, when the gearbox with a larger gear ratio is located in the healthy limb side, this

system can work well with a small demand for supplementary energy; however, when the gearbox with a larger gear ratio is located in the impaired limb side, it cannot work ideally, especially in a passive
exercise application. However, if the designed master and slave units can exchange the physical position flexibly, this scheme will be more preferable for rehabilitation robots.

2) The system has a perfect motion tracking ability. However, the errors are relatively large at the point of changing the rotational direction. This problem is considered to be resolved by improving the control program.
3) In the experiment, encoders were employed to detect the speeds of the master and slave motors, and to calculate requirements of supplementary energy. Actually, we can formulate the relationship between supplementary energy and closed-loop current and apply the formulation along with detected current to calculate the required supplementary energy in practical applications. Hence, the system will do not require any sensors. We can further improve the system's performance, and can enhance the potential for an extensive application in control fields.
4) The load-bearing capability of the experimental platform is not enough to drive an impaired limb especially with extremely weak motor function. Therefore, in real applications, the gearboxes with a larger gear ratio should be adopted to increase load-bearing capability. However, the gear ratio should not be too large, because a larger gear ratio will increase the difficulty of driving the impaired limb, and may destroy the gearbox in the slave site where the motor/gear unit is required to be back drivable during the passive training. Gearboxes with the gear ratio of 51 or 66 (Planetary Gearhead GP 32 A) may be able to provide enough driving force to drive the forearm limb to perform a flexion/extension action. The feasibility of these gear ratios should be confirmed with testing experiments in the next step. If the gear ratio is too large to work well in the back driving mode, motors with a relatively high power can be adopted to replace the ones used in this experiment.
5) In order to make the system more suitable for the application in rehabilitation robots, a new system with multi-DOF mechanism will be developed using a multi-motor combination in the future work.

6. Conclusion

The mater-slave control system has several characteristics that make it suitable for application in wearable rehabilitation robots. First, the system realizes force sensing without a force sensor. The patient can feel the resistant or active force of the impaired limb and adjust the input force accordingly to accomplish the movement. Second, the system achieves a kind of energy recycling. Therefore, a lightweight battery will probably supply enough power for the system. This may address the power problem, and relatively reduce the weight of the wearable robots, further make it favorable for the operator wearing a portable robot so as to move around freely in future applications. Third, the equivalent configuration of the master and slave sites enables the system to realize bilateral control, and eliminates the direction limitation for the master and slave manipulator. Furthermore, in the active-assistive exercise or the active-resistive exercise, the assistive force or the resistant force provided by the healthy limb can be regulated in time under the action of motion consciousness, because the motion is controlled by the patient himself/herself. For example, when the patient want to change the motion trajectory, he/she will make a preparation for changing the force direction of the healthy limb, and making a timely regulation to assist or resist the motion of the unhealthy limb. Otherwise, this scheme also has a great potential for applications in the fields of micro-manipulation, micro-assembly and medical surgery assistant.

7. References

Burgar, C.G.; Lum, P.S.; Shor, P.C. & Van Der Loos, H.F.M. (2000). Development of robots for rehabilitation therapy: the Palo Alto VA/Stanford experience, *Journal of Rehabilitation Research and Development,* Vol. 37, No. 6/2000, pp. 663–673

Gang, S. & Shuxiang, G. (2006). Development of an active self-assisted rehabilitation simulator for upper limbs, *Proceedings of the World Congress on Intelligent Control and Automation (WCICA),* Vol. 2, 2006, pp. 9444-9448

Hogan, N. & Krebs, H.I. (2004). Interactive robots for neuro-rehabilitation. *Restorative Neurology and Neuroscience,* Vol. 22, No. 3, 5/2004, pp. 349–358, 0922-6028

Hogan, N.; Krebs, H.I.; Rohrer, B.; Palazzolo, J.J.; Dipietro, L.; Fasoli, S.E.; Stein, J. & Volpe, B.T. (2006). Motions or muscles? Some behavioral factors underlying robotic assistance of motor recovery, *Journal of Rehabilitation Research and Development,* Vol. 43, No. 5, pp. 605–618

Jack, D.; Boian, R.; Merians, A.S.; Tremaine, M.; Burdea, G.C.; Adamovich, S.V.; Recce, M. & Poizner, H. (2001). Virtual reality enhanced stroke rehabilitation, *IEEE Transactions on Neural Systems and Rehabilitation Engineering,* Vol. 9, No. 3, 9/2001, pp. 308–318

Jeong, Y.; Kim, Y.K.; Kim, K. & Park, J.-O. (2001). Design and control of a wearable robot, *Robot and Human Communication - Proceedings of the IEEE International Workshop,* pp. 636-641

Krebs, H.I.; Volpe, B.T.; Aisen, M.L. & Hogan, N. (2000). Increasing productivity and quality of care: Robot-aided neuro-rehabilitation, *Journal of Rehabilitation Research and Development,* Vol. 37, No. 6/2000, pp. 639–652

Li, H. & Song, A. (2008). Force assistant master-slave telerehabilitation robotic system, *Journal of Southeast University,* Vol. 24, No. 1, 3/2008, pp. 42-45

Lum, P.S.; Burgar, C.G.; Shor, P.C.; Majmundar, M. & Van der Loos, M. (2002). Robot-assisted movement training compared with conventional therapy techniques for the rehabilitation of upper-limb motor function after stroke, *Arch Phys Med Rehabil,* Vol. 83, No. 7, pp. 952–959

Lum, P.S.; Burgar, C.G. & Shor, P.C. (2004). Evidence for improved muscle activation patterns after retraining of reaching movements with the MIME robotic system in subjects with post-stroke hemiparesis, *IEEE Transactions on Neural Systems and Rehabilitation Engineering,* Vol. 12, No. 2, 6/2004, pp. 186–194

Nef, T.; Mihelj, M.& Riener, R. (2007). ARMin: A robot for patient-cooperative arm therapy, *Medical and Biological Engineering and Computing,* Vol. 45, No. 9, 9/2007, pp. 887–900

Peng, Q.; Park, H.-S. & Zhang, L.-Q. (2005). A low-cost portable tele-rehabilitation system for the treatment and assessment of the elbow deformity of stroke patients, *Proceedings of the 2005 IEEE 9th International Conference on Rehabilitation Robotics,* Vol. 28, 7/2005, pp. 149–151

Reinkensmeyer, D.J.; Dewald, J.P.A. & Rymer, W.Z. (1999). Guidance-based quantification of arm impairment following brain injury: a pilot study, *IEEE Transactions on Rehabilitation Engineering*, Vol. 7, No. 1, 3/1999, pp. 1–11,

Reinkensmeyer, D.J.; Kahn, L.E.; Averbuch, M.; McKenna-Cole, A.; Schmit, B.D. & Zev Rymer, W. (2000). Understanding and treating arm movement impairment after chronic brain injury: progress with the ARM Guide, *Journal of Rehabilitation Research and Development*, Vol. 37, No. 6/2000, pp. 653–662

Reinkensmeyer, D.J.; Hogan, N.; Krebs, H.I.; Lehman, S.L. & Lum, P.S. (2000). Rehabilitators, Robots and Guides: New Tools for Neurological Rehabilitation, *Biomechanics and Neural Control of Posture and Movement*, pp. 516–533

Song, G. & Guo, S. (2006). Development of a novel tele-rehabilitation system, *IEEE International Conference on Robotics and Biomimetics, ROBIO*, 17-20 Dec.2006, pp. 785–789

Statistics Bureau, Japan. (2007). Statistical handbook of Japan. Available: *http://www.stat.go.jp/english/data/handbook/c02cont.htm*

van Exel, N.J.A.; Koopmanschap, M.A.; Scholte op Reimer, W.; Niessen, L.W. & Huijsman, R. (2005). Cost-effectiveness of integrated stroke services, *QJM-J. Association of Physician*, Vol. 98, No. 6, 1/2005, pp. 415-425

Zheng, H.; Davies, R.; Zhou, H.; Hammerton, J.; Mawson, S.J.; Ware, P.M.; Black, N.D. & Harris, N.D. (2006). SMART project: Application of emerging information and communication technology to home-based rehabilitation for stroke patients, *Int.J. Disability and Human Development*, Vol. 5, No. 3, 7/2006, pp. 271-276

Zhou, H.; Stone, T.; Hu, H. & Harris, N. (2008). Use of multiple wearable inertial sensors in upper limb motion tracking, *Medical Engineering and Physics*, Vol. 30, No. 1, pp. 123-133

2

Reactive Robot Control with Hybrid Operational Techniques in a Seaport Container Terminal Considering the Reliability

Satoshi Hoshino and Jun Ota
Tokyo Institute of Technology and The University of Tokyo
Japan

1. Introduction

In a seaport container terminal system, there are various types of container-handling and transport machines (Guenter, 2005). Machine reliability is a particularly serious concern due to the fact that the system is subjected to salt erosion. The reliability of an item is expressed by the probability that the item will perform its required function under given conditions for a stated time interval (Birolini, 2007).

In research that deals with flexible manufacturing systems (FMSs), system reliability has already been investigated from the viewpoint of the endurance and fault tolerance of robots or machines (Beamon, 1998) (Sun, 1994). Savsar has described the importance of preventive and corrective maintenance for system reliability (Savsar, 2005). On the other hand, few investigations of seaport container terminal systems consider reliability; one of these studies by Hoshino et al. deals with the reliability design of intelligent machines, i.e., operating robots (Hoshino & Ota, 2007), and another, by Bruzzone et al., deals with container logistics node design (Bruzzone et al., 2007). However, although operating robots are preventively maintained on the basis of the confidence level, robot failure has not been considered at all (Hoshino & Ota, 2007).

In an actual seaport container terminal system, in order to minimize the loss of operating efficiency even when a robot undergoes maintenance, a large number of robots are readied and used on the assumption that operating robots fail fortuitously. However, such a policy increases the required number of robots, and, therefore, the initial investment. Therefore, in this paper, we approach this issue from the system management aspect.

Fig.1 shows a horizontal transportation system with automated guided vehicles (AGVs), namely, the AGV transportation system in one berth. A seaport container terminal generally consists of several berths what are arranged along a wharf. The effectiveness of the system has been shown compared to the vertical one by controlling the operating robots including the AGV efficiently (Hoshino et al., 2007). Thus, we address and manage the horizontal system considering efficient maintenance of the operating robots.

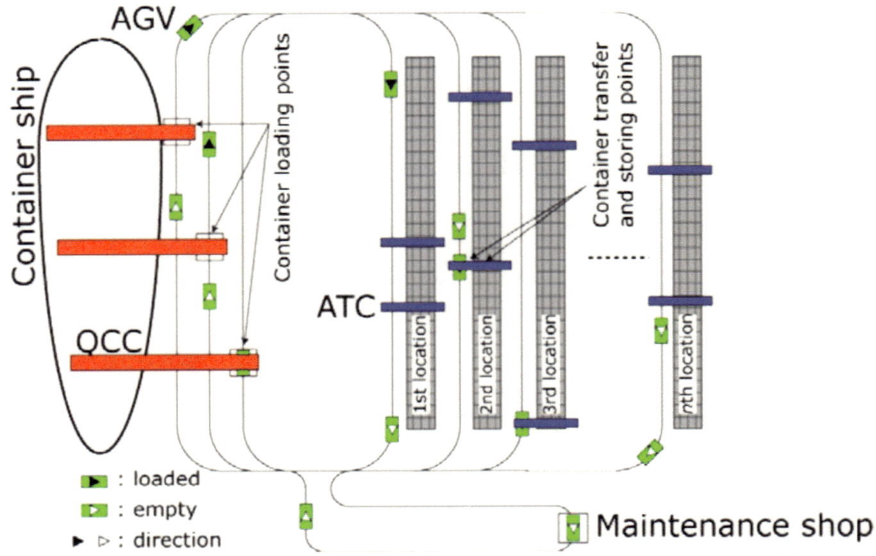

Fig. 1. Horizontal AGV transportation system in a seaport container terminal (top view)

2. Seaport Container Terminal

2.1 Horizontal AGV transportation system

In the horizontal AGV transportation system shown in Fig.1, quay container cranes (QCCs) at the quay side, automated transfer cranes (ATCs) at the container storage yard side, and AGVs for container transport between the quay and yard sides are in operation. In this paper, we refer to the AGV and ATC as operating robots. Since each robot has a radio communication device, the robots are able to share their information with neighbors based on the distributed blackboard, namely, 'sign-board' model (Wang, 1994). A container storage location consists of a 320 [TEU (Twenty-foot Equivalent Unit)] container space. There are three QCCs at the quay side in a general berth, and two ATCs of different sizes are operating at one location.

While Qiu, Hsu, and Zeng have focused on a transportation system with a bidirectional path layout to take into account inter-berth operations (Qiu & Hsu, 2001) (Zeng and Hsu, 2008), in this paper, we focus on a unidirectional path layout because the layout is suitable for more conflict-free container routing even if a simple and feasible routing rule for the system automation is applied. Thus, we do not address a multiple berth scenario, such as a traffic pattern of distributing into and gathering from different berths.

2.2 Container-handling operation

We limit container movement to one-way flow, i.e., from the quay side to the yard side in the course of container loading, transport, transfer, and storing operations as follows:

1) A QCC loads a container from the container ship to the AGV.
2) The AGV transports the container from the quay side to a destination location in the container storage yard.

3) Right after the AGV goes into an adjacent yard lane to the container storing location, an ATC in an idle state is called by the AGV.
4) The AGV begins container transfer to the ATC after the ATC arrives at the container transferring position.
5) The AGV that has completed the container transferring goes back to a QCC.
6) The ATC to which the container has been transferred stores it at the storage position; it then becomes an idle state for a next operation.

The effectiveness of the container assignment and order scheduling methods has been shown by the authors (Hoshino et al., 2005) (Hoshino et al., 2006). Thus, the operating robots perform container-handling tasks as follows: regardless of the operational state, the tasks are equally given from three QCCs; in other words, containers are equally loaded onto the AGVs by the QCCs. In addition, the containers are equally assigned to each location in the container storage yard. An execution order of the tasks is scheduled so that the total moving distance of the ATCs is minimized.

3. Challenges

Fig.2 shows the container-handling simulation result with the AGVs and ATCs. Fig.2(a) indicates the throughput of an ideal system in which, although the operating robots are not maintained preventively, they do not fail at all. Fig.2(b) indicates the throughput of a system in which preventive maintenance of the operating robots and corrective maintenance for a failed robot are done. Here, the mean time between failures (MTBFs) of the AGV and ATC in the simulation (Fig.2(b)) are 50 and 40 hours, respectively. From the results shown in Fig.2(a), it is evident that the throughput increases as the number of AGVs and ATCs increases, and, then, the throughput converges at 130 [TEU/hour]. On the other hand, from the results shown in Fig.2(b), it is evident that the maximum throughput is less than 120 [TEU/hour]; sometimes the throughput does not converge. In addition, it is clear that the throughput decreases significantly due to the maintenance activity.

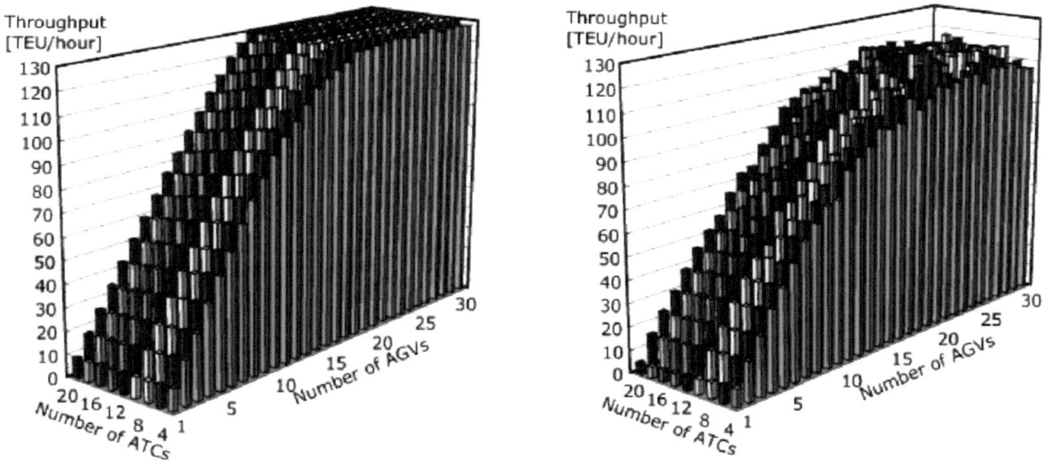

(a) Ideal system without any maintenance

(b) Actual system with preventive and corrective maintenance.

Fig. 2. Container-handling simulation result: system throughputs

These results denote that the system shown in Fig.2(b) is insufficient for a system in which the operating robots have to be maintained in consideration of robot reliability. Hence, for the realization of efficient and flexible container handling, we address the following challenge:

- Even in a case in which a robot has to be maintained due to decreased operational function or failure, ideally, the system should continue operation as efficiently as possible without interruption, as shown in Fig.2(a). Hopefully, this is done by controlling other robots and preventing the system from being obstructed by the robot undergoing maintenance.

For this challenge, we focus on operational techniques in order to utilize the mutual substitutability of the operation among robots that have similar functions. We define the system operational states as follows: 1. normally operating, 2. preventive maintenance, and corrective maintenance, and develop suitable operational techniques for the three states. By applying the developed hybrid operational techniques, each robot is able to respond to the dynamically changing states 1 to 3 reactively. This is a reactive robot control system that takes reliability into account.

4. Robot Reliability

In this paper, we assume that the probability density function on the time span of a normally operating robot in the system follows an exponential distribution. Thus, the failure rate of the operating robot $(\lambda(t))$ at time t is constant (see Eq.(1)). Each operating robot, on the basis of the failure rate λ_0, fails fortuitously (corrective maintenance state). Furthermore, the confidence level $R(t)$,which is the probability that the robot has not failed by time horizon, t, is derived from Eq.(2). Therefore, based on the confidence level, each robot stops operating and enters the preventive maintenance mode when its confidence level is under a given threshold value (preventive maintenance state). In other words, we decide the robot preventive maintenance timing on the basis of $R(t)$.

$$\lambda(t) = \lambda_0 \tag{1}$$

$$R(t) = e^{-\lambda_0 t} \tag{2}$$

The MTBF of the operating robot, *MTBF,* is derived from Eq.(3). From Eq.(1), Eq.(2), and Eq.(3), the failure rate $\lambda(t)$ and confidence level $R(t)$.can be derived from the reciprocal number of the MTBF.

$$MTBF = \int_0^\infty R(t)dt = \int_0^\infty e^{-\lambda_0 t}dt = \frac{1}{\lambda_0} \tag{3}$$

Note that although we assume the constant failure rate (CFR) in the bathtub curve and use the exponential distribution as the probability density function, these are not limited in this research framework. Other distributions, e.g., normal distribution and Weibull distribution are also available under the assumption of the decreasing or increasing failure rate (DFR or IFR) as necessary.

5. Reactive Robot Control with Hybrid Operational Techniques

5.1 Operational technique in the normal state

In the normally operating state, the robots are controlled with the use of the operational technique as follows: the AGV selects the shortest lane to the destination and does not change the destination and lane while moving. The ATC has its own operation area on the location, and, thus, the ATC does not operate in another ATC operational area.

5.2 Operational technique in the preventive maintenance state

Since there are a limited number of maintainers, in this paper, only one AGV and one ATC in the preventive maintenance mode are maintained. Hence, in a case in which multiple AGVs and ATCs enter the preventive maintenance mode at the same time, it is necessary to preventively maintain the robots efficiently in order to take advantage of the mutual substitutability of the operation among robots.
As for the AGV, if an AGV is preventively maintained on every transport lane, the AGV becomes an obstacle to other AGVs. To solve this problem, we parallelized the system by providing a maintenance shop as shown in Fig.3. By doing this, the system is able to keep operating except in a case in which all AGVs are in the maintenance mode and go to the maintenance shop. Here, an AGV that arrives at the maintenance shop first is maintained according to the First-In First-Out (FIFO) rule.

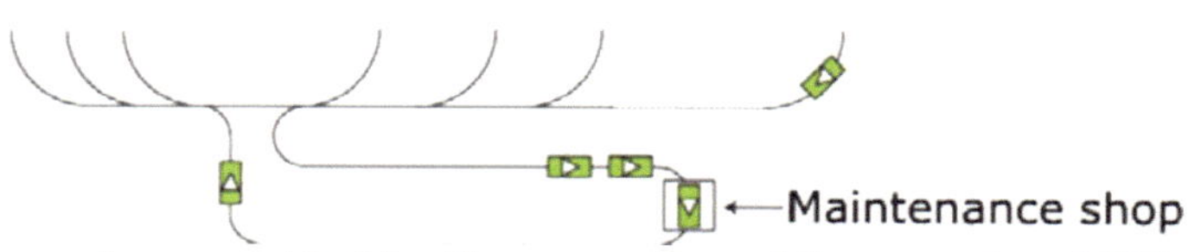

Fig. 3. Maintenance shop provided for the prevention of the operating AGVs

On the other hand, since there are two ATCs at one location, even if an ATC at the location is in the preventive maintenance mode, another ATC is able to perform its task instead by sharing their operation areas. Fig.4 shows container transfer and storing operations among the AGVs and ATCs in a case in which one ATC at the location enters the preventive maintenance mode. In Fig.4(a), two ATCs are normally operating; then, in Fig.4(b), one (small) ATC is in the preventive maintenance mode at the edge of the location. For this situation, if the other (large) ATC is in a standby state, the ATC moves to support the other's operation with the waiting AGV (see Fig.4(c)) in communication with the small ATC. However, if both ATCs at the location are in the preventive maintenance mode at the same time, the flow of incoming AGVs is disrupted on the adjacent yard lane to the location. As a result, the whole system operation might be interrupted. To solve this problem, we developed the following preventive maintenance rules:

- If there is a location where two ATCs are both in the preventive maintenance mode, one of two ATC at the location is selected for maintenance according to priority.

• If either ATC operates at every location, an ATC that enters the preventive maintenance mode first is maintained by rotation.

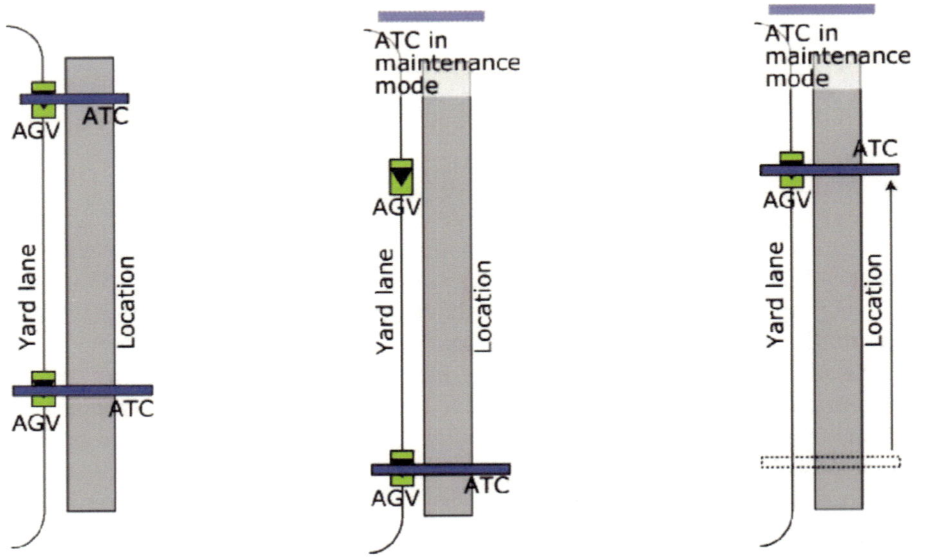

(a) Two ATCs are operating (b) Small ATC is being maintained (c) Large ATC supports operation

Fig. 4. Alternative operation performed by an operating ATC instead

5.3 Operational technique in the corrective maintenance state

It is difficult to completely prevent the accidental failure of the operating robots even if they are maintained for prevention regularly. A failed robot stops at the current position for the corrective maintenance. Therefore, as well as the operational technique in the preventive maintenance state, we consider an operational technique in the corrective maintenance state in order to take advantage of the mutual substitutability of the operation. In this paper, we focus on operational techniques for the AGV in the quay and container storage yard sides, where there are multiple lanes. In communication with each other on a communication lane, an AGV is able to identify whether any failed AGVs or ATCs exist in the quay and container storage yard sides.

5.3.1 Operational technique in the quay side

• If there is a failed AGV at a destination (QCC) or on the lane, the normally operating AGV changes the destination to another QCC closest to the current destination as a new destination according to priority and selects a new lane.

• However, if there are several QCCs that have same priority, the AGV changes the current destination to a QCC located on the yard side and selects a new lane as well in consideration of the moving distance of the AGV.

Fig.5 shows an example of the operation when an AGV fails in the quay side. Here, in the quay side, there are three QCCs operating on three quay lanes (QLs). Fig.5(a) shows that the quay side destination (QD) of an AGV moving on the (red) communication lane is QD 3.

However, the AGV notices that a failed AGV exists on QL 3 in communication with AGVs; hence, the AGV changes the destination from QD 3 to QD 2 and selects QL 2. Fig.5(b) shows a case in which, while the AGV on the communication lane is moving to destination QD 3, there are failed AGVs on QLs 3 and 2. In this case, the AGV changes the destination to QD 1 and selects QL 1.

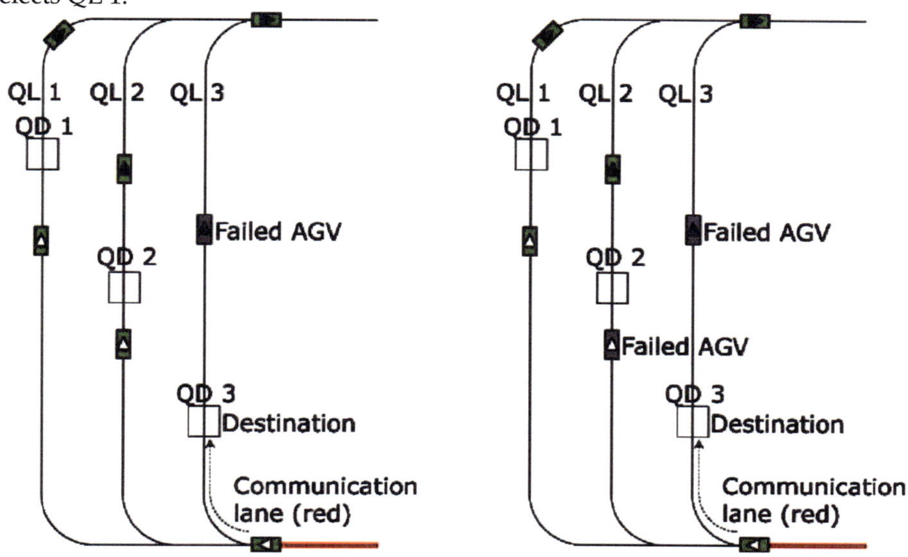

(a) Failed AGV on QL 3 (b) Failed AGVs on QLs 2 and 3
Fig. 5. Operation at the quay side in the corrective maintenance state

5.3.2 Operational technique in the yard side

- If there are failed ATCs and AGVs at a destination (location) or on the lane, the normally operating AGV changes its destination to another location closer to the current destination from the locations located on the quay side in comparison to the current destination according to priority and selects a new lane.
- However, if there are failed ATCs and AGVs at every location or on every lane located on the quay side, the normally operating AGV changes the destination to another location closer to the current destination from the locations located on the land side according to priority and selects a new lane.
- The container transfer and storing points at a location are not changed even if the destination is changed.

Fig.6 shows an example of the operation in a case in which the AGVs and ATCs failed in the container storage yard. Fig.6(a) shows that the yard side destination (YD) of an AGV moving on the (red) communication lane is YD 3, located at the adjacent 3rd location to the yard lane (YL) 3. However, the AGV notices that there is a failed AGV on YL 3 through communication with other AGVs and ATCs; hence, the AGV changes the destination from YD 3 to YD 2 from the candidates YD 1, 2, 4, and 5 and selects YL 2. In Fig.6(b), there are one failed ATC at the first location and failed AGVs on YL 3 and 2. In this case, the destination is changed to YD 4, and then YL 4 is selected as well.

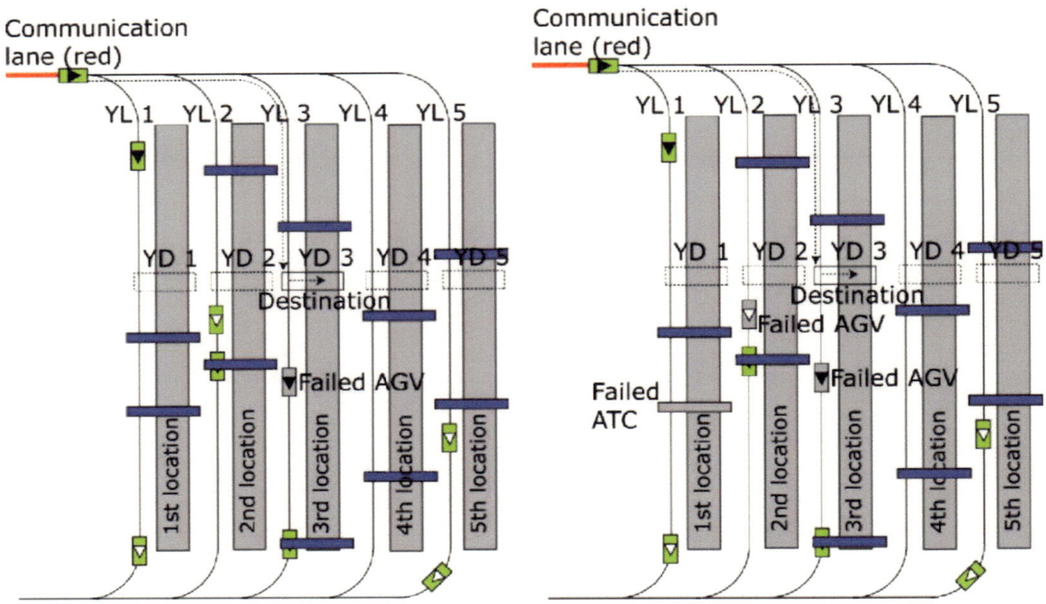

(a) Failed AGV on YL 3

(b) Failed AGVs and ATC on YLs 2 and 3 and at the 1st location

Fig. 6. Operation at the container storage yard in the corrective maintenance state

6. Simulation Experiment

6.1 Experimental condition

The MTBFs of the AGV and ATC are 50 and 40 hours, respectively. These are minimum parameters given in our previous work (Hoshino & Ota, 2007). Each operating robot is preventively maintained at time t when the confidence level is less than 0.9, that is, $R(t) < 0.9$. The $R(t)$ of a robot, which was once preventively maintained, is reset to one ($R(t) = 1.0$). Here, the initial confidence level of each operating robot at the start of a simulation is given randomly as follows: $0^{\wedge} < R(t) < 1^{\wedge}$.

As for preventive maintenance, we assume parts inspection, consumable parts replacement, and main parts replacement; thus, 0.3 to 0.5 [hour] for the AGV and 0.2 to 0.4 [hour] for the ATC are required. These preventive maintenance times are randomly determined with a uniform probability. As for the failed robots, 0.5 to 1.0 [hour] for the AGV and 0.4 to 1.0 [hour] for the ATC are required for their correction. These corrective maintenance times are also determined in a random manner with a uniform probability.

The number of containers that must be unloaded from a containership, that is, the number of tasks, is 600 [TEU]. Here, because there is a 320 [TEU] container space at one location, two locations, i.e., at least four ATCs, are needed in the system. In this experiment, we do a 10-time simulation for 10 incoming container ships. The maximum numbers of AGVs and ATCs used in the container-handling simulation are 30 and 20, respectively. As for the performance of the AGV for the container transport, the maximum traveling speeds are given as 5.56 (loaded) and 6.94 (empty) [m/s] depending on the presence of a container. The acceleration and deceleration speeds are 0.15 and 0.63 [m/s^2] regardless of the presence of a container. The maximum moving speed of the ATC is 2.5 [m/s], and the acceleration and deceleration speeds are 0.1 and 0.4 [m/s^2], respectively. The container unloading/loading time by the QCC, the container transfer time from the AGV to the ATC, and the container storing time by the ATC, which are described in 2.2, are 60, 30, and 30 seconds, respectively. To discuss the effectiveness of the proposed reactive robot control system with the developed three hybrid operational techniques, we compare the proposed system to (I) the ideal system, in which, although the operating robots are not preventively maintained, they do not fail at all with the use of the operational technique described in 5.1 (see Fig.2(a)), and (II) a system in which, although the operating robots are preventively maintained with the use of the two operational techniques described in 5.1 and 5.2, they are not efficiently controlled in the corrective maintenance state (see Fig.2(b)).

6.2 Simulation result

Fig.7 shows the comparison result of the systems on the basis of the throughput. The blue (and diamond-shaped) plot denotes the throughput of the ideal system (I); the red bar graph denotes the throughput of the system (II); and the white bar graph denotes the throughput of the proposed system, in which the operating robots are reactively controlled even in the corrective maintenance state by switching three hybrid operational techniques, described in 5.1, 5.2, and 5.3.

From the result, for the system in which the robots, which have to be maintained for the prevention and correction in consideration of the reliability, are operating, we can see that the proposed system throughput for the all combination of AGVs and ATCs is higher than the throughput of the system (II). From the results of Fig.7(d) to Fig.7(f), we obtained several higher throughputs near the ideal system throughputs. This is because the robots failed on the lanes in the quay or yard sides. In addition, the other operating robots successfully responded to the corrective maintenance state with the third operational technique. These results indicate that the robots are successfully controlled with the use of the hybrid operational techniques. On the other hand, we also obtained several throughputs near the throughputs of the system (II), e.g., as shown in Fig.7(a) with 26 AGVs. The reason for this result is that there were AGVs that failed on a single lane, such as the communication lane, and not on multiple lanes, such as the quay and yard lanes. In this case, it is needed to develop the fourth operational technique on a single lane to avoid a failed robot.

6.3 Effectiveness of the proposed system

Table 1 shows the increase of the throughput of the proposed system relative to the throughput of system (II) on the basis of the result shown in Fig.7. To discuss the effectiveness of the proposed system, the increase of the throughput is calculated after the throughput of the ideal system with a certain number of AGVs becomes nearly flat (see blue and diamond-shaped plots in Fig.7). In other words, the increase of the throughput when the number of AGVs is more than 20 in the result of Fig.7(a) and 17 in other results Fig.7(b) to Fig.7(i) is examined. In the table, "average' represents the average value of the difference between the proposed system throughput and the system throughput of (II), "max.' represents the maximum value of the difference, and "min.' represents the minimum value of the difference.

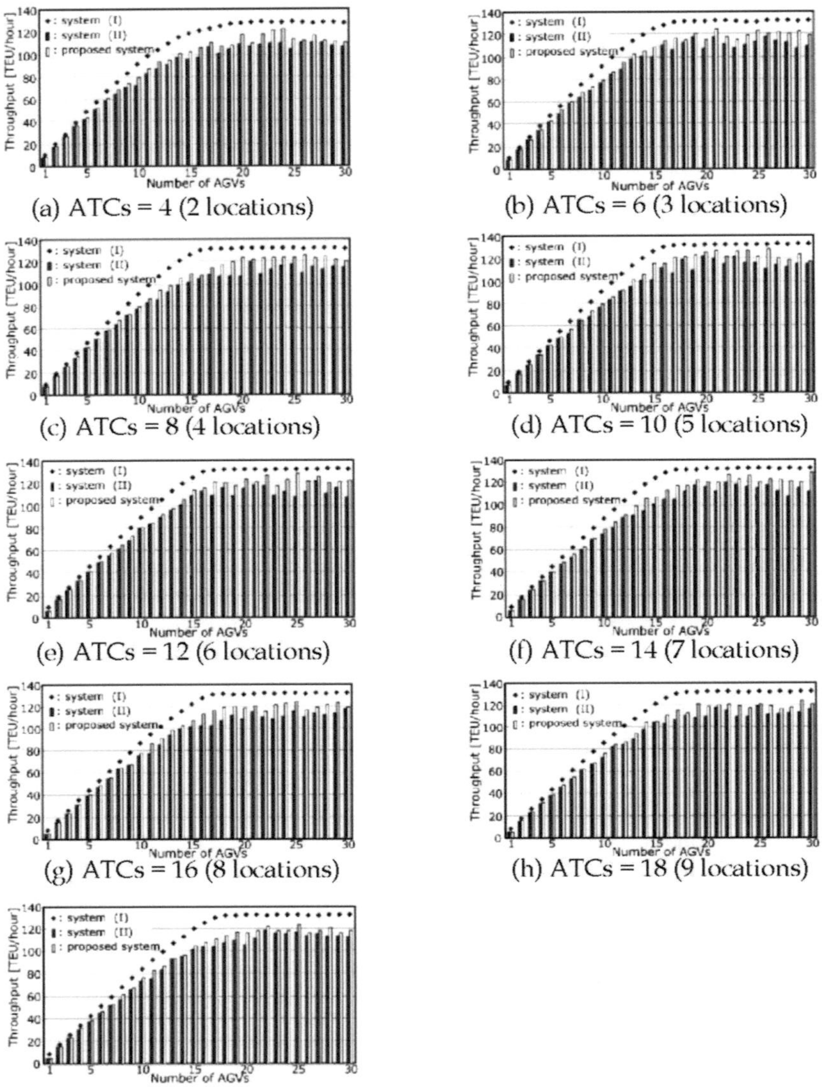

(a) ATCs = 4 (2 locations)

(b) ATCs = 6 (3 locations)

(c) ATCs = 8 (4 locations)

(d) ATCs = 10 (5 locations)

(e) ATCs = 12 (6 locations)

(f) ATCs = 14 (7 locations)

(g) ATCs = 16 (8 locations)

(h) ATCs = 18 (9 locations)

(i) ATCs = 20 (10 locations)

Fig. 7 Comparison result of systems on the basis of the throughputs

From Table 1, we can see that the increase of the proposed system throughput is 5.4 to 9.2 (average), 11.6 to 20.5 (max.), and 0.9 to 4.0 (min.). The average increase of 9.2 [TEU/hour] produces an increase of 100 [TEU] container volume within 10 hours of system operating time. From the result of the maximum value, the increase in container volume within 10 hours of system operating time was up to 200 [TEU].

# of ATCs (locations)	Average [TEU/hour]	Max. [TEU/hour]	Min. [TEU/hour]
4 (2)	6.0	12.3	0.9
6 (3)	7.5	15.0	2.6
8 (4)	9.2	16.6	1.9
10 (5)	7.5	16.7	2.1
12 (6)	9.2	20.5	2.4
14 (7)	7.8	17.2	4.0
16 (8)	8.6	14.2	1.3
18 (9)	6.4	13.0	1.0
20 (10)	5.4	11.6	2.1

Table 1. Increase of the system throughput

Furthermore, we can see that the proposed system is particularly effective when the 6 to 16 ATCs were used. This is because the number of yard lanes increases or decreases according to the number of locations in the container storage yard. In the proposed system, since the robots perform the given tasks by switching three hybrid operational techniques reactively, the AGVs could not change and select their destinations and lanes appropriately in a case in which there were few yard lanes in the yard side, e.g., four ATCs and two lanes (locations). As a result, the increase of the throughput was comparatively low. In a case in which 18 or 20 ATCs were used, i.e., there were 9 or 10 yard lanes and locations, the AGVs did not go into the yard lane successively even if an AGV or ATC failed on the yard lane or location because the tasks are assigned to each location equally for the AGVs, as described in 2.2. Hence, the increase of the throughput was low in the system with many ATCs. However, from the result that the entire throughput was higher than that of the system (II), finally, the effectiveness of the proposed system in the dynamically changing states was shown.

7. Conclusion

In this paper, we proposed a reactive robot control system with hybrid operational techniques in a seaport container terminal considering the robots' reliability. We developed operational techniques in the normal, preventive maintenance, and corrective maintenance states in order to utilize the mutual substitutability of the operation among robots. In the system, each robot was able to respond to the dynamically changing states reactively with the use of the hybrid operational techniques. Finally, for flexible and efficient container handling, we showed the effectiveness of the proposed system through a simulation experiment.

In future works, we will additionally take into account: (I) a multiple berth scenario in the systems which consist of a bidirectional path layout by developing more complex container routing rule and (II) a fluctuation of the lulls and peaks in the workload for the robots in the maintenance of them, for a highly efficient system.

8. References

Guenter, H.-O. & Kim, K.H. (2005). Container terminals and automated transport systems, Springer, ISBN 3-540-22328-2, Berlin Heidelberg Germany.

Birolini, A. (2007). Reliability engineering: theory and practice, Springer, ISBN 978-3-540-49388-4, Berlin Heidelberg Germany.

Beamon, B.M. (1998). Performance, reliability, and performability of material handling systems. *International Journal of Production Research,* Vol. 3, No. 2, pages 377-393. Sun, Y. (1994). Simulation for maintenance of an FMS: an integrated system of maintenance and decision-making. The International Journal of Advanced ManufacturingTechnology, Vol. 9, No. 1, pages 35-39.

Savsar, M. (2005). Performance analysis of an FMS operating under different failure rates and maintenance policies. *International Journal of Flexible Manufacturing Systems,* Vol. 16, No. 3, pages 229-249.

Hoshino, S. & Ota, J. (2007). Design of an automated transportation system in a seaport container terminal for the reliability of operating robots, *IEEE/RSJ International Conference on Intelligent Robots and Systems,* pages 4259-4264.

Bruzzone, A.G., Bocca, E., Longo, F., & Massei, M. (2007). Training and recruitment in logistics node design by using web-based simulation. *International Journal of Internet Manufacturing and Services,* Vol. 1, No. 1, pages 32-50.

Hoshino, S., Ota, J., Shinozaki, A., & Hashimoto, H. (2007). Hybrid design methodology and cost-effectiveness evaluation of AGV transportation systems. *IEEE Transactions on Automation Science and Engineering,* Vol. 4, No. 3, pages 360-372.

Wang, J. (1994). On sign-board based inter-robot communication in distributed robotic systems, *IEEE International Conference on Robotics and Automation,* pages 1045-1050.

Qiu, L. & Hsu, W.-J. (2001). A bi-directional path layout for conflict-free routing of AGVs. *International Journal of Production Research,* Vol. 39, No. 10, pages 2177-2195.

Zeng, J. & Hsu W.-J. (2008). Conflict-free container routing in mesh yard layouts. *Robotics and Autonomous Systems,* Vol. 56, No. 5, pages 451-460.

Hoshino, S., Ota, J., Akiko, S., & Hashimoto, H. (2005). Highly efficient AGV transportation system management using agent cooperation and container storage planning, *IEEE/RSJ International Conference on Intelligent Robots and Systems,* pages 2330-2335.

Hoshino, S., Ota, J., Akiko, S., & Hashimoto, H. (2006). Design of an AGV transportation system by considering management model in an ACT, In: *Intelligent Autonomous Systems 9,* Eds. Arai, T., Pfeifer, R., Balch, T., & Yokoi, H., pages 505-514, IOS, ISBN 978-1-58603-595-2, Amsterdam Netherlands.

3

Robust nonlinear control of a 7 DOF model-scale helicopter under wind gusts using disturbance observers

Adnan Martini[1], Frangois Leonard and Gabriel Abba
Design Engineering, Manufacturing and Control Lab., Arts et Metiers, ParisTech
France

1. Introduction

Nowadays, high levels of agility, maneuverability and capability of operating in reduced visual environments and adverse weather conditions are the new trends of helicopter design. Helicopter flight control systems should make these performance requirements achievable by improving tracking performance and disturbance rejection capability. Robustness is one of the critical issues which must be considered in the control system design for such high-performance autonomous helicopter, since any mathematical helicopter model, especially those covering large flight domains, will unavoidably have uncertainty due to the empirical representation of aerodynamic forces and moments.
Recently the control problems of unmanned scale helicopter have been attracted extensively attention of control researchers. As the helicopter can hover, it is used to implement many important flight missions such as rescue, surveillance, security operation, traffic monitoring, etc. However, helicopter, which is difficult to hover, is more complicated than other familiar control objects. Helicopter is dynamic unstable when it flights in hover mode at nearly zero forward speed. Moreover, the helicopter is open-loop unstable and most mathematical model contain a moderate-high degree of uncertainty models associated with neglected dynamics and poorly understood aeromechanical couplings. Therefore, it is very important to design a stable controller for unmanned helicopter.
Many previous works focus on (linear and nonlinear, robust, ...) control (Beji and Abichou, 2005) (Frazzoli et al., 2000) (Koo and Sastry, 1998), including a particular attention on the analysis of the stability (Mahony and Hamel, 2004), but very few works have been made on the influence of wind gusts acting on the flying system, whereas it is a crucial problem for out-door applications, especially in urban environment : as a matter of fact, if the autonomous flying system (especially when this system is relatively slight) crosses a crossroads, it can be disturbed by wind gusts and leave its trajectory, which could be critical in a highly dense urban context.
In (Martini et al., 2008), thw controllers (an approximate feedback control AFLC and approximate disturbance observer AADRC) are designed for a nonlinear model of a 7 DOF helicopter using in its approximate minimum phase model. In (Pflimlin et al., 2004), a

control strategy stabilizes the position of the flying vehicle in wind gusts environment, in spite of unknown aerodynamic efforts and is based on robust backstepping approach and estimation of the unknown aerodynamic efforts.
The purpose of this chapter is to present the stabilization (tracking) with motion planning of a 7 DOF disturbed helicopter VARIO Benzin Trainer (Lozano et al. (2005)). As a feedback control, a dynamic decoupling method obtained with an approximate minimum phase model (Koo and Sastry (1998)) is proposed AFLC. To deal with uncertainty and vertical wind gust an approximate disturbance observer is added AADRC (Martini et al. (2008)). Simulations show that AADRC is more effective than the AFLC, ie. the norm of the tracking error are lower in presence of disturbances (small body forces, air resistance and vertical wind gust). However, in the presence of nonlinear disturbances the system after linearization remains nonlinear. The observer used here overcomes easily this nonlinearities by an inner estimation of the external disturbances to impose desired stability and robustness properties on the global closed-loop system. The zero dynamics stabilizes quickly with the two controls. In section 2, a model of a disturbed helicopter is presented. In Section 3 the design and the application of two approaches of robust control for the approximate model are proposed. The study of model stability is carried out in section 4. In section 5, several simulations of the helicopter under vertical wind gust show the relevance of the two controls which are described in this work. Finally some conclusions are presented in section 6.

2. Model of the 7DOF disturbed helicopter

This section presents the nonlinear model of the disturbed helicopter VARIO Benzain Trainer starting from a non disturbed model (Vilchis, 2001). This dynamic model using the formalism of Euler-Lagrange is based on the vector of generalized coordinates q G $\mathbf{R}^7$ that can be defined by (Vilchis, 2001) :

$$q(t) = [x\ y\ z\ \varnothing\ \theta\ \psi\ \gamma]^T$$

where $\xi = [x \quad y \quad z]$ denote the position vector of the center of mass of the helicopter relative to the navigation frame for aircraft attitude problems I (Fig.2). γ is the main rotor azimuth angle. Let $(\varnothing, \theta$ and $\psi)$ denote the three Euler angles(yaw, pitch and roll angles) expressed in the body fixed frame $C = (cm, E_1, E_2, E_3)$ In addition,

we define $\eta = [\varnothing \quad \theta \quad \psi \quad \gamma]^T$, let $\Re : (cm, E_1, E_2, E_3) \rightarrow (O, E_x, E_y, E_z)$ be the rotation matrix representing the orientation of the body fixed frame C with respect to the inertial frame $\mathcal{I}$, where $\Re \in SO(3)$ is an orthogonal matrix :

$$\Re = \begin{bmatrix} C_\theta C_\varnothing & S_\psi S_\theta C_\varnothing - C_\psi S_\varnothing & C_\psi S_\theta C_\varnothing + S_\psi S_\varnothing \\ C_\theta S_\psi & S_\psi S_\theta S_\varnothing + C_\psi C_\varnothing & C_\psi S_\theta S_\varnothing - S_\psi C_\varnothing \\ -S_\theta & S_\psi C_\theta & C_\psi C_\theta \end{bmatrix} \tag{1}$$

The vector :

$$\Omega = \begin{bmatrix} \Omega_\varnothing & \Omega_\theta & \Omega_\psi & \Omega_\gamma \end{bmatrix}^T$$

denotes the angular velocity of the vehicle in the body fixed frame, which can also be written as : $\Omega = J_0 \dot{\eta}$, where $\dot{\eta} \in R^4$ denote the generalized velocities. Then :

$$\Omega = \underbrace{\begin{bmatrix} -S_\theta & 0 & 1 & 0 \\ C_\theta S_\phi & C_\phi & 0 & 0 \\ C_\theta C_\phi & -S_\phi & 0 & 0 \\ 0 & 0 & 0 & 1 \end{bmatrix}}_{J_0} \underbrace{\begin{bmatrix} \dot{\phi} \\ \dot{\theta} \\ \dot{\phi} \\ \dot{\gamma} \end{bmatrix}}_{\dot{\eta}} \tag{2}$$

This allows to define :

$$J_R^T = \begin{bmatrix} \Re & 0;0 & J_0^T \end{bmatrix}$$

In the case of VARIO helicopter, the vector of control inputs u G R^4 is given by :

$$u = \begin{bmatrix} u_1 & u_2 & u_3 & u_4 \end{bmatrix}^T$$

where :

- $u_1(m)$ is the collective pitch angle (swashplate displacement) of the main rotor and the motor power.

F_c and $\mathbf{r_c}$	Analytical Expression
fxc	$T_M u_3$
fyc	$T_M UA + T_T$
fzc	$TM + D_{VI}$
	kbu - $T_M UAZ_M$
$Tyc = Tß$	$kau + C_T + T_M u3 Z_M$
Tzc $Tfj)$	$TTXT$
$\mathbf{T_7}$	$C_m ot + CM$

Table 1. Components of simplified external forces vector (Vilchis (2001))

- $u_2(m)$ is the collective pitch angle (swashplate displacement) of the tail rotor.
- $u_3(rad)$ is the longitudinal pitch angle of the main rotor.
- $u_4(rad)$ is the lateral pitch angle of the main rotor.

We can now calculate the Euler-Lagrange equation $L = T - U$ and obtain the motion equations of helicopter :

$$\frac{d}{dt}\left[\frac{\partial L}{\partial \dot{q}_i}\right] - \frac{\partial L}{\partial q_i} = F_L + F_w \tag{3}$$

where U is the potential energy of helicopter, T is its kinetic energy and $FL = (F_c, T_c)^T$ represents the aerodynamical forces and torques applied to the helicopter at the center of mass (table 1), $F_w = (F_{raf}, T_{raf})^T$ represents the external aerodynamical forces and torques produced by the vertical wind gust.

TM, TT are the main and tail rotor thrust. Here M stands for main rotor and T for tail rotor and D_{vi} is the amplitude of the drag force created by induced velocity in the disc of the main rotor, CM and CT are the main and tail rotor drag torque, respectively. $C_{mo}t = k_{mot}$ x **ui** is the engine torque which is assumed to be proportional to the first control input. $\mathbf{ai_s}$ and bi_s are the longitudinal and lateral flapping angles of the main rotor blades, k is the blades stiffness of main rotor. ZM and XT represent the main and tail rotor center localization with respect to the center of mass, respectively.

The development of equation (3) makes it possible to obtain the following equations Vilchis (001):

$$M(q)\ddot{q}+C(q,\dot{q})\dot{q}+G(q)=Q(q,\dot{q},u,v_{raf}) \tag{4}$$

with: $M(q)\in R^{7x7}$ is the inertia matrix. $C(q,\dot{q})\in R^{7x7}$ is the Coriolis and centrifugal forces matrix, $G(q)$ represents the vector of conservative forces. $Q(q,\dot{q},u,v_{raf})=J_R^T(F_L+F_w)$ is the vector of generalized forces. $F_L=D(\dot{q},u)u+A(\dot{q})u+B(\dot{q})$ The induced gust velocity is noted v_{raf} . Let us now introduce our approach based on induced velocity. While returning to the definition of the induced

velocity in three dimensions to the main rotor disc of the helicopter (R.W.Prouty, 1995) and (Vilchis, 2001) :

$$v_1=\sqrt{-\frac{\dot{x}^2+\dot{y}^2}{2}+\sqrt{\frac{(\dot{x}^2+\dot{y}^2)^2}{4}+v_h^4}}-\frac{\dot{z}}{2}+\sqrt{\frac{\dot{z}^2}{4}+v_h^2}-v_h$$

if A is the area of the main rotor disk then :

$$v_h=\sqrt{\frac{mg+D_{vi}}{2\rho A}} \tag{5}$$

The case $\boldsymbol{X} = y = 0$ corresponds to a vertical flight :

$$v_1=v_2=-\frac{\dot{z}}{2}+\sqrt{\frac{\dot{z}^2}{4}+v_h^2} \tag{6}$$

where $\boldsymbol{v_v}$ and $\boldsymbol{vh}$ are respectively the induced velocities in the case of vertical flight and of hovering. Hereafter an ascending vertical flight with a gust velocity vraf is considered (the wind gust is vertical and it has a downward direction as the vertical induced velocity). In this case the total induced velocity becomes : vi = vv + vraf. If it is assumed that the helicopter flies at low speed, then the induced velocity in vertical flight and the induced velocity in hovering are almost the same ones : (vv « vh), so that vi = vh + vraf. Replacing this value in the force and the torque equations Fl provides the contribution STm, STt and SCm of wind on these aerodynamical actions.

$$\begin{aligned}
\delta T_M &= -c_{18}\dot{\gamma}v_{raf_v}\\
\delta T_T &= 0\\
\delta C_M &= c_{17}\dot{\gamma}v_{raf_v}u_1+c_{36}\|V\|v_{raf_v}u_3+c_{38}\dot{\gamma}v_{raf_v}u_4\\
&\quad +c_{43}\left(2v_v v_{raf_v}+v_{raf_v}^2\right)+c_{45}\|V\|v_{raf_v}+c_{46}\dot{\gamma}v_{raf_v}
\end{aligned} \tag{7}$$

Table 2 shows the change of thrust and torque of main rotor (parameter variations of the helicopter) acting on the helicopter in the presence of the vertical wind gust. These variations are calculated from a nominal position when the helicopter performs a hover flight : vrafv =0 : Tm = -77, 2 (N) and Cm = 3, 66 (N.m).

$v_{raf}(m/s)$	$\delta T_M(N)$	$\frac{\delta T_M}{T_{M_o}}\%$	$\delta C_M(N.m)$	$\frac{\delta C_M}{C_{M_o}}\%$
0,68	-14	18 %	1	27 %
3	-62	80 %	6,4	160%

Table 2. - Forces and torques variations for different vertical wind gusts

We can decompose the dynamique of (4) into two dynamics, a slowly translational dynamic and a fastly rotational dynamic, where m is the total mass of the helicopter, g is the gravitation constant.

$$\begin{aligned} m\ddot{\xi} &= \Re\left(F_c + F_{raf}\right) - mgE_z \\ M(\eta)\ddot{\eta} &= J_0^T\left(\tau_c + \tau_{raf}\right) - C(\eta,\dot{\eta})\dot{\eta} - G(\eta) \end{aligned} \tag{8}$$

3. Design and implementation of the control

3.1. Approximate feedback linearization control (AFLC)

The expressions of forces and torques (which contained 4 controls $[u_1,u_2,u_3,u_4]$) are very complex and have a strong nonlinearities. Therefore, it is appropriate to consider the main rotor thrust $\mathbf{T_M}$ and the $\tau_\phi,\tau_\theta,\tau_\psi$ couples as the new vector of control inputs. Then the real controls can be calculated. The objective of the control flight is to design an autopilot $(T_M,\tau_\phi,\tau_\theta,\tau_\psi)$ for the miniature helicopter to let the vertical, lateral, longitudinal and yaw attitude dynamics track a desired smooth trajectories $Y_d = (x_d,y_d,z_d)$ and ϕ_d: the tracking errors $e_\xi = \xi - Y_d$ and $e_\phi = \phi - \phi_d$ should converge asymptotically to zero in the presence of vertical wind gust. The calculation of the relative degrees gives : $r_1 = r_2 = r_3 = \mathbf{r_4} = 2$, the standard helicopter model have a dimension $n = 14$ and : $\mathbf{r_1} + \mathbf{r_2} + r_3 + \mathbf{r_4} = 8 < n = 14$, this implies the existence of an internal dynamics. In order to ensure system stability, we must analyze the internal stability of the system by studying the zero dynamics. A simulation study of the model (9) shows that the zero dynamics, parameterized by $\{\theta,\psi,\gamma,\dot{\theta},\dot{\psi},\dot{\gamma}\}$, is not asymptotically stable since the equilibrium point is surrounded by a family of periodic orbits. It can also be shown that exact state-space linearization fails to transforme the system into a linear and controllable system. Hence, it is impossible to fully linearize the nonlinear system (Koo and Sastry, 1998). Neglecting the couplings between moments and forces, we show that the approximated system with dynamic decoupling is full state linearizable.
Starting from (8), and neglecting the small body forces ($f_{xc} = f_{yc} = 0$) and for $F_{raf} = T_{raf} = 0$, an approximate model of the dynamics of translation is obtained :

$$\ddot{\xi}_a = \frac{1}{m}\left(\Re\begin{bmatrix}0 & 0 & T_M\end{bmatrix}^T - \begin{bmatrix}0 & 0 & mg\end{bmatrix}^T\right) \tag{9}$$

In order to make the approximate model (9) completely linearizable, we will use a dynamic extension procedure. This is done by adding two integrators for the thrust control input T_M. To simplify the expressions, we propose the change of input variables :

$$\tilde{\tau} = -M(\eta)^{-1}C(\eta,\dot{\eta})\dot{\eta} - M(\eta)^{-1}G_7 + M(\eta)^{-1}J_0^T\tau_c \Rightarrow \ddot{\eta} = \tilde{\tau}$$

we will thus consider as the control inputs the vector :

$$\overline{w}_e = \begin{bmatrix} \overline{T}_M & \tilde{\tau}_{\psi} & \tilde{\tau}_{\theta} & \tilde{\tau}_{\phi} \end{bmatrix}^T$$

where $\overline{T}_M = \ddot{T}_M$.

Using the input-output feedback linearization procedure of the position £*a*, we take the third time derivative of (9):

$$\begin{bmatrix} x_a^{(3)} \\ y_a^{(3)} \\ z_a^{(3)} \\ \ddot{\phi} \end{bmatrix} = \begin{bmatrix} * \\ * \\ * \\ 0 \end{bmatrix} + \underbrace{\begin{bmatrix} * & 0 & 0 & 0 \\ * & 0 & 0 & 0 \\ * & 0 & 0 & 0 \\ 0 & 0 & 0 & 1 \end{bmatrix}}_{A_{e_1}} \begin{bmatrix} \dot{T}_M \\ \tilde{\tau}_{\psi} \\ \tilde{\tau}_{\theta} \\ \tilde{\tau}_{\phi} \end{bmatrix} \quad (10)$$

The decoupling matrix Aei has rank 2 only, and therefore is not invertible. Hereafter, we propose to use dynamic decoupling algorithm, and continue differentiating the position £a. At last, the iteration ends, since the decoupling matrix Ae(X) has full rank and is invertible (if Tm = 0). The extended system is in the following form:

$$\underbrace{\begin{bmatrix} x_a^{(4)} \\ y_a^{(4)} \\ z_a^{(4)} \\ \ddot{\phi} \end{bmatrix}}_{v_a} = \underbrace{\begin{bmatrix} b^P \\ b^{\phi} \end{bmatrix}}_{b^e} + \underbrace{\begin{bmatrix} A^P \\ A_{\phi} \end{bmatrix}}_{A_e} \underbrace{\begin{bmatrix} \ddot{T}_M \\ \tilde{\tau}_{\psi} \\ \tilde{\tau}_{\theta} \\ \tilde{\tau}_{\phi} \end{bmatrix}}_{\ddot{w}_e} \quad (11)$$

for which the vector relative degree is { 4 4 4 2 }. The state vector of our extended system can be writhen as following:

$$X = \begin{bmatrix} x & y & z & \phi & \theta & \psi & \gamma & \dot{x} & \dot{y} & \dot{z} & \dot{\phi} & \dot{\theta} & \dot{\psi} & T_M & \dot{T}_M \end{bmatrix}^T$$

Its order is 16 and its control vector is $\overline{w}_e$. We can rewrite the true system in normal form using $(\zeta,\tilde{\eta})$, where:

$$\tilde{\eta} = (\tilde{\eta}_1(X) \quad \tilde{\eta}_2(X))$$

is such that the transformation $I(\zeta,\tilde{\eta})$ --> X define a coordinated change with the particularity that **n** depend only on Z and ***n*** (Isidori (1995)). Define Zi = *xa*, Z2 = *ya*, *Zi* = *za*, Z4 = *P*, and : *Zi* = [Zi *Zi Zi*]T, ***Zi*** = *Z*4, We have a representation of the full state model helicopter (see (8)) :

$$\begin{aligned} \dot{\zeta}_1^p &= \zeta_2^p & \dot{\zeta}_1^{\phi} &= \dot{\zeta}_2^{\phi} \\ \dot{\zeta}_2^p &= \zeta_3^p + H(X,u,v_{raf}) & \dot{\zeta}_2^{\phi} &= b^{\phi} + A^{\phi}\overline{w}_e \\ \dot{\zeta}_3^p &= \zeta_4^p & \dot{\tilde{\eta}} &= f(\zeta,\tilde{\eta},v_{raf}) \\ \dot{\zeta}_4^p &= b^p + A^p\overline{w}_e \end{aligned} \quad (12)$$

in which $H(X,u,v_{raf})$ represents the small body forces and the vertical wind gust forces and torques:

$$H(X,u,v_{raf}) = \ddot{\xi} - \ddot{\xi}_a = \frac{T_M}{m}\Re\begin{bmatrix} u_3 & u_4 + \dfrac{T_T}{T_M} & \dfrac{F_{raf}}{T_M} \end{bmatrix}^T$$

It appears that the sum of relative degree of our extended system is 14, while its size is n_e = *dim(X)* = *16.* There is a difference of 2, which corresponds to the dynamics of the main rotor which is free and which creates a dynamics of order 2, but it is stable (by simulation). It may be noted that this persistent zero dynamics does not exist in the helicopter studied in (Koo and Sastry, 1998) (Mahony et al., 1999) ((Frazzoli et al., 2000) because our helicopter has 7DOF. We can then use the approximate system, the following control who linearizes it for the new controls v_a. We obtain the following equations : $\varpi_e = -A_e^{-1} b_e + A_e^{-1} v_a$ where:

$$v_a = \left[v_\xi^p \;\; v_\phi \right]^T$$

We can then apply the following tracking control law v_a for the approximate system (11), and the true system (12) :

$$\begin{aligned} v_\xi^p &= Y_d^{(4)} - \lambda_3\left(\zeta_4^p - Y_d^{(3)}\right) - \lambda_2\left(\zeta_3^p - \ddot{Y}_d\right) - \lambda_1\left(\zeta_1^p - Y_d\right) \\ v_\phi &= \phi_d^2 - \lambda_5\left(\zeta_2^\phi - \dot{\phi}_d\right) - \lambda_4\left(\zeta_1^\phi - \phi_d\right) \end{aligned} \tag{13}$$

3.2. Active and Approached Disturbance Rejection Control (AADRC)

In this work, a methodology of generic design is proposed to treat the disturbance. A second order system described by the following equation is considered (Han, 1999; Hou et al., 2001)

$$\ddot{y} = f(y, \dot{y}, d) + bu, \tag{14}$$

where f (.) represents the dynamics of the model and the disturbance, d is the input of unknown disturbance, u is the control input, and y is the measured output. An alternative method is presented by (Han, 1999) as follows. The system in (14) is initially increased:

$$\dot{x}_1 = x_2, \quad \dot{x}_2 = x_3 + bu, \quad \dot{x}_3 = \dot{f} \tag{15}$$

where *xi* = *y*, *xi* = *y*, x3 = *f (y,y,d)*. *f (.)* is treated as an increased state. Here f and *f* are unknown. By considering $f(y, \dot{y}, d)$ as a state, it can be estimated with a state estimator. Han in (Han, 1999) proposed a nonlinear observer for (15):

$$\begin{aligned} \dot{\hat{x}} &= A\hat{x} + Bu + Lg(e, \alpha, \delta), \\ \hat{y} &= C\hat{x}, \end{aligned} \tag{16}$$

where:

$$C = \begin{bmatrix} 1 & 0 & 0 \end{bmatrix} \tag{17}$$

and:

$$A = \begin{bmatrix} 0 & 1 & 0 \\ 0 & 0 & 1 \\ 0 & 0 & 0 \end{bmatrix}; B = \begin{bmatrix} 0 \\ b \\ 0 \end{bmatrix}; L = \begin{bmatrix} L_1 \\ L_2 \\ L_3 \end{bmatrix} \tag{18}$$

The observer error is ***e*** = ***y*** - ***xx i*** and:

$$g_i(e, \alpha_i, \delta)\big|_{i=1,2,3} = \begin{cases} |e|\alpha_i \, sign(e) & |e| > \delta \\ \dfrac{e}{\delta^1 - \alpha_i} & |e| \le \delta \end{cases} \qquad \delta > 0. \tag{19}$$

The observer is reduced to the following set of state equations, and is called extended state observer (*ESO*):

$$\begin{cases} \dot{\hat{x}}_1 = \hat{x}_2 + L_1 g_1(e,\alpha,\delta), \\ \dot{\hat{x}}_2 = \hat{x}_3 + L_2 g_2(e,\alpha,\delta) + bu \\ \dot{\hat{x}}_3 = L_3 g_3(e,\alpha,\delta). \end{cases} \tag{20}$$

The dynamics of translation of the approximate helicopter has an order four (11) : $Y_a^{(4)} = \hat{v}_{\hat{\zeta}_1^p}$, with:

$$Y_a = \begin{bmatrix} \zeta_1^1 & \zeta_1^2 & \zeta_1^3 \end{bmatrix}^T$$

While the full system (12) can be put in the following form:

$$Y_a^{(4)} = f\left(Y_a, \dot{Y}_a, Y_a^{(2)}, Y_a^{(3)}, d\right) + \hat{v}_{\hat{\zeta}_1^p} \tag{21}$$

Here d represents $H(X,u,v_{raf})$ and the model uncertainty. Then the system (21) is estimated with the observer:

$$\begin{aligned} \dot{\hat{\zeta}}_1^p &= \hat{\zeta}_2^p + L_{1i} g_{1i}(\hat{e}_{1i}, \alpha_{1i}, \delta_{1i}) \\ \dot{\hat{\zeta}}_2^p &= \hat{\zeta}_3^p + L_{2i} g_{2i}(\hat{e}_{1i}, \alpha_{2i}, \delta_{2i}) \\ \dot{\hat{\zeta}}_3^p &= \hat{\zeta}_4^p + L_{3i} g_{3i}(\hat{e}_{1i}, \alpha_{3i}, \delta_{3i}) \\ \dot{\hat{\zeta}}_4^p &= \hat{\zeta}_1^p + \hat{\zeta}_5^p + L_{4i} g_{4i}(\hat{e}_{4i}, \alpha_{4i}, \delta_{4i}) \\ \dot{\hat{\zeta}}_5^p &= L_{5i} g_{5i}(\hat{e}_{5i}, \alpha_{5i}, \delta_{5i}) \end{aligned} \tag{22}$$

and for Ø-dynamics:

$$\begin{cases} \dot{\hat{\phi}}_1 = \hat{\phi}_2 + L_{16} g_{16}(\hat{e}_4, \alpha_4, \delta_4) \\ \dot{\hat{\phi}}_2 = \hat{\phi}_3 + L_{17} g_{17}(\hat{e}_4, \alpha_4, \delta_4) + \hat{v}_{\phi} \\ \dot{\hat{\phi}}_3 = L_{18} g_{18}(\hat{e}_4, \alpha_4, \delta_4) \end{cases} \tag{23}$$

where $(i = x, y, z)$, and:

$$\hat{v}_a = \begin{bmatrix} \hat{v}_{\hat{\zeta}_1^p} & \hat{v}_{\phi} \end{bmatrix}^T$$

The observer error is $\hat{e}_{1x} = \zeta \int_1^1 - \hat{\zeta} \int_1^1, \hat{e}_{1z} = \zeta \int_1^2, \hat{e}_{1z} = \zeta \int_1^3 - \hat{\zeta} \int_1^3 and \hat{e}_4 = \zeta \int_1^{\phi} - \hat{\phi}_1$and (4 = (* — 4>$i$. The controls are then defined as follows (PD control):

$$\hat{v}_{\hat{\zeta}_1^p} = Y_d^{(4)} - \lambda_3\left(\hat{\zeta}_4^p - Y_d^{(3)}\right) - \lambda_2\left(\hat{\zeta}_3^p - \ddot{Y}_d\right) - \lambda_1\left(\hat{\zeta}_2^p - \dot{Y}_d\right) - \lambda_0\left(\hat{\zeta}_1^p - Y_d\right) - \hat{\xi}_5^p$$

$$\hat{v}_{\phi} = \phi_d^2 - \lambda_5\left(\hat{\phi}_2 - \dot{\phi}_d\right) - \lambda_4\left(\hat{\phi}_1 - \phi_d\right) - \hat{\phi}_3$$

The control signal $\hat{v}_a$takes into account of the terms which depend on the observer $\hat{\zeta}_1^1 ... \hat{\phi}_1$ The fifth part $\left(\hat{\zeta}_5^1, \hat{\zeta}_5^2, \hat{\zeta}_5^3\right)$ and $\hat{\phi}_3$ which also comes from the observer, is added to eliminate the effect of disturbance in this system (21) and in ϕ – dynamics.

4. Stability Analysis

In this section, the stability of the perturbed helicopter (12) controlled using observer based control law (24) is considered. To simplify this study, the demonstration is done to one input and one output (Hauser et al., 1992) and the result is applicable for other outputs. By defining the trajectory error

$$\begin{bmatrix} e_1 \\ e_2 \\ e_3 \\ e_4 \end{bmatrix} = \begin{bmatrix} \zeta_1 \\ \zeta_2 \\ \zeta_3 \\ \zeta_4 \end{bmatrix} - \begin{bmatrix} Y_d \\ \dot{Y}_d \\ \ddot{Y}_d \\ Y_d^{(3)} \end{bmatrix} \tag{25}$$

we can write:

$$\begin{aligned} \dot{e} &= Ae + B\hat{e} + e\varphi\left(X, \hat{\zeta}, v_{raf}\right) \\ \dot{\eta} &= f(\zeta, \eta) \end{aligned} \tag{26}$$

Where $\boldsymbol{A}$ is a stable matrix determined by pole placement:

$$A = \begin{bmatrix} 0 & 1 & 0 & 0 \\ 0 & 0 & 1 & 0 \\ 0 & 0 & 0 & 1 \\ -\lambda_0 & -\lambda_1 & -\lambda_2 & -\lambda_3 \end{bmatrix}$$

and:

$$A = \begin{bmatrix} 0 & 0 & 0 & 0 & 0 \\ 0 & 0 & 0 & 0 & 0 \\ 0 & 0 & 0 & 0 & 0 \\ -\lambda_0 & -\lambda_1 & -\lambda_2 & -\lambda_3 & 1 \end{bmatrix}$$

$\hat{e}$ is the observer error:

$$\begin{bmatrix} \hat{e}_1 & \hat{e}_2 & \hat{e}_3 & \hat{e}_4 & \hat{e}_5 \end{bmatrix}^T = \begin{bmatrix} \zeta_1 & \zeta_2 & \zeta_3 & \zeta_4 & \zeta_5 \end{bmatrix}^T - \begin{bmatrix} \hat{\zeta}_1 & \hat{\zeta}_2 & \hat{\zeta}_3 & \hat{\zeta}_4 & \hat{\zeta}_5 \end{bmatrix}^T$$

Morever:

$$e\varphi\left(X, \hat{\zeta}, v_{raf}\right) = \begin{bmatrix} 0 & H\left(X, \hat{\zeta}, v_{raf}\right) & 0 & 0 \end{bmatrix}^T$$

In this equation:

$$H\left(X, \hat{\zeta}, v_{raf}\right) = e\rho\left(X, v_{raf}\right) u_a\left(X, \hat{\zeta}\right)$$

where:

$$u_a(X) = \begin{bmatrix} \ddot{T}_M & \tilde{\tau}_\psi & \tilde{\tau}_\theta & \tilde{\tau}_\phi \end{bmatrix}$$

$\boldsymbol{e}$ is a positive coupling constant. To simplify the study, we consider the case of a linear observer:

$$\begin{bmatrix}\dot{\hat{e}}_1\\ \dot{\hat{e}}_2\\ \dot{\hat{e}}_3\\ \dot{\hat{e}}_4\\ \dot{\hat{e}}_5\end{bmatrix}=\begin{bmatrix}-L_1 & 1 & 0 & 0 & 0\\ -L_2 & 0 & 1 & 0 & 0\\ -L_3 & 0 & 0 & 1 & 0\\ -L_4 & 0 & 0 & 0 & 1\\ -L_5 & 0 & 0 & 0 & 0\end{bmatrix}\begin{bmatrix}\hat{e}_1\\ \hat{e}_2\\ \hat{e}_3\\ \hat{e}_4\\ \hat{e}_5\end{bmatrix}+\underbrace{\begin{bmatrix}0\\ H\left(X,\hat{\zeta},v_{raf}\right)\\ 0\\ 0\\ p\left(X,\hat{\zeta},v_{raf}\right)\end{bmatrix}}_{e\hat{\varphi}\left(X,\hat{\zeta},v_{raf}\right)} \quad (27)$$

Where $p\left(X,\zeta,v_{raf}\right)=\ \dot{\zeta}_5$ represents the disturbance that comes from the system error and the coupling term $H\left(X,\hat{\zeta},v_{raf}\right)$ we can write: $\dot{\hat{e}}=\hat{A}\hat{e}+e\hat{\varphi}\left(X,\hat{\zeta},v_{raf}\right)$ Where A^V is a stable matrix determined by pole placement.

Theorem : Suppose that:

- The zero dynamics of the approximate system (12) are locally exponentially stable and:
- The functions $H\left(X,\hat{\zeta},v_{raf}\right)$ and $p\left(X,\hat{\zeta},v_{raf}\right)$ are locally Lipschitz and continuous.
- Then for e small and for desired trajectories with sufficiently small values and derivatives $\left(Y_d,\dot{Y}_d,\dots,Y_d^{(4)}\right)$ the states of the system (12) and the states of the observer (22, 23) will be bounded and the tracking error:

$$\|e\|=\|\zeta-Y_d\|\le ke \text{ pour } k<\infty \quad (28)$$

Proof : Since the zero dynamics of approximate model are assumed to be exponentially stable, a conservative Lyapunov theorem implies the existence of a Lyapunov function $V_1(\tilde{\eta})$ for the system $\dot{\tilde{\eta}}=f(0,\tilde{\eta})$ satisfying:

$$\begin{aligned}&k_1\|\tilde{\eta}\|^2\le V_1(\tilde{\eta})\le k_2\|\tilde{\eta}\|^2\\ &\frac{\partial V_1}{\partial\tilde{\eta}}f(0,\tilde{\eta},0)\le -k_3\|\tilde{\eta}\|^2\\ &\left\|\frac{\partial V_1}{\partial\tilde{\eta}}\right\|\le k_4\|\tilde{\eta}\|\end{aligned} \quad (29)$$

for some positive constants k_1,k_2,k_3 and k_4. We first show that $e,\hat{e},\tilde{\eta}$ are bounded. To this end, consider as a Lyapunov function for the error system ((26) and (27)):

$$V(e,\hat{e},\tilde{\eta})=e^TPe+\delta\hat{e}^T\hat{P}\hat{e}+\mu V_1(\tilde{\eta}) \quad (30)$$

where P, $\hat{P}$ > 0 are chosen so that : A^T P + PA = -I and $\hat{A}^T\hat{P}+\hat{P}\hat{A}=-I$ (possible since A and A are Hurwitz), p and S are a positive constants to be determined later. Note that, by assumption, Yd and its first derivatives are bounded : $\|\zeta\|\le\|e\|+b_d,\|\hat{\zeta}\|\le\|\hat{e}\|+\|e\|+b_d$ where $\|Y_d\|\le b_d$. The functions, $f(\zeta,\eta)$ and $\hat{\varphi}\left(X,\hat{\zeta},v_{raf}\right)$ are locally Lipschitz with $\hat{\varphi}(0,0,0)=0$ et $\hat{\varphi}_i(0,0,0)\hat{v}_a(0)=0$, we have:

$$\begin{aligned}&\left\|f\left(\zeta^1,\tilde{\eta}^1,v_{raf}{}^1\right)-f\left(\zeta^2,\tilde{\eta}^2,v_{raf}{}^2\right)\right\|\le l_q\left(\left\|\zeta^1-\zeta^2\right\|+\left\|\tilde{\eta}^1-\tilde{\eta}^2\right\|+\left\|v_{raf}^1+v_{raf}^2\right\|\right)\\ &\left\|2P\varphi\left(X,\zeta,v_{raf}\right)\right\|\le l_u\left(\|X\|+\left\|\zeta\right\|+\left\|v_{raf}\right\|\right)\\ &\left\|2\hat{P}\hat{\varphi}\left(X,\zeta,v_{raf}\right)\right\|\le\hat{l}_u\left(\|X\|+\left\|\hat{\zeta}\right\|+\left\|v_{raf}\right\|\right)\end{aligned}$$

with l_q, l_u and l_u 3 positive reals. X is a locally diffeomorphism of $(\zeta,\tilde{\eta})$ namely that exists l_x is such that:

$$\|X\| \le l_x\left(\|\zeta\| + \|\tilde{\eta}\|\right)$$

Using these bounds and the properties of $V_1(.)$, we have:

$$\begin{aligned}&\frac{\partial V_1}{\partial \tilde{\eta}} f\left(\zeta,\tilde{\eta},v_{raf}\right) = \frac{\partial V_1}{\partial \tilde{\eta}} f(0,\tilde{\eta},0) + \frac{\partial V_1}{\partial \tilde{\eta}} f\left(\zeta,\tilde{\eta},v_{raf}\right) - \frac{\partial V_1}{\partial \tilde{\eta}} f(0,\tilde{\eta},0)\\ &\le -k_3\|\tilde{\eta}\|^2 + k_4 l_q \|\tilde{\eta}\|\left(\|e\| + b_d + \|v_{raf}\|\right)\end{aligned} \tag{32}$$

Taking the derivative of $V(.,.)$ along the trajectory, we find:

$$\begin{aligned}V &= e^T P e + \delta \hat{e}^T \hat{T} \hat{e} +_1 (\tilde{\eta})\\ \dot{V} &= \dot{e}^T P e + e^T P \dot{e} + \delta \dot{\hat{e}}^T \hat{P} \hat{e} + \delta \hat{e}^T \hat{P} \dot{\hat{e}} + \mu \tfrac{\partial V_1}{\partial \eta} f(\zeta,\tilde{\eta})\\ &\le -[\tfrac{3}{8} - \epsilon\{l_u(2+l_x) - \tfrac{\delta}{2}\hat{l}_u(1+l_x)\}]\|e\|^2 + \tfrac{1}{2} - \delta +\\ &+ \epsilon\{\delta l_x(\tfrac{1}{2}l_u + 1) + \tfrac{5}{2}\hat{l}_u\} + 2(\|B\|\|P\|)^2\\ &-[\tfrac{3}{4}\mu k_3 - \tfrac{1}{2}\delta\epsilon l_u l_x - 4(\epsilon l_u l_x + \mu k_4 l_q)^2]\|\eta\|^2 -\\ &+ (2\epsilon l_u(1+l_x)b_d)^2 + \mu\tfrac{(k_4 l_q b_D)^2}{k_3} + \tfrac{1}{2}[\delta\epsilon b_d(l_u l_x + \hat{l}_u)]^2 +\\ &+ \epsilon(l_u + \delta\hat{l}_u)\|v_{raf}\|^2 + \tfrac{1}{2}\mu k_4 l_q \|v_{raf}\|^2\end{aligned}$$

Define:

$$\begin{aligned}&\mu_0 = \frac{k_3}{32\left(l_u l_x + k_4 l_q\right)^2}\\ &\delta = 1 + 4\left(\|B\|\|P\|\right)^2\\ &e_1 \le \frac{1}{8l_u(2+l_x)}\\ &e_2 \le \frac{-\frac{1}{2} + \frac{\delta}{2} - 2\left(\|B\|\|P\|\right)^2}{\delta\left[\frac{1}{2}l_u l_x + l_x + 2\hat{l}_u + l_u\right]}\\ &e_3 \le \frac{\mu}{4\delta l_u l_x}\left[k_3 + k_4 l_q\right]\end{aligned} \tag{33}$$

Then, for all $\mu \le \mu_0$ and all $e < min(\boldsymbol{\mu}, ei, ei, e\text{-}l)$ and for $\delta \ge \delta_0$, we have:

$$\begin{aligned}\dot{V} \le &-\frac{\|e\|^2}{4} - \frac{\delta}{2}\|\hat{e}\|^2 - \frac{\mu k_3\|\eta\|^2}{2} + \left[\mu\frac{\left(k_4 l_q\right)^2}{k_3} + \left[2el_u(1+\right.\right.\\ &\left.+l_x)\right]^2 + \frac{1}{2}\left[\delta e\left(l_u l_x + \hat{l}_u\right)\right]b_d^2 + e\left(l_u + \delta\hat{l}_u\right)\|v_{raf}\|^2\\ &+\frac{1}{2}\mu k_4 l_q\|v_{raf}\|^2\end{aligned}$$

Thus, $\dot{V} < 0$ whenever $\|e\|,\|\hat{e}\|$ and $\|\tilde{\eta}\|$ is large which implies that $\|\hat{e}\|,\|e\|$ and $\|\tilde{\eta}\|$ and, hence, $\|\zeta\|,\|\hat{\zeta}\|$ and $\|X\|$ are bounded. The above analysis is valid in a neighborhood of the origin. By choosing b_d and e sufficiently small and with appropriate initial conditions, we can guarantee the state will remain in a small neighborhood, which implies that the effect of the disturbance on the closed-loop can be attenuated.

5. Results in simulation

In this section, simulations are presented to illustrate the performance and robustness of proposed controls laws when applied to the full helicopter model with the small body forces, air resistance and with the influence of vertical wind gust in the case of stabilization and trajectory tracking. The regulations parameters values used concern the dynamics model of VARIO 23cc helicopter Vilchis (2001). The initial conditions are:
$\xi_0 = (2 \quad 3 \quad -0.2), \dot{\xi}_0 = 0, \phi_0 = \frac{\pi}{30}$ and $\dot{\phi}_0 = 0$. The initial value adopted for the main rotor thrust force is ***TM*** = *-73.5N*. It is exactly equal to the main rotor thrust force required for the helicopter to perform a hover flight:

- for ***AFLC***: The gains values $(\lambda_0, \lambda_1, \lambda_2, \lambda_3)$ and (λ_4, λ_5) are calculated by pole placement, 4 poles in -10 for the translational dynamics and 2 poles in -5 to **p.**
- for the AADRC: The bandwidths chosen to the observer: $w_{0x} = w_{0y} = w_{0z} = 50\text{rad/s}$, $w_{0\phi} = 25\text{rad/s}$. The choice of α= *0.5* and δ= *0.1* for x, y and z, and α= *0.5* and $\boldsymbol{\delta}$ = *0.02* for ϕ.

The induced gust velocity operating on the principal rotor is chosen as:
$v_{raf} = v_{gm} \sin(2.1 t_d) \;\; if \;\; 28 \le t \le 33$, where $t_d = t - 28$, the value of 2.1 represents $\frac{2\pi V}{L_u}$ where **V** in ***m/s*** is the height rise speed of the helicopter (V = 0.5 m/s) and v_{gm} = *0.3 m/S* is the gust density. This density corresponds to an average vertical wind gust, and L_u = *1.5m* is its length (see Fig.4). We propose a simple trajectory to verify the applicability of designed controls, the trajectories are presented in Fig.2.
Now, we show some results that we have obtained through simulations. The both controls (with or without an observer) manage to stabilize $\theta, \dot{\theta}, \gamma, \dot{\gamma}$ and $\dot{\gamma}$, which are free. The difference between the two controls appears in Fig.3 where the tracking errors are less significant by using the *AADRC* than *AFLC.* The *AADRC* compensates quickly the effect of the disturbance and the small body forces that destabilize the system.

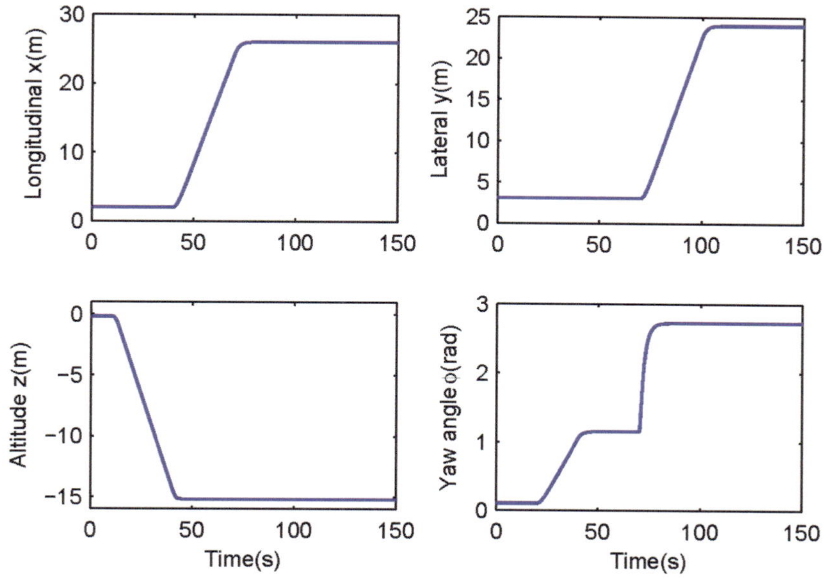

Fig. 2. Desired trajectories

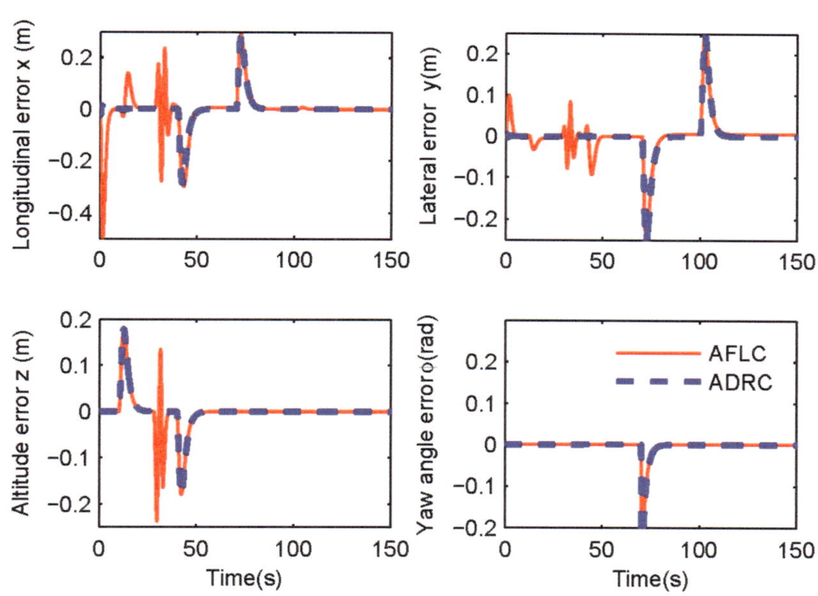

Fig. 3. Tracking errors

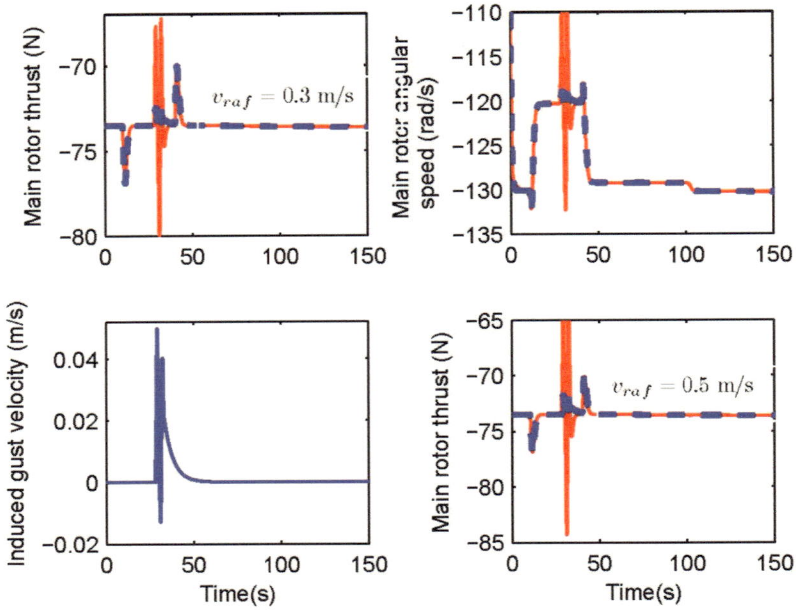

Fig. 4. Variations of T_M, $\dot{\gamma}$ and v_{raf}

One can observe that $\dot{\gamma} \to -130 rad/s$ (which correspond to hover flight). One can also notice that the main rotor angular speed is similar for the two controls as illustrated in Fig.4, but the $AADRC$ compensates quickly the effect of the disturbance.
One can see in Fig.4 that the main rotor thrust converges to values that compensate the helicopter weight, the drag force and the effect of the disturbance on the helicopter. The $AADRC$ allows the main rotor thrust T_M to be less away from its balance position than the other control. If one keeps the same parameters of adjustment for the two controls and using a larger vertical wind gust $(v_{raf} = 0.5 m/s)$ (we have a *PD* controls), we find that the $AADRC$ give better results than the $AFLC$ (see Fig.4).

6. Conclusion

In this chapter, we presented a perturbed nonlinear model of a 7DOF helicopter. As a feedback control, a dynamic decoupling method obtained with an approximate minimum phase model is proposed *AFLC*. To deal with uncertainty and vertical wind gust a disturbance observer is added *AADRC*. Simulations show that the *AADRC* is more effective than the *AFLC,* ie the tracking error are less important in presence of disturbance (small body forces, air resistance and vertical wind gust). However, in the presence of nonlinear disturbances the system after linearization remains nonlinear. The observer used here overcomes easily this nonlinearities by an inner estimation of the external disturbances to impose desired stability and robustness properties on the global closed-loop system. The zeros dynamics stabilizes quickly with the two controls.

Another investigation is planned by using the *AADRC* to stabilize the model of yaw angular displacement of a Tiny CP3 mini-helicopter mounted on experiment platform. We began by the identification setup and we finished with the experimental validation of the AADRC to stabilize the helicopter in presence of lateral wind gust. We find that the identified model is very close to measured model, and the validation results of the AADRC shows the efficiency of proposed control.

7. References

Beji, L. and A. Abichou (2005). Trajectory generation and tracking of a mini-rotorcraft. ***Proceedings of the 2005 IEEE International Conference on Robotics and Automation, Spain,*** 2618-2623.

Frazzoli, E., M. Dahleh, and E. Feron (2000). Trajectory tracking control design for autonomous helicopters using a backstepping algorithm. ***Proceedings of the American Control Conference, Chicago, Illinois,*** 4102-4107.

Han, J. (1999). Nonlinear design methods for control systems. Beijing, China : *The Proc of the 14th IF A C World Congress.*

Hauser, J., S. Sastry, and G. Meyer (1992). Nonlinear control design for slightly non-minimum phase systems : Applications to v/stol aircraft. ***Automatica*** *28*(4), 665-679.

Hou, Y., F. J. Z. Gao, and B. Boulter (2001). Active disturbance rejection control for web tension regulation. *Proceedings of the 40th IEEE Conference on Decision and Control, Orlando, Florida USA,* 4974-4979.

Isidori, A. (1995). ***Nonlinear Control Systems*** (3rd edition ed.). Berlin : SpringerVerlag.

Koo, T. and S. Sastry (1998). Output tracking control design of a helicopter model based on approximate linearization. ***The 37th Conference on Decision and Control, Florida, USA. Vol. 4,*** 3636-3640.

Lozano, R., P. Castillo, and A. Dzul (2005). ***Modeling and control of mini-flying machines.*** Springer Verlag London.

Mahony, R. and T. Hamel (2004). Robust trajectory tracking for a scale model autonomous helicopter. *Int. J. Robust Nonlinear Control 14,* 1035-1059.

Mahony, R., T. Hamel, and A. Dzul (1999). Hover control via approximate lyapunov control for a model helicopter. ***The Conference on Decision and Control, Phoenix, Arizona, USA,*** 533-544.

Martini, A., F. Léonard, and G. Abba (2008). Robust nonlinear control and stability analysis of a 7dof model-scale helicopter under wind gust. In ***IEEE/RSJ, IROS, International Conference of Intelligent Robots and Systems,*** pp. 354-359. NICE, France.

Pflimlin, J., P. Soures, and T. Hamel (2004). Hovering flight stabilization in wind gusts for ducted fan uav. ***Proc. 43 rd IEEE Conference on Decision and Control, Atlantis, Paradise Island, The Bahamas. Vol. 4,*** 3491- 3496.

R. W. Prouty (1995). ***Helicopter Performance, Stability, and Control.*** Krieger.

Vilchis, A. (2001). ***Modélisation et Commande d'Hélicoptère.*** Ph. D. thesis, Institut National Polytechnique de Grenoble.

4

Pursuit-Evasion Games in Presence of Obstacles in Unknown Environments: towards an optimal pursuit strategy *

C. Giovannangeli, M. Heymann and E. Rivlin
Computer Science Department
Haifa, 32000, Israel

Abstract

In this paper, we will incrementally build a complete pursuit algorithm to deal with a 2-players PEG in presence of a single unknown convex obstacle. We will first provide a sufficient condition to achieve capture without disappearance. Then, we will solve the *circular obstacle problem*, a particular problem highlighting a necessary trade-off between surveillance and capture. Next, the *pole problem*, as a generalization, of the convex obstacle problem will be tackled. The solution and the corresponding strategies will be detailed. A quasi-optimal pursuit strategy as regards the time to capture will be provided for the pole problem, and then transposed for the more general *convex obstacle problem*. For the cases leading to the evader victory in the pole problem, a last strategy allowing a maximal deviation of the line of disappearance will be added to complete to our pursuit algorithm. Finally, our complete pursuit algorithm will be adapted to use a heuristic minimization method instead of the strategy suggested by the resolution of the pole problem for the cases leading to the pursuer victory. Different heuristics, one being an approximation of the solution of the pole problem, will be compared with respect to the size of the capture basin and will highlight thez interest of our pursuit algorithm.

1. Introduction

Continuous differential games have been widely studied since the pioneering work of Issacs Isaacs (1965). In particular, pursuit-evasion games (PEGs) have received a great deal of attention, particularly in free spaces for problems such as the missile guidance Basar and Olsder. (1982); Espiau et al. (1992); Hájek (1975); Hutchinson et al. (1996); Isaacs (1965); Song and Um; (1996). Interesting recent works include the notion of *forward reachable sets* (related to maneuverability of the pursuer) for a team of pursuers against a fast moving evader Chung and Furukawa (2006); Chung et al. (2006). In contrast, PEGs in cluttered unknown environments, where obstacles imply specific movement constraints of the pursuer for maintaining visibility and the possibility for the evader to hide, represent a more recent problem, for which a definitive solution has not yet been found. The problem has been split into several classes. A first class of problems is addressed when the evader is not yet visible. Two major issues can be

*This work is supported by the Visiontrain project.
We thank P. Gaussier for allowing the use of the neural development platform for robotics *Promethe*.

discussed: the first consists in developing algorithms in order to find one or several static or dynamic evaders, in an environment either known or unknown, with either a single pursuer or a team of pursuers Chen et al. (2005); Gerkey et al. (2006); LaValle et al. (1997b); Park et al. (2001); Sachs et al. (2004); Suzuki and Yamashita (1992). These approaches suggest that before tracking an evader, efficient solutions to find it should be proposed. The second question, often referred as the Art Gallery Problem, consist in the efficient control of a team of robots so that every part of the environment could be visible by at least one pursuer, thus avoiding the intrusion of a robber in the art gallery Chvatal (1975); Gonzalez-Banos and Latombe (2001); O'Rourke (1983; 1987; 1998); Shermer (1992). Similar works focus on the problem of the positioning a minimum number of captors (movement captors or simply cameras) in the art gallery in order to remove invisible part of the environment. This first class of problems addressed the seeking the evader: what should be done when the evader is not yet visible.
Another major category of problems, that particularly interests us, arises as soon as the target(s) is/are visible. Approaches depend on the relative capabilities of the players, their relative knowledge, their objectives, and the number of pursuers and evaders the mission scenarios consider. A first historical question was raised by David Gal known as the Lion and Man Problem: a man (evader) and a lion (pursuer) are moving with the same speed within the non negative quadrant of the plane. In Sgall (2001), a solution that the author claims to be nearly optimal is proposed, consisting for the lion (whom coordinates are initially greater than the man's coordinates) in aligning himself between the man's position and a particular reference point (the center of the smallest circle, touching its initial position and both axis of the Euclidian space). More generally, interesting solutions for 2-players PEGs have been proposed when the map of the environment is known by the pursuer. The evader may be predictable LaValle et al. (1997a) or more interestingly unpredictable Isler et al. (2004); Murrieta-Cid et al. (2002; 2004; 2003). Thanks to the knowledge of the environment, scenarios mostly focus on maintaining the visibility of the target. Recent solutions rely on the use of a graph of mutual visibility: the environment is first subdivided into regions, and a graph that describes the visibility of each region by the others Murrieta-Cid et al. (2008). A NP-hard method based on this mutual visibility graph is proposed to provide a sufficient condition for maintaining the visibility. Another interesting study highlights that situations in which the target can never be captured Cheung (2005) may exist, even when the evader speed is smaller than the pursuer speed. Scenarios in which the pursuer has to stay at a fixed distance from the evader has also been tackled Muppirala et al. (2005).
In this paper, we consider a minimalist 2-players PEG in presence of obstacles, by assuming that: a) the map of the environment is also unknown; b) the exact positions of the pursuer and the evader are unknown. The mapping of the obstacles is not aimed since: 1) mapping is a classical and well documented problem, 2) PEGs in known environments is also well documented (but not definitively solved), 3) we assume that none of the opponents had time to do the mapping before the conflict, 4) we assume that the game ends upon the capture or the disappearance of the evader, 5) we obviously agree that the use of the map can be helpful for further pursuits in the same environment, 6) we hope in the future to deal with dynamic obstacles (moving obstacles, obstacles shape changes due to a non planar ground ...).
The problem of PEGs in unknown cluttered environment has not been extensively studied: provided solutions mostly aim at maintaining visibility in a classical indoor environment. In Gonzalez-Banos et al. (2002); Lee et al. (2002), the method is based on the minimization of a heuristic called escape risk. A more recent work proposed a better heuristic based on an approximated computation of what is called the vantage time Bandyopadhyay et al. (2006).

Interestingly, the authors point out that trying to maximize the instantaneous visibility, as done in Gonzalez-Banos et al. (2002); Lee et al. (2002), actually increases the latter probability of the evader disappearance, as opposed to a better balancing between closing the distance to the obstacle and maintaining visibility, which seems to offer a better global behavior of the algorithm.

Note that the capture as a termination mode in these previous studies was not considered; the sole objective was to maintain the visibility of the target as long as possible. Moreover, theses studies did not consider the evader as smart, resulting in a problem description without antagonist goals for the pursuer and the evader.

Hence, an interesting facet of this paper is to consider 2-player PEGs in presence of obstacles as an antagonist game, in which the evader is at least as smart as the pursuer and both of them have antagonist objectives. As Isaacs early said: *"... Difficulty of the problems when - and such is the essence of game theory - there are two opponents with conflicting aims and each is to make the best possible decisions understanding and taking into account that his antagonist is doing the same... If we seek conflicting objectives - and only such cases are of interest - the situation assumes something on the nature of the game."*

A last approach that should be cited before entering the heart of the paper is the use of genetic algorithm, inspired by evolutionist neuro-ethological data about the development of pursuit and evasion capabilities among the animal species Cliff and Miller (1996); Miller and Cliff (1994), in order to incrementally generate populations of pursuers and evaders that progress in parallel (ie: that co-evaluate) Choi et al. (2004); Eaton et al. (2002); Nitschke (2003).

In the following, we will incrementally build a complete pursuit algorithm to deal with a 2-players PEG in presence of a single unknown convex obstacle. We will first provide a sufficient condition to achieve capture without disappearance, based on the properties of the famous parallel pursuit. Then, we will solve the *circular obstacle problem*, a particular problem in which the evader, initially located on a circular obstacle, tries to hide behind it. The pursuer, initially located on the tangent to the obstacle crossing the evader position, has to capture the evader in minimum time or at least has to maximally delay its disappearance. Next, the *pole problem*, as a generalization of the convex obstacle problem, will be tackled. In this problem, the pursuer tries to capture the evader and the evader tries to rotate the line of sight in order to create a contact with a pole (corresponding to a point). The solution and the corresponding strategies will be detailed. A quasi-optimal pursuit strategy as regards the time to capture will be provided for the pole problem, and then transposed for the more general *convex obstacle problem*. For the cases leading to the evader victory in the pole problem, a last strategy allowing to maximally rotating the line of disappearance will be added to complete to our pursuit algorithm. Finally, our complete pursuit algorithm will be adapted to use a heuristic minimization method instead of the strategy suggested by the resolution of the pole problem for the cases leading to the pursuer victory. Different heuristics (inspired from the literature, or proposed here, one being an approximation of the solution of the pole problem) will be compared with respect to the size of the capture basin (initial conditions of the pursuer leading to capture without disappearance), highlighting the performance enhancement allowed by our solutions.

2. Interest of the convex obstacle problem

In this paper, a particular subset of the possible games, the *convex obstacle problem*, will be investigated. The rules are the following:

Rule 1: The map of the environment is initially unknown.

Rule 2: The pursuer is faster than the evader.

Rule 3: Each player knows the maximal speed of the other player.

Rule 4: The environment contains a single convex obstacle

Rule 5: The pursuer wins if it captures the evader in finite time while avoiding its disappearance

Rule 6: The evader wins if it succeeds in hiding or if it infinitely delays the capture.

The rule 1 has already been justified previously. The rule 2 is classical in PEGs since if the evader is faster or as fast as the pursuer, as a general rule, it can evade easily [1]. The rule 3 is also used since it largely extends the methods that can be developed. Moreover, the speed of an antagonist can be continuously estimated.

However, at first sight, one can wonder why the disappearance as a termination mode in an environment containing a single convex obstacle is interesting (rules 4, 5, 6). Indeed, in such an environment, even if the evader disappears for a while, the pursuer will eventually see it again and capture it by simply executing the following procedure: first it reaches the disappearance point and then it turns around the obstacle along its boundary. If the evader also moves along the boundary of the obstacle, the pursuer will obviously capture it. Otherwise, the pursuer can move along the obstacle boundary until being on a line orthogonal to the boundary of the obstacle crossing the position of the evader. In such a situation, capture is guaranteed without future disappearance by many pursuit strategies since the capture region is likely to not be altered by the obstacle.

Even if the obstacle is not convex but is simply such that each point of its boundary can be seen from at least one point outside the convex hull of the obstacle (let us call this kind obstacle a *nookless* obstacle), we could prove that capture is guaranteed. Indeed, after disappearance, if the pursuer simply follow the convex hull of the obstacle (which is the shortest path that allow to see all the points on the boundary of a nookless obstacle), either the evader has not entered the convex hull so the pursuer can always reach a position such that it is on the line orthogonal to the convex hull crossing the evader position (similar to the previous problem that consider a convex obstacle), or the evader has entered the convex hull. In this later case, the pursuer can always recover the sight of the evader by simply moving along the convex hull. Once the sight is recovered, the pursuer may use some strategies to prevent the evader to exit the convex hull (the problem becomes closer to a Lion and Man Problem for which solutions exist).

So, why should the disappearance be considered as a termination mode in the case of a single convex or even a nooklees obstacle. In presence of a single non-nookless obstacle or several obstacles, once the evader has disappeared, there is no deterministic guaranty to recover its sight. Indeed, when the pursuer sees a nook in the obstacle, either it enters the nook but the evader may simply have followed the convex hull, or the pursuer follows the convex hull but the evader may simply have entered the nook and may hide in a region that is not seeable from the convex hull. The same dilemma occurs with several obstacles: the pursuer can never know if the evader has turned around the obstacle behind which it has disappeared or if it is hidden behind another obstacle. To solve such situations, several pursuers seem to be required.

That is why it is very important to not lose the sight of the evader, and this is why the disappearance as a termination mode is very important, even if there is only a single convex

[1] Actually, in *Lion and Man problems* Sgall (2001), the evader can be captured even if its speed is the same as the pursuer speed thanks to a line of sight pursuit for which the reference point is well chosen.

obstacle. An efficient pursuer in such a game will largely reduce the probability to face non-deterministic situations as described above in a more general case. Hence, the case of a PEG in presence of a single obstacle can be reduced to a PEG with a single convex obstacle in order to gain insight about the general problem. Moreover, although the convex obstacle problem is the simplest 2-players PEGs in presence of unknown obstacles, an optimal solution has not yet been found.

3. Sufficient capture condition under visibility constraint

In this section, a general sufficient condition that guaranty capture thanks to the properties of the famous parallel pursuit will be established. The region, where this condition holds, covers the major part of the environment.
Assume for a moment the absence of obstacles. The BSR (Boundary of Safe Region) is defined as the frontier of the region in which the evader E is able to go without being captured, what ever the pursuer P does. If the pursuer is faster than the evader, the classical BSR of the evader involved in a PEG in a free 2D space (no obstacles) is defined by an Apollonius circle Isaacs (1965); Nahin (2007); Petrosjan (1993). This definition is evader-centered. We define here the *pursuit region* related to a particular strategy as the set of positions that can be reached by the pursuer during the game when using a particular strategy. We define also the *capture region* related to a particular strategy as the set of positions where the capture can occur. Obviously, the capture region is included in the pursuit region. Finally, we introduce a short terminology about specific geometric objects such as disappearance vertex, line of disappearance and line of sight (see fig. 1.a)

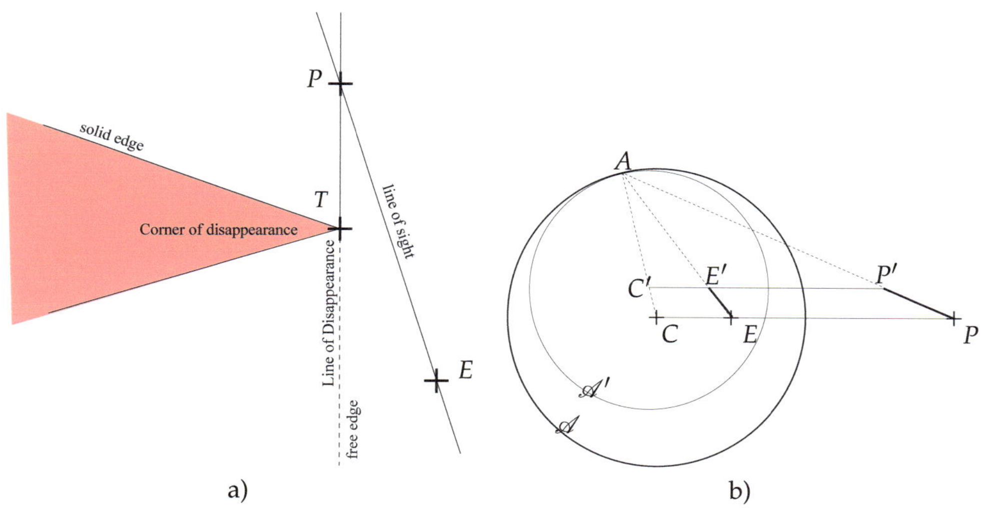

Fig. 1. a) Terminology: the line (PT) is the line of disappearance (i.e.: the tangent to the obstacle crossing the pursuer position), the line (PE) is the line of sight, and T is the disappearance vertex. b) Illustration of the Apollonius circle $\mathscr{A}$ for $\gamma = 4$ (the pursuer is twice faster than the evader), $E : (0,0)$ and $P : (6,0)$. $R = 4$ and $C : (-2,0)$. The Apollonius pursuit is equivalent to a parallel pursuit (all the lines of sight are parallel). Note that $\mathscr{A}'$ is included in $\mathscr{A}'$ and ththe two circle intersects in A.

3.1 Apollonius pursuit properties

Let consider a PEG in the 2D plan with no obstacles, involving a single purser faster than a single evader. The following convention will be used:

- Points in the space are noted with capital letters (such as the point A).
- The coordinates of a point A are noted (x_a, y_a) and (r_a, θ_a) in a polar coordinates system.
- A vector between the origin of the coordinates system and a point A is noted $\boldsymbol{a}$.
- A vector between two points A and B will be noted $\overrightarrow{AB}$ but also $\boldsymbol{b} - \boldsymbol{a}$.
- The angle of a vector $\overrightarrow{AB}$ is noted θ_{AB}.
- $\|.\|$ is the Euclidian 2d-norm.
- The distance between two points A and B can be noted AB but also $\|\boldsymbol{b} - \boldsymbol{a}\|$.
- Geometrical objects are noted with calligraphically written letters (such as the circle $\mathscr{C}$).

The following notations and relations will be used:

- $\boldsymbol{p}$ is the position of the pursuer.
- $\boldsymbol{e}$ is the position of the evader.
- v_e is the maximal speed of the pursuer.
- v_p is the maximal speed of the evader.
- $v_e < v_p$, meaning that the pursuer is faster than the evader.
- $\gamma = k^2 = (\frac{v_p}{v_e})^2$ the square of the ratio k of the pursuer speed above the evader speed.
- $\gamma > 1$ since the pursuer is faster than the evader.

Let us remind some basics results about 2-players PEGs assuming straight line motion of the evader. To capture in minimum time, the optimal pursuit strategy is obviously a straight line motion towards the closest point of capture (a point of capture is such that the time to arrive to this point is the same for both the antagonists). If the evader adopts a straight line motion, the locus of interception $\mathscr{A}$ is the set of points $X : (x, y)$ such that $\frac{\|\boldsymbol{e}-\boldsymbol{x}\|}{v_e} = \frac{\|\boldsymbol{p}-\boldsymbol{x}\|}{v_p}$, more recognizable as:

$$k.\|\boldsymbol{e} - \boldsymbol{x}\| = \|\boldsymbol{p} - \boldsymbol{x}\| \tag{1}$$

$\mathscr{A}$ is an Apollonius circle with E and P as references points and k as parameter (eq. 1 is precisely the definition of an Apollonius circle). Such a circle can be noted $\mathscr{C}(E, P, k)$ The following expression are implied by the equation 1:

$$\begin{aligned} \left(\frac{\|\boldsymbol{e}-\boldsymbol{x}\|}{v_e}\right)^2 &= \left(\frac{\|\boldsymbol{p}-\boldsymbol{x}\|}{v_p}\right)^2 \\ \gamma.\|\boldsymbol{e} - \boldsymbol{x}\|^2 - \|\boldsymbol{p} - \boldsymbol{x}\|^2 &= 0 \end{aligned} \tag{2}$$

With a few substitutions and arrangements, it follows the equation of the Apollonius circle centered on C with the radius R:

$$\begin{cases} \boldsymbol{c} = \dfrac{\gamma.}{\gamma - 1}.(\boldsymbol{e} - \boldsymbol{p}) \\ R^2 = \dfrac{\gamma}{(\gamma - 1)^2}.\|\boldsymbol{e} - \boldsymbol{p}\|^2 \end{cases} \tag{3}$$

C is obviously aligned with E and P, and its the radius R only depends on the distance $\|\boldsymbol{e}-\boldsymbol{p}\|$ between the evader and the pursuer. Thus, C, located on the extension of the segment $[PE]$, can be expressed as $\boldsymbol{c}=\boldsymbol{e}-\frac{1}{\gamma-1}(\boldsymbol{p}-\boldsymbol{e})$. We finally note that the distance between C and P only depends on the distance between E and P as follow (this result will be used later):

$$\|\boldsymbol{c}-\boldsymbol{p}\| = \sqrt{\gamma}.R = \frac{\gamma}{\gamma-1}\|\boldsymbol{e}-\boldsymbol{p}\| \tag{4}$$

The fig. 1.b illustrates the circle $\mathscr{A}$ for a given γ and for the given initial positions of the pursuer and the evader.

If the evader trajectory is a straight line toward a point A of the circle $\mathscr{A}$, there is no better strategy for the pursuer than going also to the point A, since it will go to A in straight line at its maximal speed. This strategy, often called Apollonius pursuit, is time-optimal for straight line motion. Any other pursuer movement will allow the evader to travel a distance greater than $\|\boldsymbol{e}-\boldsymbol{a}\|$.

A well known properties of the Apollonius pursuit is that any line (EP) during the game is parallel to the initial one. Indeed, as highlighted by the fig. 1.b, assume the evader has moved from E to E'. Let ρ be the ratio of the segment $[EA]$ that has been traveled by going from E to E' ($\frac{\|\boldsymbol{e}'-\boldsymbol{e}\|}{\|\boldsymbol{a}-\boldsymbol{e}\|}=\rho$). During the same time, the pursuer has moved to P', and obviously: $\frac{\|\boldsymbol{e}'-\boldsymbol{e}\|}{v_e}=\frac{\|\boldsymbol{p}'-\boldsymbol{p}\|}{v_p}$. The point A being of the Apollonius circle, it follows that: $\frac{\|\boldsymbol{a}-\boldsymbol{e}\|}{v_e}=\frac{\|\boldsymbol{a}-\boldsymbol{p}\|}{v_p}$.

By dividing the two previous equality and with a few arrangement, it follows that:

$$\frac{\|\boldsymbol{e}'-\boldsymbol{a}\|}{\|\boldsymbol{e}-\boldsymbol{a}\|}=\frac{\|\boldsymbol{p}'-\boldsymbol{a}\|}{\|\boldsymbol{p}-\boldsymbol{a}\|}=1-\rho$$

The intercept theorem (or Thales theorem) implies that the line (EP) and $(E'P')$ are parallel. Hence, the Apollonius pursuit is more generally called the parallel pursuit, for antagonists that do not move in straight line during the game.

We introduced here the name Π-strategy to refer to the optimal parallel pursuit, the one continuously minimizing the distance $\|\boldsymbol{e}-\boldsymbol{p}\|$ (the notation Π-strategy is used in Petrosjan (1993)). The Π-strategy ensures that the pursuer will capture the evader inside the circle $\mathscr{A}$, whatever the evader does. This point and other properties of the Π-strategy is reminded in the followings.

3.2 Properties of the Π-strategy

If the evader does not move in straight line, the application Π depends on the current evader velocity:

$$\begin{array}{rccc} \Pi: & \mathbb{R}^6 & \rightarrow & \mathbb{R}^2 \\ & (E,\overrightarrow{v_e},P) & \mapsto & \overrightarrow{v_p} \end{array}$$

It is well known that, in free space (absence of obstacle), if the pursuer is faster than the evader, then the Π-strategy guaranties the capture of the evader inside the initial Apollonius circle $\mathscr{A}$ in finite time without disappearance. Moreover, the Apollonius circle $\mathscr{A}$ is the BSR of the evader (i.e.: the intersection of the capture regions of all the pursuit strategies).

Let us prove the first point. The Π-strategy will first be proved to allow for the evader capture inside the initial Apollonius circle $\mathscr{A}$ (ie: this will prove that the Apollonius circle $\mathscr{A}$ is the capture region related to the Π-strategy whatever the evader does). To prove this point, note that adopting the Π-strategy implies that the new Apollonius circle after an infinitesimal move of the evader and the pursuer is fully included in the initial Apollonius circle $\mathscr{A}$. Then, an upper

bound of the time to achieve the capture can be computed, by noting that the Π-strategy is at the equilibrium as regard a min-max approach. Finally, it will be reminded (thought it is trivial) that if the pursuer does not adopt the Π-strategy, the evader may be captured outside the circle, implying that the Apollonius circle is the BSR of the evader.
Let E' and P' be the point reached by the pursuer and the evader after an infinitesimal duration:

$$\boldsymbol{e}' = \boldsymbol{e} + v_e.dt$$
$$\boldsymbol{p}' = \boldsymbol{p} + v_p.dt$$

Let us call $\mathscr{A}'$ the new Apollonius circle centered on C' with radius R' related to the new positions P' and E' of the antagonists. As previously, $\rho = \frac{\|\boldsymbol{e}'-\boldsymbol{e}\|}{\|\boldsymbol{a}-\boldsymbol{e}\|} = \frac{\|\boldsymbol{p}'-\boldsymbol{p}\|}{\|\boldsymbol{a}-\boldsymbol{p}\|}$ is the ratio of the segment $[EA]$ and $[PA]$ respectively traveled by the evader and the pursuer by going respectively from E to E' and from P to P'. The coordinates E' and P' can be expressed as:

$$\boldsymbol{e}' = \boldsymbol{e} + \rho.(\boldsymbol{a} - \boldsymbol{e}) \quad (5)$$
$$\boldsymbol{p}' = \boldsymbol{p} + \rho.(\boldsymbol{a} - \boldsymbol{p}) \quad (6)$$

By inserting these expression in the definition of the center and the radius of the Apollonius circle, and with a few arrangements, it follows the equation of the circle $\mathscr{A}'$:

$$\begin{cases} \boldsymbol{c}' = \boldsymbol{c} + \rho.(\boldsymbol{a} - \boldsymbol{c}) \\ R' = (1 - \rho).R \end{cases} \quad (7)$$

We have shown here that the center C' of the circle $\mathscr{A}'$ belongs to the segment $[CA]$, which is a radius of $\mathscr{A}$. Obsviously, the point A belongs to the new circle $\mathscr{A}'$ since it is still located at the same time of travel from the antagonists. Hence $R' = \|\boldsymbol{a} - \boldsymbol{c}'\|$.

We now have to prove that $\mathscr{A}'$ is fully included in $\mathscr{A}$. Actually, we have to show that the two circles have at most a single intersection point which is precisely A. Let us provide a geometrical proof (see fig. 1.b for the illustration): consider two circles $\mathscr{A}$ and $\mathscr{A}'$ centered respectively on C and C'. The center $C' \neq C$ is located on a radius $[CA]$ with A a point of $\mathscr{A}$. The two circles intersect at least in A. Let $A' \neq A$ be a point of $\mathscr{A}$. To prove the full inclusion of $\mathscr{A}'$ in $\mathscr{A}$, we have to prove that $C'A' > C'A$.
If $CC'A'$ is a triangle then $CC' + C'A' > CA'$. Since $CA' = CA = CC' + C'A = CA$, then $CC' + C'A' > CC' + C'A$. Hence, we have $C'A' > C'A$. If $CC'A'$ is not a triangle, as $A' \neq A$, $[AA']$ is a diameter of $\mathscr{A}$ implying that $C'A' = CC' + CA'$. As $CC' \neq 0$, it is trivial that $C'A' > CA'$ which conclude the proof.
Of course, if the evader does not travel at maximal speed, the new positions will be such that the new maximal Apollonius circle (taking the maximal speed into account) is also included in the initial one.
Indeed, on the fig. 1.b, if the evader would not have moved at its maximal speed, the new pursuer position would be closer to the new evader position. The pursuer would actually aims a point $\widetilde{A}$ located on the segment $[EA]$. Indeed, for two Apollonius circles $\mathscr{A}$ and $\widetilde{\mathscr{A}}$ sharing the same reference points E and P but with two different speed ratios, respectively k and $\widetilde{k}$ such that $\widetilde{k} > k > 1$ ($\widetilde{k}$ corresponds precisely to the Apollonius circle for an evader moving slower than v_e), all the points on the Apollonius circle with the parameter $\widetilde{k}$ (the higher) are

inside the other Apollonius circle. Note first that if one point of $\widetilde{\mathscr{A}}$ is inside $\mathscr{A}$, all the points of $\widetilde{\mathscr{A}}$ are inside $\mathscr{A}$, since the two circles cannot intersect (an intersection means that a single point is at two different distance ratios from E and P, which is impossible). Second, let O and $\widetilde{O}$ be the intersection of the segment $[EP]$ with the circles $\mathscr{A}$ and $\widetilde{\mathscr{A}}$ respectively. It is clear that $\widetilde{O}$ belongs to the segment $[EO]$ since $\|\boldsymbol{e} - \tilde{\boldsymbol{o}}\| = \frac{1}{\tilde{k}+1}\|\boldsymbol{e} - p\| < \frac{1}{k+1}\|\boldsymbol{e} - p\| = \|\boldsymbol{e} - \boldsymbol{o}\|$. As $\widetilde{O}$, a point of $\widetilde{\mathscr{A}}$ is inside $\mathscr{A}$, $\widetilde{\mathscr{A}}$ is inside $\mathscr{A}$. Thus, if the evader does not move at its maximal speed, the pursuer aims a point $\widetilde{A}$ located on the segment $[EA]$. The new pursuer position noted $\widetilde{P}'$ thus belongs to the segment $[E'P']$.
The new maximal Apollonius circle is obviously included in the initial one. Indeed, for two Apollonius circles $\mathscr{A}'$ and $\widetilde{\mathscr{A}}'$ sharing the inner reference point E' and the same speed ratio $k > 1$, but such that the outer reference points, respectively P' and $\widetilde{P}'$ are different: $\widetilde{P}'$ belongs to $[E'P']$ (the two circles $\mathscr{A}'$ and $\widetilde{\mathscr{A}}$ correspond precisely to the maximal Apollonius circles after an infinitesimal movement of the evader respectively at maximal speed and at a slower speed), $\widetilde{\mathscr{A}}'$ is inside $\mathscr{A}'$. Note first that if one point of $\widetilde{\mathscr{A}}'$ is inside $\mathscr{A}'$, all the points of $\widetilde{\mathscr{A}}'$ are inside $\mathscr{A}'$, since the two circles can not intersect (an intersection means that a single point is a the same distance ratio from E' and two different points P' and $\widetilde{P}'$ belonging to $[E'P']$, which is impossible). Second, let O' and $\widetilde{O}'$ be the intersection of the segment $[E'P']$ with the circles $\mathscr{A}'$ and $\widetilde{\mathscr{A}}'$ respectively. It is clear that $\widetilde{O}'$ belong to the segment $[E'O']$ since $\|\boldsymbol{e}' - \boldsymbol{o}'\| = \frac{1}{k+1}\|\boldsymbol{e}' - \boldsymbol{p}'\| > \frac{1}{k+1}\|\boldsymbol{e}' - \tilde{\boldsymbol{p}}'\| = \|\boldsymbol{e} - \tilde{\boldsymbol{o}}'\|$. As $\widetilde{O}'$, a point of $\widetilde{\mathscr{A}}'$ is inside $\mathscr{A}'$, $\widetilde{\mathscr{A}}'$ is inside $\mathscr{A}'$. Since $\widetilde{\mathscr{A}}' \subset \mathscr{A}' \subset \mathscr{A}$, the new Apollonius circle after an infinitesimal movement in inside the initial one whatever the evader does.
Moreover, as the only intersection of the circle $\mathscr{A}$ and $\mathscr{A}'$ is precisely the point A aimed by the evader, it is obvious that as soon as the evader does not travel in straight line at its maximal speed, it will allow the pursuer to capture it closer to its initial position. Indeed, if the evader change its direction of motion at time $t > 0$ even at maximal speed, the new Apollonius circle will no longer have any contact point with the initial circle $\mathscr{A}$. Hence, for any point E_1 inside $\mathscr{A}$ reached by the evader while the pursuer has reached P_1, the greatest distance between the evader and the pursuer (the distance $\|\boldsymbol{e_1} - \boldsymbol{p_1}\|$) is obtained for a straight line motion of the evader at maximal speed.

The capture occurs in finite time, since a bound to the time to capture exists. As the time to capture is linear with the traveled distance, resulting form the integration of the infinitesimal movements of the pursuer and the evader, let us first compute the movement of the evader that maximizes $\|\boldsymbol{e}' - \boldsymbol{p}'\| = (1-\rho).\|\boldsymbol{e} - \boldsymbol{p}\|$ after an infinitesimal movement. Note that this direction minimizes ρ. We also have that $\rho = \frac{dt.v_e}{\|\boldsymbol{e}-\boldsymbol{a}\|}$. The point A that minimizes ρ also maximizes $\|\boldsymbol{e} - \boldsymbol{a}\|$. Let us express A in a different manner as before:

$$\begin{cases} x_k^a = x_k^c + R_k^c.cos(\alpha) \\ y_k^a = y_k^c + R_k^c.sin(\alpha) \end{cases} \tag{8}$$

We now look for the $\alpha^* \in [0, 2\pi[$ that maximizes the distance $\|\boldsymbol{e} - \boldsymbol{a}\|$:

$$\alpha^* = \underset{\alpha\in[0,2\pi[}{\arg\max}\left(\boldsymbol{e} - \boldsymbol{a}\right) = \underset{\alpha\in[0,2\pi[}{\arg\max}\left(\boldsymbol{e} - \boldsymbol{a}\right)^2$$

With a few arrangements, the problem becomes:

$$\alpha^* = \underset{\alpha\in[0,2\pi[}{\arg\max} \Big((x_c - x_e)cos(\alpha) + \quad (9)$$

$$(y_c - y_e)sin(\alpha)\Big) \quad (10)$$

By studying the variation of this function with respect to α, we have that $\alpha^* = \theta_{EC}$ where θ_{EC} is the direction of the vector $\overrightarrow{EC}$. Hence, the strategy of the evader in order to maximizes the future distance to the pursuer after infinitesimal movement is simply to run away (θ_{EC} being precisely the opposite direction of the pursuer along the line of sight). In parallel, the worst evader strategy is to go toward the pursuer, since the direction of the pursuer $-\theta_{EC}$ also minimizes the future distance between the antagonists. In both case, the Π-strategy leads the pursuer to simply aim an optimal evader like in a pure pursuit strategy, known as the optimal pursuit (against any motion of the evader). The Π-strategy is time optimal for any straight line motion of the evader but also against the optimal evasion strategy (which is a straight line motion). The Π-strategy respects the equilibrium of the min-max approach.
The maximal time to capture t^* corresponds to the optimal value of a PEG involving 2 players with simple motion in free space:

$$t^* = \frac{\|\boldsymbol{p} - \boldsymbol{e}\|}{v_p - v_e}$$

The Π-strategy allows for the evader capture in finite time.
To finally prove that the capture region of the Π-strategy is the BSR of the evader, it is sufficient to notice two facts: first, there exists a strategy, the Π-strategy, that allows for the capture inside the initial Apollonius circle $\mathscr{A}$. Second, if the evader travels in straight line, any other strategy different from the Π-strategy will allow the evader to go outside $\mathscr{A}$. The Apollonius circle is the BSR of the evader.
Finally, let S_r and S_l be the points such that the lines (PS_r) and (PS_l) are the right and left tangent lines to the circle $\mathscr{A}$ starting from P. The union of the triangle PS_rS_l and the circle $\mathscr{A}$ represents the pursuit region (the set of all the pursuer-evader positions during the game) related to the Π-strategy.

3.3 A sufficient condition to guaranty capture without disappearance

Our goal is to provide here a general sufficient condition to guaranty capture under visibility constraint. For convenience, we adopt the same terminology as used in Bandyopadhyay et al. (2006); Gonzalez-Banos et al. (2002); Lee et al. (2002): the set of points that are visible from the pursuer at time t defines a region called the *visibility region* The visibility region is composed by both *solid edges* and *free edges*. A solid edge represents an observed part of the physical obstacles of the environment as opposed to a free edge, which is caused by an occlusion (see fig. 1.a) and is aligned with the pursuer position. In order to hide, the evader must cross a free edge. Any point of a free edge is called an escape point. All the points belonging to the free edges are potential escape points. The disappearance corresponds to the intersection of the light of sight with an obstacle.
An obvious capture condition under visibility constraint is the following:

Condition 3.1. *If the Apollonius circle $\mathscr{A}$ does not intersect neither any free edge, nor any obstacle, then the capture is guaranteed without disappearance by adopting the Π-strategy.*

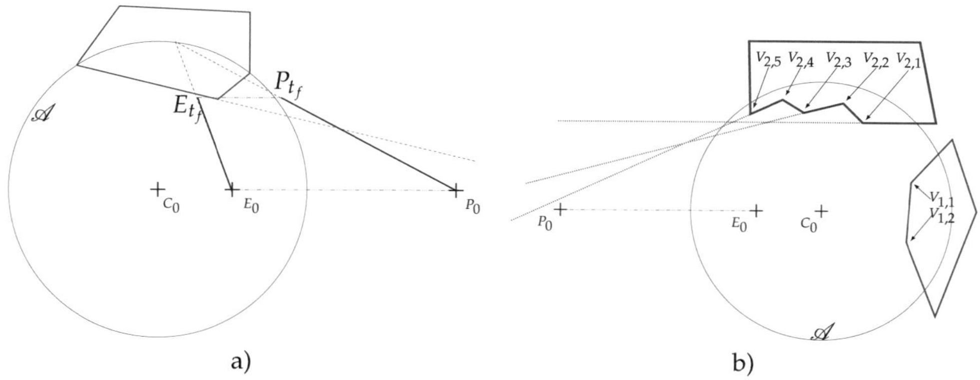

Fig. 2. a) Illustration of the creation of a free edge during the game that did not exist at the beginning of the game: at the beginning, the pursuer sees all the obstacle boundaries (blue polygon) that belong to the pursuit region. Here, $\gamma = 4$: the pursuer is twice faster than the evader. There clearly exists an evasion strategy that will break the line of sight with the obstacle at time t_f if the pursuer adopts the Π-strategy. b) Identification of the vertices that can break the line of sight if the pursuer adopts the Π-strategy. For each vertex inside the Apollonius circle, consider a point E' at a very small distance on the right of the vertex. If the evader can reach E' (i.e. the segment $[EE']$ does not intersect any obstacle edge), the Π-strategy fails. In this example, only the vertices $V_{2,3}$ and $V_{2,5}$ prevents the Π-strategy to capture the evader without disappearance. For all the other vertices, it is clear that E' is inside an obstacle, implying that the evader is not able to reach these escape points.

Indeed, the absence of free edge in the Apollonius circle implies the absence of obstacle in the part of the pursuit region which is outside the Apollonius circle. Since the Apollonius circle does not intersect any obstacle, the pursuit region of the Π-strategy is empty. Everything is thus as a PEG in free space if the pursuer adopts the Π-strategy since none of the possible segment $[E'P']$ can intersect an obstacle.
At first sight, one could think that if the initial Apollonius circle does not intersect any free edge, then capture is guaranteed. The fig. 2.a illustrates an example without any free edge intersecting the initial Apollonius circle, illustrating anyway a movement of the evader that will lead to break the line of sight if the pursuer adopts the Π-strategy. Nevertheless, such situations only happen for particular obstacle shapes that intersect the Apollonius circle. Hence, it is possible to refine the capture condition by refining which kind of obstacles is allowed in the capture region.
Our general sufficient condition to guaranty capture without disappearance is the following:

Condition 3.2. *If the Apollonius circle does not intersect any free edge and if the shape of the obstacles inside the Apollonius circle can not lead to break the line of sight if the pursuer adopts the Π-strategy, then capture is guaranteed without disappearance by adopting the Π-strategy.*

We propose here a simple method to verify if the evader is able to hide from a pursuer using the Π-strategy or not. To simplify, rotate and translate the initial coordinate system such that the new purser position is the origin of the new coordinates system, the line of sight becomes the abscise, and the abscise of E is positive (translation of a vector $-\boldsymbol{p}$ and rotation of an angle

$-\theta_{PE}$ with θ_{PE} the orientation of the vector $\overrightarrow{PE}$ in the initial coordinates system). The figure 2.b is drawn after this transformation. Then, for each vertex $V : (x_v, y_v)$ of the obstacle inside the Apollonius circle, let E' be the point at an arbitrarily small distance ϵ on the right of the vertex V: $x_{e'} = x_v + \epsilon$ and $y_{e'} = y_v$. If E' is inside the obstacle, the evader cannot use the vertex V to hide from a pursuer using the Π-strategy since it would need to cross an obstacle edge. Hence, capture without disappearance is guaranteed by adopting the Π-strategy, if for all obstacle vertices V_k inside the Apollonius circle and all the related E'_k, none of the segments $[EE'_k]$ intersects any obstacle edge. For example, in the figure 2.b, the vertices $\{V_{2,3}, V_{2,5}\}$ prevent to verify this condition, thus prevent to guaranty capture.
In the following, as soon as the condition 3.2 holds, the pursuer will adopt the Π-strategy to terminate the game.

3.4 Region of adoption of the Π-strategy

Given a convex obstacle and a position of the pursuer, let us compute the set of initial evader positions such that the Π-strategy guaranties capture, thanks to the condition thanks 3.2 (refer to fig 4.a).
First, note that, for a convex obstacle, the two contact points of the left and right tangents to the obstacle starting from the pursuer position are the only disappearance vertices. Moreover all the points between the left and right lines of disappearance are visible from P. If the evader is between the two lines of disappearance, and if its time to go to a given disappearance vertex is greater than the time for the pursuer to go to the same vertex, then the Π-strategy guaranties capture.

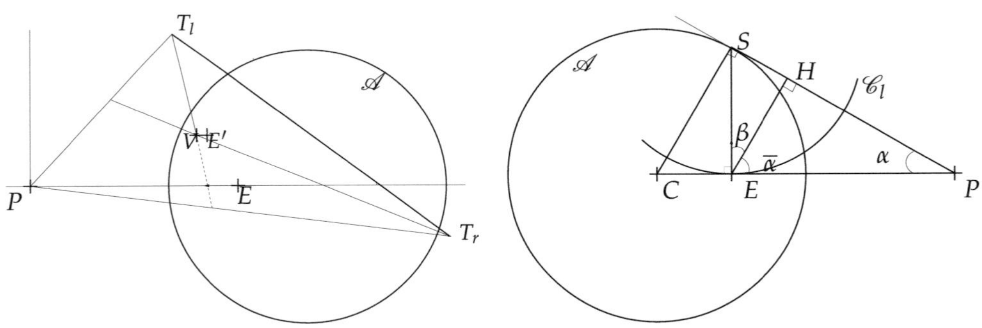

Fig. 3. a) If the evader is between the left and right line of disappearance and if the disappearance vertices do not belongs to the Apollonius circle, then, any disappearance point E' is included in the triangle $T_l V T_r$ (V being the obstacle point creating the occlusion with the pursuer), hence is inside the obstacle, which is impossible. The Π-strategy allows for the capture. b) Computation of the minimal distance between the evader and the line of disappearance, in order to guaranty that the Apollonius circle does not contain any free edge. Here, the line (PS) is the line of disappearance and the disappearance vertex T is assumed to belong to the segment $[PS]$. We demonstrate is the text that the distance EH is linear with respect to the distance PH. The line (ES) is perpendicular to the line (EP) ($\beta = \alpha$). This information helps to determine the set of the position of the evader such that no free edge intersects the Apollonius circle for a given obstacle and a given pursuer's position, when the evader is not between the two lines of disappearance (see fig. 4.a.a).

Indeed, the first point of the condition 3.2 is verifyied because there is no free edge inside the initial Apolonius circle. Moreover, the obstacle being convex, the vertices belonging to the Apollonius circle cannot break the line of sight if the pursuer adopts the Π-strategy (second point of the condition 3.2). This can be demonstrated by noting that for any future possible position of the evader E' inside the Apollonius circle and the corresponding position P' of the pursuer, an occlusion implies the presence of a point V of the obstacle between E' and P', which is impossible due to the convexity of the obstacle. Indeed, perform first the translation of a vector $-\boldsymbol{p}$ and the rotation of an angle $-\theta_{PE}$ (see fig. 3.a) in order to simplify. E being between the two tangents (PT_l) and (PT_r), it follows that $0 \leq \theta_{PT_l} \leq \pi$ and $-\pi \leq \theta_{PT_r} \leq 0$. The obstacle being convex, all the visible points of the obstacle between the two tangents belong to the triangle PT_rT_l. Consider a possible disappearance point E' in the Apollonius circle. If E' (with a positive abscise) is not in the triangle PT_rT_l, E' is not a disappearance point since it is clear that there is not any point of the obstacle (all belonging to the triangle PT_rT_l) between E' and all the possible P' on the left of E'. Hence, E' being a disappearance point, there exists a point V of the obstacle between E' and P': $y_v = y_{e'}$ and $x_v = x_{e'} - \epsilon$ with $\epsilon > 0$. V inside the triangle PT_rT_l implies $\theta_{PT_r} \leq \theta_{PV} \leq \theta_{PT_l}$. It is obvious by construction that the point E' is inside the triangle T_lVT_r since $\theta_{PT_r} < \theta_{VT_r} < (\theta_{VE'} = 0) < \theta_{VT_l} < \theta_{PT_l}$ and E' located in the same half-plan as P (the left one) relatively to the line (T_lT_r). The obstacle being convex, the segments $[T_lV]$ and $[VT_r]$ belong to the obstacle: hence the triangle T_lVT_r to the obstacle, which is impossible since E', a point of this triangle, is, by essence of a disappearance point, outside the obstacle.

To sum-up, if the evader is between the two line of disappearance and if the two vertices of disappearance are outside the Apollonius circle, the Π-strategy guaranties capture without disappearance. In practice, the verification of the second point of the condition 3.2 requires to be checked only if the evader is not between the the two lines of disappearance.

Second, if the evader is not between the two lines of disappearance, what are the positions the set of evader positions such that the Apollonius circle is tangent to a free edge? Note first that if the evader can arrive to a disappearance vertex T before the pursuer ($k.ET < PT$), this vertex belongs to the Apollonius pursuit and the capture cannot be guaranteed[2]. Otherwise, the fig 3.b helps us to compute the minimal distance between the evader and a free edge to guaranty capture under visibility constraint by adopting the Π-strategy. In fig 3.b, (PS) is the line of disappearance, H is the projection of E on the line (PS) and EH is then the distance between the line of disappearance and the evader. The disappearance vertex T belongs to the segment $[PS]$, otherwise the circle would be tangent to the line of disappearance but not tangent with the corresponding free edge (each vertex inside the Apollonius circle should be verified to lead or not to a future line of sight occlusion). The center of the Apollonius circle is noted C and, of course, the line (CS) and (PS) are perpendicular.

We are looking for an expression of the distance EH with respect to the distance HP. The Pythagor theorem also implies that $EH^2 = ES^2 - HS^2$. As the line (EH) and (CS) are parallel and due to the Thales theorem, note that:

$$\frac{CP}{EP} = \frac{PS}{HP} = \frac{CS}{EH} = \frac{\gamma}{\gamma - 1}$$

S being on the Apollonius circle, it follows that:

$$ES^2 = \frac{PS^2}{\gamma} = \frac{\gamma}{(\gamma-1)^2}.HP^2$$

[2] Until the end of the section, the distance between two points A and B will simply be noted AB.

H belonging to $[PS]$ we have:

$$HS^2 = (PS - HP)^2 = \frac{1}{(\gamma-1)^2}.HP^2$$

Hence:

$$EH = HP.\frac{1}{\sqrt{\gamma-1}}$$

The frontier between the evader positions such that the Apollonius circle intersects the line of disappearance or does not is a line (d_l) starting from P. The angle α between this line and the line of disappearance is constant:

$$\alpha = \tan^{-1}\left(\frac{1}{\sqrt{\gamma-1}}\right)$$

Note also that the line (ES) and (EP) are perpendicular. Indeed, in the rectangular triangle PHE, the angle $\widehat{PEH} = \overline{\alpha} = \frac{\pi}{2} - \alpha$. In the rectangular triangle SEH, the angle $\beta = \widehat{SEH}$ is such that

$$\beta = \tan^{-1}\left(\frac{SH}{EH}\right)$$

Since $SH = \frac{1}{\gamma-1}.HP$ and $EH = \frac{1}{\sqrt{\gamma-1}}.HP$, it follows:

$$\beta = \tan^{-1}\left(\frac{1}{\sqrt{\gamma-1}}\right) = \alpha$$

Thus, (ES) and (EP) are perpendicular since $\widehat{PES} = \alpha + \overline{\alpha} = \frac{\pi}{2}$.
Building of the set of the evader positions such that the condition 3.2 holds is now trivial. Indeed, in the fig 3.b, assume that $S = T$ (S is the disappearance vertex T). The circle centered on $S = T$ and crossing E is noted $\mathscr{C}_l$ in fig.3.b and is such that $PS = k.ES$ (or $PT = k.ET$). As (EP) is perpendicular with (ES) (hence with (ET)), the angle α is such that the line (d_l) is tangent to the circle $\mathscr{C}_l$. The contact point between the circle $\mathscr{C}_l$ and its tangent starting from P is the starting point of the frontier between the evader positions such that the Apollonius circle is tangent to the related free edge.
The evader positions such that the condition 3.2 holds are drawn as the white region in fig 4.a. For the proposed obstacle, the Π-strategy allows for the capture in the whole region where the Apollonius circle includes a part of the obstacle but no free edges. In the following, we will focus on the strategy to adopt when the evader position does not belongs to the region where the Π-strategy guaranties capture.

4. The circular obstacle problem

In order to gain insight about what should be done if the condition 3.2 does not hold, the circular obstacle problem defined in fig. 4.b will be investigated. The solving of this game will highlight the existence of a necessary trade-off between maximizing visibility and minimizing the time to capture.
In this game, the evader moves along the boundary of a circular obstacle $\mathscr{C}_e$ (the radius of $\mathscr{C}_e$ is R_e and C is the center). The pursuer is initially located on the tangent to $\mathscr{C}_e$ crossing the evader position. The pursuer tries to capture the evader as fast as possible while maintaining its visibility, or at least it tries to delay the evader disappearance as long as possible. The evader

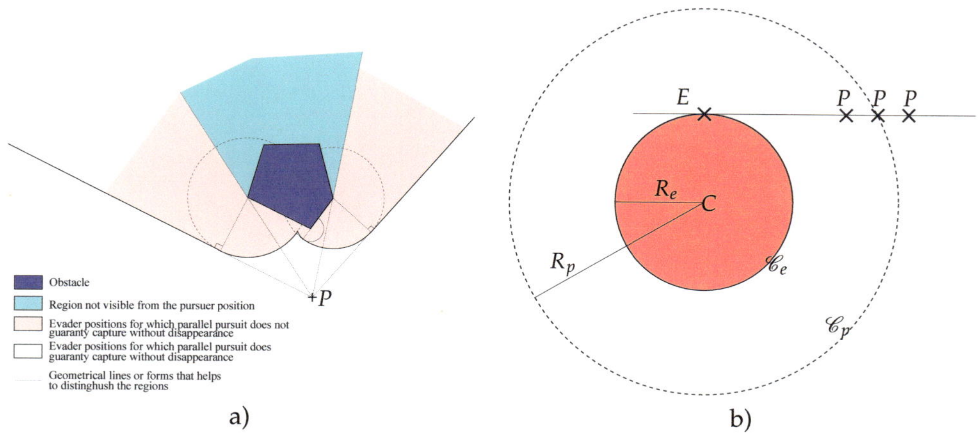

Fig. 4. a) For a given obstacle (convex) and a given position of the pursuer, the figure distinguishes the set of the evader positions such that the Π-strategy guaranties capture without disappearance (region not colored) from the positions for which a free edge intersect the Apollonius circle (region colored in light pink). The two dotted circles are the evader positions such that the time to go to the disappearance vertex is equal for both the antagonists. b) The circular obstacle problem: The evader is moving along the boundary of a circular obstacle $\mathscr{C}_e$, centered on O with the radius R_e. $\mathscr{C}_p$ is another circle centered on O with a radius R_p defined such that $\frac{R_p}{R_e} = \frac{v_p}{v_e} = k$ (the speeds are constant). The pursuer is initially located on the tangent of $\mathscr{C}_e$ touching the evader position. It tries to capture the evader in minimum time while maintaining its visibility (or at least it tries to delay the time to disappearance as long as possible). This corresponds to stay on the tangent, as it will be shown. What is its trajectory, if it starts at a distance $r(0)$? Or at least, what are the kinematics equations of its trajectory.

is initially on the boundary of the obstacle. Hence, moving along the boundary is obviously optimal in order to disappear since this movement maximally deviate the half-plane from which the evader is visible. Assume that $\mathscr{C}_p$ is another circle centered on C with a radius R_p defined such that $\frac{R_p}{R_e} = \frac{v_p}{v_e} = k$ (the speeds are constant, as usual). Hence, $\omega_e = \frac{v_e}{R_e} = \frac{v_p}{R_p}$ is then the angular speed of the evader. The pursuer's speed is $v_p = \omega_e.R_p$.

From the pursuer point of view, minimizing the time to capture (or maximizing the time of visibility maintenance if the capture is impossible) while maintaining the visibility is equivalent to stay on the tangent. Indeed, the fig 5.a illustrate how the pursuer can consume its velocity $v_p.dt$ depending on its distance r to the center C (to simplify the notation, the coordinate of the pursuer P in the polar coordinates system centered on C are r and θ, respectively the radius and the angle of the point P). The evader performs an infinitesimal angular movement $d\phi_e = \omega_e.dt$ between t and $t + dt$. Let $\mathscr{C}_v$ be the circle corresponding to the locus of the possible pursuer positions after an infinitesimal movement. With respect to r, the circles centered on P on the fig 5.a represent the possible positions $\mathscr{C}_v$ the pursuer can reach by consuming its velocity $v_p.dt$.

First, it is clear that the only solution for the pursuer to maintain the evader visibility is to aim a point on the circle $\mathscr{C}_v$ that will be in the half-plane from which the evader is visible at $t + dt$. Second, among all the choices, the best local choice to either capture as fast as possible or at

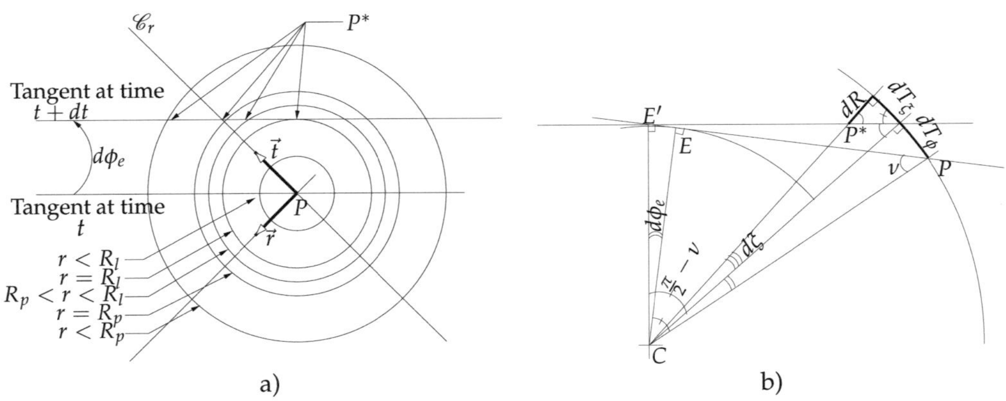

Fig. 5. a) The circular obstacle problem: For an infinitesimal rotation of the evader ($d\phi_e$), how can the pursuer consume its velocity $v_p.dt$ when located at a distance r from the rotation center C. $\vec{r}$ and $\vec{t}$ are the radial and tangential unitary vectors of the polar coordinate system centered on C. The five circles ($\mathscr{C}_v$) centered on P represents the possible positions the pursuer can reach with respect to r, by consuming its velocity $v_p.dt$. From inside to outside: $r > R_l$ implies that the circle $\mathscr{C}_v$ does not intersect the new tangent (instantaneous disappearance as soon as the evader moves); $r = R_l$ is the limit case after which disappearance is instantaneous; $R_p < r < R_l$ implies that the pursuer can maintain the visibility until $r = R_l$ (where the disappearance occurs); $r = R_p$ implies that the pursuer can infinitely maintain the visibility but can not close the distance to the evader (infinite game duration); and $r < R_p$ implies that the pursuer is able to decrease the distance to the evader while maintaining visibility (capture is guaranteed in finite time). b) Decomposition of the pursuer velocity for an infinitesimal movement of the evader.

least maintain the visibility as long as possible is to aim the point which is the closer to the future evader position: this point is actually the intersection of the circle $\mathscr{C}_v$ with the future tangent that minimizes r at time $t + dt$ (hence minimizing its distance L to the evader since $L^2 = r^2 - R_e^2$ as long as the pursuer is on the tangent line to the obstacle). Three cases are possible, with respect to r and to the number of intersections of the circle $\mathscr{C}_v$ with the future tangent:

- $r > R_l$: no intersection point exists: the disappearance is instantaneous as soon as the evader moves.
- $r = R_l$: there exists a single intersection: this corresponds to the limit case, after which the disappearance is instantaneous. Indeed, in this case, as the radius r decreases, the next situation will correspond to the first case.
- $r < R_l$: two intersection points exist: in this case, the pursuer must aim the point P^* that minimizes r at time $t + dt$ (the left one in the figure 5.a). Let us call r^* the radius of P^*. Three sub-cases are possible:
  - $r^* < r$: the pursuer close the distance to the evader ($\dot{r} < 0$) and the capture will eventually occurs in finite time.
  - $r^* = r$: the pursuer stay on the circle with the radius r ($\dot{r} = 0$), and the game duration will be infinite.

- $r^* > r$: the pursuer gets away from the evader ($\dot{r} > 0$): it can only maintain visibility by increasing the distance to the evader, which will eventually disappear.

Let us express the radial and tangential components of the pursuer speed in order to locally minimize the distance to the evader under visibility constraint. The fig 5.b illustrates the different variables to solve the problem, considering an infinitesimal angular movement $d\phi_e$ of the evader.
The pursuer movement can be decomposed into one radial and two tangential components (P^* is the aimed position at $t + dt$):

- $dT_\phi = r.d\phi_e$: the tangential component in order to maintain visibility while remaining at the same distance from the evader.
- $dT_\xi = r.d\xi$: the tangential component in order to reach the line (CP^*) after having performed dT_ϕ.
- dR: the radial component in order to reach the point P^* after having perform dT_ϕ and dT_ξ.

The infinitesimal velocity vector is expressed as follows:

$$\overrightarrow{v_p}.dt = -dR.\overrightarrow{r} + (dT_\phi + dT_\xi)\overrightarrow{t} \tag{11}$$

with $\overrightarrow{r}$ and $\overrightarrow{t}$ the radial and tangential unitary vectors of the polar coordinate system centered on C.
Let us compute dT_ξ as a function of dR. The angle $\nu = (\overrightarrow{PE}, \overrightarrow{PC})$ is really helpful, since $\nu = \sin^{-1}(\frac{R_e}{r})$, and $\tan(\nu) = \frac{dT_\xi}{dR}$ by construction (see fig 5.b). We deduce that:

$$\begin{aligned} dT_\xi &= dR.\tan(\nu) \\ &= \tan(\sin^{-1}(\frac{R_e}{r})) \\ &= \frac{R_e}{\sqrt{r^2 - R_e^2}}.dR \end{aligned} \tag{12}$$

The pursuer velocity can now be expressed as follows:

$$\overrightarrow{v_p} = \dot{r}.\overrightarrow{r} + (\omega_e.r - \dot{r}.\frac{R_e}{\sqrt{r^2 - R_e^2}})\overrightarrow{t} \tag{13}$$

This expression of the pursuer velocity allows to remain on the tangent (valid for P^* and also for the second intersection point of the future tangent with the circle $\mathscr{C}_v$) and is valid in any case if of course $r \leq R_l$ (remember that R_l is the limit case).
The pursuer speed being constant, we obtain the following differential equation (norm of the pursuer velocity):

$$\dot{r}^2.\frac{r^2}{r^2 - {R_e}^2} + \dot{r}.\frac{2.r\omega_e.R_e}{\sqrt{r^2 - R_e^2}} + {\omega_e}^2.r^2 - {v_p}^2 = 0 \tag{14}$$

This equation is quadratic and admits two expressions of $\dot{r}$ (noted $\dot{r}_-$ and $\dot{r}_+$ such that $\dot{r}_- < \dot{r}_+$), corresponding to the two intersections of the future tangent with the locus $\mathscr{C}_v$ of the future pursuer position:

$$\begin{aligned}\dot{r}_- &= \frac{\omega_e.\sqrt{r^2 - {R_e}^2}}{r}.(R_e - \sqrt{R_p^2 + R_e^2 - r^2}) \\ \dot{r}_+ &= \frac{\omega_e.\sqrt{r^2 - {R_e}^2}}{r}.(R_e + \sqrt{R_p^2 + R_e^2 - r^2}) \end{aligned} \quad (15)$$

Reaching P^* obviously corresponds to use the smallest expression $\dot{r}_-$. Moreover, the radius R_l (the limit case for which the circle $\mathscr{C}_v$ has a single intersection with the future tangent) can be computed easily since this is the one for which $\dot{r}_- = \dot{r}_+$:

$$\begin{aligned} & \dot{r}_- = \dot{r}_+ \\ \leftrightarrow \quad & \sqrt{R_p^2 + R_e^2 - R_l^2} = -\sqrt{R_p^2 + R_e^2 - R_l^2} \\ \leftrightarrow \quad & R_l = \sqrt{R_e^2 + R_p^2} \end{aligned} \quad (16)$$

The kinematics equation of the pursuer trajectory, consisting in locally minimizing the distance to the evader under the visibility constraint for the circular obstacle problem is:

$$\begin{cases} \dot{r} = \dfrac{\omega_e.\sqrt{r^2 - {R_e}^2}}{r}.(R_e - \sqrt{R_p^2 + R_e^2 - r^2}) \\ \dot{\theta} = \omega_e - \dfrac{\dot{r}.R_e}{r.\sqrt{r^2 - R_e^2}} \end{cases} \quad (17)$$

Unfortunately, the pursuer trajectory cannot be expressed thanks to classical known functions. The fig. 6.a and 6.b shows the course of the game for an initial position of the pursuer close to but inside the limit circle $\mathscr{C}_p$ ($r(0) < R_p$), on the limit circle ($r(0) = R_p$), and close to but outside the limit circle ($r(0) > R_p$). The trajectories are generated with a numerical solver of differential equation (ode45) provided by Matlab®. Each time step can be seen as a new initial condition, so these trajectories contains almost all the trajectories for initial conditions such that $r(0) \leq R_l$.

The resolution of this game is interesting for at least three reasons: the first one is that most of the methods for visibility maintenance in known environment provided until now assumed a polygonal environment in order to decompose it into a finite number of sub-regions. In the case of a circular obstacle, the number of regions would be infinite and the known methods cannot be applied. The second reason is that this game clearly illustrates the trade-off between fast capture and visibility maintenance in PEGs in presence of obstacles, if the visibility maintenance is a hard constraint of the game. The last one is that this resolution gives insight on what should be done in unknown environment: it seems that doing the minimal but necessary effort to maintain visibility and consuming the spare power in reducing the distance to the future evader position is a relevant strategy, actually locally optimal. The only constraint is to estimate what would be this minimal necessary effort for visibility maintenance.

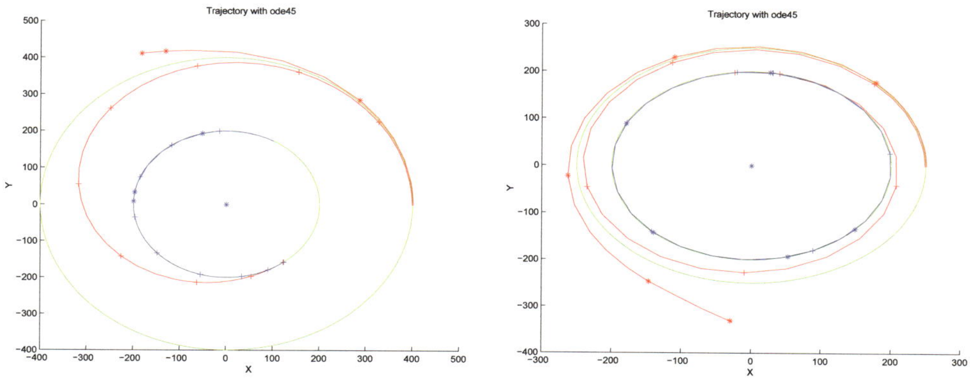

a) b)

Fig. 6. The circular obstacle problem. a) Here, $R_p = 400$, $v_p = 4$, $R_e = 200$, $v_e = 2$. $r(0) = 399$ for the inner trajectory and $r(0) = 401$ for the outter one. The inner green circle is the obstacle ($\mathscr{C}_e$), and the outer green circle is the limit circle $\mathscr{C}_p$ (These circle correspond to the infinite trajectories of the pursuer and the evader when $r(0) = R_p$). The red and the blue trajectories are respectively the trajectory of the pursuer and the evader. The crosses and the star on the trajectories, plotted at the same time step, help to verify that the pursuer is always on the tangent to the circle, touching the evader position. b) Here, $R_p = 250$, $v_p = 2.5$, $R_e = 200$, $v_e = 2$. $r(0) = 249$ for the inner trajectory and $r(0) = 251$ for the outer one.

5. A 2-person PEG biased by a single unknown convex obstacle: Construction of an algorithm

5.1 The pole problem

Let us consider some given initial condition for the convex obstacle problem. As illustrated by the fig. 1.a from the pursuer point of view, the evader will try to hide by crossing the line of disappearance forward the segment $[PT]$ (by crossing the free edge). This line of disappearance can be seen as a stick, anchored on the fixed disappearance vertex T, and such that the pursuer controls its orientation.

As the pursuer does not know the shape of the obstacle outside its visibility region, the worst case would be an extremely sharp obstacle. Hence, a simplification of the convex obstacle problem is to consider the disappearance vertex T as a simple pole or a punctual obstacle. T is now taken as the center of a polar coordinate system as illustrated in the fig. 7. The position of the evader and the pursuer are now respectively noted (r_e, θ_e) and (r_p, θ_p).

The pole problem is defined as follows:

- The evader wins if it can change the sign of the angle α or if it can arrive to the pole before the pursuer ($r_e = 0$) where a final infinitesimal move terminate the game.
- The pursuer wins if it can arrive to the pole before the evader ($r_p = 0$ and $r_p < r_e$) while maintaining the sign of α.

Obviously, if $\frac{r_p}{k} \geq r_e$, then the evader wins whatever the pursuer does by simply going toward the pole ($\dot{r_e} = -v_e$ and $\dot{\theta_e} = 0$). On the fig. 7, for the drawn position of the pursuer and $k = 2$, the initial positions of the evader such that $\frac{r_p}{k} \geq r_e$ is the colored semi-circle (the problem is symmetrical for $\alpha < 0$).

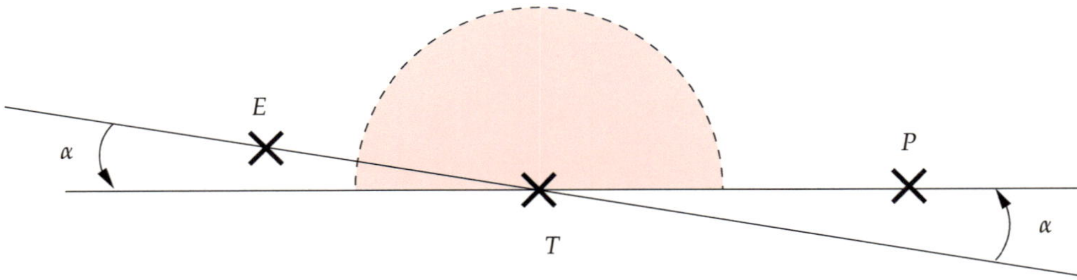

Fig. 7. The pole problem: an approximation of the convex obstacle problem. T is the disappearance vertex, which can be seen as a simple pole by the pursuer in a worst case scenario. Let α be the angle between the lines $(ET)$and (EP). The evader wins the game if it can change the sign of the angle α or if it can arrive to the pole before the pursuer. The pursuer wins if it can avoid the evader to win, and if it can arrive to the pole before the evader. The colored semi-circle represents the positions such that the evader can arrive to the pole before the pursuer by simply moving in straight line (the radius is $\frac{r_p}{k}$ and with $k = 2$). Inside the semi-circle, the evader wins, and otherwise the pursuer wins as shown in the text.

If $\frac{r_p}{k} < r_e$, whatever the evader does, it will be shown there exists a pursuit strategy that guaranties the pursuer victory. Suppose that initially $\alpha > 0$ as in the fig. 7. In order to avoid the evader disappearance, the pursuer must maintain $\alpha > 0$ and it is sufficient to arrive at the pole before the evader to ensure the victory. It is only sufficient because at a given moment of the game, the capture may be guaranteed by adopting the Π-strategy (condition 3.2). To preserve the sign of α, a simple strategy is to rotate at the same angular speed as the evader and to spend the spare power of the velocity vector in decreasing the distance to the pole (let us call this strategy the α-invariant strategy). The kinematics equation of the pursuer adopting the α-invariant strategy is:

$$\begin{cases} \dot{\theta_p} = & \dot{\theta_e} \\ \dot{r_p} = & -\sqrt{{v_p}^2 - {r_p}^2 . \dot{\theta_p}^2} \end{cases} \tag{18}$$

Let us show that the α-invariant enables the pursuer to arrive to the pole in finite time. We have the following relations (with $\dot{r_p} < 0$, and $\dot{\theta}_e > 0$ as in the figure 7):

$$\begin{aligned} v_p^2 &= \dot{r_p}^2 + {r_p}^2 . \dot{\theta_e}^2 \\ v_e^2 &= \frac{{v_p}^2}{k^2} = \dot{r_e}^2 + {r_e}^2 . \dot{\theta_e}^2 \end{aligned}$$

Assume that $\frac{r_p}{k} < r_e$ (it is at least true for $t = 0$), we deduce that:

$$\begin{aligned} {r_e}^2 &> \frac{{r_p}^2}{k^2} \\ {r_e}^2.{\dot{\theta}_e}^2 &> \frac{{r_p}^2.{\dot{\theta}_e}^2}{k^2} \\ \frac{{v_p}^2}{k^2} - {\dot{r_e}}^2 &> \frac{{v_p}^2}{k^2} - \frac{{\dot{r_p}}^2}{k^2} \\ {\dot{r_e}}^2 &< \frac{{\dot{r_p}}^2}{k^2} \end{aligned} \tag{19}$$

Whatever the sign of $\dot{r_e}$, since $\dot{r_p} < 0$, we have the following relation: $\frac{\dot{r_p}}{k} < \dot{r_e}$.
Let us express the derivative at time t as a limit:

$$\begin{aligned} \dot{r_e}(t) &= \lim_{dt \to 0} \frac{r_e(t+dt) - r_e(t)}{dt} \\ \dot{r_p}(t) &= \lim_{dt \to 0} \frac{r_p(t+dt) - r_p(t)}{dt} \end{aligned}$$

Assume $\frac{\dot{r_p}(t)}{k} < \dot{r_e}(t)$ and $\frac{r_p(t)}{k} < r_e(t)$ (at least true for $t = 0$), it follows that

$$\begin{aligned} \lim_{dt \to 0} \frac{r_e(t+dt) - r_e(t)}{dt} &> \lim_{dt \to 0} \frac{r_p(t+dt) - r_p(t)}{k.dt} \\ \lim_{dt \to 0} \frac{r_e(t+dt)}{dt} - \frac{r_p(t)}{k.dt} &> \lim_{dt \to 0} \frac{r_e(t+dt) - r_e(t)}{dt} \end{aligned}$$

Hence,

$$r_e(t+dt) > \frac{r_p(t+dt)}{k}$$

We have shown that, if $r_e > k.r_p$, the α-invariant strategy implies that $\frac{\dot{r_p}}{k} < \dot{r_e}$. We have shown that if $r_e(t) > k.r_p(t)$ and $\frac{\dot{r_p}(t)}{k} < \dot{r_e}(t)$, then $r_e(t+dt) > k.r_p(t+dt)$. Hence, if the pursuer uses the α-invariant strategy and if initially $r_e(0) > k.r_p(0)$, then $r_e(t) > k.r_p(t)$ for all $t > 0$.
Let t^* be the instant such that $r_p(t^*) = 0$ (the pursuer arrives at the pole). It is clear that the evader is not yet arrived at the pole since $r_e(t^*) > k.r_p(t^*) = 0$.
It finally has to be proved that the pursuer is able to arrive at the pole in finite time. Consider the following relation:

$$r_p = m.k.r_e$$

with $m > 0$ a temporal function. If we can show that the $\dot{r_p}$ admit a negative upper bound, the pursuer is obviously able to arrive to the pole. Let us compute the upper limit of $\dot{r_p}$:

$$\begin{aligned} \max \dot{r_p} &= -\sqrt{{v_p}^2 - {r_p}^2.\max(\theta_e)^2} \\ &= -\sqrt{k^2.{v_e}^2 - {r_p}^2 \frac{{v_e}^2}{{r_e}^2}} \\ &= -k.v_e.\sqrt{(1-m^2)} \end{aligned} \tag{20}$$

Now, if we can prove that $\dot{m} < 0$, then we would have proved that whatever the evader does, the α-strategy enables the pursuer to decrease faster and faster the distance to the pole, guarantying to arrive to the pole in finite time.

$$\dot{m} = \frac{1}{k}.\frac{\dot{r_p}r_e - \dot{r_e}r_p}{r_e^2} \tag{21}$$

The sign of $\dot{m}$ depends on $\dot{r_p}r_e - \dot{r_e}r_p$. If $\dot{r_e} > 0$, it is clear that $\dot{m} < 0$ since it is known that $\dot{r_p} < 0$. If $\dot{r_e} < 0$, it follows $\dot{r_p} < k\dot{r_e}$ and that $k.r_e - r_p > 0$, we have:

$$\begin{aligned} \dot{r_p}r_e - \dot{r_e}r_p &< \dot{r_e}(k.r_e - r_p) \\ &< 0 \end{aligned} \tag{22}$$

In any case, $\dot{m} < 0$, which means that the pursuer is able, even in worst case, to converge faster and faster to the pole.
A last remarks is that α can be arbitrarily small without changing the solution of the problem. Hence, a faster pursuit strategy than the α-invariant strategy is to aim the pole as long as $|\alpha| > \epsilon$, with ϵ a security margin arbitrarily chosen, and to use the α-invariant strategy when $|\alpha| = \epsilon$. We call this strategy the α-minimal strategy which is defined as follow:

$$\begin{aligned} &if\ |\alpha| > \epsilon && \begin{cases} \dot{\theta_p} = & 0 \\ \dot{r_p} = & v_p \end{cases} \\ &and\ otherwise && \begin{cases} \dot{\theta_p} = & \theta_e \\ \dot{r_p} = & -\sqrt{v_p{}^2 - r_p{}^2.\dot{\theta_e}^2} \end{cases} \end{aligned} \tag{23}$$

This strategy is locally quasi-optimal since it does the minimal necessary effort to maintain visibility and maximally reduce the distance to the victory position. Interestingly, going toward the pole is equivalent to aim a future position of an evader trying to hide by crossing the free edge. It corresponds to an adaptive proportional pursuit such that the reference point is not a future position of the evader but rather a potential escape point, being precisely on the line of disappearance. The solution of the pole problem is the following:

- If $r_p < k.r_e$ and if the pursuer adopts the α-minimal strategy (or even the α-invariant strategy), the capture without disappearance is guaranteed in finite time.
- If $r_p \geq k.r_e$ and if the evader goes directly towards the pole, disappearance is guaranteed.

5.2 From the pole problem to the convex obstacle problem

The difference between the pole problem and the convex obstacle problem is that the antagonists can not rotate indefinitely around the pole. Indeed, the point T belongs to two obstacle edges. As a consequence, on one hand the α-minimal strategy no longer guaranties capture if $r_e > \frac{r_p}{k}$, but only guaranties to see the next obstacle edge without disappearance. On the other hand $r_e \leq \frac{r_p}{k}$ does no guaranty the evader disappearance. Indeed, if $r_e \leq \frac{r_p}{k}$, the pursuer can rotate the line of disappearance by performing a tangential movement in order to hope to see

the hidden part of the obstacle before the evader disappearance. Two cases must be considered according the position of the orthogonal projection H of the evader position on the line of disappearance (PT).
If the projection H of the position of the evader on the line of disappearance is not forward $[PT]$ (see fig. 8.a), then the evader may disappear by simply reaching the point T. The best the pursuer can do is to spent the time required for the evader to go to T in maximally rotating the line of disappearance. Consider a circle centered on P, with the radius $k.r_e$ (the distance the pursuer can travel while the evader tries to reach the point T). To maximally deviate the line of disappearance, the pursuer must aim the tangent line to this circle crossing the point T, as illustrated on the fig 8.a. For a given position r_e of the evader, the maximal deviation of the line of disappearance is the angle $\delta^* = \sin^{-1}(\frac{k.r_e}{r_p})$ and the distance between the pursuer and the point T will be $r_p^+ = \sqrt{r_p{}^2 - k^2.r_e{}^2}$ at the end of the complete movement. The velocity vector of the pursuer must form an angle $\overline{\delta^*} = \cos^{-1}(\frac{k.r_e}{r_p})$ with the line (PT). Of course, as soon as a new obstacle edge becomes visible, the pursuer has to wonder about its new strategy according to what it sees. Let us call ω the angle between the current line of disappearance and the next visible obstacle edge. If $\omega > \delta^*$, the evader is able disappear whatever the pursuer does. If $\omega \leq \delta^*$, the game will continue since the pursuer will see a new obstacle edge before the evader disappearance. In the following, this strategy will be called the $MD - LoD$ strategy (standing for Maximal Deviation of the Line of Disappearance).
If H is located on the extension of the segment $[PT]$ (see fig. 8.b), the problem is more complex. Indeed, there may exist a future free edge the evader can reach during the game. These future free edges lie outside the pursuer visibility region, and form an angle δ with the line (PT). A particular future line of disappearance is noted (d_δ).
Note that the possible angle δ such that the evader can reach the line d_δ are bounded: $\delta + \alpha \leq \frac{\pi}{2}$ ($\alpha = \widehat{ETP'}$, with P' the symmetrical point of P with respect to T), because reaching a line such that $\delta + \alpha > \frac{\pi}{2}$ is equivalent to reach the point T and is also equivalent to reach the line d_δ with $\delta = \frac{pi}{2} - \alpha$.
In order to hide by crossing a line (d_δ), the best evader motion is to aim its own projection on this line. Let D_δ be this projection. Let $\overline{D_\delta}$ be the projection of P on the line (d_δ). Going straight to $\overline{D_\delta}$ would then be the best solution for the pursuer to avoid the disappearance of the evader. Hence, the evader looks for a line such that $k.\|e - d_\delta\| - \|p - \overline{d_\delta}\| < 0$ (the time for the evader to go the D_δ is smaller than the time for the pursuer to go to $\overline{D_\delta}$). The evader looks for a δ above which $k.r_e.sin(\alpha + \delta) - r_p.sin(\delta) < 0$.
First, if $k.r_e > r_p$ (region A if the fig. 8.b), such a line does not exist. Indeed:

$$k.r_e > r_p \tag{24}$$
$$k.r_e.sin(\delta) > r_p.sin(\delta) \tag{25}$$
$$k.r_e.sin(\delta + \alpha) > r_p.sin(\delta) \tag{26}$$
$$k.r_e.sin(\delta + \alpha) - r_p.sin(\delta) > 0 \tag{27}$$
$$\tag{28}$$

Hence, if the evader is in the region A, the pursuer will use the α-minimal strategy as suggested by the resolution of the pole problem in order to maintain the evader visibility until seeing a new vertex that "deals new cards".

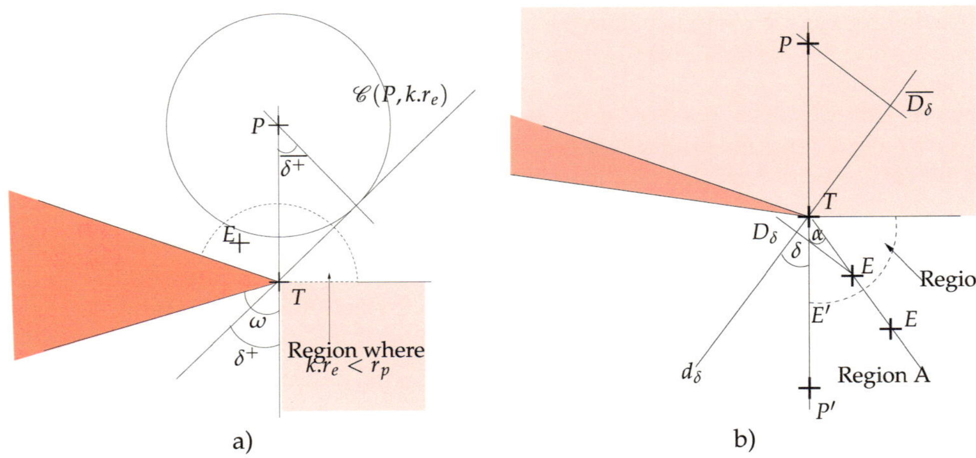

Fig. 8. a) How can the pursuer maximally rotate the line of disappearance if the $r_e < \frac{r_p}{k}$ and if the projection H of E on the line (PT) belongs to the segment $[PT]$ (the light pink region is not considered for the moment). The pursuer must draw the circle centered on its position with a radius $k.r_e$ (corresponding to the position it can reach in straight line while the evader is going to T) and must aim the contact point of the tangent to the circle crossing the point T (this tangent is the maximally rotated line of disappearance the pursuer can create while the evader goes to T). More simply, its velocity vector must from an angle $\overline{\delta^+} = cos^{-1}(\frac{k.r_e}{r_p})$ with the line (PT). In this example, whatever the pursuer does, the evader is able to disappear by reaching the point T because $\omega > \delta^+$. If $\omega \leq \delta^+$, the game would continue whatever the evader does. b) What happens if the projection H of E on the line (PT) is on the extension of the segment $[PT]$ (the light pink region has been treated previously). If the evader is in the region A ($k.r_e > r_p$), it cannot reach any line d_δ before the pursuer. The pursuer will adopt a α-minimal strategy until it sees a new disappearance vertex. If the evader is in the region B ($k.r_e \leq r_p$), there always exist a line d_δ the evader can reach before the pursuer, at least for $\delta = \frac{\pi}{2} - \alpha$ which corresponds for the evader to reach the point T. The pursuer can only deviate the line of disappearance to hope to see the hidden part of the obstacle before the evader reaches a line of disappearance.

Second, if $\frac{r_p}{k} > r_e$ (region B of the fig 8.b), everything becomes drastically more complex. Hence, we decided to not tackle the case in this article. In the following, the pursuer will adopt the α-minimal strategy if the evader is in the region B of the fig 8.b, but we are aware that this case should be considered very thoroughly and carefully in order to determine an efficient strategy. We are also aware that the α-minimal strategy may lead to the evader disappearance in this region.

5.3 An incrementally built pursuit algorithm for the convex obstacle problem

Incrementally, a complete pursuit strategy, which combined three strategies and determine which one is the most relevant according to the current situation, has been built:

- Π-strategy: if the condition 3.2 holds, the pursuer will adopt the Π-strategy to conclude the game. All the games the pursuer can win will finish by the adoption of the Π-strategy.
- MD-LoD strategy: if the projection of the evader position H on the line (PT) is not on the extension of the segment [PT] and if $\frac{r_p}{k} > r_e$, the pursuer uses the MD-LoD strategy to maximally deviate the line of disappearance in order to hope to get the sight of the next hidden edge of the obstacle. If the two possible vertices of disappearance (the left one and the right one) verify this condition, the pursuer should deviate the line of disappearance for which $t_p - t_e$ is the higher, t_p and t_e being the time to reach a given disappearance vertex for respectively the pursuer and the evader.
- α-minimal strategy: If the projection H on the closest disappearance line if forward the segment $[PT]$, then the α-minimal strategy will be used as suggested by the resolution of the pole problem. Yet, if $r_e < \frac{r_p}{k}$, we noticed that a better strategy, which is not under the scope of this article, is likely to exist and should be built.

The figure 9 illustrates on a given example (a given position of the pursuer and a given convex) which strategy is used according to the evader positions. In the following, the concept underlying the α-minimal strategy will be compared with other heuristics that has been proposed in the literature or that appears relevant.

6. Heuristic comparison

In this section, our pursuit algorithm, especially the interest of the α-minimal strategy, will be evaluated. A measure to compare the efficiency of different algorithms is proposed: the

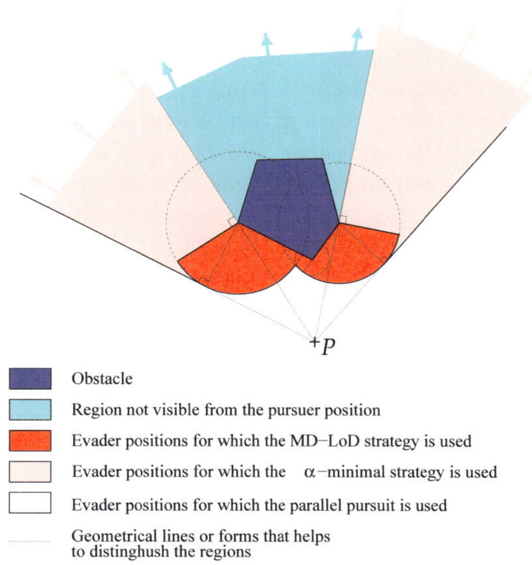

Fig. 9. Strategy used with respect to the evader position, for a fixe position of the pursuer and a given convex obstacle.

size of the capture basins. The proposed methods in the literature for the problem of visibility maintenance are based on heuristics. Our experiment will consist in building the capture basin of our pursuit algorithm if a particular heuristic is used instead of the α-minimal strategy. Heuristics inspired by Bandyopadhyay et al. (2006); Gonzalez-Banos et al. (2002), as well as other simple heuristics that appears relevant for the problem, one approximating the α-minimal strategy, will be compared. We choose to not directly implement the α-minimal strategy because it requires more information that the other heuristics (the evader angular speed related to the disappearance vertex). Anyway, it will be shown that the heuristic approximating the α-minimal strategy largely outperforms the other heuristics. A last important point concerns the evader strategy. Although this article does not deal with evasion, we need that the pursuer plays against a relatively smart evader. In this section, an evasion strategy (which we do not claim to be optimal but simply smart) will be proposed and uses against the pursuer in the experiments.

6.1 Capture and evasion basin

In order to compare pursuit heuristics, a measure is needed. The method proposed here is somehow inspired by the dynamical systems theory. Let the couple evader-pursuer be a coupled dynamical system. As each opponent state is completely determined by 3 coordinates (x, y, θ) (or 2 for simple motion without any constraints on the turning rate), the dynamical system is defined by 6 dimensions. The topology of the obstacles corresponds to a high dimensional set of parameters. An important criterion that can be taken into account to justify that a pursuit algorithm is better than another one is the volume of the capture basin (ie: the set of initial conditions such that the pursuer eventually wins the game): the wider the capture basin, the better the pursuit algorithm for this environment. An optimal algorithm should be such that all the capture basins related to other algorithms are included in the capture basin of this optimal algorithm for any convex obstacles.
As it is particularly difficult to represent such a basin for a 2-players PEG (at least 4 dimensions), and even more difficult to analyze it, the heuristic comparison will be reduced in the followings to the building of capture basins, assuming that a set of initial conditions is fixed (the initial state of the evader). Then, for a given paving of the environment, we build the capture basin of each heuristic in 2 dimensions. For example, in the circular obstacle problem, the capture basin related to the optimal pursuit-evasion strategy is the ring defined by the set of points (r, θ) such that $R_e \leq r < R_p$ and the evasion basin is obviously the set of points such that $r > R_p$.

6.2 List of the variables

Before describing the different heuristics that will be compared, let us first give the list of the variables on which they rely. Then we will define the evader strategy. The obstacle is a polygon (or at least a segment) G when it is seen by the pursuer and G^E when it is seen by the evader (see fig 10).

- T_r and T_l are the two vertices of the polygon G such that the lines (PT_r) and (PT_l) are the right and left tangents to the polygon G starting from P (the two lines of disappearance). The two free edges correspond to the extension of the segments $[PT_l]$ and $[PT_r]$.
- H_r and H_l are respectively the projections of the evader on the line of disappearance (PT_r) and (PT_l).
- r_r and r_l are the distances between the pursuer and the vertices T_r and T_l respectively: $r_r = \|\boldsymbol{p} - \boldsymbol{t_r}\|$ and $r_l = \|\boldsymbol{p} - \boldsymbol{t_l}\|$.

- r'_r and r'_r: if H_r (resp. H_l) is forward $[PT_r]$ (resp. forward $[PT_l]$), we introduce $r'_r = \|\boldsymbol{t_r} - \boldsymbol{h_r}\|$ (resp. $r'_l = \|\boldsymbol{t_l} - \boldsymbol{h_l}\|$) the distance between T_r and H_r (resp. between T_l and H_l).
- h_r and h_l are the distances the evader has to travel in order to hide by crossing (PT_r) and (PT_l) forward $[PT_r]$ and forward $[PT_l)$ respectively. If the path to hide is a broken line, the distance must be computed accordingly.
- l: distance between the line of sight and the disappearance vertex.

To summarize, with the subscript $x = \{r, l\}$ specifying that either the right or left disappearance vertex is considered, r_x is the distance between the pursuer P and the disappearance vertex T_x, h_x is the distance the evader must travel to reach the closest disappearance point on the line of disappearance (PT_x) (the point H_x if H_x is forward $[PT_x]$ or the point T_x otherwise) and r'_x is the distance between the disappearance vertex T_x and the projection H of the evader on the extension of the segment $[PT_x]$ (if the projection on the free edge does not exist because H_x is not forward $[PT_x]$, then $r'_x = 0$).

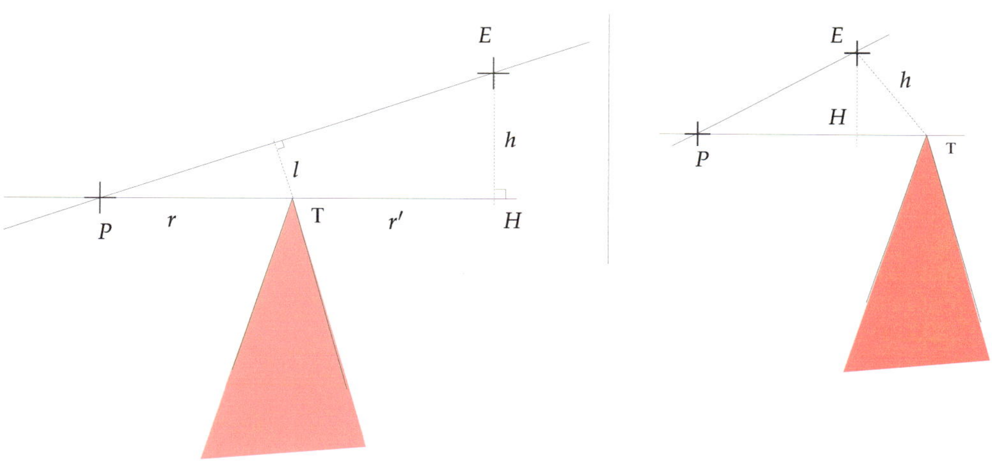

Fig. 10. Distances and points used for the computation of the heuristics by the pursuer. Here, we consider the left line of disappearance. $r = \|\boldsymbol{p} - \boldsymbol{t}\|$, wherever the evader is. As regard the evader E, if its projection H on the line (PT) is not forward the segment $[PT]$, $h = \|\boldsymbol{e} - \boldsymbol{t}\|$ and $r\prime = 0$. If H is forward $[PT]$, $h = \|\boldsymbol{e} - \boldsymbol{h}\|$, $r\prime_r = \|T - H\|$ and finally l is the distance between the line of disappearance vertex and the line of sight. These definitions hold for both the right and left line of disappearance. Hence each variable can be noted with a subscript $x = \{r, l\}$ specifying which line is considered: (r_x, r'_x, h_x).

6.3 Heuristics-based pursuit algorithm under visibility constraint

In order to equitably compare the different heuristic and to tackle the problem of a real game where the players have opposite objectives, we first need a smart evader. Evasion by hiding is not a trivial problem. An obvious local strategy for the evader is to locally aim the most secure disappearance point (H_x or T_x) or at least to run-away in order to delay an unpreventable capture. Evasion strategies not being under the scope of this article, the algorithm to choose the most secure disappearance point is not provided here.

6.3.1 Foreword

The problem of a 2-players PEG in an unknown cluttered environment has recently been tackled. The proposed solutions consist in locally minimizing either the *escape risk* Gonzalez-Banos et al. (2002); Lee et al. (2002) or the *vantage time* Bandyopadhyay et al. (2006). The sole problem of the surveillance was addressed in these works: the termination modes were either the duration of the game or on the disappearance of the evader. Interestingly, the vantage time minimization (let us call this approach VTM) seems to outperform the escape risk minimization (ERM). The authors have highlighted that the surveillance is enhanced by a good balancing between the radial movement (the movement towards the disappearance point) and the tangential movement (orthogonal to the line of disappearance). The ERM gives a too high influence to the tangential movement, and increase the latter probability for a smart evader to escape. On the contrary, the VTM give a higher influence to the radial movement. By an early decrease of the distance to the disappearance vertex, the influence of the future tangential movement is higher and allows for longer visibility maintenance. Here, we claim that the most interesting balancing between the radial and the tangential components of the velocity actually corresponds to a minimal necessary effort in visibility maintenance in order to maximally close the distance to the disappearance vertex (α-minimal strategy). As the evader must aim the disappearance line in order to disappear, closing the distance to the disappearance vertex somehow corresponds to for the pursuer to move towards a future position of an evader that would try to hide.

6.3.2 List of the heuristics

As previously said, the minimization of the different heuristics will be used instead of the α-minimal strategy in our global pursuit algorithm. One of these heuristics leads to a pursuit behavior that is very close to the α-minimal strategy.

The first heuristic H_{ER} is inspired by the escape risk function proposed in Gonzalez-Banos et al. (2002); Lee et al. (2002):

$$H'_{ER} = \max_{x \in \{r,l\}} \left(\frac{v_e . r_x}{v_p . h_x}\right)$$

Step after step, the pursuer should choose to move in order to minimize H_{ER}. An average among all the free edges could have been used instead of the max operator as in Gonzalez-Banos et al. (2002); Lee et al. (2002) but the resulting behavior would leads the pursuer to equilibrated the escape risk among all the free edge influences, whereas the max operator leads to prior focus on the riskiest free edge. A preferable method is to estimate the most critical free edge $x^* = \{r,l\}$ (it is trivial in the region where the heuristic minimization is used as illustrated by the fig. 9). Hence, the following heuristic H_{ER} is equivalent to the heuristic H'_{ER}:

$$H_{ER} = \frac{r}{h}$$

with $r = r_{x^*}$ and $h = h_{x^*}$. In the following, we will use this more simple formalism ($r' = r'_{x^*}$). The constant $\frac{v_e}{v_p}$ is removed, since it has no influence on the local minimization.

The second heuristic inspired by Bandyopadhyay et al. (2006) aims at reducing the vantage time, which corresponds to the time required to push the evader in the area such that the distance to hide is greater than the distance to avoid hiding (assuming that the current evader velocity will not change). The authors proposed an approximated computation of this time. They first estimate the vertex T_x behind which the evader tries to hide (equivalent to find x^*). Here, the most critical escape path $x^* \in \{r,l\}$ is first computed. The velocity vector

$v_r \vec{r} + v_t \vec{t}$ ($\vec{r}$ and $\vec{t}$ are the unit vectors in the tangential and the radial direction) that minimizes the vantage time, also minimizes the risk defined as:

$$H_{VT} = \frac{r - h}{v_r + v_t(r'/r) - v_e}$$

The authors deduced that the correct velocity vector is: $(v_p/\sqrt{r^2 + r'^2})(r\vec{r} + r'\vec{t})$, by differentiating H_{VT} with respect to v_t and v_r.
A third heuristic we introduce here simply compares the distance the pursuer has to travel to avoid hiding (by reaching the vertex that may break the line of sight) with the distance for the evader to reach the related free edge. $x^* \in \{r, l\}$ is first computed. This heuristic (let it be called *spatial hidability*) is the following:

$$H_{SH} = r - h$$

Note that this heuristic is simpler but very close to H_{VT}, since the vantage time estimation results from the integration of the expression $r - h$.
We also proposed a forth heuristic that compares the time needed by the pursuer to avoid hiding with the time for the evader to reach the related free edge (let us call it *temporal hidability*), knowing $x^* \in \{r, l\}$:

$$H_{TH} = r - k.h$$

Finally, we propose a last heuristic which approximates the α-minimal strategy. $x^* \in \{r, l\}$ is first computed. As long as the distance l between the line of sight and the disappearance vertex T is greater than a given security distance l_0, the heuristic minimization should lead the pursuer to aim the disappearance vertex. When the distance l become smaller than l_0, the heuristic minimization should lead the pursuer to use part of its velocity to perform a tangential movement. The following heuristic called H_{AMA} (standing for Alpha-Minimal Approximation) provides such a behavior:

$$H_{AMA} = r - \left(\frac{l_0}{l}\right)^n .h$$

with $n > 1$ that can be adapted ($n = 2$ in the following to delay the beginning of the tangential movement). If l is greater than l_0, $\left(\frac{l_0}{l}\right)^n .h$ is negligible as compared with r and the pursuer will aim the disappearance vertex. If l becomes smaller than l_0, r becomes negligible as compared with $\left(\frac{l_0}{l}\right)^n .h$ leading the pursuer to perform a tangential movement.
For comparison, the direction of movement for each heuristic, computed by differentiating the heuristic, is given in the table 1. We note that the heuristics H_{VT} and H_{SH} have the same gradient direction.
In the following, the capture basin of our pursuit algorithm, embedding each one of the proposed heuristic instead of the α-minimal strategy will be compared for different obstacle shapes against the smart evasion strategy introduced previously.

6.4 Results in a virtual environment

In fig 11.a, 11.b, and 11.c, the capture basins of our pursuit algorithm, using the minimization each of the heuristics H_{ER}, H_{TM}, H_{SH}, H_{VT} and H_{AMA}, are displayed for different obstacles. The capture basin of the pure pursuit is also displayed. The 2 dimensions of the capture basin correspond to the initial positions (x, y) of the pursuer from which it achieves the capture of the evader always starting at the same position $(60, 75)$ (the length unit is the meter). The

Algo.	Expression	radial	component	tangential	component
ERM	$\frac{r}{h}$	$-r$	$\overrightarrow{r}$	$\frac{r^2}{h}.r'$	$\overrightarrow{t}$
THM	$r - k.h$	$-r$	$\overrightarrow{r}$	$k.r'$	$\overrightarrow{t}$
VTM	$\frac{r-h}{v_r+v_t(r'/r)-v_e}$	$-r$	$\overrightarrow{r}$	r'	$\overrightarrow{t}$
SHM	$k - h$	$-r$	$\overrightarrow{r}$	r'	$\overrightarrow{t}$
AHAM$_{n=2}$	$r - \left(\frac{l_0}{l}\right)^2.h$	$-r$	$\overrightarrow{r}$	$\left(\frac{l_0}{l}\right)^2.r'$	$\overrightarrow{t}$

Table 1. Direction of the gradient of the proposed heuristics 11.

speed are $v_e = 2$ and $v_p = 4\ m.s^{-1}$. As foreseen, the best algorithm is undoubtedly the AHA-minimization. The related capture basin includes almost all the other capture basins. Inspired from the solving of the pole problem, this strategy leads the pursuer to aim the disappearance vertex as fast as possible while minimally counter-balancing the movements of the evader when the line of sight and the line of disappearance are very close: the direction of the movement is such that the pursuer does not change the orientation of the disappearance line excepted in order to compensate the evader tangential move when the disappearance is imminent. The fast reaching of the disappearance vertex allows for easier visibility maintenance because the required leverage to compensate the evader tangential movement will be minimal when the disappearance becomes imminent.
Moreover, by aiming the disappearance vertex, the pursuer performs an adaptive proportional navigation since it aims a future position of the evader (obviously, the evader aims a point on the disappearance line). The resulting behavior is between a pure pursuit and the Π-strategy: the pursuer moves along the shortest path to the potential points of capture.
The fig 11.c provides an example of a game in our 2D virtual environment: The evader uses the strategy described previously and the pursuer uses our pursuit algorithm with the AMA heuristic minimization (the one approximating the α-minimal strategy). In this situation, only the AMA minimization allows for the capture.

7. Conclusion

In this article, the problem of pursuit under visibility constraint in an unknown cluttered environment has been tackled. First, a sufficient condition of capture in the presence of unknown obstacles has been established. The Π-strategy consisting in an optimal parallel pursuit guaranties the capture without disappearance if the capture region (the Apollonius circle) does not contains any free edge and if the obstacles included in the capture region can not break the future line of sight. We then wonder what should be done in other situations.
We first solved the circular obstacle problem, a particular game in which the evader moves along the boundary of a circular obstacle and the pursuer is initially located on tangent line to the obstacle touching the evader position. The resolution highlighted that, under visibility constraint, the pursuit algorithm that locally optimizes the time to capture leads in parallel to perform the minimal necessary effort in maintaining the visibility. The pole problem has then been investigated. This game is an approximation of an extremely sharp obstacle vertex. The solution showed that the pursuer wins if it can arrive to the pole before the evader, by simply compensating the rotation of the line of sight with a rotation of the line of disappearance. Otherwise, the evader wins by simply reaching the pole. This has led us to propose a pursuit strategy called the α-minimal strategy consisting in moving towards the disappearance vertex

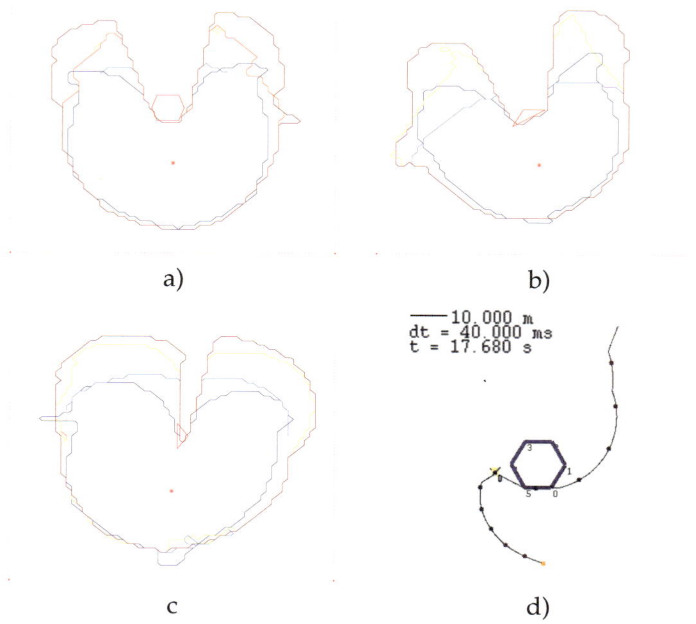

Fig. 11. a,b,c) Capture basin of each algorithm: The position of the evader is fixed (the red crosses). The red polygon is the obstacle. The dark blue basin correspond to a simple pure pursuit, the light blue one to the ERM, the cyan one to the THM, the orange and yellow one (almost superimposed) to the SHM and the VTM respectively, and the maroon one to the AMAM. Expected for particular cases due to the non-optimal behavior of the evader, the capture basin of the strategy approximating the α-minimal strategy includes all the other capture basins. d) Illustration of the AMA heuristic minimization that approximates the α-minimal strategy. The pursuer aims the disappearance vertex as long as possible and begins to perform a tangential movement when disappearance becomes imminent. The other heuristic do not allow for the evader visibility maintenance.

as fast as possible while preventing the imminent evader disappearance by a minimal compensation of the line of sight rotation. In the general case of a convex obstacle, this strategy guaranties the pursuer to see the next obstacle vertex without evader disappearance in the region where it wins the pole problem. In the region where the evader wins the pole problem, the generalization to the case of a convex obstacle is harder. If the projection of the evader on the line of disappearance is not on the related free edge, we established a pursuit behavior that aims at maximally rotating the line of disappearance before the evader disappearance, in order to hope to see the hidden part of the obstacle. This strategy called MD-LoD, standing for maximal Deviation of the Line of Disappearance, allows extending the capture basin in particular situations. If the projection of the evader position belongs to a free edge, the analysis becomes much more difficult and was not under the scope of this article.
Incrementally, a pursuit algorithm has been built. It combines the Π-strategy if it guaranties capture without disappearance, the $MD - LoD$ strategy when the evader is able to arrive to the disappearance vertex before the pursuer (and if it projection of the line of sight does not

belongs to a free edge) and the α-minimal strategy otherwise. Finally, we compared the capture basins of our pursuit algorithm modified such that the minimization of a given heuristic is used instead of the α-minimal strategy. Two of these heuristics were inspired by previous heuristics found of the literature (escape risk Gonzalez-Banos et al. (2002); Lee et al. (2002) and vantage time Bandyopadhyay et al. (2006)), two of them appeared relevant to the problem (spatial and temporal hidability) and the last one was built to approximate the α-minimal strategy. As foreseen, the strategy consisting in closing the distance to the disappearance vertex as fast as possible and doing the minimal necessary effort to maintain visibility extends the capture basin.
All along the article, even though the building of an evasion strategy was not addressed, the evader has always been considered as intelligent. For the simulation, we propose a geometrical method to locally aim the most secure instantaneous escape point or to run away if disappearance is impossible. In particular situation, it is clear that a better evasion strategy exists as highlighted by the pole problem.
In future work, it will be important to provide more global evasion strategies in order to evaluate how far the one we proposed is from an optimal and to imagine the possible evolutions of our algorithm. The concepts underlying the building of our pursuit algorithm, especially the α-minimal strategy and the sufficient capture condition we established based on the properties of the Π-strategy, should be also considered to tackle the problem of several unknown non-convex obstacles. Based on the insight provided by this study, it is also possible to investigate new pursuit concepts involving several pursuers in presence of multiple obstacles, not necessarily nookless.

8. References

Bandyopadhyay, T., Li, Y., Ang, M., and Hsu, D. (2006). A greedy strategy for tracking a locally predictable target among obstacles. In *IEEE Int. Conf. on Robotics and Automation ICRA06*, pages 2342–2347.

Basar, T. and Olsder., G. J. (1982). *Dynamic Noncooperative Game Theory*. Academic Press, London/New York, January 1995 (reedition of 1982).

Chen, Y.-C., Qi, H., and Liu, X. (2005). Mas-based pursuit-evasion algorithm under unknown environment. In *Int. Conf. on Machine Learning and Cybernetics*, volume 1, pages 265–269.

Cheung, W. A. (2005). *Constrained Pursuit-Evasion Problems in the Plane*. PhD thesis, the University of British Colombia.

Choi, H.-L., Ryu, H., Tahk, M.-J., and Bang, H. (2004). A co-evolutionary method for pursuit-evasion games with non-zero lethal radii. *Engineering Optimization*, 36(18):19–36.

Chung, C. and Furukawa, T. (2006). Time-optimal control of pursuers using forward reachable sets. In *3rd Int. Conf. on Computational Intelligence, Robotics and Automation CIRA06*.

Chung, C., Furukawa, T., and Goktogan, A. (2006). Coordinated control for capturing a highly maneuverable evader using forward reachable sets. In *IEEE Int. Conf. on Robotics and Automation ICRA06*, pages 1336–1341.

Chvatal, V. (1975). A combinatorial theorem in plane geometry. *Journal of Computorial Theory (B)*, 18:39–41.

Cliff, D. and Miller, G. F. (1996). Co-evolution of pursuit and evasion ii: Simulation methods and results. In *From Animals to Animats 4: Proceedings of the Fourth Int. Conf. on Simulation of Adaptive Behavior*, pages 506–514.

Eaton, M., McMillan, M., and Tuohy, M. (2002). Pursuit-evasion using evolutionary algorithms in an immersive three-dimensional environment. In *IEEE Int. Conf. on Systems, Man and Cybernetics SMC02*, volume 2, pages 348– 353.

Espiau, B., Chaumette, F., and Rives, P. (1992). A new approach to visual servoing in robotic. *IEEE Trans. on Robotics and Automation*, 8(3):313–326.

Gerkey, B. P., Thrun, S., and Gordon, G. (2006). Visibility-based pursuit-evasion with limited field of view source. *Int. Journal of Robotics Research*, 25(4):299 – 315.

Gonzalez-Banos, H. and Latombe, J. (2001). A randomized art-gallery algorithm for sensor placement. In *17th ACM Symp. on Computational Geometry (SoCG'01)*, pages 232–240.

Gonzalez-Banos, H., Lee, C.-Y., and Latombe, J.-C. (2002). Real-time combinatorial tracking of a target moving unpredictably among obstacles. In *IEEE Int. Conf. on Robotics and Automation ICRA02*, volume 2, pages 1683 – 1690.

Hájek, O. (1975). Pursuit games. *Mathematics in Science and Engineering*, 120.

Hutchinson, S. A., Hager, G. D., and Corke., P. I. (1996). A tutorial on visual servo control. *IEEE Trans. Robot. Automat.*, 12(5):651–670.

Isaacs, R. (1965). *Differential Games : A Mathematical Theory With Applications To Warfare And Pursuit, Control And Optimization*. Wiley.

Isler, V., Kannan, S., and Khanna, S. (2004). Locating and capturing an evader in a polygonal environment. In *Workshop on Algorithmic Foundations of Robotics (WAFR'04)*.

LaValle, S., Gonzalez-Banos, H., Becker, C., and Latombe, J.-C. (1997a). Motion strategies for maintaining visibility of a moving target. In *Int. Conf. on Robotics and Automation*, volume 1, pages 731 – 736.

LaValle, S. M., Lin, D., Guibas, L. J., Latombe, J.-C., and Motwani, R. (1997b). Finding an unpredictable target in a workspace with obstacles. In *IEEE Int. Conf. on Robotics and Automation ICRA97*, pages 737–742.

Lee, C.-Y., Gonzalez-Banos, H., and Latombe, J.-C. (2002). Real-time tracking of an unpredictable target amidst unknown obstacles. In *7th Int. Conf. on Control, Automation, Robotics and Vision ICARCV02*, volume 2, pages 596– 601.

Miller, G. F. and Cliff, D. (1994). Co-evolution of pursuit and evasion: Biological and game-theoretic foundations. *Cognitive Science Research Paper*, 311.

Muppirala, T., Hutchinson, S., and Murrieta-Cid, R. (2005). Optimal motion strategies based on critical events to maintain visibility of a moving target. In *IEEE Int. Conf. on Robotics and Automation ICRA05*, pages 3826 – 3831.

Murrieta-Cid, R., González-Banos, H., and on Robotics and, B. T. I. P. I. I. C. (2002). A reactive motion planner to maintain visibility of unpredictable targets. In *IEEE Int. Conf. on Robotics and Automation ICRA02*.

Murrieta-Cid, R., Monroy, R., Hutchinson, S., and Laumond, J.-P. (2008). A complexity result for the pursuit-evasion game of maintaining visibility of a moving evader. In *IEEE Int. Conf. on Robotics and Automation ICRA08*, pages 2657 – 2664.

Murrieta-Cid, R., Sarmiento, A., Bhattacharya, S., and Hutchinson, S. (2004). Maintaining visibility of a moving target at a fixed distance: the case of observer bounded speed. In *IEEE Int. Conf. on Robotics and Automation ICRA04*, volume 1, pages 479 – 484.

Murrieta-Cid, R., Sarmiento, A., and Hutchinson, S. (2003). On the existence of a strategy to maintain a moving target within the sensing range of an observer reacting with delay. In *Proceedings. 2003 IEEE/RSJ Int. Conf. on Intelligent Robots and Systems (IROS 2003)*, volume 2, pages 1184 – 1191.

Nahin, P. J. (2007). *Chases and Escapes: The Mathematics of Pursuit and Evasion*. Princeton University Press.

Nitschke, G. (2003). Co-evolution of cooperation in a pursuit evasion game. In *Intl. Conf. on Intelligent Robots and Systems IROS03*, volume 2, pages 2037 – 2042.

O'Rourke, J. (1983). Galleries need fewer mobile guards: A variation on chvatal's theorem. *Geometriae Dedicata*, 14:273–283.

O'Rourke, J. (1987). *Art Gallery Theorems and Algorithms*. Oxford University Press, London, New York.

O'Rourke, J. (1998). *Advances in Discrete and Computational Geometry*, chapter Open problems in the combinatorics of visibility and illumination, pages 237–243. eds. B. Chazelle and J. E. Goodman and R. Pollack, (Contemporary Mathematics) American Mathematical Society, Providence.

Park, S.-M., Lee, J.-H., and Chwa, K.-Y. (2001). Visibility-based pursuit-evasion in a polygonal region by a searcher. *Lecture Notes in Computer Science, Springer-Verlag*, 2076:456–468.

Petrosjan, L. A. (1993). *Differential Games of Pursuit*. World Scientific Publishing Company.

Sachs, S., Rajko, S., and LaValle, S. M. (2004). Visibility-based pursuit-evasion in an unknown planar environment. *Int. Journal of Robotics Research*, 23(1):3–26.

Sgall, J. (2001). A solution of david gale's man and lion problem. *Theoretical Comput. Sci.*, 259(1-2):663–670.

Shermer, T. (1992). Recent results in art galleries. In *Proceedings of the IEEE*, volume 80(9), pages 1384–1399.

Song, T. L. and Um;, T. Y. (1996). Practical guidance for homing missiles with bearings-only measurements. *IEEE Transactions on Aerospace and Electronic Systems*, 32(1):434 – 443.

Suzuki, I. and Yamashita, M. (1992). Searching for a mobile intruder in a polygonal region. *SIAM J. Computing*, 21(5):863–888.

5

Motion Planning by Integration of Multiple Policies for Complex Assembly Tasks

Natsuki Yamanobe[1], Hiromitsu Fujii[2], Tamio Arai[2] and Ryuichi Ueda[2]
[1]*National Institute of Advanced Industrial Science and Technology (AIST), JAPAN*
[2]*The University of Tokyo, JAPAN*

1. Introduction

Robotic assembly has been an active area of manipulation research for several decades. However, almost all assembly tasks, especially complex ones, still need to be performed manually in industrial manufacturing. The difficulty in planning appropriate motion is a major hurdle to robotic assembly.

In assembly tasks, manipulated objects come into contact with the environment. Thus, force control techniques are required for successfully achieving operations by regulating the reaction forces and dealing with uncertainties such as the position errors of robots or objects. Under force control, a robot's responsiveness to the reaction forces is determined by force control parameters. Therefore, planning assembly motions requires designing appropriate force control parameters. Many studies have investigated simple assembly tasks such as peg-in-hole, and some knowledge of appropriate force control parameters for the tasks has been obtained by detailed geometric analysis (Whitney, 1982). However, the types of parameters that would be effective for other assembly tasks are still unknown.

Here, it should be noted that the efficiency is always required in industrial application. Therefore, force control parameters that can achieve successful operations with a short time are highly desirable. However, it is difficult to estimate the cycle time, which is the time taken to complete an operation, analytically. Currently, designers have to tune the control parameters by trial and error according to their experiences and understanding of the target tasks. In addition, for complex assembly, such as insertion of complex-shaped objects, a robot's responsiveness to the reaction forces is needs to be changed according to the task state. Since tuning force control parameters with determining task conditions for switching parameters by trial and error imposes a very heavy burden on designers, complex assembly has been left for human workers.

Several approaches to designing appropriate force control parameters have been presented. They can be classified as follows: (a) analytical approaches, (b) experimental approaches, and (c) learning approaches based on human skill. In the analytical approaches, the necessary and sufficient conditions for force control parameters that will enable successful operations are derived by geometric analysis of the target tasks (e.g., Schimmels, 1997, Huang & Schimmels, 2003). However, the analytical approaches cannot be utilized for obtaining the parameters to achieve operations efficiently since the cycle time cannot be estimated analytically. Further, it is difficult to derive these necessary or sufficient conditions by geometric analysis for complex shaped objects. In the experimental approaches, optimal control parameters are obtained by learning or by explorations based on the results of iterative trials (e.g., Simons, 1982, Gullapalli et al., 1994). In these approaches, the cycle time is measurable because operations are performed either actually or virtually. Thus, some design methods that consider the cycle time have been proposed (Hirai et al., 1996, Wei & Newman, 2002). However, Hirai et al. only dealt with simple planar parts mating operations, and the method presented by Wei and Newman was applicable only to a special parallel robot. In addition, these approaches cannot be applied to complex assembly since it is too time-consuming to explore both parameter values and task conditions for switching parameters. In the last approaches based on human skill, the relationship between the reaction forces and the appropriate motions are obtained from the results of human demonstration (e.g., Skubic & Volz, 2000, Suzuki et al., 2006). Although some studies on these approaches have addressed the complex assembly that needs some switching of parameters, they cannot always guarantee the accomplishment of tasks because of the differences in body structure between human demonstrators and robots. Above all, relying on human skill is not always the best solution to increasing the task efficiency. Therefore, there is no method for planning assembly motions that can consider the task efficiency and have the applicability to complex assembly.

From another point of view, a complex assembly motion consists of some basic assembly motions like insertion or parts matting motions. Basic assembly motions can be accomplished with fixed force control parameters; therefore, it is relatively simple to program them. In addition, there are many types of control policies and task knowledge that are applicable to planning complex assembly motions: programs previously coded for similar tasks; human demonstration data; and the expertise of designers regarding the task, the robot, and the work environment.

Therefore, we adopt a step by step approach in order to plan complex assembly motions required in industrial applications. First, a method for basic assembly motion has been presented in order to design appropriate force control parameters that can efficiently achieve operations (Yamanobe et al, 2004). Then, based on the results, a policy integration method has been proposed in order to generate complex assembly motions by utilizing multiple policies such as basic assembly motions (Yamanobe et al, 2008). In this paper, we present these methods and show the simulation results in order to demonstrate the effectiveness of them.

This paper will proceed in the following way: Section 2 explains the problem tackled in this paper. In Section 3, a parameter designing method for basic assembly motion is firstly shown. In Section 4, a method for planning robot motions by utilizing multiple policies is then presented. In Section 5, the proposed methods are applied to clutch assembly. Basic assembly motions that constitute the clutch assembly motion are first obtained based on the method explained in Section 3, and the simulation results of integrating them are shown. Finally, Section 6 concludes this paper.

2. Problem Definition

In assembly tasks, the next action is determined on the basis of observable information, such as the current position of the robot, the reaction forces, and the robot's responsiveness; and information of the manipulated objects obtained in advance. Therefore, we assume that assembly tasks can be approximated by Markov decision processes (MDPs) (Sutton & Barto, 1998).

The problem considered in this paper is then formalized as follows:

- States $S = \{s_i | i = 1, \cdots, N_s\}$: A robot belongs to a state s in the discrete state space, S . A set of goal states, $S_{goal} \subset S$, is settled.
- Actions $\mathcal{A} = \{a_j | j = 1, \cdots, N_a\}$: The robot achieves the task by choosing an action, a, from a set of actions, $\mathcal{A}$, at every time step. A control policy for assembly tasks is defined as a sequence of force control parameters. Thus, the actions are represented as a set of force control parameters. While only one action is applied for basic assembly: $N_a = 1$, several actions need to be provided and swiched according to the states for achieving complex assembly: $N_a > 1$.
- State transition probabilities $\mathcal{P}^a_{ss'}$: State transition probability depends only on the previous state and the action taken. $\mathcal{P}^a_{ss'}$ denotes the probability that the robot reaches s' after it moves with a from s .
- Rewards $\mathcal{R}^a_{ss'} \in R$: $\mathcal{R}^a_{ss'}$ denotes the expected value of the immediate evaluation given to the state transition from s to s' by taking a . The robot aims to maximize the sum of rewards until it reaches a goal state. An appropriate motion is defined as the motion that can achieve a task efficiently. Hence, a negative value, namely, a penalty that is proportional to the time required for a taken action, is given as the immediate reward at each time step.

In addition, this paper presumes that the robot is under damping control, which is described as follows:

$$v_{ref} = v_0 + A f_{out} , \tag{1}$$

where $v_{ref} \in R^6$ is the reference velocity applied to the robot; $v_0 \in R^6$ is the nominal velocity; $A \in R^{6\times6}$ is the admittance matrix; and $f_{out} \in R^6$ is the reaction force acting on the object from the environment. Both the nominal velocity v_0 and the admittance matrix A are damping control parameters and determine robot motions. The admittance matrix determines how the reference velocity should be modified according to the reaction force, and the nominal velocity describes the motion of the robot in free space.

3. Method of Designing Force Control Parameters for Basic Assembly

In order to obtain effective policies for basic assembly motions, a method of designing force control parameters that can reduce the cycle time has been proposed (Yamanobe et al., 2004). An experimental approach is adopted so as to evaluate the cycle time; and the parameter design method through iterative operations is formulated as a nonlinear constrained optimization problem as follows:

$$\begin{aligned} \text{minimize}: \quad & V(p) \\ \text{subject to}: \quad & p \in C, \end{aligned} \tag{2}$$

where $V(\boldsymbol{p})$ is the objective function that is equal to the cycle time; $\boldsymbol{p}$ is a vector that consists of optimized parameters; and C is a set of optimized parameters that satisfy certain constraints, which are conditions that must be fulfilled to ensure successful motions. Here, the optimized parameters are damping control parameters, such as the admittance matrix $\boldsymbol{A}$ and the nominal velocity v_0.

A difficulty in this optimization problem is that it is impossible to calculate the derivatives of the objective function with respect to the optimized parameters since the cycle time is obtained only through trials. Therefore, we used a direct search technique: a combination of the downhill simplex method and simulated annealing (Press et al., 1992).

This method can deal with various assembly motions accomplished with fixed force control parameters. In addition, specific conditions desired for a particular operation can be easily considered by adding to the constraints of the optimization. Some effective policies for basic assembly motions, such as insertion motion and search motion, were obtained based on this method; the detailed results are shown in Section 5.

4. Motion Planning by Integration of Multiple Policies

In order to plan complex assembly motions, we have proposed a method for integrating several basic assembly motions and task knowledge that are effective for task achievement (Yamanobe et al. 2006) (Fig. 1). In our method, we represent a control policy for robots with a state action map, which denotes a look-up table connecting a state of a robot and its surroundings to its actions. Owing to the simplicity of the map, we can handle various policies and knowledge that exists in different forms using only one format, i.e., a state action map. The effective policies are selected and represented in a map by designers, and a new policy for the target task is efficiently constructed based on them. Here, it is difficult to determine the conditions for effectively applying the policies to the task. In some states, the applied policies would conflict with others and fail to achieve the task. Our method develops a robot motion by modifying the applied policies for the states in which they result in a failure.

4.1 Related works

On existing policies exploitation, several studies have been conducted especially in reinforcement learning in order to quickly learn motions for new tasks. Thrun and Mitchell proposed lifelong learning (Thrun & Mitchell, 1995). In this approach, the invariant policy of individual tasks and environments is learned in advance and employed as a bias so as to accelerate the learning of motions for new tasks. Tanaka and Yamamura presented a similar idea and applied it to a simple navigation task on a grid world (Tanaka & Yamamura, 2003). The past learning experiences are stored as the mean and the deviation of the value functions obtained for each task which indicates the goodness of a state or an action. Minato and Asada showed a method for transforming a policy learned in the previous tasks into a new one by modifying it partially (Minato & Asada, 1998). Although these approaches can acquire a policy that is common to a class of tasks and improve their learning performance by applying it to a new task in the class, only one type of policy is utilized in these methods.

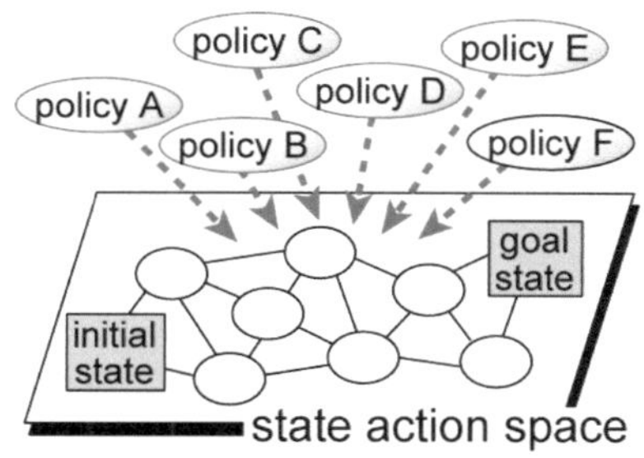

Fig. 1. Robot motion obtained by the integration of multiple policies

In the case of multiple-policy applications, Lin proposed a learning method to use various human demonstration data as informative training examples for complex navigation tasks (Lin, 1991). However, this method cannot deal with false teaching data. Sutton et al. defined a sequence of actions that is effective for task as an option; they then presented an approach to increase the learning speed by using options interchangeably with primitive actions in the reinforcement learning framework (Sutton et al., 1999). This approach can modify the unsuitable parts of options in the learning process and, therefore, integrate multiple options. This approach is similar to our methodology. However, the usable policy is limited to a sequence of actions. The advantage of our method is to be easily able to deal with various types of existing policies and knowledge.

4.2 Method for integrating multiple policies

As described above, the basic idea of our method is as follows: first, all applied policies are written in a state action map; after that, a new policy for the target task is constructed by partially modifying the applied policies.

Applied policies, such as policies for basic assembly motions, are selected by designers and represented in a state action map. The states in which each policy is represented are also determined by designers. Knowledge for the target task defines the state space and rewards, and sets a priority among the applied policies. When multiple policies are represented on a map, the map includes

states in which no policy is applied,

states in which multiple policies are written, and

states in which the actions following the applied policies fail to achieve the task.

We define the last-named states as "failing states." In order to obtain a new policy that is feasible for the target task, the following processes are required.

- Policy definition according to the applied policies.
- Selection of failing states.
- Policy modification for the failing states.

Let us explain each procedure in the following sub-sections.

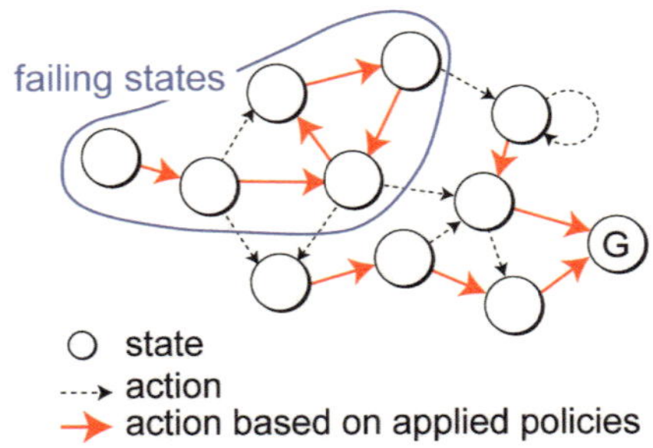

Fig. 2. Failing states

4.2.1 Policy Exploration Based on Applied Policies

$s_{\text{policy}} \in S$ represents the state in which policies are introduced. A set of actions available at s_{policy}, $\mathcal{A}_{\text{p}}(s_{\text{policy}}) \subset \mathcal{A}$, is defined as follows:

$$\mathcal{A}_{\text{p}}(s_{\text{policy}}) = \{a_{\text{p}} \in \mathcal{A}_{\text{p}}^{k}(s_{\text{policy}}) \mid k = 1, \cdots, N_{\text{p}}(s_{\text{policy}})\}, \tag{3}$$

where $\mathcal{A}_{\text{p}}^{k}(s_{\text{policy}})$ is a set of actions based on policy k at s_{policy} and $N_{\text{p}}(s_{\text{policy}})$ is the number of policies applied to s_{policy}. At the state in which no policy is applied, $s_{\overline{\text{policy}}}$, the robot can take all actions involved in $\mathcal{A}$. The new policy for the target task is efficiently decided on the basis of these actions limited by the applied policies. An optimal control policy can maximize the state value, $V(s)$, that is defined as the expected sum of the rewards from a state s to a goal state. The new policy is explored while estimating the state value function, V, based on dynamic programming (DP) (Bellman, 1957) or reinforcement learning (Sutton & Barto, 1998).

4.2.2 Failing States Selection

If actions are limited by the applied policies, the robot might fail to perform the task at some states. The failing states, S_{fail}, are defined as the states from which the robot cannot reach a goal state only with the actions implemented based on the applied policies. Fig. 2 shows an example of failing states.
In failing states, state transitions are infinitely repeated. Since a penalty is given for each action, the state value at a failing state, $V(s_{\text{fail}})$, decreases. Hence, we select the failing states by using the decrease in the state values. First, a state $\tilde{s}_{\text{fail}}$ with a value $V(\tilde{s}_{\text{fail}})$ that is lower than $V_{\min}$ is found. $V_{\min}$ is the threshold value. Then, S_{fail} is defined as a set of $\tilde{s}_{\text{fail}}$ and the states that the robot can reach from $\tilde{s}_{\text{fail}}$ according to the actions limited by the applied policies.

4.2.3 Policy Modification

In order to correct the infinite state transitions in the range of the failing states, the applied policies need to be modified partially. In particular, the actions that are available in the failing states are changed from the actions limited by the policies, $\mathcal{A}_{\text{p}}(s_{\text{policy}})$, into the normal actions, $\mathcal{A}$, that are available for the robot. Then, the new policy is explored again only for the failing states. By repeating these processes until no failing state is selected, we can efficiently obtain a new policy that is not optimal but feasible for the whole target task.

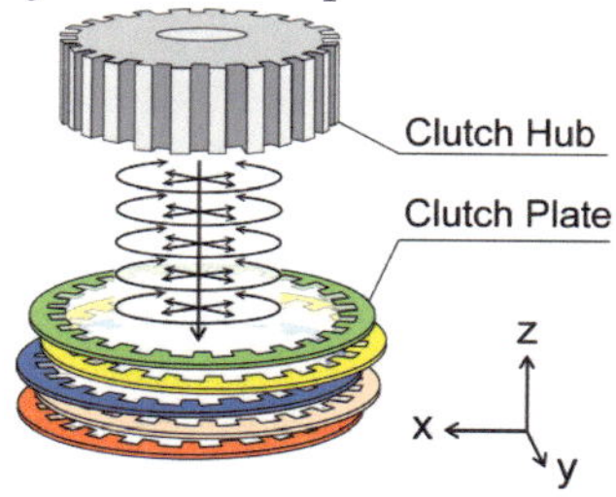

Fig. 3. Clutch assembly

5. Application of Policy Integration Method to Complex Assembly

The proposed method for the integration of multiple policies is applied to clutch assembly (Fig. 3) in order to demonstrate its validity in complex assembly.

Clutch assembly is a complicated assembly task, in which a splined clutch hub is inserted through a series of movable toothed clutch plates. Since the clutch plates can move in the horizontal plane and rotate about the vertical, the plates are nonconcentric and have random phase angles before the clutch hub is inserted. In order to efficiently execute the task, a search motion for matching the centerline and the phase angle of the hub to those of each plate is required in addition to a simple insertion motion. However, the task is achieved by only search motion in practical applications because it is difficult to perceive that the teeth on the hub become engaged with the proper grooves on the plate.

In this section, the appropriate motion for clutch assembly is developed by integrating the policies for insertion motion and search motion.

5.1 Simulator for clutch assembly

We utilize a simulator for integrating multiple policies as well as the optimization for basic assembly motions in order to avoid problems such as the occurrence of a crash when an operation results in a failure during policy exploration and the deterioration of objects and/or a robot on account of iterative operations. Although a modelling error might be a problem in a simulation, this problem can be overcome by developing a simulator based on preliminary experiments.

In this subsection, the simulator used in this paper is explained. The simulator consists of a physical model and a control system model. The physical model has been developed using LMS DADS that is mechanical analysis software. This model expresses the work environment in which operations are performed and is composed of the manipulated object and the assembled objects. For the simulator of clutch assembly, the physical model consists of a clutch hub as the manipulated object, clutch plates as the assembled objects, and the housing that holds the clutch plates.

The control system model has been developed using MATLAB Simulink. In this model, the mechanical compliance and the control system of the robot are expressed. A schematic view of the simulator is shown in Fig. 4. The position of the manipulated object, $x_{\mathrm{object}} \in R^6$, and the reaction force acting on the object, f_{out}, constitute the output from the physical model and are fed into the control system model. The reference velocity of the robot v_{ref} is calculated from the damping control law (eq. 1) by the damping controller. The position controller of the robot is modeled as a second-order system. The robot is modeled as a rigid body, and its mechanical compliance is described as a spring and a damper between the end-effector of the robot and the manipulated object. Based on the position controller and the robot's mechanical compliance, the position of the robot, $x_{robot} \in R^6$, is written as follows:

$$\begin{aligned} M_s[\ddot{x}_{robot} + 2\zeta\omega_n(\dot{x}_{robot} - v_{ref}) + \omega_n{}^2(x_{robot} - v_{ref}\Delta t)] &= D_e(\dot{x}_{object} - \dot{x}_{robot}) + K_e(x_{object} - x_{robot}) \\ &= -f_{in}, \end{aligned} \tag{4}$$

where $M_s \in R^{6\times6}$ is the inertia matrix; ζ and ω_n are the damping coefficient and the natural frequency of the second-order system, respectively; Δt is the sampling time in the controller; $D_e \in R^{6\times6}$ and $K_e \in R^{6\times6}$ are the damping matrix and the stiffness matrix between the robot and the manipulated object, respectively. f_{in} is the force acting on the manipulated object from the robot through the spring and the damper and fed into the physical model for actuating the object.

In oder to obtain data for the simulator, preliminary experiments: measurement of the stiffness of the robot, K_e, and clutch assembly, were performed using a 6 DOF (degree of freedom) manipulator, FANUC M-16i. A clutch consisting of five clutch plates was used in the experiments of clutch assembly. Each clutch plate has 45 teeth and is 0.8 [mm] thick; the distance between adjacent plates is 3.75 [mm]. The plates are contained within a fixed subassembly, and they can move independently in the horizontal plane ±1 [mm] and rotate about the vertical. The clutch hub is 95 [mm] in diameter and 35 [mm] in height. The height of each of the teeth is 5 [mm]. It is possible to represent the actual tasks by adjusting the parameters in the simulator. Thus, the parameters of the control system model and the coefficient of kinetic friction in the physical model were determined by trial and error in order to obtain simulation results that are close to the experimental results.

5.2 Acquisition of policies for basic assembly motions

Using the method for optimizing force control parameter, which is presented in Section 3, appropriate policies for insertion motion and search motion are obtained.

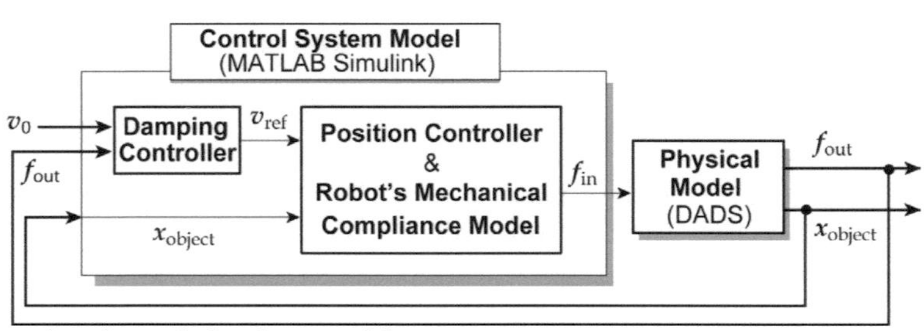

Fig. 4. Schematic view of clutch assembly

5.2.1 Policy acquisition for insertion motion

A policy for insertion motion, i.e. appropriate force control parameters, is obtained on the basis of cylindrical peg-in-hole tasks. A simulator for peg-in-hole tasks was first developed only by changing the physical model in the simulator for clutch assembly. Then, the optimization of force control parameters was performed by considering the following constraints.

Stability conditions

We consider the stability of the control system in the case where the manipulated object is in contact with the assembled object. When the manipulated object is constrained, eq. 4 can be discretely expressed as follows:

$$\boldsymbol{M}_{\mathrm{s}}\left[\left(\frac{x_{\mathrm{robot}(i+1)}-2x_{\mathrm{robot}(i)}+x_{\mathrm{robot}(i-1)}}{\Delta t^2}\right)+2\zeta\omega_n\left(\frac{x_{\mathrm{robot}(i)}-x_{\mathrm{robot}(i-1)}}{\Delta t}-v_{\mathrm{ref}(i)}\right)+\omega_n{}^2\left(x_{\mathrm{robot}(i)}-x_{\mathrm{ref}(i)}\right)\right]$$
$$=\boldsymbol{K}_{\mathrm{e}}\left(x_{\mathrm{object}}-x_{\mathrm{robot}(i)}\right)-\boldsymbol{D}_{\mathrm{e}}\left(\frac{x_{\mathrm{robot}(i)}-x_{\mathrm{robot}(i-1)}}{\Delta t}\right), \tag{5}$$

where, x_{object} is constant; $\dot{x}_{\mathrm{object}}=0$; and $x_{\mathrm{robot}(i)}$ is the position of the robot at $t=i\Delta t$. Using eq. 5 and considering the delay of the reaction force information from the force sensor, we can discretely express the damping control law (eq. 1) as follows:

$$\begin{aligned} v_{\mathrm{ref}(i)} &= \frac{x_{\mathrm{object}}-x_{\mathrm{robot}(i)}}{\Delta t} \\ &= v_0+\boldsymbol{A}\left[\boldsymbol{K}_{\mathrm{e}}\left(x_{\mathrm{object}}-x_{\mathrm{robot}(i-2)}\right)-\boldsymbol{D}_{\mathrm{e}}\left(\frac{x_{\mathrm{robot}(i-2)}-x_{\mathrm{robot}(i-3)}}{\Delta t}\right)\right], \end{aligned} \tag{6}$$

Let $\boldsymbol{X}_{\mathrm{robot}(i)}=\left(x_{\mathrm{robot}(i)}, x_{\mathrm{robot}(i-1)}, x_{\mathrm{robot}(i-2)}, x_{\mathrm{robot}(i-3)}\right)^T \in R^{24}$; eq. 5 and eq. 6 can then be rewritten as follows:

$$\begin{aligned} \boldsymbol{X}_{\mathrm{robot}(i+1)} &= \begin{bmatrix} \boldsymbol{W}_{11} & \boldsymbol{W}_{12} & \boldsymbol{W}_{13} & \boldsymbol{W}_{14} \\ \boldsymbol{I} & \boldsymbol{O} & \boldsymbol{O} & \boldsymbol{O} \\ \boldsymbol{O} & \boldsymbol{I} & \boldsymbol{O} & \boldsymbol{O} \\ \boldsymbol{O} & \boldsymbol{O} & \boldsymbol{I} & \boldsymbol{O} \end{bmatrix} \boldsymbol{X}_{\mathrm{robot}(i)}+\boldsymbol{C} \\ &:= \boldsymbol{W}\boldsymbol{X}_{\mathrm{robot}(i)}+\boldsymbol{C}, \end{aligned} \tag{7}$$

where

$$\boldsymbol{W}_{11}=\boldsymbol{M}_{\mathrm{s}}^{-1}\left[2\boldsymbol{M}_{\mathrm{s}}-\Delta t\left(2\zeta\omega_n\boldsymbol{M}_{\mathrm{s}}+\boldsymbol{D}_{\mathrm{e}}+\Delta t\boldsymbol{K}_{\mathrm{e}}\right)\right],$$
$$\boldsymbol{W}_{12}=\boldsymbol{M}_{\mathrm{s}}^{-1}\left[\Delta t\left(2\zeta\omega_n\boldsymbol{M}_{\mathrm{s}}+\boldsymbol{D}_{\mathrm{e}}\right)-\boldsymbol{M}_{\mathrm{s}}\right],$$
$$\boldsymbol{W}_{13}=-\Delta t\omega_n\boldsymbol{A}\left(\Delta t\omega_n+2\zeta\right)\left(\Delta t\boldsymbol{K}_{\mathrm{e}}+\boldsymbol{D}_{\mathrm{e}}\right),$$
$$\boldsymbol{W}_{14}=-\Delta t\omega_n\boldsymbol{A}\boldsymbol{D}_{\mathrm{e}}\left(\Delta t\omega_n+2\zeta\right),$$
$$\boldsymbol{C}=\left[\Delta t^2\boldsymbol{M}_{\mathrm{s}}^{-1}\left[\omega_n\boldsymbol{M}_{\mathrm{s}}\left(\Delta t\omega_n+2\zeta\right)\left(v_0+\boldsymbol{A}\boldsymbol{K}_{\mathrm{e}}x_{\mathrm{object}}\right)+\boldsymbol{K}_{\mathrm{e}}x_{\mathrm{object}}\right], \boldsymbol{O}, \boldsymbol{O}, \boldsymbol{O}\right]^T,$$

and $\boldsymbol{I}\in R^{6\times 6}$ is the identity matrix.

The series $\boldsymbol{X}_{\mathrm{robot}(i)}$ must converge to a certain value in order to ensure the stability of the control system. Therefore, the stability condition can be theoretically described as $|\lambda_j|<1$, where λ_j is each of the eigenvalues of $\boldsymbol{W}$. Here, the state of the control system gets close to the instability as the maximum value of $|\lambda_j|$, λ_{max}, becomes large. Then, λ_{max} can be used as a value that evaluates the instability of the system. Considering the modeling error of the simulator, we define the stability condition in the optimization as:

$$\lambda_{max} < 0.99 \tag{8}$$

Condition for nominal velocity

To achieve insertion, the z-element of the nominal velocity v_{0z} must be negative.

Limitation of the reaction force

The rating of the force sensor bounds the allowable reaction force. We define the limit of the reaction force as the rating: 294 [N] and 29.4 [Nm].

If the given parameters cannot satisfy these above constraints, the simulation is stopped. The operation is regarded as a failure, and a very large value is assigned to the objective function. Namely, we use a penalty function to consider these constraints.

Here it should be noted that, if the optimization of the control parameters were performed simply, the obtained parameters would be too specialized in a specific initial condition. Thus, simulations are performed with possible errors in the initial position of the peg in order to deal with the various errors. Let the maximal value of the possible position error of the peg be 1 [mm] and that of rotation error be 1 [deg]. Six kinds of position errors are considered here: along the x-axis, along the y-axis (positive, negative), rotation error around the x-axis (positive, negative), and rotation error around the y-axis. Simulation of the peg-in-hole is performed with the above errors in the initial position of the peg; then, the mean value of the six kinds of cycle time obtained from the simulation with each error is defined as the objective function.

The admittance matrix $\boldsymbol{A}$ was defined as a diagonal matrix; the robot was position-controlled only around the insertion axis, i.e. z-axis, and the x-axis and the y-axis were treated equally because of cylindrical peg-in-hole tasks. The optimization was performed for the optimized control parameters, $\boldsymbol{p} = (a_{x-y}, a_z, a_{rx-ry}, v_{0z})^T$, and the damping control parameters are thus expressed as follows:

$$\begin{aligned} \boldsymbol{A} &= \mathrm{diag}(a_{x-y}, a_{x-y}, a_z, a_{rx-ry}, a_{rx-ry}, 0), \\ v_0 &= (0, 0, v_{0z}, 0, 0, 0)^T. \end{aligned} \tag{9}$$

In Fig. 5, the results for the optimization are presented. The horizontal axis in Fig. 5 represents the number of the simplex deformation in the optimization, which denotes the progress of the parameter exploration. The values of each element of the vector $\boldsymbol{p}$ at the best and worst points of the simplex are plotted. Similarly, the objective values at the best and worst points of the simplex are also plotted.

As shown in the bottom-left figure in Fig. 5, the objective value decreased as the optimization proceeded. Around the deformation count 80, the objective values at the worst point of the simplex were huge. That is because the parameters at the points broke the condition for the reaction force. The value of a_{rx-ry} shown in the middle-left figure in Fig. 5 became large for dealing with the orientation errors quickly. As shown in the top-right and the middle-right figures in Fig. 5, the magnitude of a_z and v_{0z} changed interacting each other. The peg is inserted more quickly, as v_{0z} increases. However, the larger the value of the initial velocity is, the larger the magnitude of the reaction force becomes. Thus, the value of a_z changed in order to keep the reaction force from violating its constraint. As shown in these results, we obtained the appropriate force control parameters that can achieve insertion motions with short cycle time and handle various possible errors.

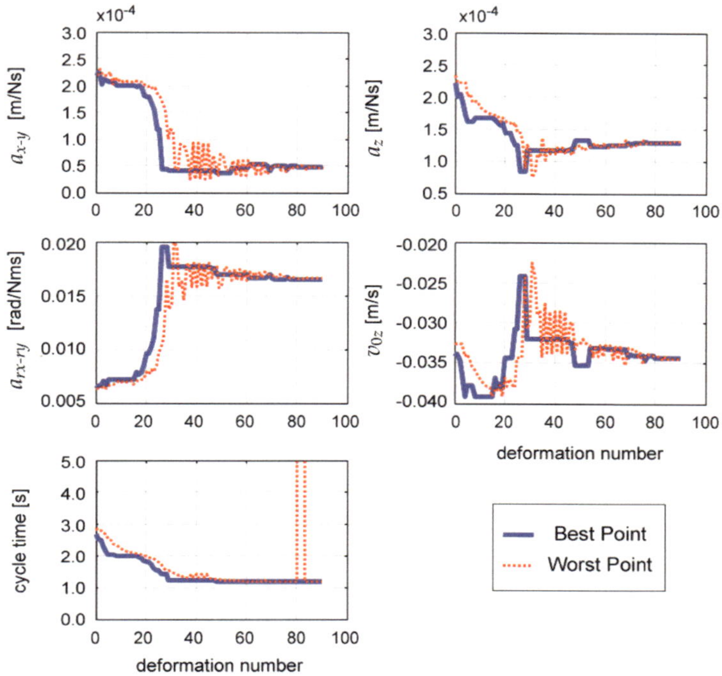

Fig. 5. Results of optimization for insertion motion

The simulation whose results are shown in Fig. 5 took about 189[h] using a Windows PC with Pentium 4 CPU running at 2.8[GHz].

5.2.2 Policy acquisition for search motion

A policy for search motion is acquired on the basis of the clutch assembly performed in the preliminary experiments.

In the search motion, a cyclic motion in the horizontal plane is performed while pressing the assembled object in order to engage the manipulated object with it. Each clutch plate can move in the x-y plane and rotate about the z-axis in the clutch assembly. The cyclic motion along the x-axis and the y-axis as well as around the z-axis was adopted and was achieved by reversing the nominal velocity v_0 when the hub goes beyond the search area $\mathbf{R} = (\pm R_x, \pm R_y, \pm R_{rz})^T = (\pm 1\,[\text{mm}], \pm 1\,[\text{mm}], \pm 4\,[\text{deg}])^T$. The elements of the nominal velocity, which are related to the cyclic motion, are v_{0x}, v_{0y}, and v_{0rz}; they were determined as follows:

$$(v_{0x}, v_{0y}, v_{0rz})^T = k_c \boldsymbol{v}_{bc}, \tag{10}$$

where k_c is the coefficient of the cyclic motion velocity, and $v_{bc} \in R^3$ is the base velocity of the cyclic motion defined so as to cover the entire search area $\boldsymbol{R}$. In order to achieve this motion when the clutch hub is constrained by the clutch plates, the target force $f_t \in R^6$, which is the force applied by the manipulated object on the environment in the steady state, should be defined appropriately. Here, the target force f_t is expressed from eq. 1 as follows:

$$v_0 + \boldsymbol{A} f_t = 0 . \tag{11}$$

Let $v_{bc} = (2.25 \times 10^{-3}\ [\mathrm{m/s}], 7.54 \times 10^{-4}\ [\mathrm{m/s}], 0.92\ [\mathrm{rad/s}])^T$ and $f_{tc} = (f_{tx}, f_{ty}, f_{trz})^T = (49\,[\mathrm{N}], 49\,[\mathrm{N}], 7.8\,[\mathrm{Nm}])^T$, which is the elements of the target force related to the cyclic motion, based on the experiences. The admittance matrix $\boldsymbol{A}$ was defined as a diagonal matrix. The manipulator was position-controlled for the directions that were not relevant to the cyclic motion and the pressing: around the x-axis and the y-axis. Therefore, the vector of the optimized control parameters $\boldsymbol{p}$ was defined as $\boldsymbol{p} = (k_c, a_z, v_{0z})^T$, and the damping control parameters are then expressed as follows:

$$\begin{aligned} \boldsymbol{A} &= \mathrm{diag}(a_x, a_y, a_z, 0, 0, a_{rz}), \\ v_0 &= (v_{0x}, v_{0y}, v_{0z}, 0, 0, v_{0rz})^T, \end{aligned} \tag{12}$$

where v_{0x}, v_{0y}, v_{0rz} is calculated from eq. 10 with k_c and v_{bc}, and a_x, a_y, a_{rz} is determined using eq. 11 with f_{tc}.

The optimization for search motion was executed with the same constraints considered in that for the insertion motion: stability condition, condition for nominal velocity, and limitation of the reaction force. In addition, in order to deal with various arrangements of the clutch plates, we divided the clutch assembly into two phases: insertion to the first plate from free space and insertion to the other clutch plates. Simulations were only performed for the insertion through the first and the fifth plate with the possible errors of the clutch plate, which are presented in Table 1. The mean value of the cycle time obtained from the simulation through each plate and with each error was defined as the objective function as well as the optimization for the insertion motion.

The results of this optimization are presented in Fig. 6. The horizontal axis represents the number of the simplex deformation in the optimization. The vertical axis represents the values of each element of the vector $\boldsymbol{p}$ and the objective values at the best and worst points of the simplex.

Position error (along x-axis)	Phase angle error
+0.4 [mm]	+1 [deg]
+0.4 [mm]	-1 [deg]
-0.4 [mm]	+1 [deg]
-0.4 [mm]	-1 [deg]

Table 1. Position/phase angle error of the clutch plate in the parameter optimization

As shown in the bottom-right figure in Fig. 6, the objective value decreased as the optimization proceeded. It shows that the parameters that can achieve the task with short cycle time were obtained. The value of k_c plotted in the top-left figure grew large as the optimization proceeded, and the value of a_z and the absolute value of v_{0z} got smaller as interacting each other. The decrease in a_z causes the stiff force control along the insertion axis. Due to the increase in k_c, which led to the increase of the cyclic motion velocity, the stiff force control was desired in order to insert the clutch hub effectively when the teeth on the hub engaged with the grooves on the plates. In addition, since the stiff force control tends to cause large reaction force, v_{0z} changed interactively in order to keep the reaction force from violating its constraints.

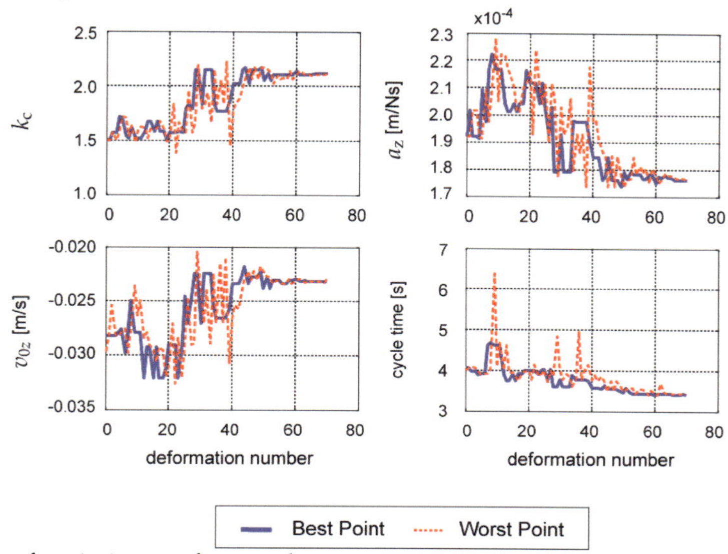

Fig. 6. Results of optimization for search motion

The cyclic motion and the pushing force need to be determined appropriately in the clutch assembly. For example, when the velocity of the cyclic motion is too high with soft force control along the insertion axis, the teeth on the clutch hub will fail to engage with the grooves on the clutch plate. When the clutch hub pushes the clutch plate with too large force, the pressed plate tends to move along with the hub. As shown in the above results, the obtained force control parameters through the optimization had a good balance between the velocity and the force and can deal with various plate arrangements.

The simulation whose results are shown in Fig. 6 took about 39[h] using a Windows PC with Pentium 4 CPU running at 2.8[GHz].

5.3 Integration of policies for insertion and search motions

An appropriate policy for the clutch assembly is constructed by integrating the policies for insertion and search motions that are obtained in the previous subsection. The limitation of the reaction forces is utilized as task knowledge and defines the states in which the reaction forces exceed their limit as terminal states of the task.

5.3.1 State space

A state space of assembly tasks is constructed by the current position of the robot, the reaction forces, and the robot's responsiveness. However, the number of states becomes enormous if all kinds of states are addressed. In clutch assembly, the reaction force along the insertion axis, $f_{\mathrm{out}\,z} \in R^6$, is the most effective state for recognizing that the teeth on the clutch hub becomes engaged with the proper grooves on the clutch plate. Therefore, the state space was confined as follows:

$$s = s\left(f_{\mathrm{out}\,z}, d\right) \tag{13}$$

The reaction force $f_{\mathrm{out}\,z}$ was segmented into 62 states between the lower limit and the upper limit of it. The robot's responsiveness d was divided into two: $d_{\mathrm{insertion}}$ which is the responsiveness for the insertion motion and d_{search} which is that for the search motion.

5.3.2 Actions

The policies of insertion and search motions were applied to the whole state space. A set of implemented actions was defined as follows:

$$\mathcal{A}\left(s_{\mathrm{policy}}\right) = \left\{\boldsymbol{a}_{\mathrm{insertion}}, \boldsymbol{a}_{\mathrm{search}}\right\}, \tag{14}$$

where $\boldsymbol{a}_{\mathrm{insertion}}$ and $\boldsymbol{a}_{\mathrm{search}}$ are the damping control parameters that can effectively perform the insertion and search motions, respectively. When an action is selected, the damping control parameters expressed by the action are applied to the damping controller of the simulator.

In damping control, the target force, which is the force applied by the manipulated object on the environment in the steady state, is defined by the applied damping control parameters. The target force along the insertion axis of the insertion motion is about twice of that of the search motion, and the admittance along the insertion axis of the insertion motion is smaller than that of the search motion. Namely, the robot strongly presses the assembled object when the insertion motion is applied, compared with when the search motion is done.

5.3.3 Method for exploring new policy

In assembly tasks, it is difficult to obtain the model of the target task, i.e. to calculate the state transition probability $\mathcal{P}_{ss'}^{a}$ beforehand because of the uncertainties like the position errors of the robot and the friction between manipulated objects. Therefore, we adopted Q-learning (Sutton & Barto, 1998) in order to explore a new policy. Q-learning is one of the reinforcement learning techniques and can construct a policy without the model of the target task. The goal states were defined as the states in which the clutch assembly is successfully achieved. The error states were also determined as the states in which the reaction forces exceed their limit. If the robot reaches the error states, the simulation is stopped, and the task is regarded as a failure. In order to reduce calculation time for obtaining a new policy, the three clutch plate model was applied.

Rewards:

The robot selects and executes an action at each sampling time of its control system, and the sampling time Δt is 0.004 [s]. Thus, a reward −0.004 was given at each step. In addition, a penalty −4 was given when the task results in a failure.

Learning parameters:

	Position error (along x-axis)	Phase angle error
Type1	Alternately: ±0.5 [mm]	Alternately: +4, 0 [deg]
Type2	Alternately: ±0.5 [mm]	All plates: +4 [deg]
Type3	Alternately: ±0.5 [mm]	Alternately: ±2 [deg]
Type4	All plates: +0.5 [mm]	Alternately: +4, 0 [deg]
Type5	All plates: +0.5 [mm]	All plates: +4 [deg]

Table 2. Initial position/phase angle of the clutch plates

Type1	Type2	Type3	Type4	Type5	mean
1.18 [sec]	1.18 [sec]	1.16 [sec]	0.98 [sec]	0.75 [sec]	1.05 [sec]

Table 3. Cycle time of the clutch assembly based only on the search motion

The ε-greedy method was used for making experiences and in which actions are conducted randomly at the rate ε. The parameter of ε-greedy method and the learning parameters of Q-learning, α, were adopted as $\varepsilon = 0.1$ and $\alpha = 0.1$, respectively.
Uncertainties of the task:
A new policy for clutch assembly needs to handle various plate arrangements. Thus, simulations were performed against the five kinds of arrangements of the clutch plates described in Table 2. As a reference, Table 3 shows the cycle time obtained by the simulations with each plate arrangement in Table 2 based only on the policy of search motion.

5.3.4 Simulation results of integrating the policies for insertion and search motions

A new policy for the clutch assembly was developed by integrating two policies: for insertion motion and search motion, based on the conditions that are mentioned above. The simulation results are presented in Fig. 7. The horizontal axis represents the learning step. The vertical axis shows the average cycle time of ten trials. Here, the average is obtained without including the results of tasks failed. As shown in Fig. 7, the average cycle time was slightly shortened as the learning proceeded. In addition, the cycle time was extremely reduced compared with that obtained based only on the search motion presented in Table 3. It shows that integrating the insertion motion into the search motion is effective to the clutch assembly.
In Fig. 8, a result of the clutch assembly using the state action map obtained after 600 trials is shown. The last graph shows the action taken at each sampling time during the task. As shown in Fig. 8, the clutch hub was inserted through each clutch plate with a cyclic search motion and the insertion motion. The values of z and $f_{\mathrm{out}z}$ represent the fitting of the hub in each plate. When the teeth of the clutch hub is engaged with that of each clutch plate, the reaction force $f_{\mathrm{out}z}$ becomes small since only the frictional force is acting on the hub. From the result of the selected action, the policy of insertion motion was continuously selected while $f_{\mathrm{out}z}$ was almost zero. In fact, the effective policy was taken with perceiving the engagement of the objects based on the obtained state action map. Compared to the result based only on the search motion (Fig. 9), the hub is quickly inserted by selecting the policy of insertion motion. **Error! Reference source not found.**Fig. 9 presents the cycle time of the clutch assembly against the plate arrangements in Table 2 using the obtained state action. It

shows that the obtained new policy can also deal with various arrangements of the clutch plates.

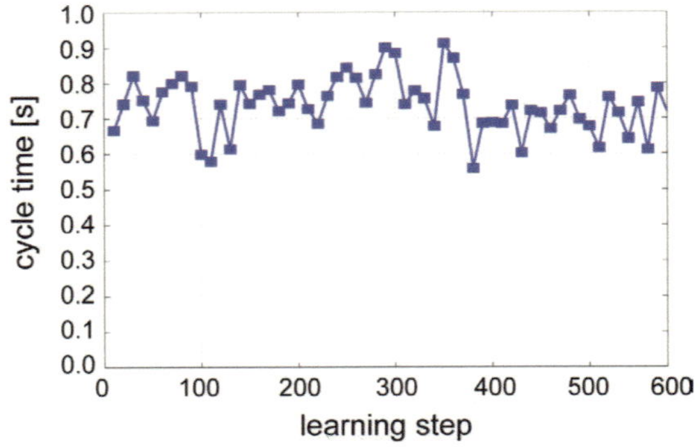

Fig. 7. Change of the cycle time through the learning

On the other hand, the insertion motion and the search motion were selected at random while the reaction force f_{outz} was acting on the clutch hub. In the insertion motion, the clutch hub strongly presses each clutch plate compared with the case of the search motion. If the insertion motion is chosen when the position and the phase angle of the hub are not matched to those of the objective plate, the task might result in a failure by the excessive reaction forces. Actually the reaction force in Fig. 8 is larger than that in Fig. 9. The learning was not converged after 600 trials, and the robot failed to perform the task because of the excessive reaction forces once in three times on average. The insertion motion and the search motion are similar motion in a certain part. In addition, the clutch assembly can be accomplished only with the search motion. It is hard to differentiate these motions while constructing a new policy. In future work, the reward setting should be improved, and a learning method also needs to be developed for obtaining a new policy efficiently based on similar policies.

6. Conclusion

In this paper, we proposed a step by step approach to planning robot motions for complex assembly tasks. It can be assumed that a complex assembly motion consists of some basic assembly motions, which can be accomplished with fixed force control parameters. Therefore, we firstly proposed an optimization method for basic assembly motions to obtain appropriate force control parameters that can efficiently achieve operations. Based on the results, we then proposed a policy integration method in order to generate complex assembly motions by utilizing optimized basic assembly motions.

These methods were applied to clutch assembly in this paper. Using the parameter optimization method, effective policies for insertion and search motions were obtained; and then, a new control policy for clutch assembly was developed by integrating these basic assembly motions. Based on the new policy obtained by the policy integration method, the effective motion was chosen with perceiving the engagement of the clutch hub and the clutch plates. The result shows that the proposed approach makes it possible to generate complex assembly motions that need some switching of force control parameters according to task states.

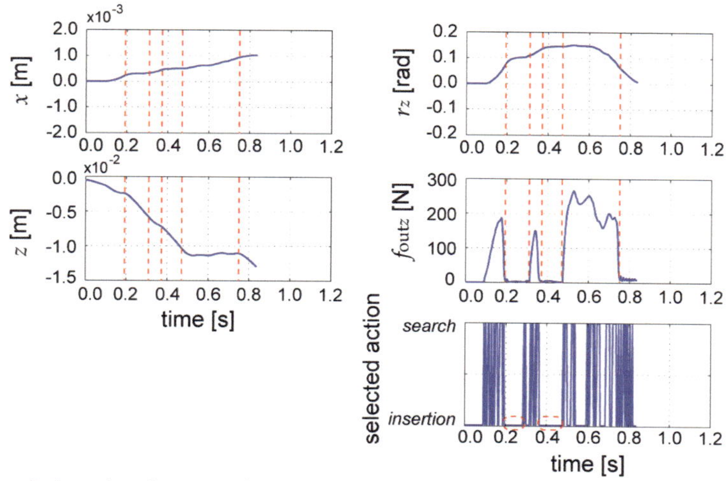

Fig. 8. Result of the clutch assembly based on the state action map obtained after 600 iterations

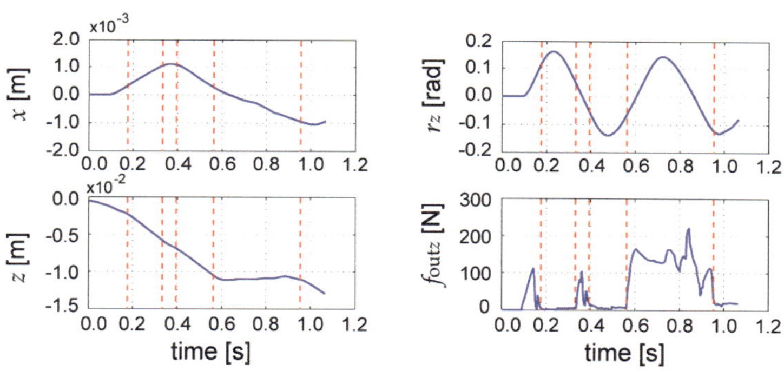

Fig. 9. Result of the clutch assembly based only on the search motion with the same clutch plate arrangement with that used in Fig. 8.

It is one of the future works to develop a learning method in order to efficiently generate a new policy based on similar policies. Although the approach was applied only to the clutch assembly in this paper, it will be applied to the other complex assembly tasks for more evaluation.

7. References

Bellman, R. (1957). *Dynamic Programming*, Princeton University Press

Gullapalli, V.; Franklin, J.A. & Benbrahim, H. (1994). Acquiring robot skills via reinforcement learning, *IEEE Control Systems*, Vol. 14, No. 1, pp. 13–24

Hirai, S.; Inatsugi, T. & Iwata, K. (1996). Learning of admittance matrix elements for manipulative operations, *Proceeding of IEEE/RSJ International Conference on Intelligent Robots and Systems*, pp. 763–768

Huang, S. & Schimmels, J.M. (2003). Sufficient conditions used in admittance selection for force-guided assembly of polygonal parts, *IEEE Transaction on Robotics and Automation*, Vol. 19, No. 4, pp. 737–742

Lin, L-J (1991). Programming Robots Using Reinforcement Learning and Teaching, *Proceeding on the Ninth National Conference on Artificial Intelligence*, pp. 781–786

Minato, T. & Asada, M. (1998). Environmental Change Adaptation for Mobile Robot Navigation, *Proceeding of IEEE/RSJ International Conference on Intelligent Robots and Systems*, pp. 1859–1864

Press W.H.; Teukolsky, S.A.; Vetterling, W.T. & Flannery, B.P. (1992). *Numerical Recipes in C: The Art of Scientific Computing, 2nd ed.*, pp. 394–455, Cambridge University Press, ISBN-10: 0521750334

Schimmels, J.M. (1997). A linear space of admittance control laws that guarantees force-assembly, *IEEE Transaction on Robotics and Automation*, Vol. 13, No. 5, pp. 656–667

Simons, J.; Van Brussel, H.; De Schutter, J. & Verhaert, J. (1982). A self-learning automation with variable resolution for high precision assembly by industrial robots, *IEEE Transaction on Automatic Control*, Vol. AC–27, No. 5, pp. 1109–113

Skubic, M. & Volz, R.A. (2000). Acquiring Robust, Force-Based Assembly Skills from Human Demonstration, *IEEE Transaction on Robotics and Automation*, Vol. 16, No. 6, pp. 772–781

Sutton, R.S. & Barto, A.G. (1998). *Reinforcement Learning: An Introduction*, The MIT Press, ISBN-10: 0262193981

Sutton R.S.; Precup, D. & Singh, S. (1999). Between MDPs and semi-MDPs: A framework for temporal abstraction in reinforcement learning, *Artificial Intelligence*, Vol. 112, pp. 181–211

Suzuki, T.; Inagaki, S. & Yamada, N. (2006). Human Skill Modelling Based on Stochastic Switched Dynamics, *Proceeding of IFAC Conference on Analysis and Design of Hybrid Systems*, pp. 64–70

Tanaka, F. & Yamamura, M. (2003). Multitask Reinforcement Learning on the Distribution of MDPs, *Proceeding of IEEE International Symposium on Computational Intelligence in Robotics and Automation*, pp. 1108–1113

Thrun, S. & Mitchell, T.M. (1995). Lifelong Robot Learning, *Robotics and Autonomous Systems*, Vol. 15, pp. 25–46

Wei, J. & Newman, W.S. (2002). Improving robotic assembly performance through autonomous exploration, *Proceeding of IEEE International Conference on Robotics and Automation*, pp. 3303–3308

Whitney, D.E. (1982). Quasi-static assembly of compliantly supported rigid parts, *Transaction of ASME, Journal of Dynamic Systems, Measurement and Control*, Vol. 104, pp. 65–77

Yamanobe, N.; Maeda, Y. & Arai, T. (2004). Designing of Damping Control Parameters for Peg-in-Hole Considering Cycle Time, *Proceeding of IEEE International Conference on Robotics and Automation*, pp. 1129-1134

Yamanobe, N.; Fujii, H.; Arai, T. & Ueda, R. (2008). Motion Generation for Clutch Assembly by Integration of Multiple Existing Policies, *Proceeding of IEEE/RSJ International Conference on Intelligent Robots and Systems*, pp. 3218–3223

6

Robotic Strategies to Assist Pilots in Landing and Takeoff of Helicopters on Ships and Offshore

Alexandre Campos, Jacqueline Quintero, Roque Saltarén,
Manuel Ferre and Rafael Aracil
Universidad Politécnica de Madrid
Spain

1. Introduction

A large number of maritime structures used for commercial, exploration and military purposes are equipped with a helicopter landing deck or winching area, respectively. Offshore flight operations are a highly complex and specialized process. It requires high levels of training, competence and skill to plan a flight, land and takeoff from an offshore installation even in good weather flying conditions. These operations can be seriously degraded by environmental effects that may be present around installations, vessels and their helidecks. The effects of wind turbulence from topsides structures or hot gases cause difficult landings and temporary loss of aircraft control by the pilot. Hot exhaust plumes from gas turbines on an installation can affect helicopter performance. The power output of the turbine engines on a helicopter is reduced as the air temperature at the intake increases. Rapid temperature changes can also induce engine surge and even compressor stall or flameout. The low cloud at helideck and other weather conditions affect visibility. The wind and wave induced dynamic motions on helidecks, on floating installations and vessels which represent a potential hazard to helicopter operations.
Therefore, development of new equipment and technologies are required in order to ease helicopter crew with the execution of flight operation. Aiming at assists pilots in these flight operations, we propose a new approach using an active helideck. Active-helideck based on Stewart-Gough platform is designed, developed, computational simulated and tested as a helicopter floating helideck. The objective of this testbed is to show the advantages of helicopters that use an active helideck upon landing on and taking off from ships or from offshore structures. Active-helideck compensates simulated movements of a ship at sea. The main goal of this study is to maintain the robot's end effector (helideck) in a quasi-static position in accordance to an absolute inertial frame. Compensation is carried out through the coordinate action of its six prismatic actuators in function of an inertial measurement unit. Moreover, the simulation of the sea movement is done by a parallel robot called ship platform with three degrees of freedom. A Computed Aided Engineering (CAE) system is used for modeling, simulation and analysis before construction testbed. The ship platform is

built with a vertical oscillation along the z axis, *i. e.* heave, and rotates on remaining axes, *i. e.* roll and pitch. Active helideck is able to compensate simulated movements by considering the ship as an inertial frame as observed in the experiment.
Firstly, this chapter presents the different strategies used for assist crew in flight operation. Next, the testbed is described. Follow, kinematics model and control strategy for the testbed are explained. Then, it includes part of the results obtained through the simulation of testbed. It is then followed by the experimental procedure and results. Finally, conclusions are pointed out.

2. Assistance strategies

In order to assist crew in difficult maneuvers or tasks different strategies have been implemented. The wind based problems are identified through a wind-tunnel model study using flow visualization. The results determine appropriate position of the helideck, obstructions or exhausts, considered in design phase. Additionally, the designed helideck position should minimize its motion. Problem analysis result in operating limits being imposed by the helicopter operators.
Mathematical models for ship movements are formulated (Fossen, 2002), (Matusiak, 2002). Reliable information on weather and helideck motions needs to be provided at the flight aiming at outset maximum opportunity to be completed safely and efficiently or abort flight operations. Wave measurement (using pressure sensors, horizontal radars, vertically operating distance or velocity meter) and analysis with fast communication systems have been developed (Pajala, 2002). Moreover, by means of a satellite (*e.g.* ERS-1 European Remote-Sensing Satellite-1), wind and wave fields, sea-surface temperature, ocean tides, iceberg monitoring, oil and pollution detection can be determined. Satellite measurements combined with wave model gives real time wave information for sea operation (He & Zinping, 1997), (Pajala, 2002). Furthermore, an aircraft/ship dynamic interface analysis simulation software package has been developed in order to analyze dynamic interface that exists between ships and embarked aircraft. It has been extensively validated and applied (Linn & Langlois, 2006), (Langlois & LaRosa, 2003). Aircraft performance data is scheduled in the flight manual and/or the operations manual which enables flight crew to accommodate the varying ambient conditions and operate in such a way that the helicopter has sufficient space and engine performance to approach land and take off from helidecks in securely.
Computer simulations are some approaches carried out for studying this case. Oh et al. design an autopilot for autonomous helicopter landing on a rocking ship through a tether in helideck (target) tracking (Oh et al., 2005), (Oh et al., 2006). In this case, a controller is considered to ensure the landing of a helicopter on a ship based on measurement of angle between the helicopter/ship and the cable. The tether permits divide the problem of landing in two control problems, position control and attitude-altitude control. In the first, when the helicopter is far the destination, the rotation reference signals are used to control the translation of the helicopter. In the second, due to coupling of the translation of the helicopter to the rotation through the tether, the translation reference rates are created to achieve a desired behavior attitude and altitude.
Dalamagkidis *et al.* design a gimbaled landing platform on an unmanned ground vehicle (UGV) (Figure 1). Instead of a ship, an UGV is used in order to transport miniature vertical takeoff and landing (VTOL) (Dalamagkidis et al., 2006). A two-axes gimbaled is used for

ensuring a horizontal landing and free from vibrations platform for the VTOL, even during the movement of the UGV. The inertial measure unit (IMU) of UGV provides the heading and elevation of the platform. Then the platform is first rotated around the heading axis until the elevation axis is horizontal and after heading has been changed, the platform will be rotated around its new elevation vector, until the heading vector is vertical; the platform assumes a horizontal pose.

Fig. 1. 3D representation of the landing platform installed on UGV top (Dalamagkidis et al., 2006)

Fig. 2. The Stewart Platform is installed on MORTIMER (Graf & Dillmann, 1997)

Before Dalamagkidis a similar configuration was used, but the platform mounted on the vehicle is used to aceleration compensation of it. Graf and Dillmann used a Stewart platform installed on the mobile robot so that abrupt vehicle movements and accelerations affect the object in a smooth way in x- and y-direction. The Stewart platform is moved so that the gravity create a contrarily directed acceleration (Graf & Dillmann, 1997), (Graf & Dillmann, 1999) as is show in the Fig 2.

The controller of the stewart platform is connected directly to the controller of the mobile system to get the latest acceleration values all the time. The acceleration vector of the robot is inverted and transformed into the coordinate system of the platform. A double integration is realized to get position and orientation data for the platform. The inverse kinematics deliver the length of the legs.
With similar objectives Dan and et al. mounted at the top of the mobile base is the planar 3RRR manipulator which serves as compensation device (Decker et al., 2001), (Dang & Ebert-Uphoff, 2004). The compensation platform moves in three degrees of freedom (DOFs): translation and rotation in the vertical plane (see Figure 3). They decided for a three degrees-of-freedom motion planning algorithm, based on a combination of flexible local optimization scheme based on feedback algorithms (FMPA) and a so called Pendulum Algorithm (3DOF-FMPA). The pendulum algorithm emulates the response of a virtual pendulum when subjected to linear accelerations applied at its pivot point. The object is attached to the virtual pendulum. This pendulum approach is motivated by the fact that the free swinging motion of a pendulum automatically minimizes the lateral force acting on the object. For this application, the linear acceleration of the mobile robot is mapped to the linear acceleration of the pendulum's pivot point. The resulting linear motion of the pendulum's tray is mapped to the linear acceleration of the parallel platform.

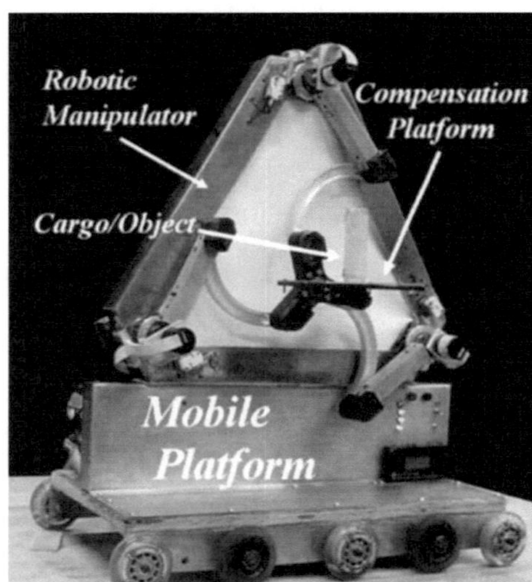

Fig. 3. Testbed consisting of mobile base with acceleration compensation platform acting in vertical plane (Dang & Ebert-Uphoff, 2004)

At present chapter, a new approach using an *active helideck* is intorduced. Its aim is to compensate possible movements of a landing platform caused by wind and/or waves. A 6-dof parallel robot has been designed, simulated and built (see Figure 4). Its objective is to test active helideck on a moving structure. A ship at sea movement simulator named *ship platform* has been designed and built. The more important offshore structure movements: roll, pitch and heave (Fossen, 2002), are considered; therefore a *3* dof parallel platform (Cheng, 1994), two of rotation and one of movement along the vertical axis have been used as a ship platform.

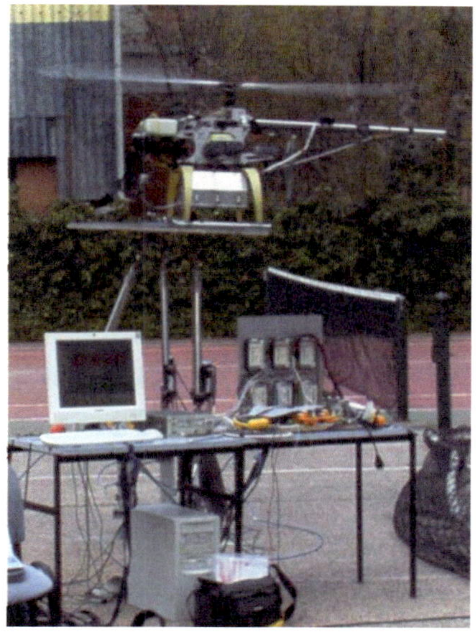

Fig. 4. Testbed: active helideck (AH) and ship platform (SP)

3. Testbed Description

In order to compensate helideck movement with the wind and wave effects on the ship's hull, as well as simulating this movement in the testbed, some dimension considerations have taken into account. The position amplitude is 0.1 m, orientation amplitude is 6° and wave period is 10 s. Orientation angle and the wave period were taken from RMS values of general operability limiting criteria for ships (Nordforsk, 1987). On the other hand, a big ship has been considered in order not to take into account scaling factors in terms of size and mass/inertia properties or transients due to helicopter landing/taking off for this testbed.

The testbed in this study is based on parallel robot architecture, *i.e.* a closed kinematics chain. A 6 dof Stewart-Gough platform is selected for the active helideck (Stewart, 1966), (Merlet, 2000) as shown in Figure 5.

Stewart Platforms are mainly used for flight or driving simulators (Baret, 1978) and for machine tools (Pierrot et al., 2001), (Thomas et al., 2002). The advantages of Stewart Platforms are the excellent weight and payload capability ratio and precise positioning (Merlet, 2000). A Stewart platform is commonly used in obtaining acceleration, but it can also generate an anti acceleration if the base platform is moving (Graf & Dilmann, 1997); application of Stewart platforms compensate acceleration of a moving robot. However, only vibrations of relatively minimal amplitude are considered (Graf & Dilmann, 1997), (Geng & Haynes, 1993), (Cobb et al., 1999), (Vaillon et al., 1999). A greater amplitude is considered in vibration control simulation of a Stewart-Gough platform on flexible suspension (Cheng et al., 2003), where the platform compensates the movements of a radio telescope mounted at a suspending structure. In such case, expected wind-induced vibration can be as high as 0.5 m in position and 3° in orientation. In this given case, the reaction force caused by the motion of the stabilized platform will lead to perturbation on the base platform as the base platform is not fixed on the ground. It will then induce vibration on the whole system.

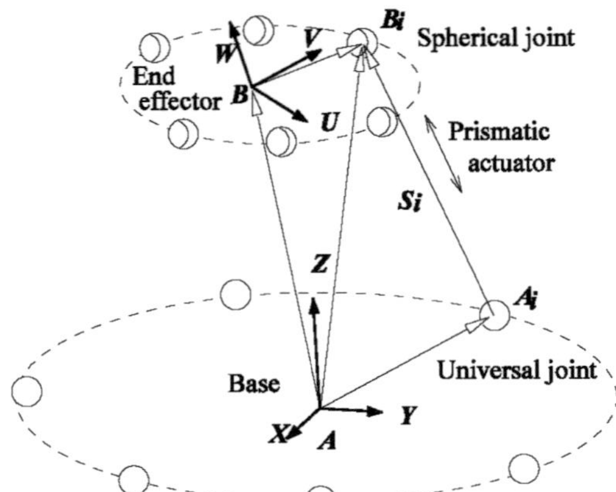

Fig. 5. Schematic architecture of the 6-dof parallel robot used as active helideck

Aiming at simulating the ship movement, the closed kinematic chain used is the 3-dof parallel manipulator shown in Figure 6.
In this chapter, the proposed alternative is a testbed compounded by two main parts, see Figure 4, the *ship platform* from now on named **SP,** *i. e.* a ship at sea movement simulator, and the *active helideck* from now on named **AH,** *i. e.* a Stewart-Gough platform.

3.1 Active Helideck

The active Helideck **(AH)** is a Stewart-Gough platform *6-UPS* (Universal, Prismatic, Spherical) whose end effector (helideck) is connected to the base, *i. e.* ***SP*** end effector, through six prismatic electrical actuators. Prismatic actuators are connected to the end effector by spherical joints and to the base by universal joints. Ball screws actuated by electric brush-less motors have been used considering prismatic movement.

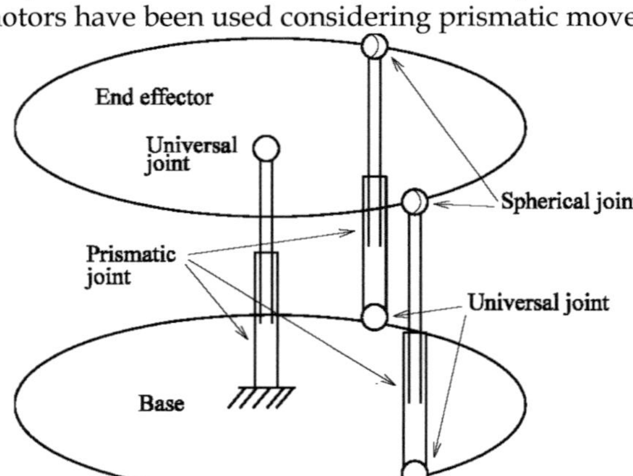

Fig. 6. Schematic architecture of the 3-dof parallel robot used as ship movement simulator

3.2 Ship Platform

The **SP** requires that its platform end effector (base of **AH)** moves along vertical axis *(Z)*, taking into account ship movement *heave,* rotation around axes *X* and *Y* with respect to its base, and also *roll* and *pitch* from ship movements. Therefore, a 3dof parallel platform is constructed which consists of four pneumatic prismatic actuators respectively controlled by proportional valves:

- Two pneumatic cylinders fixed to the base that are connected by a universal joint to the end effector (ship hull). These cylinders move simultaneously in order to control height, *i. e.* heave of the **SP** end effector. These two cylinders are cinematically considered as one central cylinder.
- Two pneumatic cylinders provide the rotational movements, *i. e.* roll and pitch; these cylinders, called external cylinders, are connected by universal joints to the base and by spherical joints to the end effector.

Therefore, **AH** compensates the movement of the **SP** in order to maintain the deck, placed on its end effector, in a quasi-static position. The constructed testbed is shown in Figure 7. Two sensors are installed on the testbed in order to obtain the information from the environment which allow actuating the **AH** conveniently. An inertial measurement unit (IMU) is located at the **SP** end effector in order to get its instantaneous orientation, through Euler parameters. An ultrasonic length sensor is used to quantify SP height that has to be compensated in order to keep the helideck in a quasi-static height.

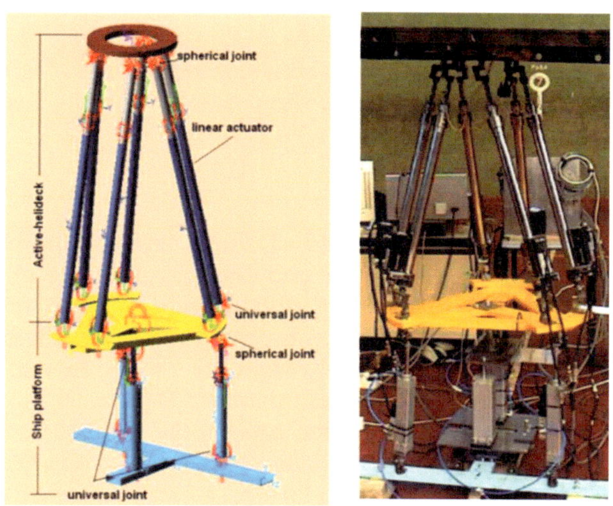

Fig. 7. Testbed for computational simulation and for real testing

4. Kinematic model

Using the concepts of the parallel robot inverse kinematics, control signals can be obtained in order to determine AH coordinate movements according SP behavior. This inverse kinematics calculates the limb lengths, which is reached through the active prismatic joints (see Figure 5), given the SP location (position and orientation).

4.1 Mathematical Tools

Two Cartesian coordinate systems are attached to base and moving platform in order to calculate parallel robot inverse kinematics. Frames *A(X, Y, Z)* and *B(U, V, W)* respectively as shown in Figure 5.

Each joint location in the base is described by the position vector A_i in the coordinate frames A. Similarly, each joint location in the end effector is described by the position vector B_j in the coordinate frame *B*. Therefore, the *i*-th limb length is the magnitude of the vector directed from A_i to B_i, *i. e.* S_i:

$$d_i = |S_i| = B_i - A_i \tag{1}$$

In order to carry out the vectorial subtraction of (1), the vectors must be expressed in the same reference frame. A point *p* expressed in reference frame *B* may be expressed in reference frame *A*, using the matrix of homogeneous transformation (Tsai, 1999):

$$A_{T_B} = \begin{bmatrix} A_{R_{B(3x3)}} & A_{q(3x1)} \\ \gamma_{(1x3)} & \rho_{(1x1)} \end{bmatrix} \tag{2}$$

Then using the matrix of homogeneous transformation, the position in the new reference frame is obtained by:

$$A_p = {}^A T_B {}^B p \tag{3}$$

$${}^A p = {}^A q + {}^A R_B {}^B p \tag{4}$$

where, ${}^A\mathbf{p} = [\mathbf{p_x} \ \mathbf{p_y}' \ \mathbf{p_z}' \ \mathbf{1}]$ is the position, in homogeneous coordinates, expressed in reference frame *A* and ${}^B\mathbf{p} = [\mathbf{p_u}' \ \mathbf{p_v} \ \mathbf{p_w}' \ \mathbf{1}]$ is the position in homogeneous coordinates as expressed in reference frame *B*.

4.2 Testbed kinematic model

Three reference frames are selected in the testbed: global reference frame G, located at the **SP** base; reference frame *A* located at the **SP** end effector (or **AH** base), and the reference frame *B* located at **AH** end effector, see Figure 8. Initially these frames are oriented according the IMU axes as ruled by: X axis yields along magnetic north direction and Z axis aligns to gravitational force direction. The IMU is attached to the **SP** end effector. It is used to measure its orientation through the four Euler parameters (e_0, e_1, e_3).

The coordinates of the universal joint axes that connect the prismatic actuators at **AH** end effector are determined in a Cartesian coordinates system relative to reference frame *B*. The points where the **AH** actuators are attached to their base (A_i) are specified on reference frame *A*. Additionally, the points where the **SP** pneumatic cylinders are attached to the **AH** base A_i are specified on reference frame *A* too. Finally, the points G_i where pneumatic cylinders are attached to **SP** base (ground) are specified on reference frame *G*, as shown in Figure 8.

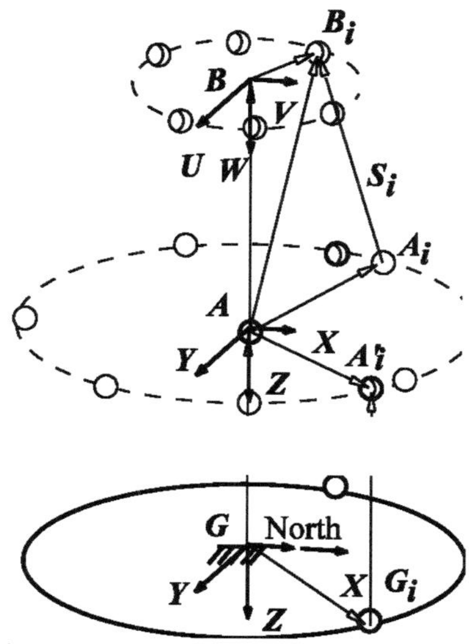

Fig. 8. Reference frames fixed at experimental platform

5. Control strategy

The objective of this control strategy is to determine **SP** and **AH** displacements; in such a way that **SP** represents the sea movements that have to be compensated by **AH** aiming at a quasi-static deck. A schematic diagram for the experimental platform is presented in Figure 9.

5.1 Ship Platform Strategies of Movement

Ship movements are described through sinusoidal displacements for the pistons. It is important to notice that the central cylinder determines the ship hull heave. The combination of central and external cylinders determines the ship hull pitch and roll. The displacement of the **SP** central cylinder is sinusoidal around a initial position and independent from the two external cylinders. This displacement causes the desired heave for the **SP.** External cylinders move follow a sinusoidal displacement too. The **SP** control loop movement is closed with the IMU attached in the **SP** end effector and with the ultrasonic length sensor which measures the **SP** end effector height. This loop for the SP movement is closed with the IMU attached in the SP end effector and with the ultrasonic length sensor which measures the SP end effector height.

The rotation matrix is obtained by using values provided by IMU, the reference frame A origin position is [0, 0, z] with respect to reference frame G, where z is read by a ultrasonic sensor, and the origin position of the reference frame B, *i. e.* the desired deck height, has to be well-known and it is indeed a variable to control. Through the inverse kinematics, effective displacements that must be reached to obtain the desired total displacement of the actuator are calculated in accordance to current position and orientation of the SP. The sinusoidal movements of the external actuators are out of phase in order to provide movement around X axis, *i. e.* pitch. The central piston is out of phase in order to obtain oscillations around the y axis, *i. e.* roll.

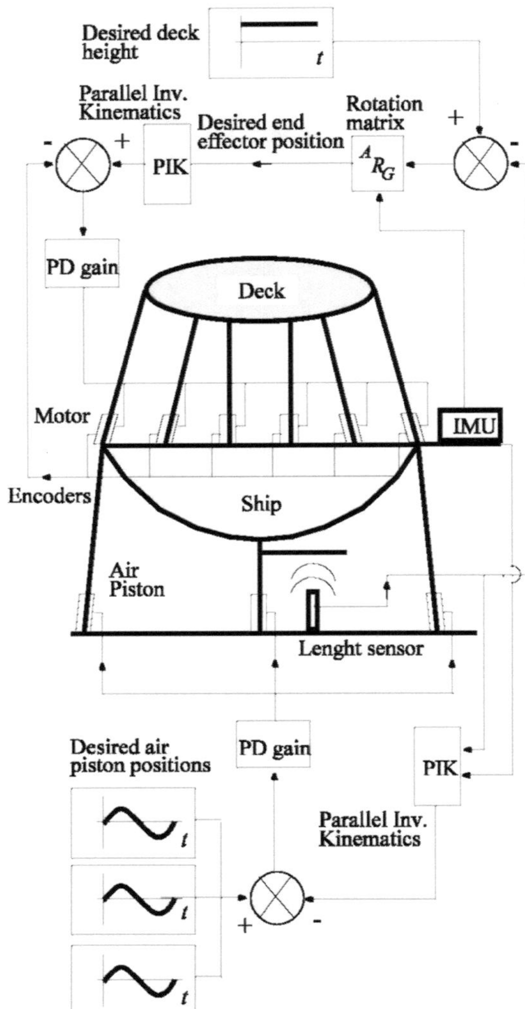

Fig. 9. Testbed schematic diagram

5.2 Active Helideck Strategies of Movement

The helideck end effector is required to remain in a constant position and orientation with respect to the inertial frame G. Aiming at such, it is then necessary to calculate the **AH** actuator displacements according to the instantaneous **SP** position and orientation. Nevertheless, applying the equations of inverse kinematics, **AH** end effector position has to be determined with respect to the reference frame G. Using Eq. (2), transformation of coordinates are calculated:

$$^{A}p_{f} = \begin{bmatrix} ^{A}R_{G} & ^{A}R_{G} \\ [0\ 0\ 0] & 1 \end{bmatrix} \tag{5}$$

where,

$^{G}p_{f}$ is the helideck end effector position with respect to inertial reference frame G. This value

is maintained constant for any movement of inferior platform.
$^{A}p_{f}$ is the position of the frame *G* with respect to frame *A*. This position is obtained by:

$$^{A}R_{G} = {}^{A}R_{G}\left(-{}^{G}p_{A}\right) \tag{6}$$

$^{A}R_{G}$ is the rotation matrix of frame *G* respect to frame *A*. This it is obtained transposing the rotation matrix $^{G}R_{A}$.
Finally, **AH** actuator lengths are calculated by using parallel inverse kinematics (IPK).

5.3 Position Control

Control stage is based on a dSPACE tm DS1103 card. DS1103 is a digital signal processor (DSP) on a PC platform where the controller can be easily programmed. The control law is introduced in C environment, then compiled and downloaded to DS1103 card using real time tools included in the Control desk package (dSpace, 2001). The controller card captures voltage signals from inertial measure unit, length sensor and motor encoders. It also calculates every three millisecond new control signal and feeds the motor amplifiers and the air piston proportional valves.
Control position for both platforms *(i. e.* **SP** and **AH)** is performed through proportional derivative PD controllers so as to assure stability even with errors in steady state. These PD gains for electric motors and for pneumatic proportional valves are tuned experimentally. It is important to consider that due compressive air properties, air pistons' behavior is different when moving up or moving down. Therefore, two different PD gains are set to each pneumatic piston.

6. Computational simulation

Prior to the implementation of the first active helideck prototype, a Computer Aided Engineering (CAE) system to simulate the testbed is used. The CAE system employed permits to build models of mechanical systems and to simulate the full-motion behavior of the models. A powerful numerical analysis automatically solves the equations of motion for kinematic, static and dynamic simulations, that lets animate and optimize these models. In this work, the computational simulation confirms kinematics of the robot, its workspace and predicts the behavior of the system before the proposed control algorithms. Primitive geometries of software are used to model the true prototype, as shown in Figure 7. Measures position and orientation of the **AH** base are used to simulate the sensors. A movement for each of the linear actuators is imposed according sea movement and the resulting **SP** motion effector is validated. According to the obtained measurements, the necessary displacement for each linear actuator of the helideck is estimated. Therefore, the **AH** prismatic joints, are actuated based on this estimation. Through this simulation, we verify that the required displacements for the **AH** prismatic joints fulfill the conditions to keep the helideck in a quasi-static position, using a PD control strategy.
Simplified models were also made to determine the height to which should be placed the **AH** end effector, in order to optimize the necessary forces on the actuators. The results of the simulations are summarized in Figure 10, where the maximum required force versus the desired (initial) height is plotted. For different heights, we examined the required forces in

each actuator, for a full cycle of movement. The initial position of **AH** end effector is chosen (so that the maximum force required is minimal 1142 mm respect its base).

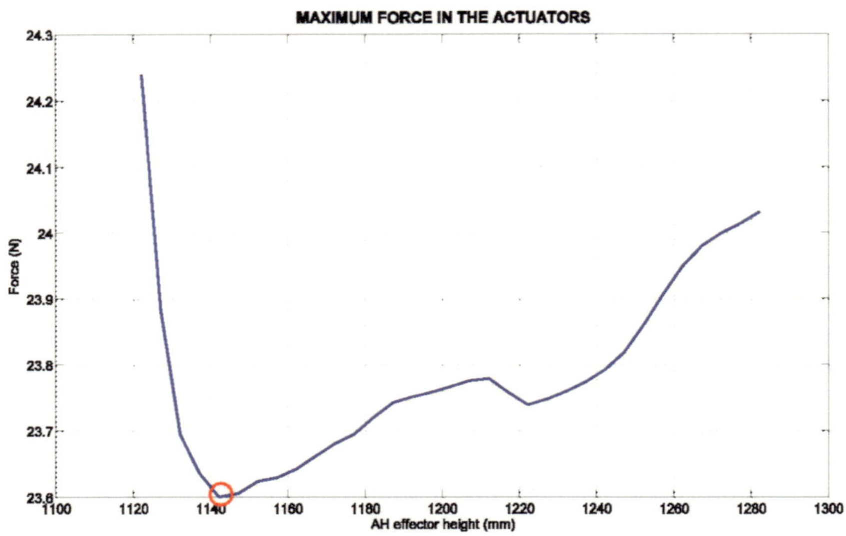

Fig. 10. Maximum force exerted on the linear actuators

7. Experimental test

Tests have been carried out on the experimental platform in the laboratory and then in field as have been shown in sequence of Figure 11. This test is required to verify that the helideck (**AH** end effector) remains quasi-static in spite of **SP** movements.

Initially, a trajectory planner is used for moving the helideck to a position that allows the actuator movements aiming at compensating the inferior platform movements at any moment. During this time the inferior platform pneumatic actuators are locked. Then, external pneumatic cylinders are driven in order to carry out **SP** to a horizontal position. At that moment **AH** control is activated to compensate the **SP** movement. Then, all the pneumatic cylinders are moved to the middle of their strokes and begin a sinusoidal movement. This movement causes a variation in inferior platform orientation and position, which is measured by the sensors. From the sensors information, the processor calculates the **AH** actuators' displacement needed to hold its fixed location with respect to the inertial frame. Using this information, the actuators are energized through the amplifiers. After that the helicopter approaches and lands on compensated helideck.

Finally, the position and orientation of helideck and inferior platform are compared in order to observe the compensation and the deck's desired behavior.

Fig. 11. Helicopter approaching and landing on Helideck

7.1 Experimental Results

From the experimental results, the position and orientation variation between helideck and inertial frame can be observed.

The orientation is described through Euler parameters (e_0, e_1, e_2 and e_3), where e_0 is associated to the rotation angle around the axis whose *X*, *Y* and *Z* components are associated ej, e_2, and e_3.

From Figure 12, **SP** presents an angular oscillation around *X* (roll) and *Y* (pitch) axis representing the sea movement. Additionally, the helideck orientation remains quasi-static in relation to the inertial frame in spite of the orientation variation of its base, *i. e.* the **SP.** It is important to highlight that the desired reference Euler parameters, for a horizontal deck position, are $e_1 = 1$ and $e_1 = e_2 = e_3 = 0$.

In Figure 13 is shown the helideck displacement with respect to **SP** aiming at remaining quasi-static with respect to inertial frame as desired. **AH** has to compensate the vertical movements as well. Last plot of Figs. 13 shows the helideck behavior for a given wave input, represented by inferior platform vertical movement. It is worth mentioning that the noise in the signal is caused by the use of the ultrasonic length sensor.

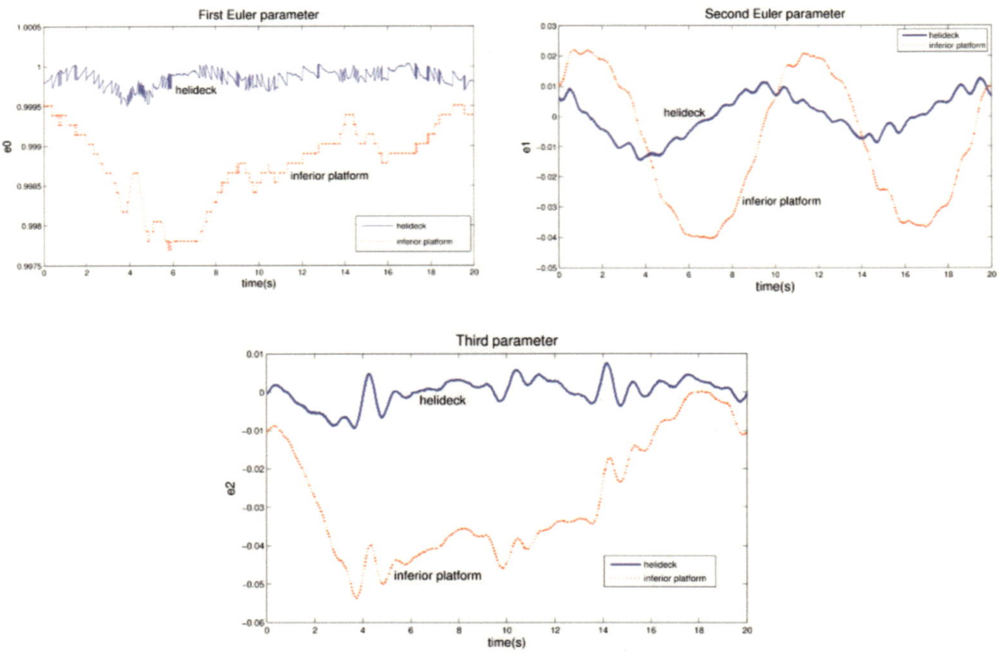

Fig. 12. Euler parameters (*e0*, *e1*, *e2*) for helideck and SP

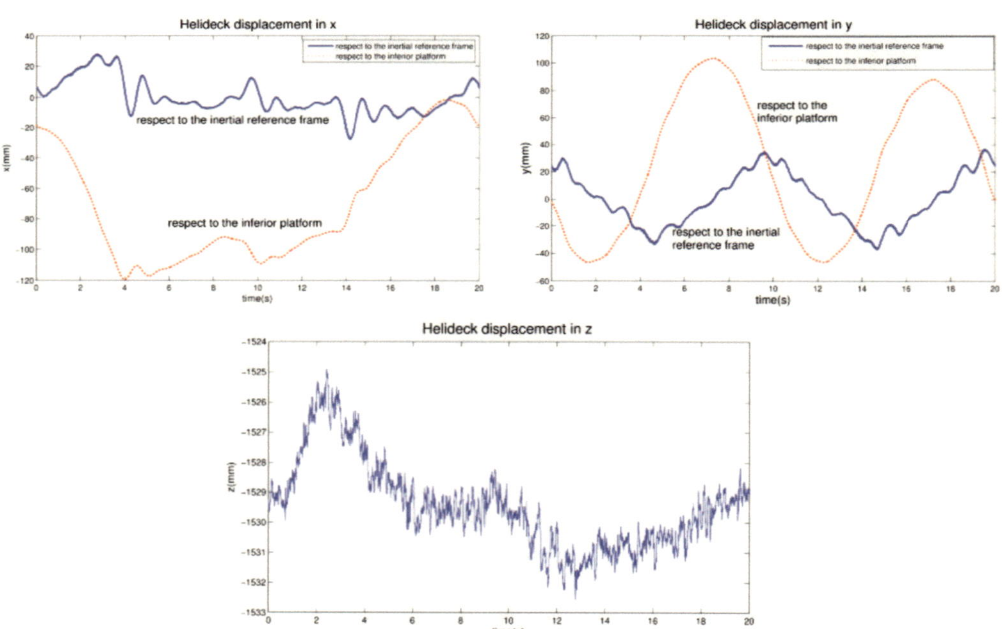

Fig. 13. *X*, *Y*, y *Z* position coordinates for helideck with respect to SP and to inertial frame

8. Conclusions and future works

Several solution strategies have been proposed to assist helicopter pilots en offshore operations (landing and taking off). These strategies are in general based on measurement, mathematical models and computational simulation.
A testbed based on Stewart-Gough platform has been used, built and tested as helicopter deck. It was observed through computational simulation and experimental test that the active helideck is able to compensate the simulated movement provided by the sea to a ship or off-shore structure in order to ease helicopter landing and take-off maneuvers, considering a big enough ship with inertial frame properties.
For the testbed, a simulation analysis has determined the most suitable height of the **AH** end effector. This is next to the minimum height required to allow compensation of the movement.
The Stewart-Gough platform used in this study can still be improved with a larger base diameter so that the desired helideck angles and height could be attained. However, active helideck inertia due the ship platform motion could decrease.
A dimensional analysis for ship mass/inertia could be done in order to closely simulate a real ship's behavior.

9. References

Baret, M. (1978). Six Degrees of Freedom Large Motion System for Flight Simulators, Piloted Aircraft Environment Simulation Techniques, *Proceedings of AGARD Conference No. 249, Piloted Aircraft Environment Simulation Techniques*, pp. 22-1/22-7, Bruxelles.

Cheng, H. (1994). Real-Time Manipulation of a Hybrid Serial-and-Parallel Driven Redundant Industrial Manipulator. *J. of Dynamic Systems Measurement and Control-Transactions of the ASME*, Vol. 116, No. 4, pp. 687-70.

Cheng, G.; Ren, Y. & Dai, S. (2003). Vibration Control of Gough-Stewart Platform on Flexible Suspension. *J. of Robotics and Automation, IEEE*, Vol. 19, No. 3, pp. 489-493.

Cobb R.; Sullivan J.; Das A.; Davis L.; Hyde T.; Davis T.; Rahman Z. & Spanos J. (1999). Vibration Isolation and Suppression System for Precision Payloads in Space. *J. of Smart Materials and Structures*, Vol. 8, No. 6, pp. 798-812.

Dalamagkidis, K.; Ioannou, S.; Valavanis, K. & Stefanakos, E. (2006). A Mobile Landing Platform for Miniature Vertical Take-off and Landing Vehicles, *Proceedings of the 14th Mediterranean Conference on Control and Automation – MED '06, IEEE*, pp. 1-6 Barcelona.

Dang, A. X. & Ebert-Uphoff, I. (2004). Active Acceleration Compensation for Transport Vehicles Carrying Delicate Objects. *IEE Transactions on robotics*, Vol. 20, No.l 5, Oct. 2004, pp. 830-839.

Decker, M. W.; Dang A. X. & EbertUphoff Imme. Motion Planning for Active Acceleration Compensation. *Proc. IEEE Int. Conf. Robotics and Automation*, , pp. 1257-1264, Seoul, Korea, May 21-26, 2001.

dSpace GmbH (2001). *DS1103 PPC Control Board Hardware Reference*, Germany.

Ford, T.; Hardesty, M. & Bobye, M. (2005). Helicopter Ship Board Landing System, *Proceedings of the 18th International Technical Meeting of the Satellite Division, ION GNSS*, pp. 979-988, Long Beach, CA.

Fossen, T. (2002). *Marine Control Systems: Guidance, Navigation and Control of Ships, Rigs and Underwater Vehicles*, Marine Cybernetics AS, Trondheim, Norway.

Geng, Z. & Haynes, L. (1993). 6-Degree-of-Freedom Active Vibration Isolation Using a Stewart Platform Mechanism. *J. of Robotic Systems*, Vol. 10, No. 5, pp. 725-744.

Graf, R. & Dillmann, R. (1997). Active Acceleration Compensation Using a Stewart-Platform on a Mobile Robot, *Proceedings of the 2nd Euromicro Workshop on Advanced Mobile Robots*, pp. 59-64, Brescia, Italy.

Graf, R. & Dillmann, R. (1999). Acceleration compensation using a Stewart platform on a mobile robot, *Third European Workshop on Advanced Mobile Robots (Eurobot' 99)*, pp. 17 - 24, 6-8 Sept. 1999.

He, Y. & Jinping , Z. (1997). Ocean Wave Spectrum Reconstruction from e s-1 Satellite Scatterometer Data, *Proceedings of Geoscience and Remote Sensing, 1997 – IGARSS '97, IEEE*, Remote Sensing - A Scientific Vision for Sustainable Development, Vol. 1 of, pp. 356-358.

Langlois, R. G. & LaRosa M. (2003). T. A. R., Development, Validation, and Application of the Dynaface Helicopter/ship Dynamic Interface Simulation Software Package, *Proceedings of the Summer Computer Simulation Conference, SCSC (Ed.)*, pp. 167-176.

Linn, D. & Langlois, R. (2006). Development and Experimental Validation of a Shipboard Helicopter on-Deck Maneuvering Simulation. *J. of Aircraft*, Vol. 43, No. 4, pp. 895-906.

Matusiak, J. (2002). Two-Stage Approach to Determination of Large Amplitude Motions of a Rigid Ship in Waves, *Proceedings of the 15th Nordic Seminar in Computational Mechanics - NSCM15*, pp. 1 - 10, Aalborg, Denmark.

Merlet, J. (2000). *Parallel Robots*, Kluwer Academic Publisher.

Nordforsk (1987). *Assessment of a Ship Performance in a Seaway*, The Nordic Cooperative Project, Sea Keeping Performance of Ships. Trondheim, Norway: Marintek.

Oh, S.; Pota, H.; Pathak, K. & Agrawal, S. (2005). Autonomous Helicopter Landing on a Moving Platform Using a Tether, *Proceedings of ICRA*, pp. 3960-3965, Barcelona.

Oh, S.; Pathak, K.; Agrawal, S.; Pota, H. & Garratt, M. (2006). Approaches for a Tether-Guided Landing of an Autonomous Helicopter. *IEEE Transactions on Robotics*, Vol. 22, No. 3, june 2006, pp. 536-544.

Pajala, P. (2002). *Wave Measurements*, Maritime Institute of Finland, Sjkulla, Finland.

Pierrot, F.; Marquet, M.; Company, O. & Gil, T. (2001). H4 Parallel Robot: Modeling, Design and Preliminary Experiments, *Proceedings of ICRA, IEEE*, pp. 3256-3261, Seoul.

Stewart D. (1996). A Platform with Six Degrees of Freedom, *Proceedings of the Institution of Mechanical Engineers*, Vol. 180, No. 15, pp. 371-386.

Thomas, U.; Maciuszek, I. & Wahl, F. (2002). A Unified Notation for Serial, Parallel, and Hybrid Kinematic Structures, *Proceedings of ICRA*, Vol. 3, IEEE, pp. 2868-2873, Washington.

Tsai, L.-W. (1999). *Robot Analysis: the Mechanics of Serial and Parallel Manipulators*, John Wiley & Sons, New York.

Vaillon, L.; Petitjean, B.; Frapard, B. & Lebihan, D. (1999). Active Isolation in Space Truss Structures: from Concept to Implementation. *J. of Smart Materials and Structures*, Vol. 8, No. 6, pp. 781-790.

7

Optimality Principles and Motion Planning of Human-Like Reaching Movements

Mikhail M. Svinin*, Igor A. Goncharenko**,
Shigeyuki Hosoe* and Yoshihito Osada*
*The Institute of Physical and Chemical Research, RIKEN, Japan
**3D Incorporated, Yokohama, Japan

Abstract
The paper deals with modeling of human-like reaching movements. Several issues are under study. First, we consider a model of unconstrained reaching movements that corresponds to the minimization of control effort. It is shown that this model can be represented by the well-known Beta function. The representation can be used for the construction of fractional order models and also for modeling of asymmetric velocity profiles. Next, we address the formation of boundary conditions in a natural way. From the mathematical point of view, the structure of the optimal solution is defined not only by the form of the optimality criterion but also by the boundary conditions of the optimization task. The natural boundary conditions, defined in this part of the paper, can also be used in modeling asymmetric velocity profiles. Finally, addressing the modeling of reaching movements with bounded control actions, we consider the minimum time formulation of the optimization problem and (for the n-th order integrator) find its analytical solution.

1. Introduction

Understanding the trajectory formation in reaching movements is a very important research problem in computational neuroscience and modern robotics. From the viewpoint of control theory the problem can be attacked from two directions, one deals with the open-loop control and employs optimization approaches (Flash et al., 2003; Kawato, 1999) while the other is based on the sensory-motor feedback control (Arimoto et al., 2005; Tsuji et al., 2002). These two research lines are quite different but both important because, in our opinion, it is the blending of the feedforward and feedback control that defines the appearance of reaching movements.

Indeed, experiments show that for relatively slow movements the velocity profiles are left-skewed (Milner and Ijaz, 1990; Nagasaki, 1989), a feature that is well-captured by the feedback control. On the other hand, for relatively fast movements the velocity profile of the hand trajectory tends to a symmetric[1] bell-shaped form (Abend et al., 1982; Morasso, 1981), that can be captured by the feedforward control (Flash and Hogan, 1985). Thus, the resulting movement can be considered as the superposition of a major ballistic (feedforward) component and a

[1] Experiments also show that for very fast movements the velocity profiles can be even slightly skewed to the right (Nagasaki, 1989; Wiegner and Wierbicka, 1992).

corrective (feedback) component (Berthier, 1996; Milner, 1992). How specifically the blending of these components works is one the greatest secrets of the CNS.
In this paper we will deal with the feedforward control of reaching movements. The reasoning behind this scheme is that in fast reaching movements the effective use of feedback is nearly impossible due to the delays in neural circuits and muscle activation. The feedforward control is often modeled via optimization theory. In the optimization approach, the trajectory of the human arm is predicted by minimizing, over the movement time T, an integral performance index $\mathcal{J}$ subject to boundary conditions imposed on start and end points. The performance index can be formulated in the joint space (Soechting and Lacquaniti, 1981; Uno et al., 1989) or in the task space normally associated with the human hand (Flash and Hogan, 1985; Ohta et al., 2004). Depending on whether the arm dynamics is taken into account or not, the performance index is defined in the motion (Flash and Hogan, 1985; Soechting and Lacquaniti, 1981) or force domain (Svinin et al., 2005; Uno et al., 1989).
Several modeling issues are under study in this paper. First, in Section 2 we consider a generalization of the classical minimum jerk model of the free-space reaching movements, relate it to the lowest polynomial approach, and show that the solution is given by the regularized incomplete Beta function. Next, in Section 3, we address the boundary conditions of the optimization task. Since the optimal solution is defined not only by the optimality criterion but also by the boundary conditions, the latter is not not less important than the former. However, the role of the boundary conditions is rarely discussed in the literature on reaching movements. Then, in Section 4, we address the formation of reaching movements with bounded control actions and consider the minimum time formulation of the optimization problem. Finally, conclusions are summarized in Section 5.

2. Minimum effort model

In our analysis, we will consider a simplified, one-dimensional model of human movements where the configuration dependence of the human arm is ignored and the motion considered at the hand level. For this model one can formulate a generalized criterion of optimality minimizing the squared n-th derivative of the hand position x over the movement time:

$$\mathcal{J} = \frac{1}{2}\int_0^T \left(\frac{d^n x}{dt^n}\right)^2 dt. \tag{1}$$

From the control theoretical point of view this is the minimum effort criterion for the controlled system $x^{(n)}(t) = u$, describing the hand movement in the free space. Here, u is the control input, and n is the order of the motor dynamics associated with the hand. This order implies $2n$ boundary conditions necessary for the correct formulation of the optimization problem. The state vector is composed of the hand position and its first $n-1$ derivatives. Assuming that the hand is at rest in the beginning and in the end of the reaching movement, one assigns

$$x(0)=0, \quad \dot{x}(0)=0, \quad \ddot{x}(0)=0, \ldots, x^{(n-1)}(0)=0, \tag{2}$$

$$x(T)=L, \quad \dot{x}(T)=0, \quad \ddot{x}(T)=0, \ldots, x^{(n-1)}(T)=0, \tag{3}$$

where L stands for the length of reaching movement.
The optimization problem—minimization of the criterion (1) under the boundary conditions (2,3)—is often raised in the prediction of human reaching movements, normally for $n = 3$ (the

minimum hand jerk model (Flash and Hogan, 1985)) and occasionally for $n = 4$ (the minimum snap model (Flash, 1983; Wiegner and Wierbicka, 1992)). Comparing to these models, the minimum hand acceleration model ($n = 2$) is often rejected on the ground that it produces non-zero initial and final accelerations, which is considered to be incompatible with experimental data (Stein et al., 1986).

A general formula for the optimal velocity profiles has been suggested in (Piazzi and Visioli, 2000; Richardson and Flash, 2002). Here, we present a formal derivation, establish the solution, and recognize it as a famous special mathematical function. To pose the optimization problem in non-dimensional settings, introduce the following change of variables: $x(t) = L\,y(\tau(t))$ where $\tau = t/T,\ \tau \in [0,1]$. In the new variables y and τ we have the criterion

$$\mathcal{J} = \frac{1}{2}\int_0^1 \left(\frac{d^n y}{d\tau^n}\right)^2 d\tau, \tag{4}$$

and the boundary conditions

$$y(0) = 0, \quad y'(0) = 0,\, y''(0) = 0, \ldots, y^{(n-1)}(0) = 0, \tag{5}$$

$$y(1) = 1, \quad y'(1) = 0,\, y''(1) = 0, \ldots, y^{(n-1)}(1) = 0, \tag{6}$$

where $y^{(k)}(\tau) = T^k x^{(k)}(t)/L,\ k = 1, \ldots, n-1$. As can be easily shown, the Euler-Lagrange equation for the problem under consideration is $y^{(2n)}(\tau) = 0$. The general solution for this equation is given by the polynomial $y(\tau) = \sum_{i=0}^{2n-1} c_i\,\tau^i$, where the coefficients c_i are established from the boundary conditions (5,6). Thus, the optimization problem under consideration is equivalent to the construction of the lowest order polynomial satisfying the boundary conditions (5,6).

To clarify the analytical structure of the solution, we recast the optimization task as the optimal control problem. Let us introduce the state vector $\boldsymbol{y} = \{y, y', \ldots, y^{(n-1)}\}^{\mathrm{T}}$ and define the control input $u = y^{(n)}$. The state dynamics are the n-th order integrator $\dot{\boldsymbol{y}} = \boldsymbol{A}\boldsymbol{y} + \boldsymbol{b}u$, where $\boldsymbol{b} = \{0, \ldots, 0, 1\}^{\mathrm{T}}$, and the elements of the matrix $\boldsymbol{A}$ are defined as

$$[\boldsymbol{A}]_{ij} = \begin{cases} 1 & \text{if } j = i+1, \\ 0 & \text{otherwise.} \end{cases} \tag{7}$$

The analytical solution for the minimum effort control problem, seeking the control minimizing $\mathcal{J} = \frac{1}{2}\int_0^1 u^2\, d\tau$ for the dynamic system $\dot{\boldsymbol{y}} = \boldsymbol{A}\boldsymbol{y} + \boldsymbol{b}u$ and the boundary conditions $\boldsymbol{y}(0) = \boldsymbol{y}_0,\ \boldsymbol{y}(1) = \boldsymbol{y}_1$, is well-established in the control literature (Luenberger, 1969). It can be represented as

$$\boldsymbol{y}(\tau) = e^{\boldsymbol{A}\tau}\left(\left\{\boldsymbol{I} - \boldsymbol{W}(\tau)\boldsymbol{W}^{-1}(1)\right\}\boldsymbol{y}_0 + \boldsymbol{W}(\tau)\boldsymbol{W}^{-1}(1)e^{-\boldsymbol{A}}\boldsymbol{y}_1\right), \tag{8}$$

where

$$\boldsymbol{W}(\tau) = \int_0^{\tau} e^{-\boldsymbol{A}s}\boldsymbol{b}\boldsymbol{b}^{\mathrm{T}}e^{-\boldsymbol{A}^{\mathrm{T}}s}ds. \tag{9}$$

For the n-th order integrator the matrix exponent is defined as

$$\left[e^{\boldsymbol{A}\tau}\right]_{ij} = \begin{cases} \dfrac{\tau^{j-i}}{(j-i)!} & \text{if } j \geq i, \\ 0 & \text{if } j < i, \end{cases} \tag{10}$$

and by direct calculations one finds

$$[\boldsymbol{W}(\tau)]_{ij} = \frac{(-1)^{i+j}\tau^{2n+1-i-j}}{(2n+1-i-j)(n-i)!(n-j)!}. \tag{11}$$

As shown in (Rose and Bronson, 1969), for the n-th order integrator the matrix $\boldsymbol{W}(\tau)$ can be decomposed as $\boldsymbol{W}(\tau) = \tau \boldsymbol{P}(-\tau)\boldsymbol{H}\boldsymbol{P}(-\tau)$, where

$$[\boldsymbol{P}(\tau)]_{ij} = \begin{cases} \dfrac{\tau^{n-j}}{(n-i)!} & \text{if } i = j, \\ 0 & \text{otherwise.} \end{cases} \tag{12}$$

and $\boldsymbol{H}$ is the Hilbert matrix

$$[\boldsymbol{H}]_{ij} = \frac{1}{(2n+1-i-j)}, \tag{13}$$

the inverse of which is known to be

$$\left[\boldsymbol{H}^{-1}\right]_{ij} = \frac{(-1)^{i+j}(2n-i)!(2n-j)!}{(2n+1-i-j)\{(n-i)!(n-j)!\}^2(i-1)!(j-1)!}. \tag{14}$$

The elements of the inverse matrix $\boldsymbol{W}^{-1}(\tau) = \tau^{-1}\boldsymbol{P}^{-1}(-\tau)\boldsymbol{H}^{-1}\boldsymbol{P}^{-1}(-\tau)$ can now be easily established:

$$\left[\boldsymbol{W}^{-1}(\tau)\right]_{ij} = \frac{(2n-i)!(2n-j)!}{(2n+1-i-j)(n-i)!(n-j)!(i-1)!(j-1)!\tau^{2n+1-i-j}}. \tag{15}$$

For the rest-to-rest movements $\boldsymbol{y}_0 = \boldsymbol{0}$, $\boldsymbol{y}_1 = \{1, 0 \ldots, 0\}^{\mathsf{T}}$, and $e^{-\boldsymbol{A}}\boldsymbol{y}_1 = \boldsymbol{y}_1$. Therefore, $\boldsymbol{y}(\tau) = e^{\boldsymbol{A}\tau}\boldsymbol{W}(\tau)\boldsymbol{W}^{-1}(1)\boldsymbol{y}_1$. The components of the vector $\boldsymbol{y}(\tau)$, which are the derivatives of the corresponding order of $y(\tau)$, are found from (10,11,15) by direct computation:

$$y^{(i-1)}(\tau) = \sum_{j=i}^{n}\sum_{s=1}^{n} \frac{(-1)^{j+s}(2n-1)!(2n-s-1)!\tau^{2n+1-i-s}}{(j-i)!(2n+1-s-j)\{(n-s)!\}^2(n-j)!(n-1)!(s-1)!}. \tag{16}$$

To find the velocity $y'(\tau)$, we set $i = 2$ in (16) and take into account that

$$\begin{aligned} \sum_{j=2}^{n} \frac{(-1)^j}{(j-2)!(2n+1-s-j)(n-j)!} &= -\frac{1}{(s-n-1)(s-n-2)\ldots(s-n-(n-1))} \\ &= \frac{(-1)^n\Gamma(1+n-s)}{\Gamma(2n-s)} = \frac{(-1)^n(n-s)!}{(2n-s-1)!}, \end{aligned} \tag{17}$$

where Γ is the usual Gamma function. Then, after some simple manipulations using the binomial expansion for $(1-\tau)^{n-1}$, one obtains

$$y'(\tau) = \frac{(2n-1)!\tau^{n-1}}{(n-1)!}\sum_{s=1}^{n}\frac{(-1)^{n+s}\tau^{n-s}}{(n-s)!(s-1)!} = \frac{\tau^{n-1}(1-\tau)^{n-1}}{B(n,n)}, \tag{18}$$

where

$$B(n,n) = \frac{(n-1)!(n-1)!}{(2n-1)!} \tag{19}$$

is the symmetric Beta function. Therefore, the solution $y(\tau)$ can be formally represented as

$$y(\tau) = \frac{\int_0^\tau p^{n-1}(1-p)^{n-1}dp}{B(n,n)} \triangleq \frac{B(\tau;n,n)}{B(n,n)} \triangleq \bar{B}(\tau;n,n). \quad (20)$$

This expression is known as the regularized incomplete Beta function (Abramowitz and Stegun, 1972). While it is widely used in many fields, it is best known for its applications in statistics (Dutka, 1981). The basic properties of the Beta distribution are well-known (Abramowitz and Stegun, 1972). In particular, the velocity $y'(\tau)$ is symmetric with respect to the middle point $\tau = 1/2$, the position $y(1/2) = 1/2$, and all even derivatives $y^{(2k)}(1/2) = 0$ for $k = 1,2,3\ldots$.

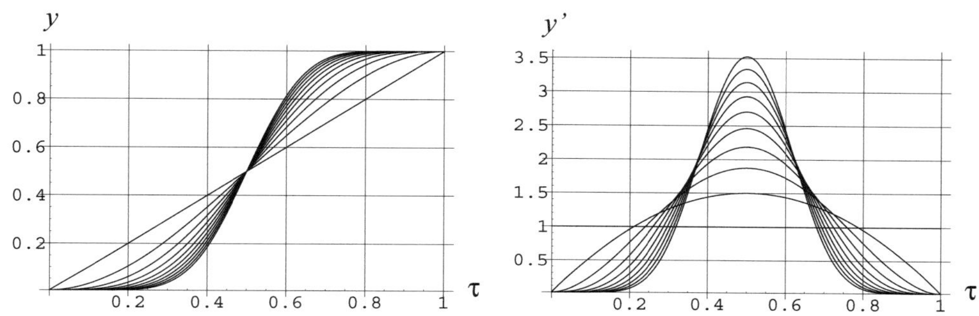

Fig. 1. Normalized position and velocity profiles in the minimum effort model for $n = 1,2,\ldots,10$.

The first 10 solutions for $y(\tau)$ and $y'(\tau)$ are plotted in Figure 1. While the solutions $y_n(\tau)$ are continuous functions of τ for any fixed n, in the limiting case of $n \to \infty$ the solution is discontinuous. In this case the velocity is the unit impulse function $y'_\infty(\tau) = \delta(\tau - 1/2)$, $\int_0^1 y'_\infty(\tau)d\tau = 1$. The position $y_\infty(\tau)$ is the unit step function, which is unity for $\tau > 1/2$ and is 0 for $\tau < 1/2$.

Several comments are in order.

1°. It is noticed in (Flash and Hogan, 1985) that while, in general, the minimum jerk model ($n = 3$) is a good candidate for mimicking human-like movements, the minimum snap model ($n = 4$) sometimes also provide a reasonable fit to the experimental data of unconstrained reaching movements. In this connection, it should be noted that the Beta function is defined not only for integer but also for real n. This suggests that the criterion (1) can be generalized to non-integer orders using, for example, Riemann-Liouville fractional integrals and derivatives (Samko et al., 1993). It also suggests that in fitting experimental data of reaching movements we can use real numbers $n \in [3,4]$ (see Figure 2). The assumption of real n sounds biologically plausible because the natural sensors and actuators in the human body are likely to deal with the fractional derivatives.

2°. The symmetry of the Beta function with respect to its last two arguments in (18,20) is explained by the symmetric placement of the boundary conditions. It can be shown that for the asymmetric placement (n_a boundary conditions at the start point and n_b at the end point, $n_a + n_b = 2n$) the solution is expressed through the asymmetric Beta function

$$y(\tau) = \frac{\int_0^\tau p^{n_a-1}(1-p)^{n_b-1}dp}{B(n_a,n_b)} \triangleq \frac{B(\tau;n_a,n_b)}{B(n_a,n_b)}, \quad (21)$$

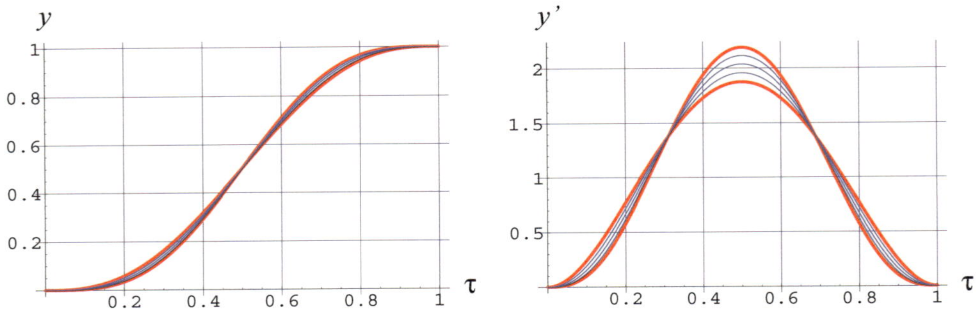

Fig. 2. Normalized position and velocity profiles for the integer (red lines: $n = 3$ and $n = 4$) and fractional (blue lines: $n = 3.25$, $n = 3.5$, and $n = 3.75$) orders.

where

$$B(n_a, n_b) = \frac{(n_a - 1)!(n_b - 1)!}{(n_a + n_b - 1)!}. \tag{22}$$

The asymmetric Beta functions are illustrated in Figure 3.

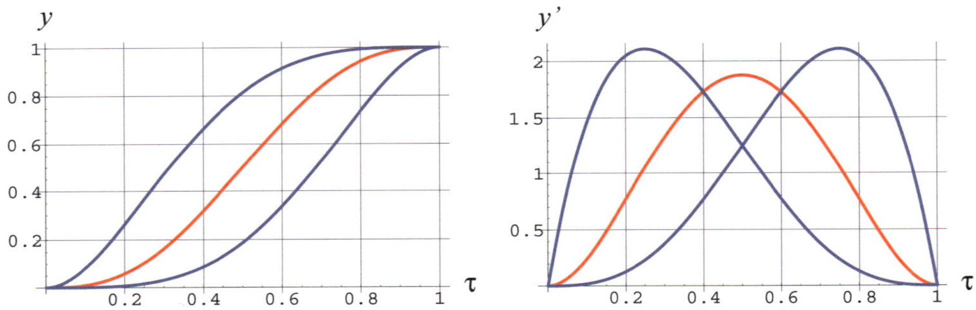

Fig. 3. Normalized position and velocity profiles given by the symmetric (red line: $B(3,3)$, center) and asymmetric (blue lines: $B(2,4)$ on the left, and $B(4,2)$ on the right) Beta functions.

Note that while the function (21) is the lowest order polynomial satisfying the asymmetric boundary conditions, it does not have a variational meaning. Strictly speaking, it does not minimize (4) because in the classical variational formulation the boundary conditions at the end-points (those that are imposed directly, as in (5,6), and those obtained automatically from the variational formulation and sometimes called the natural boundary conditions (Gelfand and Fomin, 1963)) are always placed symmetrically.

3. Natural boundary conditions

This section addresses the formation of the boundary conditions. They are an important part of the optimization problem because the optimal solution is defined not only by the form of the optimality criterion but also by the boundary conditions, and they are not less important than the criterion itself.

In the optimization problem considered in Section 2 we fixed first $n-1$ derivatives of the hand position, which, as far as human movements are concerned, might not be always plausible

from the physiological point of view[2]. Note that in experiments human subjects are normally not literally requested to produce zero end-point accelerations; they are only instructed to move comfortably from one point to another and stop there. Therefore, it does not seem reasonable to constrain the initial and final hand accelerations to zero in the optimization models. We can reliably pose the boundary conditions only for position and velocities. The remaining boundary conditions should be established differently.
Consider again the optimality criterion given by (1) but assume now that instead of (2,3) we fix only the first $k \leq n-1$ derivatives of the hand position:

$$x(0) = 0, \quad \dot{x}(0) = 0, \dots, \quad x^{(k)}(0) = 0, \tag{23}$$

$$x(T) = L, \quad \dot{x}(T) = 0, \dots, \quad x^{(k)}(T) = 0. \tag{24}$$

The variation of the functional (1) is defined as

$$\delta \mathcal{J} = \int_0^T x^{(2n)}(t)\, \delta x\, dt + \left[\sum_{i=0}^{n-1} x^{(2n-1-i)}(t)\, \delta x^{(i)}(t) \right]_0^T . \tag{25}$$

Taking into account that $\delta x^{(i)}(0) = \delta x^{(i)}(T) = 0$ for $i = 0, \dots, k$, from the necessary condition of optimality, $\delta \mathcal{J} = 0$, one obtains the Euler-Lagrange equation $x^{(2n)}(t) = 0$, and establishes the following boundary conditions:

$$x^{(n)}(0) = 0, \quad x^{(n+1)}(0) = 0, \dots, \quad x^{(2n-2-k)}(0) = 0, \tag{26}$$

$$x^{(n)}(T) = 0, \quad x^{(n+1)}(T) = 0, \dots, \quad x^{(2n-2-k)}(T) = 0, \tag{27}$$

Note that in the calculus of variations the boundary conditions (26,27), obtained from the condition $\delta \mathcal{J} = 0$, are called natural (Gelfand and Fomin, 1963). The solution of the optimization problem is the polynomial of degree $2n-1$ whose coefficients are established from the directly imposed (23,24) and the natural (26,27) boundary conditions.
Consider, for the purpose of illustration, the case of $n = 3$ and $k = 1$. In this model we fix the boundary conditions for the position and velocities of the hand, while the hand accelerations in the beginning and the end of movement are defined ŞautomaticallyŤ from the minimization of the squared hand jerk over the movement time. This differs from the conventional formulation where the boundary conditions for accelerations are set to zero. The minimum hand jerk model with the boundary conditions (23,24) and (26,27) yields

$$x = L\tau^2(2.5 - 2.5\tau^2 + \tau^3), \tag{28}$$

where $\tau = t/T$, while the conventional minimum hand jerk model gives

$$x = L\tau^3(10 - 15\tau + 6\tau^2). \tag{29}$$

Both these models produce single-phased velocity profiles. However, the maximal velocity in the former model is $1.5625L/T$, while in the latter one it is $1.875L/T$.
It should be noted that the solution (28) features non-zero initial and final acceleration jumps because in this model $\ddot{x}(0) \neq 0$ and $\ddot{x}(T) \neq 0$. To avoid possible confusion, it should be said that no coordinate jumps are physical, of course. Strictly speaking, the exact modeling of this

[2] For example, in the minimum snap model ($n = 4$) the end-point jerk are supposed to be zero. However, it is not evident that human can do the end-point control of the 3rd derivative of the hand position.

phenomenon requires more sophisticated models considered at different time scales as the coordinate (acceleration, in our case) change is considerably high in a short instance of time. However, in simplified models, such as the one considered in our paper, this behavior can be modeled by the jumps in the boundary conditions.
The assumption of zero initial and final hand accelerations can be too restrictive when dealing with fast movements in highly dynamic reaching tasks Svinin et al. (2008). For the free-space movements the applicability of the models (28) or (29) is likely to depend on the movement duration T. For relatively slow movements the CNS has enough time to nullify the end-point accelerations and this defines the model (29). For relatively fast movements the end-point accelerations are placed at will and one gets the model (28).
To unify the minimum hand jerk model (28) with natural boundary conditions for accelerations with the conventional one (29), it is reasonable to assume that the CNS controls the boundary conditions. Intuitively, the way they are controlled is changed from task to task and, in general, the more dynamic the reaching task the looser the control of the boundary conditions.
To reflect the cost of control of the boundary conditions, one can replace the optimality criterion (1) by

$$\mathcal{J} = \frac{1}{2}\int_0^T \left(\frac{d^n x}{dt^n}\right)^2 dt + \frac{1}{2}\sum_{i=0}^{n-1} w_{i0}\left(x^{(i)}(0)-\bar{x}_0^i\right)^2 + w_{iT}\left(x^{(i)}(T)-\bar{x}_T^i\right)^2. \tag{30}$$

Here, $\bar{x}_0^i$ and $\bar{x}_T^i$ are the "desired" steady state values for the derivatives of the hand trajectory at the boundary points, and $w_{i0} \geq 0$ and $w_{iT} \geq 0$ are the corresponding weight coefficients assigned by the CNS. For simplicity, we assume that $\bar{x}_0^i = 0$ for $i = 0, \ldots, n$, $\bar{x}_T^i = 0$ for $i = 1, \ldots, n$, and $\bar{x}_T^0 = L$.
Consider now the criterion of optimality given by (30). Note that now the separation of the boundary conditions into the directly imposed and the natural ones is no longer necessary, as we assume that all of them are controlled by the CNS and all of them are established from the condition $\delta\mathcal{J} = 0$. In other words, *all the boundary conditions are natural* now. It is straightforward to show that the variation of the functional (30) is defined as

$$\delta\mathcal{J} = \int_0^T x^{(2n)}(t)\,\delta x\,dt + \left[\sum_{i=0}^{n-1} x^{(2n-1-i)}(t)\,\delta x^{(i)}(t)\right]_0^T + \sum_{i=0}^{n-1} w_{i0}\left(x^{(i)}(0)-\bar{x}_0^i\right)\delta x^i(0) + w_{iT}\left(x^{(i)}(T)-\bar{x}_0^i\right)\delta x^i(T). \tag{31}$$

From $\delta\mathcal{J} = 0$ one obtains the Euler-Lagrange equation, $x^{(2n)}(t) = 0$, and establishes the natural boundary conditions. Since the variations $\delta x^i(t)$ are not restricted at the boundary points, one obtains:

$$-x^{(2n-1-i)}(0) + w_{i0}\left(x^{(i)}(0) - \bar{x}_0^i\right) = 0, \tag{32}$$

$$x^{(2n-1-i)}(T) + w_{iT}\left(x^{(i)}(T) - \bar{x}_T^i\right) = 0, \tag{33}$$

where $i = 0, \ldots, n-1$. It is worth pointing out two limiting cases. First, if $w_{i0} \to \infty$ (the CNS places a very high gain) the initial state for the i-th derivative coincides with the desired one,

as the condition (32) is reduced to $x^{(i)}(0) = \bar{x}_0^i$. If, on the other hand, the CNS cannot afford to control $x^{(i)}(0)$ then $w_{i0} \to 0$ and we get $x^{(2n-1-i)}(0) = 0$.
Let us continue the illustrative example of $n = 3$ and assign $\bar{x}_0^i = 0$, $i = 0, 1, 2$, $\bar{x}_T^0 = L$, and $\bar{x}_T^i = 0$, $i = 1, 2$. Assuming the tight control for the initial and final position and velocity, we set $w_{i0} = w_{iT} = \infty$, $i = 0, 1$. By considering, for simplicity, the symmetric assignment of the accelerations weights, define $w \triangleq w_{20} = w_{2T}$. It is not difficult to show that the optimal solution, parameterized by the acceleration weight w is defined as

$$x = \frac{L\tau^2\left(30 + 10\bar{w}\tau - 15(2+\bar{w})\tau^2 + 6(2+\bar{w})\tau^3\right)}{(12+\bar{w})}, \tag{34}$$

where $\tau = t/T$, and $\bar{w} = wT$. For $w = 0$ we get the model (28) with natural boundary conditions for accelerations, and for $w \to \infty$ we get the conventional minimum hand jerk model (29). Changing the acceleration weight from zero to infinity defines a family of curves (all in the class of the minimum hand jerk models). The velocity and acceleration profiles of this family for $L = 1, T = 1$ are shown in Fig. 4.

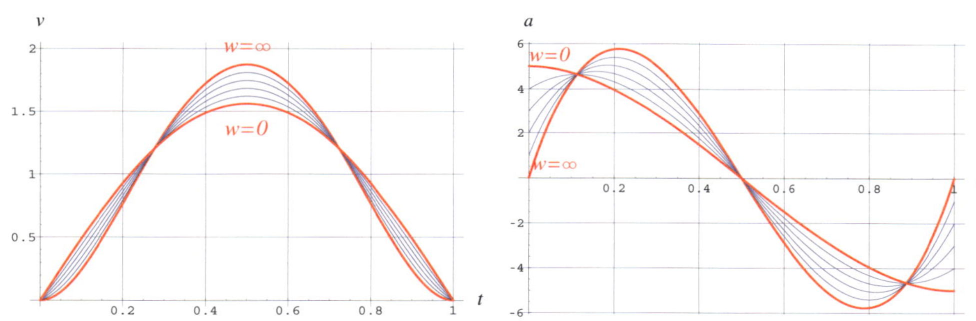

Fig. 4. Velocity and acceleration profiles for the family of curves (34).

Several comments are in order.
1°. Asymmetric velocity profiles can be obtained by asymmetric placement of the weight coefficients ($w_{i0} \neq w_{iT}$). Note also that the optimality criterion (30) can be generalized by the introduction of matrix weight coefficients.
2°. It is not clear how specifically the CNS may assign the weight coefficients. Intuitively, the human perception and control of position is better than that of velocity, that of velocity is better than that of accelerations, and so on, and it is plausible to assume that $w_{i,0} > w_{i+1,0}$ and $w_{i,T} > w_{i+1,T}$. The specific assignment of the weight coefficients must depend on how fast the reaching movement under inspection is. In this connection, it should be noted that the analysis of reaching movements is often conducted under the subtle assumption that the CNS operates without undue stress, away from the limits of neuromuscular performance (Flash and Hogan, 1985). But this assumption is not specific.
It is reported that for comfortable reaching movements in free space the movement time is in the range of [0.25, 0.8]s for the traveling distance varying from 0.2m to 0.5m (Kawato, 1995). Since the conventional minimum hand jerk model and the minimum torque change model are well established and experimentally verified for these ranges, one can presume that in such movements the boundary conditions for the hand accelerations are zero. However, if the movement time is decreased the CNS obviously cannot afford to keep the initial hand

accelerations zero. Therefore, for curtain ranges of time the reaching movements will feature the acceleration jumps (soft impacts). And if the movement duration is decreased further, the movements will feature even the velocity jumps (hard impact). One can see such impulsive movements in the exercises of karate fighters.
Obviously, the movement time T cannot be reduced arbitrarily without violating the constraints imposed by the neural and musculoskeletal hardware on the traveling distance, maximal force, and the accuracy of reaching. These constraints are ignored in the optimization problems considered in this section. Taking into account additional considerations would lead to different models with possibly different criteria of optimality (Nelson, 1983). One of these models is the minimum time model.

4. Minimum time model

In this section, the hand trajectory $x(t)$ is determined by minimizing the performance index

$$\mathcal{J} = \int_0^T dt = T, \tag{35}$$

where T is the movement time to be obtained. For the simplicity of formulation, we will consider the rest-to-rest time-optimal control where the boundary conditions are defined by (2,3) and the scalar control input is bounded as

$$|u(t)| \leq U. \tag{36}$$

For a linear controllable system, existence and uniqueness of the time-optimal solution are guaranteed, and the solution is "bang-bang" control with a finite number of switches (Hermes and LaSalle, 1969; Lee and Markus, 1967). It is easy to show that for the system under consideration (n-th order integrator) the optimal solution is symmetric about $T/2$, and the number of switches is $n+1$ ($t_0 = 0, t_1, t_2, \ldots, t_n = T$).
Assuming $L > 0$, we set $u(t) = (-1)^s U$ for the time interval $t \in [t_s, t_{s+1}]$, $s = 0, 1, \ldots, n-1$, and, by integrating $x^{(n)} = u(t)$, define the $(n-k)$-th derivative ($k = 1, \ldots, n$) of the hand position:

$$x^{(n-k)}(t) = \frac{(t-t_s)^k}{k!}(-1)^s U + \sum_{p=0}^{k-1} x^{(n-k+p)}(t_s)\frac{(t-t_s)^p}{p!}. \tag{37}$$

For $t = t_{s+1}$ one therefore has

$$x^{(n-k)}(t_{s+1}) = \frac{(t_{s+1}-t_s)^k}{k!}(-1)^s U + \sum_{p=0}^{k-1} x^{(n-k+p)}(t_s)\frac{(t_{s+1}-t_s)^p}{p!}. \tag{38}$$

By considering this recurrent relationship with the zero initial condition (2), one can represent $x^{(n-k)}(t_s)$ in the following non-recurrent form

$$x^{(n-k)}(t_s) = \frac{U}{k!}\sum_{i_1=1}^{s}\sum_{i_2=1}^{s}\cdots\sum_{i_k=1}^{s}(-1)^{\min\{i_1,i_2,\ldots,i_k\}-1}\Delta t_{i_1}\Delta t_{i_2}\ldots\Delta t_{i_k}, \tag{39}$$

where $\Delta t_i = t_i - t_{i-1}$, and $s = 1, 2, \ldots, n$. Now, substitute $x^{(n-k)}(t_n)$, $k = 1, 2, \ldots, n$, into the boundary condition (3). After some algebra, one obtains the following system of n equations

$$2\left\{\sum_{i=1}^{n-1}(-1)^i t_i^k\right\} + (-1)^n t_n^k = 0, \quad k = 1, \ldots, n-1, \tag{40}$$

$$2\left\{\sum_{i=1}^{n-1}(-1)^i t_i^n\right\} + (-1)^n t_n^n = (-1)^n n! L/U, \tag{41}$$

with respect to the unknown switching times $t_1, t_2, \ldots, t_n$. The system (40) is homogenous, and with the introduction of the non-dimensional times $\tau_i = t_i/t_n$ it is converted to the following form

$$\sum_{i=1}^{n-1}(-1)^i \tau_i^k = (-1)^{n+1}/2, \quad k = 1, \ldots, n-1. \tag{42}$$

Let us show that the solution for this system is given by

$$\tau_i = \sin^2\left(\frac{\pi i}{2n}\right), \quad i = 1, \ldots, n-1. \tag{43}$$

First, by converting (43) to the double angle representation and substituting it into (42), one obtains

$$\sum_{i=1}^{n-1}(-1)^i \tau_i^k = \sum_{i=1}^{n-1}(-1)^i \left\{\frac{1}{2}\left(1 - \cos\frac{\pi i}{n}\right)\right\}^k =$$

$$= \sum_{i=1}^{n-1}(-1)^i \frac{1}{2^k}\left(1 - \cos\frac{\pi i}{n}\right)^k = \frac{1}{2^k}\sum_{j=0}^{k}(-1)^j C_k^j \sum_{i=1}^{n-1}(-1)^i \cos^j\frac{\pi i}{n}, \tag{44}$$

where $C_p^q = p!/(q!(p-q)!)$ denotes the binomial coefficient. Next, it is known (see section 4.4.2 of (Prudnikov et al., 1986)) that

$$\sum_{k=0}^{n-1}(-1)^k \cos^m\frac{\pi k}{n} = \begin{cases} \frac{1}{2}\{1 - (-1)^{m+n}\} & \text{for } m = 0, 1, 2, \ldots, n-1 \\ \frac{n}{2^{n-1}} & \text{for } m = n \end{cases}. \tag{45}$$

Therefore, for $k < n$ we have

$$\sum_{i=1}^{n-1}(-1)^i \tau_i^k = \frac{1}{2^k}\sum_{j=0}^{k}(-1)^j C_k^j \left\{\frac{1}{2}\left[-1 - (-1)^{n+j}\right]\right\} =$$

$$= \frac{1}{2^{k+1}}\underbrace{\sum_{j=0}^{k} C_k^j(-1)^{j+1}}_{0} - \frac{(-1)^n}{2^{k+1}}\underbrace{\sum_{j=0}^{k} C_k^j}_{2^k} = -\frac{(-1)^n}{2}. \tag{46}$$

Thus, the non-dimensional switching times τ_i are defined by (43). It remains now to find the duration of movement $T \triangleq t_n$. For this purpose, rewrite (41) as

$$2\left\{\sum_{i=1}^{n-1}(-1)^i \tau_i^n\right\} + (-1)^n = (-1)^n n! \frac{L}{UT^n}. \tag{47}$$

By considering (44) for $k = n$ and taking into account (45) and (46), one obtains

$$\sum_{i=1}^{n-1}(-1)^i\tau_i^n = \frac{1}{2^n}\sum_{j=0}^{n}(-1)^j C_n^j \sum_{i=1}^{n-1}(-1)^i \cos^j\frac{\pi i}{n} =$$

$$\underbrace{\frac{1}{2^n}\sum_{j=0}^{n-1}(-1)^j C_n^j \sum_{i=1}^{n-1}(-1)^i \cos^j\frac{\pi i}{n}}_{-\frac{(-1)^n}{2}} + \frac{(-1)^n}{2^n}\underbrace{\sum_{i=1}^{n-1}(-1)^i \cos^n\frac{\pi i}{n}}_{\frac{n}{2^{n-1}}} =$$

$$-\frac{(-1)^n}{2} + \frac{(-1)^n n}{2^{2n-1}}. \quad (48)$$

Therefore, the minimal movement time is defined as

$$T = \sqrt[n]{2^{2(n-1)}(n-1)!L/U}, \quad (49)$$

and the control switching times $t_0, t_1, t_2, \ldots, t_n$ are computed as

$$t_i = \sqrt[n]{\frac{2^{2(n-1)}(n-1)!L}{U}}\sin^2\left(\frac{\pi i}{2n}\right), \quad i = 0, \ldots, n. \quad (50)$$

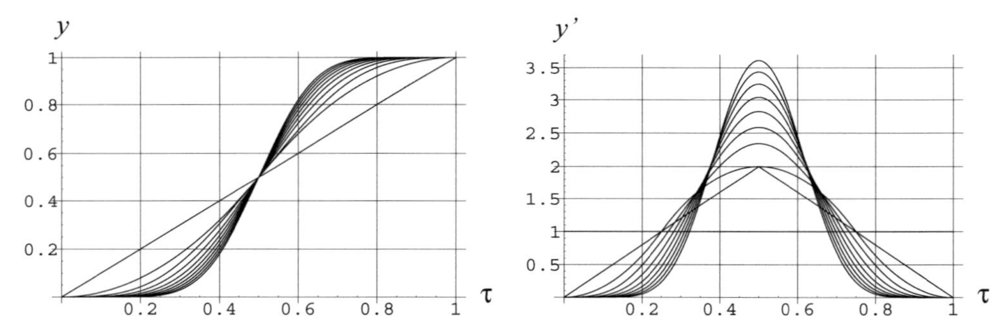

Fig. 5. Normalized position and velocity profiles in the minimum time model for $n = 1, 2, \ldots, 10$.

The first 10 solutions for the normalized position and velocity profiles in the minimum time model are plotted in Figure 5. Comparing these plots with those shown in Figure 1, one can say that qualitatively the minimum time model is similar to the minimum effort model.
Several comments are in order.

1°. It should be noted that in modeling of rapid reaching movements the minimum time model is often rejected (in favor of, say, the minimum hand jerk model) on the ground that it features acceleration jumps, which is not compatible with experimental data (Stein et al., 1986; Wiegner and Wierbicka, 1992). The rejection of the minimum time model only for this reason is premature because it is based on the fallacious comparison of the minimum time model for $n = 2$ with the minimum effort model for $n = 3$.

2°. The minimum time model for $n = 3$ does not feature acceleration jumps (the discontinuities here are at the jerk level). The velocity profile in this model looks like the typical bell-shaped pattern observed in experiments, with the normalized maximal velocity being between those in the minimum hand jerk ($n = 3$) and snap ($n = 4$) models (see Table 1). Thus, the minimum time model for $n = 3$ can be a legitimate candidate for mimicking human-like reaching movements.

model	$n=1$	$n=2$	$n=3$	$n=4$	$n=5$	$n=6$	$n=7$	$n=8$
minimum effort	1.0	1.5	1.875	2.188	2.461	2.707	2.933	3.142
minimum time	1.0	2.0	2.0	2.343	2.584	2.823	3.039	3.242

Table 1. Normalized velocity at the middle point: $\frac{L}{T}v(\frac{T}{2})$.

3°. An interesting approach to modeling of human-like reaching movements is proposed in (Ben-Itzhak and Karniel, 2008). The approach is based on the minimum acceleration model and can be linked to the minimization of kinetic energy. In the generalized form, for the controlled system $d^n x/dt^n = u(t)$ the trajectory is found by minimizing the criterion $\mathcal{J} = \frac{1}{2}\int_0^T \left(d^{n-1}x/dt^{n-1}\right)^2 dt$ under the constraints $|u(t)| \leq U$. This is a singular optimal control problem. We conjecture that this approach is equivalent to the minimum time model when the movement duration is specified by (49). This fact can be easily verified for $n = 2$ and $n = 3$ but remains to be proven for the general case.

5. Conclusions

Several issues related to modeling human-like reaching movements were under study in this paper. First, a mathematical model, based on the minimization of square derivatives of the hand position over the movement duration, has been established in the analytical form. This is a generalized model, described by the well-known Beta function. The representation can be used for the construction of fractional order models and also for modeling of asymmetric velocity profiles. Next, we addressed a natural (from the viewpoint of the calculus of variations) formation of the boundary conditions. From the mathematical point of view, the structure of the optimal solution is defined not only by the form of the optimality criterion but also by the boundary conditions of the optimization task. The natural boundary conditions, defined in our paper, can also be used in modeling asymmetric velocity profiles. Finally, addressing modeling reaching movements with bounded control actions, we have considered the minimum time formulation of the optimization problem and (for the n-th order integrator) established its analytical solution.

6. References

Abend, W., Bizzi, E., and Morasso, P. (1982). Human arm trajectory formation. *Brain*, 105:331–348.

Abramowitz, M. and Stegun, I. (1972). *Handbook of Mathematical Functions With Formulas, Graphs and Mathematical Tables*. U.S. Government Printing Office, Washington D.C., 10th edition.

Arimoto, S., Sekimoto, M., Hashiguchi, H., and Ozawa, R. (2005). Natural resolution of ill-posedness of inverse kinematics for redundant robots: A challenge to bernstein's degrees-of-freedom problem. *Advanced Robotics*, 19(4):401–434.

Ben-Itzhak, S. and Karniel, A. (2008). Minimum acceleration criterion with constraints implies bang-bang control as an underlying principle for optimal trajectories of arm reaching movements. *Neural Computation*, 20(3):779–812.

Berthier, N. (1996). Learning to reach: A mathematical model. *Developmental Psychology*, 32(5):811–823.

Dutka, J. (1981). The incomplete Beta function—a historical profile. *Archive for History of Exact Sciences*, 24:11–29.

Flash, T. (1983). *Organizing Principles Underlying The Formation of Hand Trajectories*. PhD thesis, Massachusetts Institute of Technology, Cambridge, USA.

Flash, T. and Hogan, N. (1985). The coordination of arm movements: An experimentally confirmed mathematical model. *The Journal of Neuroscience*, 5(7):1688–1703.

Flash, T., Hogan, N., and Richardson, M. (2003). Optimization principles in motor control. In Arbib, M., editor, *The Handbook of Brain Theory and Neural Networks*, pages 827–831. MIT Press, Cambridge, Massachusetts, 2nd edition.

Gelfand, I. and Fomin, S. (1963). *Calculus of Variations*. Prentice-Hall, Englewood Cliffs, New Jersey.

Hermes, H. and LaSalle, J. (1969). *Functional Analysis and Time-Optimal Control*. Academic, New York.

Kawato, M. (1995). Motor learning in the brain. *Journal of the Robotics Society of Japan*, 13(1):11–19.

Kawato, M. (1999). Internal models for motor control and trajectory planning. *Current Opinion in Neurobiology*, 9:718–727.

Lee, E. and Markus, L. (1967). *Foundations of Optimal Control Theory*. Wiley, New York.

Luenberger, D. (1969). *Optimization by Vector Space Methods*. John Wiley & Sons, New York.

Milner, T. (1992). A model for the generation of movements requiring endpoint precision. *Neuroscience*, 49(2):487–496.

Milner, T. and Ijaz, M. (1990). The effect of accuracy constraints on three-dimensional movement kinematics. *Neuroscience*, 35(2):365–374.

Morasso, P. (1981). Spatial control of arm movements. *Experimental Brain Research*, 42:223–227.

Nagasaki, H. (1989). Asymmetric velocity profiles and acceleration profiles of human arm movements. *Experimental Brain Research*, 74:319–326.

Nelson, W. (1983). Physical principles for economies of skilled movements. *Biological Cybernetics*, 46:135–147.

Ohta, K., Svinin, M., Luo, Z., Hosoe, S., and Laboissière, R. (2004). Optimal trajectory formation of constrained human arm reaching movements. *Biological Cybernetics*, 91(1):23–36.

Piazzi, A. and Visioli, A. (2000). Minimum-time system-inversion-based motion planning for residual vibration reduction. *IEEE/ASME Transactions on Mechatronics*, 5(1):12–22.

Prudnikov, A., Brychkov, Y., and Marichev, O. (1986). *Integrals and Series, Vol. 1: Elementary Functions*. Gordon & Breach, New York.

Richardson, M. and Flash, T. (2002). Comparing smooth arm movements with the two-thirds power law and the related segmented-control hypothesis. *The Journal of Neuroscience*, 22(18):8201–8211.

Rose, N. and Bronson, R. (1969). On optimal terminal control. *IEEE Transactions on Automatic Control*, 14(5):443–449.

Samko, S., Kilbas, A., and Marichev, Y. (1993). *Integrals and Derivatives of Fractional Order and Their Applications*. Gordon and Breach Science Publications.

Soechting, J. and Lacquaniti, F. (1981). Invariant characteristics of a pointing movement in man. *The Journal of Neuroscience*, 1:710–Ű720.

Stein, R., Oguztoreli, M., and Capaday, C. (1986). What is optimized in muscular movements? In Jones, N., McCartney, M., and McComas, A., editors, *Human Muscle Power*, pages 131Ű–150. Human Kinetics, Champaign, Illinois.

Svinin, M., Goncharenko, I., and Hosoe, S. (2008). On the boundary conditions in modeling of human-like reaching movements. In *Proc. IEEE/RSJ Int. Conference on Intelligent Robots and Systems, IROS'2008*, volume 1, pages 518–525, Nice, France.

Svinin, M., Masui, Y., Luo, Z., and Hosoe, S. (2005). On the dynamic version of the minimum hand jerk criterion. *Journal of Robotic Systems*, 22(11):661–676.

Tsuji, T., Tanaka, Y., Morasso, P., Sanguineti, V., and Kaneko, M. (2002). Bio-mimetic trajectory generation of robots via artificial potential field with time base generator. *IEEE Transactions on Systems, Man, and Cybernetics—Part C: Applications and Reviews*, 32(4):426–439.

Uno, Y., Kawato, M., and Suzuki, R. (1989). Formation and control of optimal trajectory in human multijoint arm movement. minimum torque-change model. *Biological Cybernetics*, 61:89–101.

Wiegner, A. and Wierbicka, M. (1992). Kinematic models and elbow flexion movements: Quantitative analysis. *Experimental Brain Research*, 88:665–673.

8

An Experimental Study of Three-Dimensional Passive Dynamic Walking with Flat Feet and Ankle Springs

Terumasa Narukawa, Kazuto Yokoyama,
Masaki Takahashi and Kazuo Yoshida
Keio University
Japan

1. Introduction

Passive dynamic bipeds were first studied by McGeer (McGeer, 1990) as inspired by a bipedal toy described in (McMahon, 1984). Passive dynamic walkers can walk down a shallow slope without actuators and controllers(McGeer, 1990; 1993). McGeer has built passive walkers that exhibit steady motion using a Poincaré map, which he called as a stride function, to analyze the gaits(McGeer, 1990; 1993). This method is quite useful and is independent of the biped model. The key idea that he examined is the stability of the entire step-to-step motion, and not the local stability at every instance. This analysis is also rather useful for actuated bipeds (Hobbelen and Wisse, 2007).
Firstly, McGeer studied two elementary passive walking models derived from a wagon wheel. One model was a rimless wheel model on a slope, and the other a synthetic wheel model on level ground as shown in Fig. 1. The motion of the models is constrained to the sagittal plane. Each model captures the fundamental mechanism of passive dynamic walking.

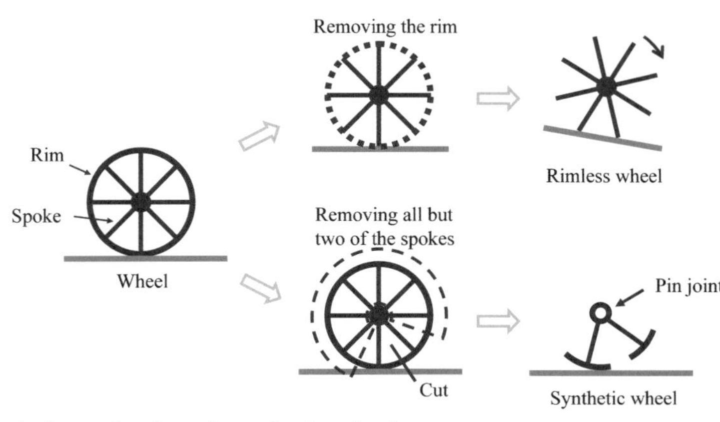

Fig. 1. Wheel, rimless wheel, and synthetic wheel.

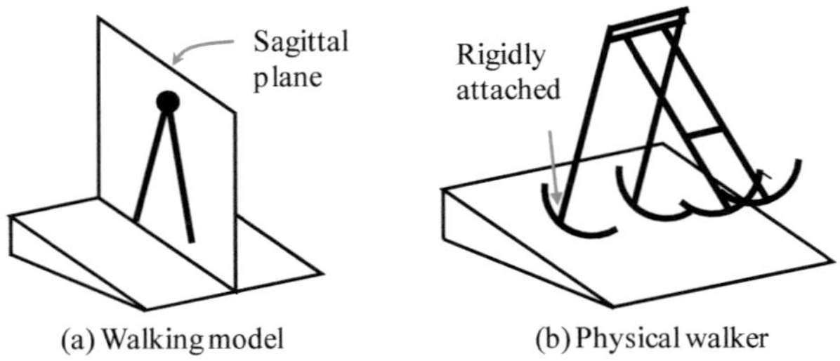

Fig. 2. 2D Passive Biped Walker.

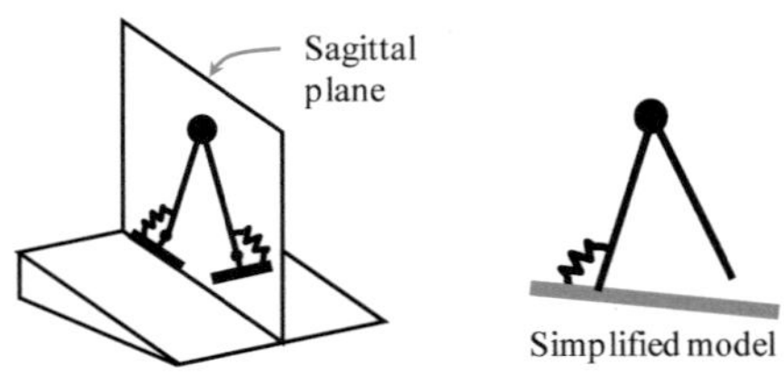

Fig. 3. A 2D straight-legged passive walker with flat feet and ankle springs (Wisse et al., 2006).

1.1 Rimless Wheel

A rimless wheel can be obtained by simply removing the rim from the wheel as shown in Fig. 1. One of the features captured by the rimless wheel is the stance leg motion which acts as an inverted pendulum motion. The other feature is the heel strike when the swing leg touches the ground. The rimless wheel has a periodic motion for a given slope angle(McGeer, 1990) whose stable region is very large(Wisse et al., 2005). If the initial rolling speed is sufficiently large and the slope angle is large enough corresponding to the relative angle between the spokes, the rimless wheel never falls forward, and converges to the equilibrium motion(McGeer, 1990; Wisse et al., 2005). This remarkable feature is used to strengthen the stability of passive walkers (Ikemata et al., 2006).

1.2 Synthetic Wheel

Passive motion of the swing leg can be explained by the synthetic wheel model(McGeer, 1990) as shown in Fig. 1. In this model, the rim was not removed. The rim was cut between the spokes, and all but two of the spokes removed. A pin joint and a large point mass were put at the hub, i.e. the hip. If the leg mass is assumed to be negligible compared to the hip mass, the swing leg motion will not disturb the stance leg motion. The stance leg rolls at a constant speed on the level floor because it is part of a wheel. McGeer showed that initial conditions exist, such that the synthetic wheel exhibits periodic motion(McGeer, 1990). The step period of the synthetic wheel is determined solely by the free pendulum period of the swing leg.

1.3 2D Passive Biped Walker

Following the study of rimless wheel and the synthetic wheel model, McGeer increased the complexity of the biped model. A kneeless passive biped walker is similar to a synthetic wheel model, however allows for variation of parameters, e.g., radius of the arc feet and location of the leg mass(McGeer, 1990). McGeer also built several physical passive walkers, with and without knees(McGeer, 1990; 1993). The motion of the McGeer's passive walkers is

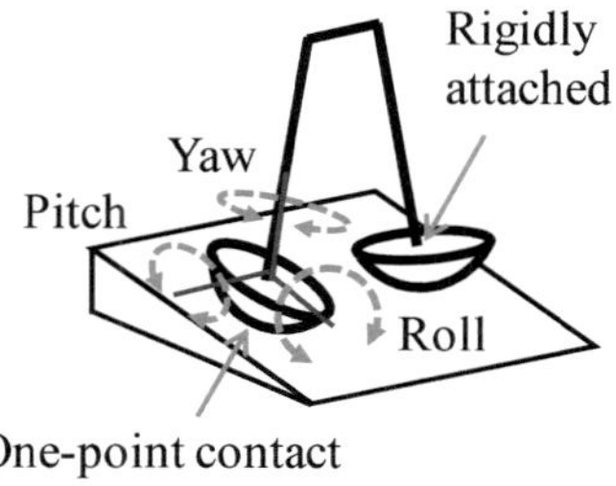

Fig. 4. A 3D passive biped walker

constrained against falling over sideways, as in numerical simulation models. The motion is only in the fore and aft, and vertical, i.e., sagittal, planes as shown in Fig. 2 (a). McGeer and his imitators used four legs, with each set of two legs connected so that they moved identically, to constrain the motion of the walker to the sagittal plane(McGeer, 1990) as indicated in Fig. 2 (b). The physical two-dimensional walkers without knees have a problem of foot scuffing at midstance. Stepping stones make clearance for the swing foot. Shortening the swing leg by lifting the feet via a lead screw mechanism with small motors is an alternative solution(McGeer, 1990). Two-dimensional passive walkers exhibit walking stability on a shallow slope.

Goswami et al. studied the compass model (Goswami et al., 1996), which consists of two straight legs connected by a frictionless hinge at the hip, devoid of actuators and control. The mass is at the hip and legs and the motion is constrained to the sagittal plane. Goswami et al. also showed that the compass model can exhibit period-doubling bifurcation, eventually leading to apparently chaotic gaits, by increasing the slope angle(Goswami et al., 1996; 1998). Garcia et al. introduced the simplest walking model (Garcia et al., 1998), which is similar to the compass model, except that the leg mass is located at the tip of the leg, and the hip mass is much larger than the foot mass. The assumption of negligible leg mass makes the motion of a swinging leg not affect the motion of the hip. The simplest passive walking model still walks stably on a shallow slope (Garcia et al., 1998), although the basin of attraction of the simplest walking model is very small (Schwab and Wisse, 2001). The simplest model exhibits period doubling bifurcations, leading to apparently chaotic gaits, as with the compass model (Garcia et al., 1998). Wisse et al. studied a 2D straight-legged passive walker with flat feet and ankle springs by simplifying the interaction of the spring and the foot (Wisse et al., 2006) as indicated in Fig. 3. The stance leg with the foot and ankle springs is modeled as a point foot with a torsional spring between the stance leg and the floor. Wisse et al. showed that arc-shaped feet rigidly connected to the legs and flat feet with ankle springs have a similar effect on the disturbance behavior in the simple 2D passive walking model (Wisse et al., 2006). Since McGeer's work, a few passive walkers that exhibit stable walking have been constructed. Garcia et al. copied McGeer's 2D kneed passive walker and performed detailed analysis of the gaits(Garcia, 1999; Garcia et al., 2000). Ikemata et al. used a stopper to maintain a constant inter-leg angle at heel strike(Ikemata et al., 2006). The stopper enabled the 2D kneed passive walker obtain high stability. The passive walker can walk on a treadmill for 35 minutes with 4010 steps.

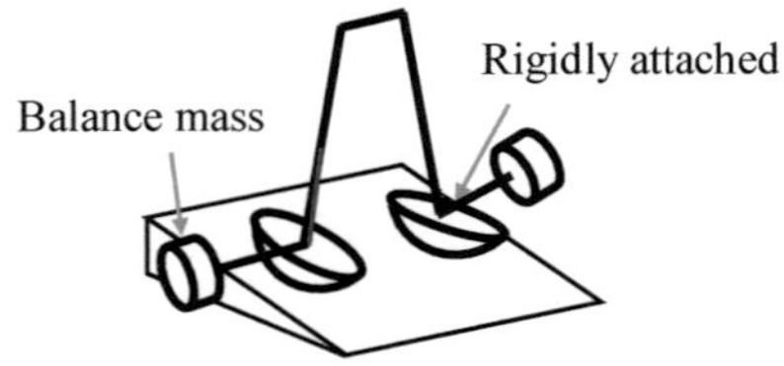

Fig. 5. A simple 3D passive biped walker with laterally extended balance bars (Coleman and Ruina, 1998).

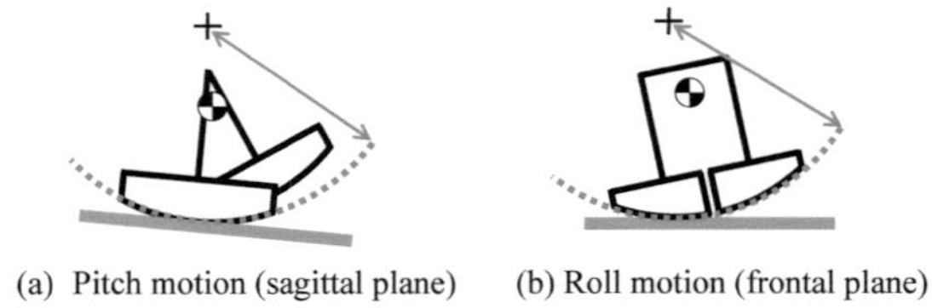

Fig. 6. A simple 3D passive biped walker with arc-shaped feet whose center of the radius of curvature is higher than the center of the total mass (Tedrake et al., 2004; Tedrake, 2004).

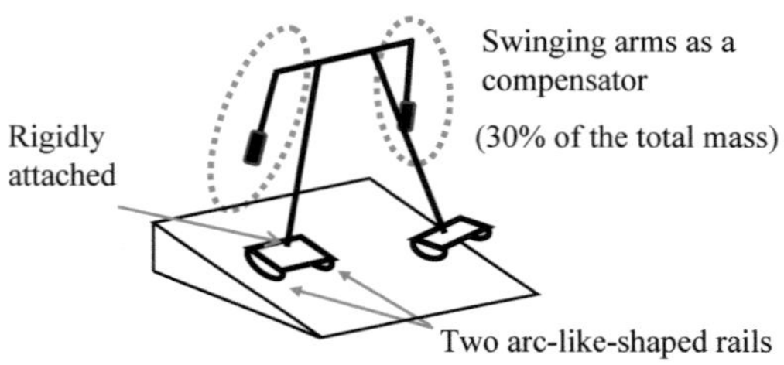

Fig. 7. The most sophisticated three-dimensional passive walker(Collins et al., 2001). It has several non human-like features, e.g., stiff ankles, heavy arms, and out and forward arm motion.

1.4 Scope of This Study

We have developed simple 3D passive walkers that can take longer steps and walk faster than other simple 3D passive walkers with arc-shaped feet (Narukawa et al., 2008; 2009a;b). The main feature of our 3D passive walkers is its foot and ankle design; it has flat feet with ankle springs instead of arc-shaped feet rigidly attached to the legs. Experimental results have shown that flat feet with ankle springs stabilize the yaw motion, and our 3D passive biped walker can take longer steps and walk faster than simple arc-footed 3D biped walkers because of its flat feet and ankle springs (Narukawa et al., 2008; 2009a).

This study investigates the effects of torsional spring stiffness on the pitch motion at the ankle joints of the developed walker. Experimental tests are performed to prove that torsional spring stiffness affects the overall motion of the walker and selecting springs with appropriate torsional spring stiffness aids in reducing the oscillating motion of the feet induced by the impact with the ground.

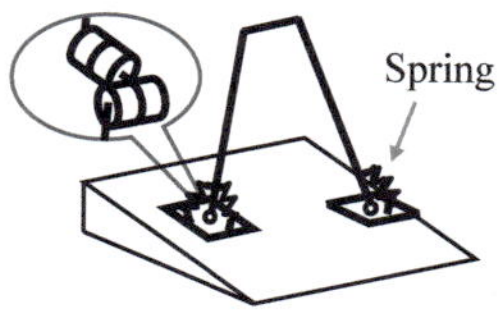

Fig. 8. Our simple 3D passive biped walker with flat feet and ankle springs (Narukawa et al., 2008; 2009a).

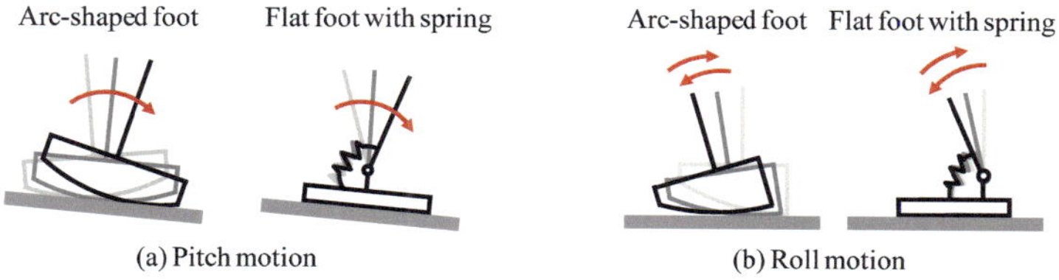

Fig. 9. Motion of a passive walker with flat feet and ankle springs compared to a passive walker with arc-shaped feet rigidly attached to the legs. For clarity, only the stance leg is shown.

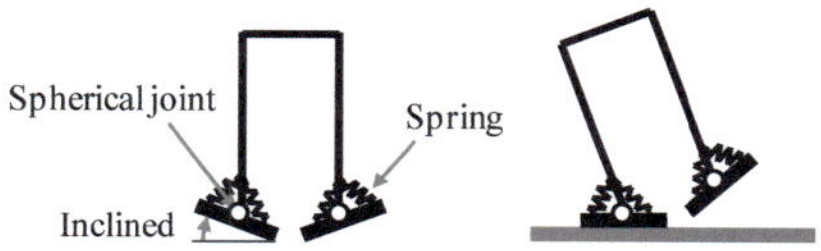

Fig. 10. A 3D passive biped walker with sprung ankles and flat feet. The initial spring tension induces an inclined roll angle (Kinugasa et al., 2008).

2. 3D Passive Biped Walker

Passive walking has been studied mainly in two dimensions, i.e., the motion is constrained to the sagittal plane e.g. (Garcia et al., 2000; 1998; Goswami et al., 1996; Ikemata et al., 2006; McGeer, 1990; Schwab and Wisse, 2001). Almost all of the physical passive walkers have arc-shaped feet rigidly connected to the legs because the arc-shaped feet can handle the disturbance behavior of the walkers better than point feet (Wisse and van Frankenhuyzen, 2003). Although 2D passive walkers demonstrate that passive leg dynamics can provide stable walking, stable 3D passive walking remains a challenge because of unstable roll and yaw motions as shown in Fig. 4. Roll is rotation about an axis in the direction of motion, and pitch is that about an axis perpendicular to that direction.

2.1 Simulation Studies

After studying 2D passive walkers, McGeer demonstrated 3D passive walking without constraining the motion to the sagittal plane (McGeer, 1993). This walker has arc-shaped feet, similar to those of 2D passive walkers. However McGeer only found unstable motion in 3D passive walking and a physical 3D passive walker was not reported. Numerical simulation of three-dimensional passive walking is usually found only with unstable motions (Coleman

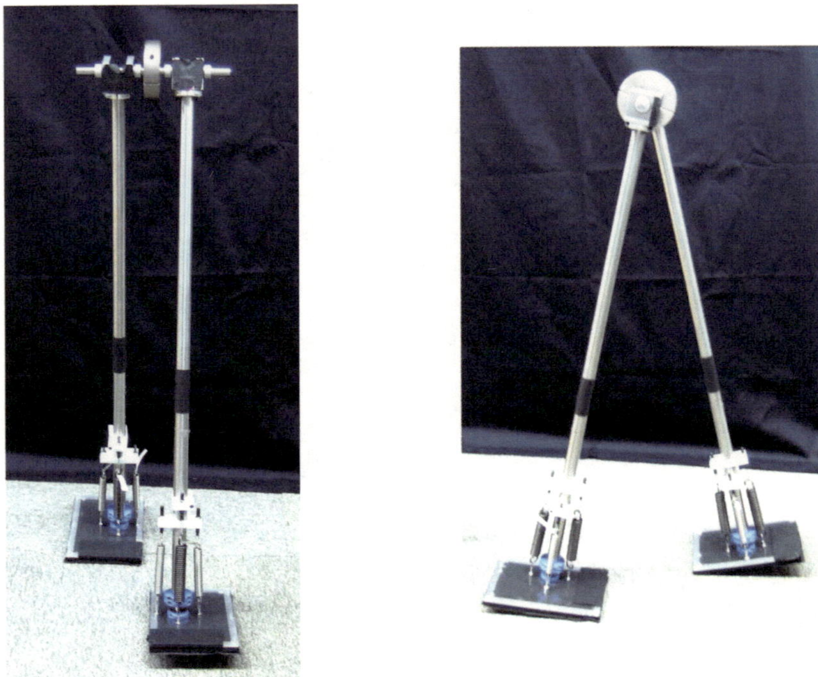

(a) Front view (b) Side view

Fig. 11. A simple 3D straight-legged passive biped walker with flat feet and ankle springs.

and Ruina, 1998; McGeer, 1993). Adolfsson et al. studied three-dimensional passive walking by changing the mechanism configuration from a planar passive walking model to a three-dimensional passive walking model (Adolfsson et al., 2001). Gait stability was investigated under parameter variations. Stable gaits of the three-dimensional model were found only when the feet were large enough to overlap. Wisse et al. proposed a 3D biped with a pelvic body as a passive-dynamic compensator for unstable yawing and rolling motion (Wisse et al., 2001). Wisse et al. studied a 3D passive walker having cylinder-shaped feet with ankle joints that kinematically couple roll to yaw (Wisse and Schwab, 2005).

2.2 Physical Passive Walkers

Coleman and Ruina built a simple two-leg passive walker with rounded feet attached rigidly to the legs, and laterally extended balance bars as shown in Fig. 5 (Coleman and Ruina, 1998). Although it cannot stand still, it can walk stably in three dimensions. However it takes only very short steps, resulting in low walking speed. Although Coleman et al. found stable motion of a 3D passive biped using an optimization method to find a set of walker parameters that produced stable walking (Coleman et al., 2001); however, the parameters are far from those of a physical prototype that exhibits stable motion. One of the simplest 3D passive biped walkers was built by Tedrake et al.(Tedrake et al., 2004; Tedrake, 2004). It has large arc-shaped feet, whose center of the radius of curvature is higher than the center of the total mass, which allows it to stand (Fig. 6). The contact with the ground occurs only at a point, and provides insufficient friction against unstable yaw motion, thus it can take

Parameter	Unit	Value	Parameter	Unit	Value
m_L	kg	0.70	l_{FF}	m	0.12
m_H	kg	0.88	l_{FL}	m	0.18
l_L	m	0.74	l_{FT}	m	0.09
l_A	m	0.03	r_L	m	0.49
l_H	m	0.09	I_{xx}	$kg \cdot m^2$	0.07
l_{FW}	m	0.08	I_{yy}	$kg \cdot m^2$	0.07

Table 1. Parameters of the physical 3D passive walker

very small strides only. The most sophisticated three-dimensional passive walker was built by Collins et al.(Collins et al., 2001) as shown in Fig. 7. They did not use simulation studies because numerical simulations of three-dimensional passive walkers are difficult, mainly due to collisions between the swing foot and the ground, and the frictional phenomena between the stance foot and the ground during gaits. Their 3D passive walker was improved by trial and error during experimental study. To reduce the unstable yaw motion, swing arms were attached to the counter side legs. Although the walker has several non human-like features, e.g., stiff ankles, heavy arms (accounting for 30 % of the total mass of the walker), and out and forward arm motion, the three-dimensional passive biped walks stably at about 0.5 m/s and exhibits impressively human-like motion.

3. 3D Passive Biped Walker with Flat Feet and Ankle Springs

We investigated simple 3D straight-legged passive walking with flat feet and ankle springs, as shown in Fig. 8, to overcome the limitations of arc-footed walkers, while maintaining mechanical simplicity (Narukawa et al., 2008; 2009a). The walker is composed of a hip, two straight legs, and two feet. The walker does not have knees and the yaw degree of freedom at the ankles so as to maintain a small number of degrees of freedom. The ankles have two degrees of freedom in roll and pitch motion. The proposed 3D passive walker do not have a compensator, such as swinging arms (Collins et al., 2001) or a pelvic body (Wisse et al., 2001), for unstable motion.
The main feature of the 3D passive walker is the flat feet with ankle springs, which enable the walker to mimic the motion of a 3D straight-legged passive walker with rigidly attached arc-shaped feet (Tedrake et al., 2004), as shown in Fig. 9, while providing sufficient friction torque against yaw. The flat feet and ankle springs allow the 3D passive walker to take long steps and walk faster than passive walkers with arc-shaped feet by stabilizing the unstable yaw motion (Narukawa et al., 2008; 2009a). Kinugasa et al. built a 3D straight-legged passive walker with flat feet and springs attached to ankles; it is 0.6 m tall (Kinugasa et al., 2008). The initial spring tension induces an inclined roll angle, as indicated in Fig. 10. Their 3D biped walks stably, but takes only very short steps of about 0.04 m.

4. Physical 3D Passive Biped Walker with Flat Feet and Ankle Springs

A physical 3D passive biped walker was constructed, as shown in Fig. 11, to investigate the proposed method(Narukawa et al., 2008; 2009a).

Parameter	Unit	Value	
		roll	pitch
a	m	0.025	0.025
b	m	0.030	0.030
h	m	0.112	0.142
s	m	0.103	0.130
k	N/m	4900	see Fig. 14

Table 2. Parameters of the foot and ankle design of the 3D passive walker

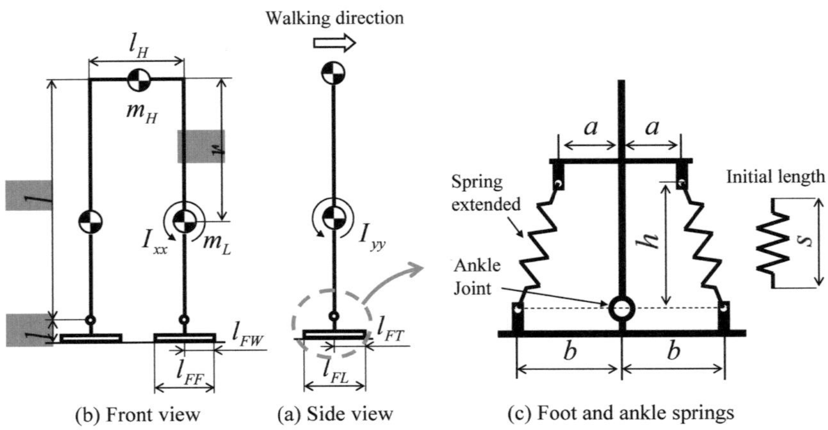

Fig. 12. Notations for physical parameters of the 3D passive walker with flat feet and ankle springs.

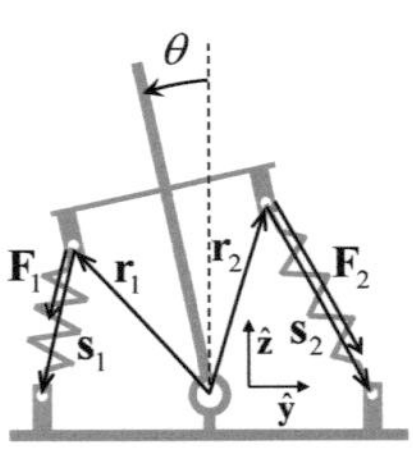

Fig. 13. A model of the foot and ankle springs. The tension springs are attached to the leg and the foot. The torque about the joint is produced by the springs.

4.1 Walker Parameters

Table 1 shows the parameter values of the physical walker. Figure 12 indicates the notations. The legs are symmetric, and the hip is assumed to be a point mass. The length of the legs, width of the hip, and mass distribution were determined as follows. First, we decided to make the legs about 0.8 m long, based on the length of human legs, other human-like passive walkers (Collins et al., 2001), and passive-based walkers (Collins and Ruina, 2005; Collins et al., 2005). The period of the swing leg motion is about 1.7 s. The width of the hip is as short as possible. Extra mass is added to the hip to increase the height of the center of mass

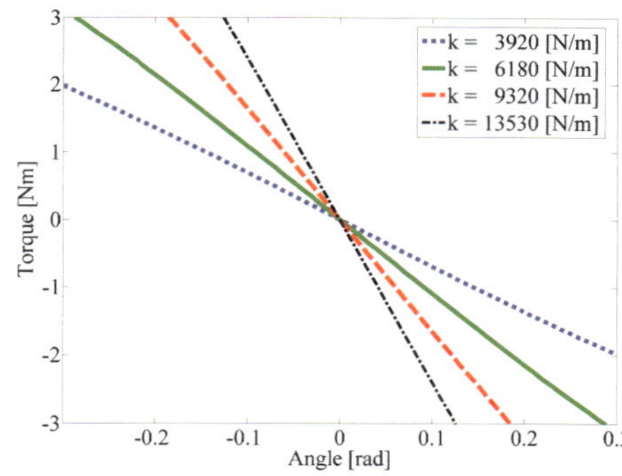

Fig. 14. Relationship between the pitch angle and the torque applied by the springs.

of the walker. The 3D passive walker weighs 2.3 kg, and its center of mass measured from the ground is about 0.5 m.

4.2 Torsional Spring Constant at the Ankle Joint

The flat foot and the leg are joined by a universal joint that has two degrees of freedom. Sponge sheets are attached to the soles of the feet to increase friction. The torsional spring effect is realized by using a pair of tension springs. Tension springs are attached to the feet and the legs to produce torque about the joint. The effect of the tension springs is simply calculated as follows(Narukawa et al., 2008; 2009a). First we assume that the roll and pitch motion are each isolated. The pitch motion is described as shown in Fig. 13. The forces produced by the springs are

$$|\mathbf{F}_1| = k\left(|\mathbf{s}_1| - s\right), \tag{1}$$

$$|\mathbf{F}_2| = k\left(|\mathbf{s}_2| - s\right) \tag{2}$$

where k is the spring constant and s is the initial length of the spring. We assume that the spring force obeys Hooke's law. The torque produced by the spring forces becomes

$$\mathbf{T} = \sum_{i=1}^{2} \mathbf{r}_i \times \mathbf{F}_i = \sum_{i=1}^{2} k\left(|\mathbf{s}_i| - s\right) \mathbf{r}_i \times \frac{\mathbf{s}_i}{|\mathbf{s}_i|}. \tag{3}$$

Figure 13 provides the necessary notation. Table 2 shows the values of the parameters of the physical walker. Then

$$\mathbf{r}_1 = \left(-h\sin\theta - a\cos\theta\right)\hat{\mathbf{y}} + \left(h\cos\theta - a\sin\theta\right)\hat{\mathbf{z}}, \tag{4}$$

$$\mathbf{r}_2 = \left(-h\sin\theta + a\cos\theta\right)\hat{\mathbf{y}} + \left(h\cos\theta + a\sin\theta\right)\hat{\mathbf{z}}, \tag{5}$$

$$\mathbf{s}_1 = -\mathbf{r}_1 - b\hat{\mathbf{y}}, \tag{6}$$

$$\mathbf{s}_2 = -\mathbf{r}_2 + b\hat{\mathbf{y}}. \tag{7}$$

$\hat{\mathbf{y}}$ and $\hat{\mathbf{z}}$ are unit vectors illustrated by Fig. 13.

The torsional spring constant for the roll motion at the stance ankle is obviously an important factor in enabling the straight-legged walker to rock adequately from side to side to avoid problematic scuffing of the swing leg and allowing it to swing forward. The study of the roll and pitch motions with ankle springs clarifies the effect of the springs so that the roll and pitch

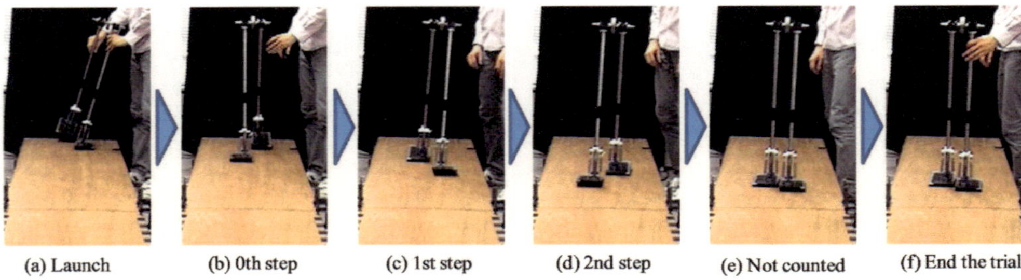

Fig. 15. Counting steps in passive walking. The first landing is counted as the zeroth step and the second landing as the first.

Successful Steps	Number of times (in 100 launch)				
	3920 N/m	6180 N/m	9320 N/m	13530 N/m	∞ N/m
0	40	12	7	12	40
1	51	68	68	70	49
2	9	14	19	12	10
3	0	3	6	5	1
4	0	3	0	1	0
Expectation	0.78	1.17	1.24	1.13	0.72

Table 3. Experimental result for each pitch angle spring constant

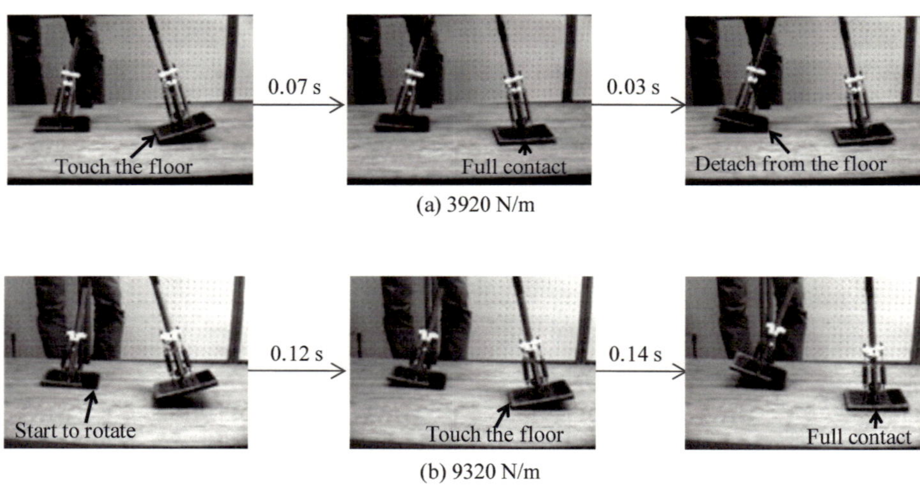

Fig. 16. Comparison of the foot motion for varying spring stiffness

motions are coordinated. From (Narukawa et al., 2009a), the value of the spring constant of the extension spring for roll motion is determined to be 4900 N/m with a torsional spring stiffness of about 9 N-m/rad. Figure 14 shows the relationship between the ankle angle and the torque applied by the springs for pitch motion when the spring constant is 3920, 6180, 9320, and 13530 N/m.

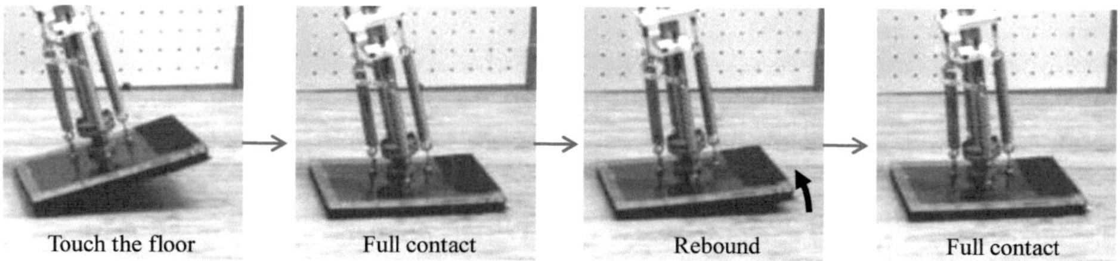

Fig. 17. Rebound at foot impact (3920 N/m)

5. Experimental Results

Experimental conditions were as follows. The slope angle was about 1.6 degrees; the slope was about 3.6 m long and 0.6 m wide. In each trial, the walker was started by hand from the top of the slope.

5.1 Successful Steps at Different Spring Constants for Pitch Motion

Figure 15 shows how we count the number of realized steps. After the launch in Fig. 15 (a), first landing is counted as the zeroth step and second landing is the first step, as shown in Fig. 15 (b) and (c). When the swing leg lands and the heel is behind the toe of the stance foot, we do not count the landing as a successful step and regard the walking as a failure, as shown in Fig. 15 (e) and (f). In some situations, we do not count a landing as a successful step even though the swing leg lands in front of the stance leg. Table 3 shows the experimental results. We changed pitch spring constant of the ankles and launched the biped walker 100 times for each settings. The spring constant ∞ [N/m] means that we locked the ankle pitch movements with wires instead of springs. However, the feet are not ideal rigid bodies because we attached sponge sheets to the soles of the feet. The result shows that a medium spring constant effectively stabilizes passive walking. High and low spring constants are ineffective because unsuccessful trials (0 step) occurred frequently, producing a low expectation of successful steps.

5.2 Foot Motion with the Ground

The spring stiffness of the pitch motion affects the foot motion with the floor. Figure 16 compares the foot motion for different spring stiffness. When the spring constant is 3920 N/m, which is equal to a torsional spring stiffness of about 7 Nm/rad, the foot of the stance leg remains in full contact with the floor until the heel of the swing leg touches the floor. Next, the front foot fully impacts the floor and rebound occurs, as indicated in Fig. 17. On the other hand, when the torsional spring stiffness is large, e.g., the spring constant is 9320 N/m, the rotation of the stance foot around the toe occurs before the swing foot touches the floor and the rebound after the full contact of the front foot is dramatically reduced. A high torsional spring stiffness for pitch motion leads to a smooth transition at the exchange of the stance leg. When the torsional spring stiffness is very large, the pitch angle is always 0, the stance foot almost always rotates around the heel or toe, and rebound does not occur.

6. Conclusions

This paper presents a simple 3D passive biped walker with flat feet and ankle springs. Experimental tests were performed to investigate the effects of torsional spring stiffness on the pitch motion at the ankle joints of the walker. When the spring stiffness is low, oscillating motion is induced by the impact of the feet with the ground. Experimental results showed that using springs with appropriate torsional spring stiffness effectively reduces the oscillating motion. The rebound of the front foot after full contact with the ground reduces dramatically with appropriate torsional spring stiffness. Appropriate stiffness enables the biped walker to walk smoothly and also stabilizes the walker. However, when the spring stiffness is either high or low, it become difficult for the walker to walk.

ACKNOWLEDGMENTS

This work was supported in part by Grant in Aid for the Global Center of Excellence Program for "Center for Education and Research of Symbiotic, Safe and Secure System Design" from the Ministry of Education, Culture, Sports, Science and Technology in Japan.

7. References

Adolfsson, J., Dankowicz, H., and Nordmark, A. (2001). 3D passive walkers: Finding periodic gaits in the presence of discontinuities. *Nonlinear Dynamics*, 24(2):205–229.

Coleman, M. J., Garcia, M., Mombaur, K., and Ruina, A. (2001). Prediction of stable walking for a toy that cannot stand. *Physical Review E*, 64(2):22901–.

Coleman, M. J. and Ruina, A. (1998). An uncontrolled walking toy that cannot stand still. *Physical Review Letters*, 80(16):3658–3661.

Collins, S. and Ruina, A. (2005). A bipedal walking robot with efficient and human-like gait. *Proceedings of the 2005 IEEE International Conference on Robotics and Automation*, pages 1983–1988.

Collins, S., Ruina, A., Tedrake, R., and Wisse, M. (2005). Efficient bipedal robots based on passive-dynamic walkers. *Science*, 307(5712):1082–1085.

Collins, S., Wisse, M., and Ruina, A. (2001). A three-dimensional passive-dynamic walking robot with two legs and knees. *International Journal of Robotics Research*, 20(7):607–615.

Garcia, M. (1999). *Stability, scaling, and chaos in passive-dynamic gait models*. PhD thesis, Cornell University.

Garcia, M., Chatterjee, A., and Ruina, A. (2000). Efficiency, speed, and scaling of two-dimensional passive-dynamic walking. *Dynamics and Stability of Systems*, 15(2):75–99.

Garcia, M., Chatterjee, A., Ruina, A., and Coleman, M. (1998). The simplest walking model: Stability, complexity, and scaling. *Journal of Biomechanical Engineering-Transactions of the ASME*, 120(2):281–288.

Goswami, A., Espiau, B., and Keramane, A. (1996). Limit cycles and their stability in a passive bipedal gait. *Proceedings of the 1996 IEEE International Conference on Robotics and Automation*, pages 246–251.

Goswami, A., Thuilot, B., and Espiau, B. (1998). A study of the passive gait of a compass-like biped robot: Symmetry and chaos. *International Journal of Robotics Research*, 17(12):1282–1301.

Hobbelen, D. G. E. and Wisse, M. (2007). Limit cycle walking. In Hackel, M., editor, *Humanoid Robots: Human-like Machines*, pages 277–294. I-Tech Education and Publishing, Vienna, Austria.

Ikemata, Y., Sano, A., and Fujimoto, H. (2006). A physical principle of gait generation and its stabilization derived from mechanism of fixed point. *Proceedings of the 2006 Conference on International Robotics and Automation*, pages 836–841.

Kinugasa, T., Kotake, K., Haruki, T., Tanaka, H., and Yoshida, K. (2008). 3D passive dynamic walker with sprung ankle and flat foot -a design method by natural frequency index without yaw and roll compensator-. *Proceedings of the 2008 JSME Conference on Robotics and Mechatronics*, pages 1P1–B12–.

Kuo, A. D. (1999). Stabilization of lateral motion in passive dynamic walking. *International Journal of Robotics Research*, 18(9):917–930.

McGeer, T. (1990). Passive dynamic walking. *International Journal of Robotics Research*, 9(2):62–82.

McGeer, T. (1993). Passive dynamic biped catalogue, 1991. In Chatila, R. and Hirzinger, G., editors, *Experimental Robotics II: The 2nd International Symposium, Toulouse, France, June 25-27 1991*, pages 465–490. Springer-Verlag.

McMahon, T. A. (1984). Mechanics of locomotion. *International Journal of Robotics Research*, 3(2):4–28.

Narukawa, T., Yokoyama, K., Takahashi, M., and Yoshida, K. (2008). A simple 3D straight-legged passive walker with flat feet and ankle springs. *IEEE/RSJ International Conference on Intelligent Robots and Systems*, pages 2952–2957.

Narukawa, T., Yokoyama, K., Takahashi, M., and Yoshida, K. (2009a). Design and construction of a simple 3D straight-legged passive walker with flat feet and ankle springs. *JSME Journal of System Design and Dynamics*, 3(1):1–12.

Narukawa, T., Takahashi, M., and Yoshida, K. (2009b). Design and Stability Analysis of a 3D Rimless Wheel with Flat Feet and Ankle Springs. *JSME Journal of System Design and Dynamics, 3(3):258-269.*

Schwab, A. L. and Wisse, M. (2001). Basin of attraction of the simplest walking model. *Proceedings of the ASME Design Engineering Technical Conference*, pages DETC2001/VIB–21363–.

Tedrake, R., Zhang, T. W., Fong, M.-F., and Seung, H. S. (2004). Actuating a simple 3D passive dynamic walker. *Proceedings of the IEEE International Conference on Robotics and Automation*, 2004(5):4656–4661.

Tedrake, R. L. (2004). *Applied Optimal Control for Dynamically Stable Legged Locomotion*. PhD thesis, Massachusetts institute of technology.

Wisse, M., Hobbelen, D. G. E., Rotteveel, R. J. J., Anderson, S. O., and Zeglin, G. J. (2006). Ankle springs instead of arc-shaped feet for passive dynamic walkers. *Proceedings of the IEEE-RAS International Conference on Humanoid Robots*, pages 110–116.

Wisse, M. and Schwab, A. L. (2005). Skateboards, bicycles, and three-dimensional biped walking machines: Velocity-dependent stability by means of lean-to-yaw coupling. *International Journal of Robotics Research*, 24(6):417–429.

Wisse, M., Schwab, A. L., van der Linde, R. Q., and van der Helm, F. C. T. (2005). How to keep from falling forward: Elementary swing leg action for passive dynamic walkers. *IEEE Transactions on Robotics*, 21(3):393–401.

Wisse, M., Schwab, A. L., and vander Linde, R. Q. (2001). A 3D passive dynamic biped with yaw and roll compensation. *Robotica*, 19:275–284.

Wisse, M. and van Frankenhuyzen, J. (2003). Design and construction of mike; a 2D autonomous biped based on passive dynamic walking. *Proceedings of the Second International Symposium on Adaptive Motion of Animals and Machines*, pages 4–8.

9

Active Knee-release Mechanism for Passive-dynamic Walking Machines

Kalin Trifonov and Shuji Hashimoto
Waseda University
Japan

1. Introduction

In this chapter we will present the design and development of two knee mechanisms. One uses permanent magnets to lock the knee in its extended position and the other features an active mechanism for releasing the passively locked knee. We will also present a comparison between the experimental results achieved with each of the two knee mechanisms.
One of the big, and still unsolved, problems in robotics is achieving efficient and stable bipedal walking. There are two main strategies used to control walking. First, the traditional approach is to control the joint-angle of every joint at all times. Crucial disadvantages of this approach are that it results in a non-efficient gait in terms of energy consumption (Collins et al., 2005), it requires complex controllers and programming, and this strategy often results in gaits that are unnatural when compared to the human gait. Second is a somewhat new strategy called passive-dynamic walking, introduced by Tad McGeer (McGeer 1990) in the late 80's, early 90's. A walker based on the passive-dynamic walking principle uses its own mechanical dynamics properties to determine its movement. Such walkers can walk down slight inclines without any actuators, sensors or controllers. The energy that is necessary in order to sustain the walking motion is provided by gravity. The force of gravity is also enough to offset the losses due to the impact of the feet on the ground and friction. The advantages of passive-dynamic walking are high-energy efficiency, simple or no control, and a human-like gait. The main disadvantage is that because they are not actively powered, they can only walk on downhill slopes. This disadvantage can be eliminated by modifying walkers to include actuators that supply the necessary power instead of gravity (Collins and Ruina, 2005; Wisse and Frankenhuyzen, 2003; Wisse, 2004). This enables them to walk not only downhill, but on level and uphill surfaces as well. This possibility greatly increases the prospects for practical application.
The knee mechanism is a major part in passive-dynamic walkers. There are several different designs that have been implemented in walkers up to now. The original walker built by McGeer uses a mechanism with suction cups that keeps the knee extended as shown in Figure 1. The drawback of the suction cups design is that it is difficult to set up and not very efficient.

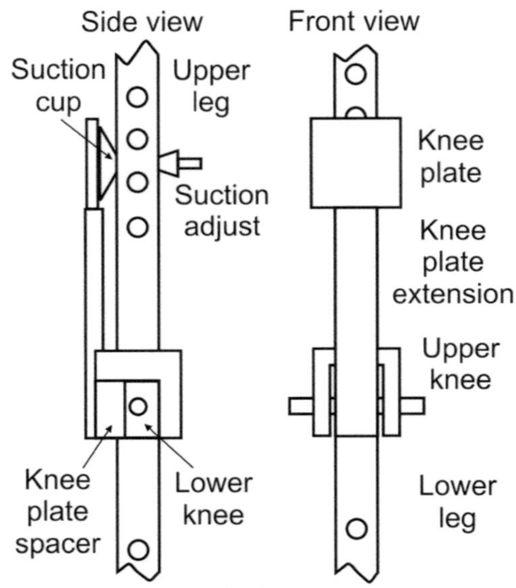

Fig. 1. Knee design with suction cups knee-locking.

Another popular design is used in the University of Delft's Mike (Wisse and Frankenhuyzen, 2003) and subsequent walkers Max and Denise (Wisse, 2004). The locking of the knee is achieved actively by McKibben muscles, which are counteracted by weak springs as shown in Fig.2. As a drawback we can mention that the McKibben muscles are not linear, and require controller that takes this feature into account. They also require a source of air.

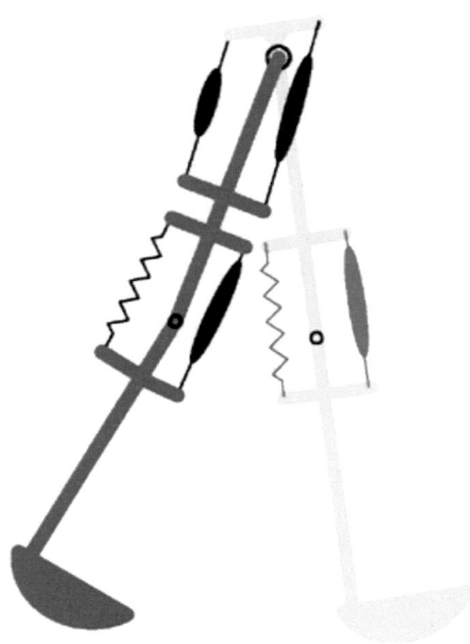

Fig. 2. Knee design with McKibben muscles knee-locking.

A third popular knee design is implemented in the Cornell powered biped (3). It features an electromagnetic release system shown in Fig.3, where (A), (B), (C), (D), (E), (F), (G) and (H) are a latch arm, a roller, a shank, a hinge, a shaft, a latch surface, a thigh and a solenoid, respectively. This design is robust and easy to control, but it is comprised of many parts, which makes it quite complicated. A similar design, where an electromagnetic clutch is used to engage or disengage a knee motor is developed by Baines (Baines, 2005).

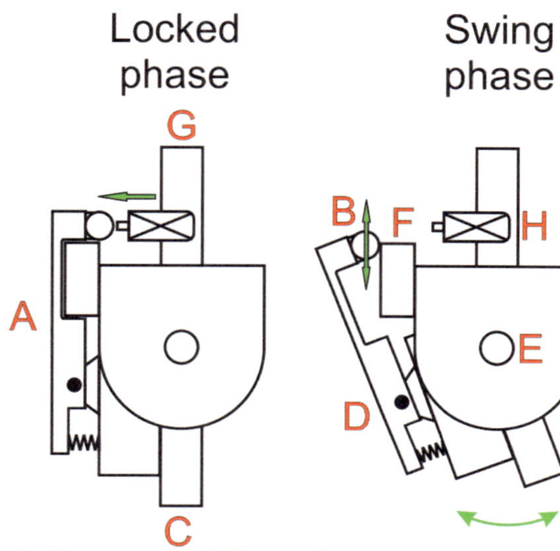

Fig. 3. Knee design with electromagnetic knee-release.

We developed our two knee locking mechanism with simplicity in mind. We wanted to understand if it was possible to develop a passive walker and a knee mechanism specifically based only on observation and experimentation without any modeling and simulations. A detailed model describing the mechanisms of generation and stabilization of a fixed point of passive walking, as well as leg-swing motion analysis of a passive-dynamic walker can be found in the research done by Prof. Sano's team at the Nagoya Institute of Technology (Ikemata et al., 2007), (Ikemata et al., 2008). Our aim was to build a mechanism that is simple, robust, and easy to use and set up. The purposes of this chapter are to present the mechanical design of the two knee mechanisms, to introduce the achieved experimental results, to make a comparison between them, and to discuss their effectiveness.

2. Knee Mechanism with Permanent Magnets

The dynamics of passive-dynamic walkers cause the swinging leg to bend and extend on its own. However, in order to achieve a stable gait, the knee must be able to swing with minimal friction, meaning minimal energy loss. Taking this into consideration, the most logical choice for the knee joint is a ball bearing. Additionally, the knee must be equipped with a knee-locking mechanism that supports the knee during its extended phase and prevents it from bending while bearing the weight of the walker.
For our walker, the knee is cut from an aluminum block and is comprised of only an upper knee, to which the aluminum lower leg is attached directly through a shaft and a pair of ball

bearings (Trifonov and Hashimoto, 2006). For the locking mechanism, we are using a knee plate spacer and a knee plate, cut from acrylic, as with the original McGeer design, but we decided to try a new approach by using magnets instead of a suction cup. We adjust the locking magnetic force by changing the distance between the magnet(s) and the steel plate. This can be achieved either by using magnets with different sizes or by using a different number of magnets. The smaller the distance is, the stronger the force. Another advantage of the magnetic lock is that it does not require physical contact between the locking parts (magnet and steel plate). In this way the material wear is reduced and the lock can be used longer without having to worry about replacing some of its parts. 3D renderings are shown in Fig. 4, where (A) is knee, (B) is knee plate, (C) is magnet(s), and (D) is a steel plate. A drawing of the knee mechanism with some main dimensions is shown in Fig. 5.

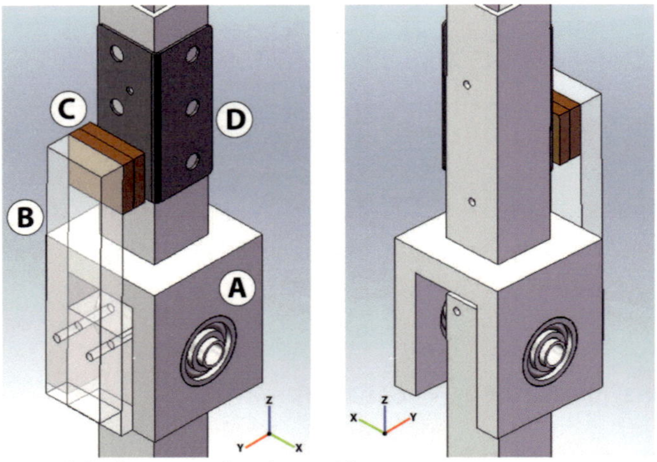

Fig. 4. 3D renderings of the knee mechanism with permanent magnets.

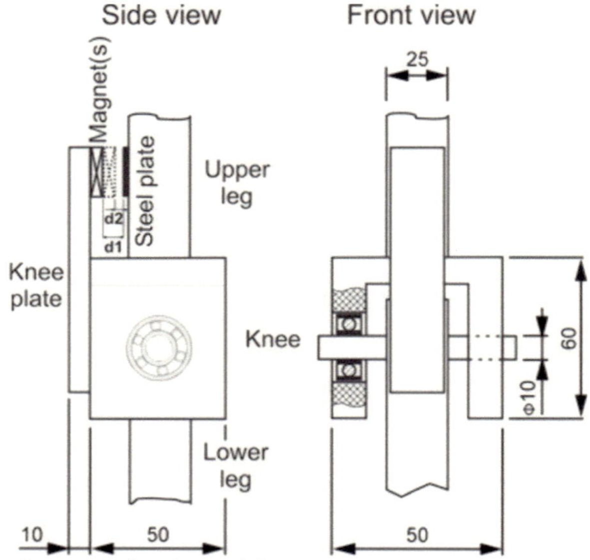

Fig. 5. Drawing of the knee mechanism with permanent magnets.

3. Knee Mechanism with an Active Release System

We designed a second newer, simpler, and lower in weight knee-locking mechanism (Trifonov and Hashimoto, 2007). The locking mechanism is constructed of acrylic, ABS, steel, and aluminum. The knee-locking mechanism consists of a knee (A), knee plate (B), locking axle (C), locking hook (D), base plate (E), and a DC motor (F) as shown in Fig. 6 and Fig. 7. Additionally, there is a switch attached to each foot of the walker, which is used to control the DC motor, but is not shown in the figure. The entire knee mechanism was designed in 3D modeling software and cut on a CAM machine. The knee is cut from aluminum, the knee plate from acrylic, the locking axle from steel, and the locking hook and the base plate are cut from ABS.

An active release system has been implemented before on a passive-dynamic walker. The Cornell powered biped (Collins and Ruina, 2005) uses an electromagnetic solenoid for the release of the passively locked knee mechanism. The advantages of our system are the much simpler design and the absence of a controller.

The locking action is done passively. As the swing leg extends before hitting the ground, the locking axle hits the front edge of the locking hook, lifting it. After the locking axle passes under the hook, it comes back down to lock the axle, effectively locking the knee itself. The locking hook is balanced by a counter weight in such a way that it comes back down to its initial position after the locking axle has lifted it. Just before the stance leg lifts from the ground and starts to swing, the foot switch comes into contact with the ground and switches to the ON position, thus turning on the power for the DC motor. This causes the motor to lift the locking hook and release the knee. Immediately after the leg lifts off the ground and starts swinging, the foot switch returns to the OFF position, cutting the power, and the locking hook returns to its initial position. The foot switch is mounted to the side of the foot plate, such that it does not influence the walking of the machine.

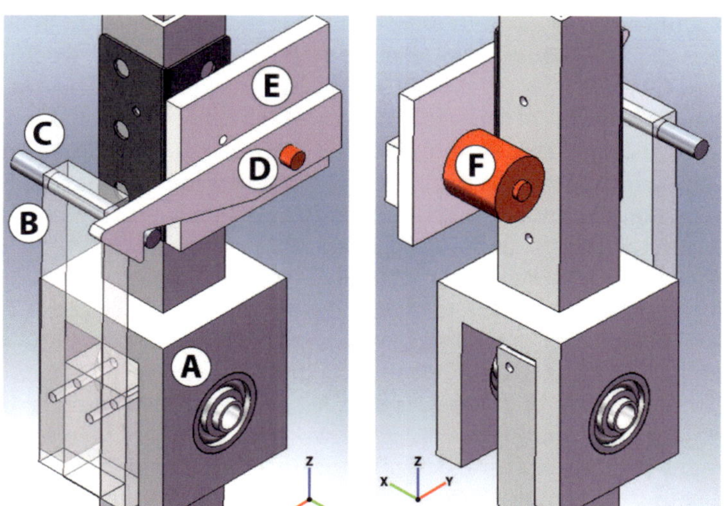

Fig. 6. 3D renderings of the knee mechanism with an active release.

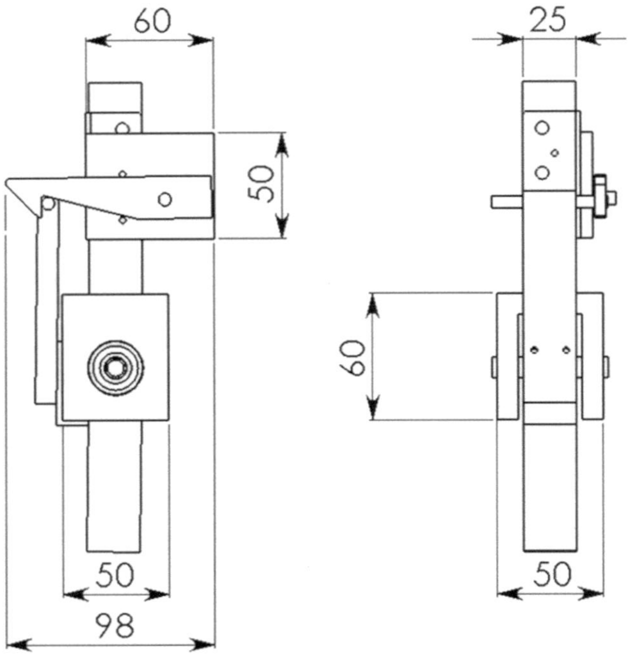

Fig. 7. Drawing of the knee mechanism with an active release.

4. Experiments and Results

To compare the two knee mechanisms, experiments were conducted with the same walker shown in Fig. 8, built from square aluminum tubes for the legs and 2mm thick steel plate for the feet (Trifonov and Hashimoto, 2006). For the thighs and lower legs, we used 2.5 by 2.5cm square aluminum tubes with lengths of 34 and 43.5cm respectively. The total height of the walker is 89cm and the radius of the feet is 12.3cm. The total weight is 4.5kg. The knees were outfitted first with the magnetic system and then with the active release one. The walker was set on a ramp, which measures 3m in length, 90cm in width, and has a 3° grade relative to the ground. The ramp is covered with a rubber mat to reduce the chance of foot slippage.
We performed several sets of a hundred trials (walks) down the ramp for both knee mechanisms and counted the steps that the walker completed each time. We denote a trial as successful if the walker manages to make five to seven steps before it exits the ramp. While five to seven steps may seem short, we postulate that after five steps, the walker has achieved a steady gait, and would ideally continue assuming a longer ramp existed. However, the impracticality of a longer ramp led us to set this number of steps as the criteria for deciding walk success. Fig. 9 shows a comparison between the two knee mechanism designs in terms of average number of steps made in each of the hundred trials.
As the results show, using the knee mechanism with active release, we can achieve a reasonable amount of successful trials. Out of a hundred trials, the walker achieved an average of forty-four successful walks with the active release system, while the magnetic approach resulted in only seven. In addition, using the active mechanism produces fewer failures than the magnetic one.

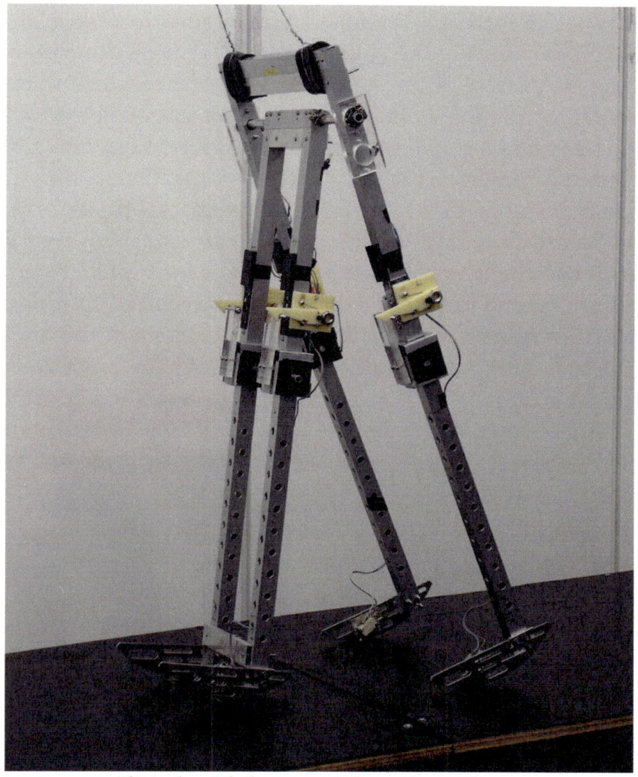

Fig. 8. Walker on the ramp, outfitted with knees with active release system.

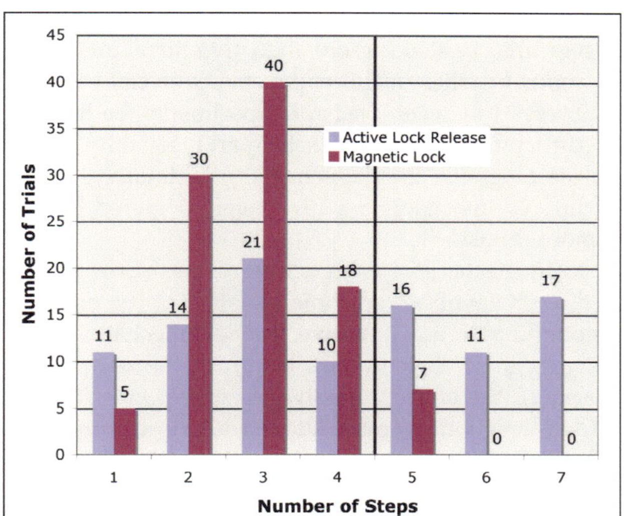

Fig. 9. Comparison between the experimental results achieved with the two mechanisms.

There are several types of reasons for a failed trial in general. One is an incorrect start of the walker by the person performing the experiments. As this is done manually, it is subjective and depends on the experience of the starter. In case of an incorrect start the walker fails on the first or the second step of the walk. If the walker is started correctly and goes beyond the first couple of steps it enters a stable gait and from this moment onwards there are two other possible reasons for failure, which may occur at any time. One is slippage of the foot against the slope, which may be attributed to dirt or other obstacles present on it. Another is failure to lock or unlock the knee. Failure to lock the knee is usually caused by the so-called knee bouncing. That is, when the knee extends too fast, the knee plate bounces off the knee, and the locking hook has no time to lock it in place. We have tried to reduce this to a minimum by adding a small 1mm rubber mat to the knee face to cushion the hit. Failure to unlock the knee is mainly due to a late attempt to do it. If the foot switch activates the DC motor after the time when the knee starts to bend, the locking axle is already applying pressure to the locking hook and it is unable to lift and release the knee. By adjusting the foot switch to activate earlier in the walking cycle we have significantly reduced the occurrence of this problem.

5. Discussion

In this chapter we presented two knee mechanism designs. One features a permanent magnet locking system and the other an active release system. We performed a series of experiments with both mechanisms mounted on the same passive-dynamic walking machine and compared the results we achieved.

The first knee mechanism was based on permanent magnets. We speculated that changing the distance between a permanent magnet and a steel plate, and hence changing the magnetic force, would be sufficient to control the release moment of the knee with this passive magnetic mechanism. The experiments showed that the walker using this mechanism was never able to make more than five steps and was only able to make a successful trial, as defined earlier in the paper, in seven out of a hundred attempts. As a result of what we observed in several sets of experiments we have reached the conclusion that it was very difficult to precisely setup and reliably uses the machine in the configuration with the magnetic knee mechanism. Ultimately, we decided to design and build a completely different mechanism, with actively powered knee release action, which is much simpler and more robust.

Our design of the knee mechanism with active release showed promising results in the experiments. Even though we observed some variation of the number of successful trials, it is obvious, that although not entirely passive, the new mechanism is more efficient in terms of the walker managing to walk the entire length of the ramp when compared with the previous design based on the entirely passive, magnetic lock. The active release approach allows the walker to achieve longer, more stable walks and is more robust and reliable. We performed several sets of a hundred trials and managed to achieve an average of forty-four successes. Using the proposed design we were also able to obtain a more even distribution between trials of five, six, and seven step walks achieved by the walker. The experimental results show that the walker, equipped with the new knee-locking mechanism makes five or more steps in a higher percentage of the trials. Also the unsuccessful trials were greatly reduced.

Our goal from the start of this research was to prove that it is possible to design a simple and usable passive-dynamic walker without any complex modeling and simulations. We wanted to see if the trial and error method would work for passive-dynamic walking where the stability range of the actual machines is very narrow. The results show clearly that there is an obvious improvement going from the first design to the second one. This means that if the work continues in the same way additional improvement is possible.
Our plans for the future include adding active control of the knee release system and building a starting mechanism for the walker in order to be able to control the release moment precisely and to reduce the failure rate even further. We are already working on adding a controller and hip motor to the walker. Our goal is set on achieving a highly successful walk on the experimental ramp and eventually on a flat surface.

6. Acknowledgement

This research was also supported by Waseda University Grant for Special Research Projects, No.2008B-094, the Grant-in-Aid for the WABOT-HOUSE Project by Gifu Prefecture, the JSPS 21st Century Center of Excellence Program, "The innovative research on symbiosis technologies for human and robots in the elderly dominated society" and the JSPS Global Center of Excellence Program, "Global Robot Academia".

7. References

A. Baines, Knee Design for a Bipedal Walking Robot Based on a Passive-Dynamic Walker. B.S. Thesis, Department of Mechanical Engineering, MIT, (2005).

S. H. Collins, A. L. Ruina, R. Tedrake, M. Wisse. Efficient bipedal robots based on passive-dynamic Walkers. Science magazine, vol. 307 pp. 1082-1085, (2005).

S. H. Collins, A. Ruina. A bipedal walking robot with efficient and human-like gait. In Proc. IEEE ICRA 2005, pp. 1983 - 1988, Barcelona, Spain, (2005).

Y. Ikemata, A. Sano, Hideo Fujimoto, Generation and Local Stabilization of Fixed Point Based on a Stability Mechanism of Passive Walking, Proc. of ICRA 2007, pp. 3218-3223, Rome, Italy (2007).

Y. Ikemata, K. Yasuhara, A. Sano, Hideo Fujimoto, A study of the leg-swing motion of passive walking, Proc. of ICRA 2008, pp. 1588-1593, Pasadena, USA, (2008).

T. McGeer. Passive dynamic walking. International Journal of Robotics Research, 9(2) pp.62-82, (1990).

K. Trifonov, S. Hashimoto, Design Improvements in Passive-Dynamic Walkers, In Proc. International Conference "Automatics and Informatics '06", Sofia, Bulgaria, pp. 35-38, (2006).

K. Trifonov, S. Hashimoto, Active knee-lock release for passive-dynamic walking machines, In Proc. IEEE Robio 2007, pp. 958-963, Sanya, China, (2007).

M. Wisse, J. V. Frankenhuyzen. Design and construction of Mike: a 2D autonomous biped based on passive dynamic walking. In Proc. 2nd Int. Symp. on AMAM, Kyoto, Japan, (2003).

M. Wisse, Essentials of dynamic walking. Analysis and design of two-legged robots. Ph.D. Thesis, Technische Universiteit Delft, (2004)

10

Simplified Human Hand Models for Manipulation Tasks

Salvador Cobos, Manuel Ferre, Rafael Aracil,
Javier Ortego and M. Ángel Sanchéz-Urán
Universidad Politécnica de Madrid
Spain

1. Introduction

The human hand is the most dexterous and versatile biomechanical device that possesses the human body, this device created by Nature during millions years of evolution represents one of the more distinctive qualities among other animals. Since the 70s and 80s, important contributions have appeared in physiological studies of the human hand (I. A. Kapandji, 1970), (I. A. Kapandji, 1981). Studies about robotic hands have made several contributions e.g. the Stanford/JPL hand (K. S. Salisbury & B. Roth, 1983), the Utah hand (S. C. Jacobsen et al., 1986), the Okada hand (T. Okada, 1982), the Belgrade/USC hand (G. Bekey et al., 1990), the UB hands (C. Melchiorri and G. Vassura 1992), (C. Melchiorri and G. Vassura 1993), the DLR hands (J. Butterfass et al., 1999), (J. Butterfass et al., 2001), the University of Tokyo hand (Y. K. Lee & I. Shimoyama, 1999), Barrett Hand (W. T. Townsend, 2000), the Robo-Naut hand by NASA (C. S. Lovchik et al., 2000), the Karlsruhe University ultra-light hand (S. Schulz et al., 2001), the GIFU hand (H. Kawasaki et al., 2001), the Shadow Dextrous Hand (Shadow Robot Company), (F. Röthling et al., 2007), a prosthetic hand (H. Yokoi et al., 2004), the DLR-HIT-Hand (H. Liu et al., 2008) and other. These devices have different kinematic configurations with respect to the number of Degrees of Freedom (DoF) controlled, number of fingers, number of joints, type of actuation, etc. This chapter describes simplified human hand models that properly represent the kinematic behaviour of the human hand in accordance with the precision and application required. The first part describes a human hand model with 24 DoF. This model represents a balance between complexity and realism. Simplified human hand (SHH) models are analyzed using the model with 24 DoF. These SHH models (1 to 24 DoF) are evaluated in accordance with the level of dexterous or power required. A Cyberglove® (Immersion) is used for the experiments carried out in this work. Kinematic constrainswere checked with the information provided by the glove. Also, this glove was used for evaluating the error of the SHH versus the full 24 DoF hand model. Finally, the experiments carried out with SHH and 24 DoF hand model compare the efficiency in grasping for circular and prismatic grasps in accordance with the application.

2. Kinematic Human Hand Model

The hand model used for this work is based on the human skeleton. The kinematic model is comprised of 19 links that imitate the corresponding human bones, and 24 degrees of freedom (DoF) that represent the joints. Links and joints are defined by $L_{i,j}$ and $\theta_{i,j}$, where i represents a finger (i=Thumb, Index, Middle, Ring or Little) and j its corresponding link or joint respectively. Two kinematic configurations are considered in this hand model, one for the thumb and another for the rest of the fingers. Therefore, the same kinematic configuration is used for the index, middle, ring and little fingers. This configuration is defined by 5 joints and 4 links: metacarpal ($L_{i,Me}$), proximal ($L_{i,P}$), middle ($L_{i,Mi}$) and distal ($L_{i,D}$) links. The joints are defined as: carpometacarpal ($\theta_{i,CMC}$), proximal interphalangeal ($\theta_{i,PIP}$) distal interphalangeal ($\theta_{i,DIP}$) and metacarpophalangeal, which is modelled by a universal joint (2 DoF) that defines the abduction/adduction (θ_{i,MCP_aa}) and flexion/extension (θ_{i,MCP_fe}) rotations. The thumb is modelled by 4 DoF and 3 links: metacarpal ($L_{T,M}$), proximal ($L_{T,P}$), and distal ($L_{T,D}$) . The thumb joints are defined as: metacarpophalangeal (θ_{T,MCP_fe}), interphalangeal ($\theta_{T,IP}$) and trapeziometacarpal, which is also defined by a universal joint that defines the abduction/adduction (θ_{T,TMC_aa}) and flexion/extension (θ_{T,TMC_fe}) respectively. Figure 1 shows details of both kinematic models. The main features of this model are: the use of 4 DoF for the thumb modelling, the inclusion of the $\theta_{i,CMC}$ joint, and the movement concatenation in the $\theta_{i,MCP}$ joint. The MCP joint of the thumb is considered to have 1 DoF (θ_{T,MCP_fe}). The $\theta_{i,CMC}$ joint allows simulation of the palm arc, the palm deformation when the hand is grasping a ball or similar objects. The MCP abduction/adduction (θ_{i,MCP_aa}) twist is defined before the MCP flexion/extension (θ_{i,MCP_fe}) aiming at simulating finger displacements in a better way. These points contribute to the definition of highly realistic hand movements and gestures. In the following sections the forward and inverse kinematic models are developed.

2.1 Forward Kinematics

Forward kinematic is used to obtain the finger tip position and orientation according to the finer joint angles. These models are obtained for both cases (thumb and the other fingers).
Model equations are calculated by means of the Denavit-Hartenberg (D-H) parameters (M.W. Spong, 2006). This convention is commonly used for mechanism and robotics modelling. Human anatomical terminology has been used to describe the hand model. Direct kinematic equations are required when virtual grasping is simulated. More specifically, a user wearing a glove that provides the human joints and forward kinematic equations determines what the fingertip positions and orientations are. This information will also be used to calculate object contact points, grasping routes, etc.

2.1.1 Forward kinematics of the index, middle, ring and little fingers

Table 1 shows the D-H parameters for index, middle, ring and little fingers. Joints represent the variables of this model that are defined by $\theta_{i,j}$. Links are defined by the parameters $a_{i,j}$ that represent the length of the bones. Parameter $d_{i,j}$ is always null since bones are aligned, and parameter $\alpha_{i,j}$ is the angle of separation of the Z_{i-1} axis and the Z_i axis, measured in a plane perpendicular to the X_i axis, utilizing the rule of the right hand.

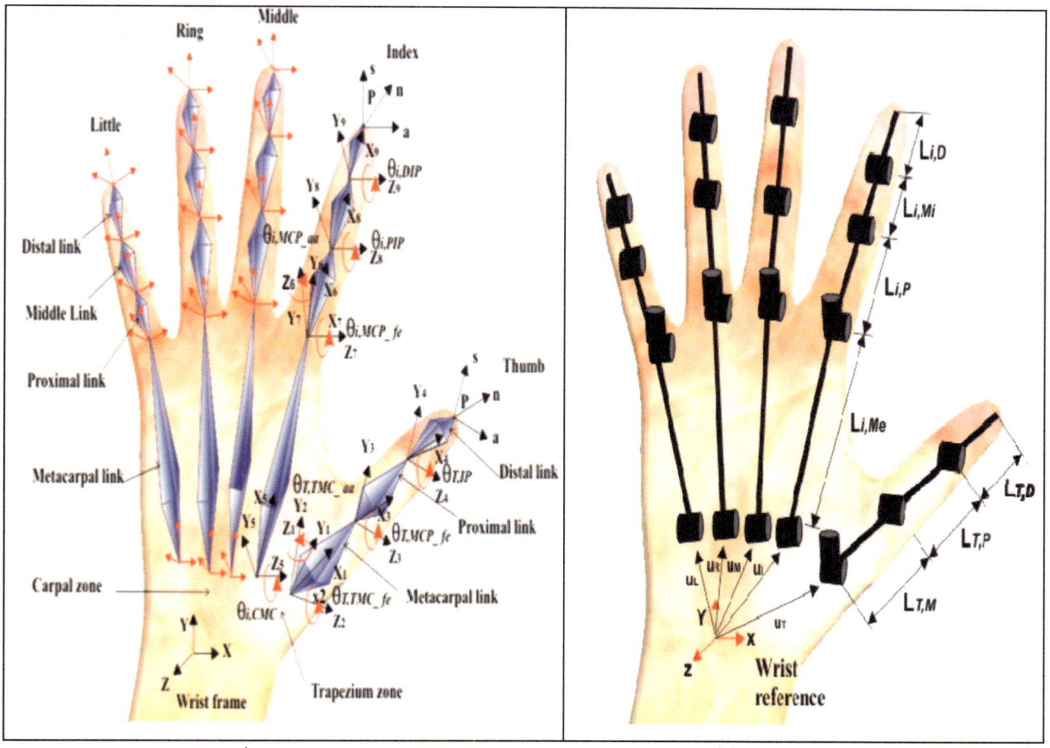

a) b)

Fig. 1. Kinematic configuration of the human hand with 19 links and 24 DoF.
Thumb is defined by 3 links ($L_{T,M}$, $L_{T,P}$ and $L_{T,D}$) and 4 DoF (θ_{T,TMC_aa}, θ_{T,TMC_fe}, θ_{T,MCP_fe} and $\theta_{T,IP}$) whereas index, middle, ring and little fingers are defined by 4 links ($L_{i,Me}$, $L_{i,P}$, $L_{i,Mi}$ and $L_{i,D}$) and 5 DoF ($\theta_{i,CMC}$, θ_{i,MCP_aa}, θ_{i,MCP_fe}, $\theta_{i,PIP}$ and $\theta_{i,DIP}$).

Joint	$\theta_{i,j}$	$d_{i,j}$	$a_{i,j}$	$\alpha_{i,j}$
1	$\theta_{i,CMC}$	0	$L_{i,Me}$	$\pi/2$
2	$\theta_{i,MCP\ aa}$	0	0	$-\pi/2$
3	$\theta_{i,MCP\ fe}$	0	$L_{i,P}$	0
4	$\theta_{i,PIP}$	0	$L_{i,Mi}$	0
5	$\theta_{i,DIP}$	0	$L_{i,D}$	0

Table 1. D-H parameters for index, middle, ring and little fingers.

The forward kinematics of these fingers is shown in equation 1, which is defined according to the parameters of Table 1.

$$p_i = {}^{-1}_{0}T_i(u_i)\cdot{}^{0}_{5}T_i(\theta_{i,j}) = {}^{-1}_{0}T_i(u_i)\cdot{}^{0}_{1}T_i(\theta_{i,CMC})\cdot{}^{1}_{2}T_i(\theta_{i,MCP\_aa})\cdot{}^{2}_{3}T_i(\theta_{i,MCP\_fe})\cdot{}^{3}_{4}T_i(\theta_{i,PIP})\cdot{}^{4}_{5}T_i(\theta_{i,DIP}) \quad (1)$$

Where terms in this equation are: p_i represents a matrix that contains position and orientation of the *i*-finger tip with respect to the center of the wrist, u_i represents the vector between the center of wrist and the corresponding *i*-finger reference frame, ${}^{0}_{5}T_i(\theta_{i,j})$ is a matrix that contains the homogeneous matrix between the i-finger reference frame and its corresponding *i*-finger tip. This matrix consists of the concatenation of matrixes that represent the contribution of each *i*-finger joint displacement ($\theta_{i,CMC}$, $\theta_{i,MCP\ aa}$, $\theta_{i,MCP\ fe}$, $\theta_{i,PIP}$, $\theta_{i,DIP}$) the translation and rotation contribution of each joint is defined by the matrix ${}^{j-1}_{j}T_i(\theta_{i,j})$.

$$ {}^{j-1}_{j}T_i(\theta_{i,j}) = \begin{pmatrix} \cos(\theta_{i,j}) & -\cos(\alpha_{i,j})\sin(\theta_{i,j}) & \sin(\alpha_{i,j})\sin(\alpha_{i,j}) & a_{i,j}\cos(\theta_{i,j}) \\ \sin(\theta_{i,j}) & \cos(\alpha_{i,j})\cos(\theta_{i,j}) & -\sin(\alpha_{i,j})\cos(\alpha_{i,j}) & a_{i,j}\sin(\theta_{i,j}) \\ 0 & \sin(\alpha_{i,j}) & \cos(\alpha_{i,j}) & 0 \\ 0 & 0 & 0 & 1 \end{pmatrix} \quad (2) $$

2.1.2 Homogeneous matrixes for the index, middle, ring and little fingers.

In order to simplify the equations, some variables are replaced (**VR**) for the corresponding variable showed in table 2.

$\theta_{i,j}$	**VR**	$a_{i,j}$	**VR**	***Link***
$\theta_{i,CMC}$	θ_5	$L_{i,Me}$	L_4	Metacarpal
$\theta_{i,MCP\ aa}$	θ_6	0	0	
$\theta_{i,MCP\ fe}$	θ_7	$L_{i,P}$	L_5	Proximal
$\theta_{i,PIP}$	θ_8	$L_{i,Mi}$	L_6	Middle
$\theta_{i,DIP}$	θ_9	$L_{i,D}$	L_7	Distal

Table 2.

$c = \cos$

$s = \sin$

$$ {}^{0}T_1(\theta_{i,CMC}) = \begin{bmatrix} c\theta_{i,CMC} & 0 & -s\theta_{i,CMC} & L_{i,Me}c\theta_{i,CMC} \\ s\theta_{i,CMC} & 0 & c\theta_{i,CMC} & L_{i,Me}s\theta_{i,CMC} \\ 0 & -1 & 0 & 0 \\ 0 & 0 & 0 & 1 \end{bmatrix} \quad (3) $$

$$ {}^{1}T_2(\theta_{i,MCP\_aa}) = \begin{bmatrix} c\theta_{i,MCP\_aa} & 0 & s\theta_{i,MCP\_aa} & 0 \\ s\theta_{i,MCP\_aa} & 0 & -c\theta_{i,MCP\_aa} & 0 \\ 0 & 1 & 0 & 0 \\ 0 & 0 & 0 & 1 \end{bmatrix} \quad (4) $$

$$ {}^{2}T_{3}(\theta_{i,MCP\_fe}) = \begin{bmatrix} c\theta_{i,MCP\_fe} & -s\theta_{i,MCP\_fe} & 0 & L_{i,p}c\theta_{i,MCP\_fe} \\ s\theta_{i,MCP\_fe} & c\theta_{i,MCP\_fe} & 0 & L_{i,p}s\theta_{i,MCP\_fe} \\ 0 & 0 & 1 & 0 \\ 0 & 0 & 0 & 1 \end{bmatrix} \tag{5} $$

$$ {}^{3}T_{4}(\theta_{i,PIP}) = \begin{bmatrix} c\theta_{i,PIP} & -s\theta_{i,PIP} & 0 & L_{i,Mi}c\theta_{i,PIP} \\ s\theta_{i,PIP} & c\theta_{i,PIP} & 0 & L_{i,Mi}s\theta_{i,PIP} \\ 0 & 0 & 1 & 0 \\ 0 & 0 & 0 & 1 \end{bmatrix} \tag{6} $$

$$ {}^{4}T_{5}(\theta_{i,DIP}) = \begin{bmatrix} c\theta_{i,DIP} & -s\theta_{i,DIP} & 0 & L_{i,D}c\theta_{i,DIP} \\ s\theta_{i,DIP} & c\theta_{i,DIP} & 0 & L_{i,D}s\theta_{i,DIP} \\ 0 & 0 & 1 & 0 \\ 0 & 0 & 0 & 1 \end{bmatrix} \tag{7} $$

$$ {}^{-1}T_{0}(u_{i}) = \begin{bmatrix} 1 & 0 & 0 & u_{i,x} \\ 0 & 1 & 0 & u_{i,y} \\ 0 & 0 & 1 & u_{i,z} \\ 0 & 0 & 0 & 1 \end{bmatrix} \tag{8} $$

$$ {}^{0}T_{5} = \begin{bmatrix} n_{2x} & s_{2x} & a_{2x} & P_{2x} \\ n_{2y} & s_{2y} & a_{2y} & P_{2y} \\ n_{2z} & s_{2z} & a_{2z} & P_{2z} \\ 0 & 0 & 0 & 1 \end{bmatrix} \tag{9} $$

Where:

$$ n_{2x} = ((c_5c_6c_7 - s_5s_7)c_8 + (-c_5c_6s_7 - s_5c_7)s_8)c_9 + (-(c_5c_6c_7 - s_5s_7)s_8 + (-c_5c_6s_7 - s_5c_7)c_8)s_9 \tag{10} $$

$$ n_{2y} = ((s_5c_6c_7 + c_5s_7)c_8 + (-s_5c_6s_7 + c_5c_7)s_8)c_9 + (-(s_5c_6c_7 + c_5s_7)s_8 + (-s_5c_6s_7 + c_5c_7)c_8)s_9 \tag{11} $$

$$ n_{2z} = (-s_6c_7c_8 + s_6s_7s_8)c_9 + (s_6c_7s_8 + s_6s_7c_8)s_9 \tag{12} $$

$$ s_{2x} = -((c_5c_6c_7 - s_5s_7)c_8 + (-c_5c_6s_7 - s_5c_7)s_8)s_9 + (-(c_5c_6c_7 - s_5s_7)s_8 + (-c_5c_6s_7 - s_5c_7)c_8)c_9 \tag{13} $$

$$ s_{2y} = -((s_5c_6c_7 + c_5s_7)c_8 + (-s_5c_6s_7 + c_5c_7)s_8)s_9 + (-(s_5c_6c_7 + c_5s_7)s_8 + (-s_5c_6s_7 + c_5c_7)c_8)c_9 \tag{14} $$

$$ s_{2z} = -(-s_6c_7c_8 + s_6s_7s_8)s_9 + (s_6c_7s_8 + s_6s_7c_8)c_9 \tag{15} $$

$$a_{2x} = c_5 s_6 \quad (16)$$

$$a_{2y} = s_5 s_6 \quad (17)$$

$$a_{2z} = c_6 \quad (18)$$

$$a_{2z} = c_6 \quad (19)$$

$$P_{2x} = ((c_5c_6c_7 - s_5s_7)c_8 + (-c_5c_6s_7 - s_5c_7)s_8)c_9L_7 + (-(c_5c_6c_7 - s_5s_7)s_8 + (-c_5c_6s_7 - s_5c_7)c_8)s_9L_7 + (c_5c_6c_7 - s_5s_7)c_8L_6 + (-c_5c_6s_7 - s_5c_7)s_8L_6 + (c_5c_6c_7 - s_5s_7)L_5 + c_5L_4 \quad (20)$$

$$P_{2y} = ((s_5c_6c_7 + c_5s_7)c_8 + (-s_5c_6s_7 + c_5c_7)s_8)c_9L_7 + (-(s_5c_6c_7 + c_5s_7)s_8 + (-s_5c_6s_7 + c_5c_7)c_8)s_9L_7 + (s_5c_6c_7 + c_5s_7)c_8L_6 + (-s_5c_6s_7 + c_5c_7)s_8L_6 + (s_5c_6c_7 + c_5s_7)L_5 + s_5L_4 \quad (21)$$

$$P_{2z} = (-s_6c_7c_8 + s_6s_7s_8)c_9L_7 + (s_6c_7s_8 + s_6s_7c_8)s_9L_7 - (s_6c_7c_8 + s_6s_7s_8)L_6 - s_6c_7L_5 \quad (22)$$

2.1.3 Forward kinematics of the thumb

Forward kinematics of the thumb is defined in a similar way. Table 3 shows the D-H parameters of the thumb.

Joint	$\theta_{T,j}$	$d_{T,j}$	$a_{T,j}$	$\alpha_{T,j}$
1	$\theta_{T,TMC\,aa}$	0	0	$\pi/2$
2	$\theta_{T,TMC\,fe}$	0	$L_{T,M}$	0
3	$\theta_{T,MCP\,fe}$	0	$L_{T,P}$	0
4	$\theta_{T,IP}$	0	$L_{T,D}$	0

Table 3. D-H Parameters for the Thumb

Equation 23 represents the direct kinematics for the thumb as follows:

$$p_T = {}^{-1}_{0}T_T(u_T)\,{}^{0}_{4}T_T(\theta_{T,j}) = {}^{-1}_{0}T_T(u_T)\,{}^{0}_{1}T_T(\theta_{T,TMC\_aa})\,{}^{1}_{2}T_T(\theta_{T,TMC\_fe})\,{}^{2}_{3}T_T(\theta_{T,MCP\_fe})\,{}^{3}_{4}T_T(\theta_{T,IP}) \quad (23)$$

Where terms in this equation are:

P_T represents a matrix that contains position and orientation of the thumb finger tip with respect to the center of the wrist, u_T represents the vector between the center of wrist and the corresponding thumb reference frame, ${}^{0}_{4}T_T(\theta_{T,j})$ is a matrix that contains the homogeneous matrix between the thumb reference frame and its finger tip. This matrix is also a concatenation of the corresponding matrixes; which are obtained by the thumb joints (θ_{T,TMC_aa}, θ_{T,TMC_fe}, $\theta_{T,MCP\,fe}$, $\theta_{T,IP}$).

2.1.4 Homogeneous matrixes for the Thumb model

In order to simplify the equations, some variables are replaced (VR) for the corresponding variable showed in table 2.

$\theta_{T,j}$	VR	$a_{T,j}$	VR	*Link*
$\theta_{T,TMC\ aa}$	θ_1	*0*	*0*	
$\theta_{T,TMC\ fe}$	θ_2	$L_{T,M}$	L_1	*Metacarpal*
$\theta_{T,MCP\ fe}$	θ_3	$L_{T,P}$	L_2	*Proximal*
$\theta_{T,IP}$	θ_4	$L_{T,D}$	L_3	*Distal*

Table 4.

$c = \cos$

$s = \sin$

$$ {}^{0}T_1(\theta_{T,TMC\_aa}) = \begin{bmatrix} c\theta_{T,TMC\_aa} & 0 & s\theta_{T,TMC\_aa} & 0 \\ s\theta_{T,TMC\_aa} & 0 & -c\theta_{T,TMC\_aa} & 0 \\ 0 & 1 & 0 & 0 \\ 0 & 0 & 0 & 1 \end{bmatrix} \tag{24} $$

$$ {}^{1}T_2(\theta_{T,TMC\_fe}) = \begin{bmatrix} c\theta_{T,TMC\_fe} & -s\theta_{T,TMC\_fe} & 0 & L_{T,M}c\theta_{T,TMC\_fe} \\ s\theta_{T,TMC\_fe} & c\theta_{T,TMC\_fe} & 0 & L_{T,M}s\theta_{T,TMC\_fe} \\ 0 & 0 & 1 & 0 \\ 0 & 0 & 0 & 1 \end{bmatrix} \tag{25} $$

$$ {}^{2}T_3(\theta_{T,MCP}) = \begin{bmatrix} c\theta_{T,MCP} & -s\theta_{T,MCP} & 0 & L_{T,P}c\theta_{T,MCP} \\ s\theta_{T,MCP} & c\theta_{T,MCP} & 0 & L_{T,P}s\theta_{T,MCP} \\ 0 & 0 & 1 & 0 \\ 0 & 0 & 0 & 1 \end{bmatrix} \tag{26} $$

$$ {}^{3}T_4(\theta_{T,IP}) = \begin{bmatrix} c\theta_{T,IP} & -s\theta_{T,IP} & 0 & L_{T,D}c\theta_{T,IP} \\ s\theta_{T,IP} & c\theta_{T,IP} & 0 & L_{T,D}s\theta_{T,IP} \\ 0 & 0 & 1 & 0 \\ 0 & 0 & 0 & 1 \end{bmatrix} \tag{27} $$

$$ {}^{-1}T_0(u_{Thumb}) = \begin{bmatrix} 1 & 0 & 0 & u_{T,x} \\ 0 & 1 & 0 & u_{T,y} \\ 0 & 0 & 1 & u_{T,z} \\ 0 & 0 & 0 & 1 \end{bmatrix} \tag{28} $$

$$ {}^{0}T_{4} = \begin{bmatrix} n_{1x} & s_{1x} & a_{1x} & P_{1x} \\ n_{1y} & s_{1y} & a_{1y} & P_{1y} \\ n_{1z} & s_{1z} & a_{1z} & P_{1z} \\ 0 & 0 & 0 & 1 \end{bmatrix} \tag{29} $$

Where:

$$ n_{1x} = (c_1c_2c_3 - c_1s_2s_3)c_4 + (-c_1c_2s_3 - c_1s_2c_3)s_4 \tag{30} $$

$$ n_{1y} = (s_1c_2c_3 - s_1s_2s_3)c_4 + (-s_1c_2s_3 - s_1s_2c_3)s_4 \tag{31} $$

$$ n_{1z} = (s_2c_3 + c_2s_3)c_4 + (-s_2s_3 + c_2c_3)s_4 \tag{32} $$

$$ s_{1x} = -(c_1c_2c_3 - c_1s_2s_3)s_4 + (-c_1c_2s_3 - c_1s_2c_3)c_4 \tag{33} $$

$$ s_{1y} = -(s_1c_2c_3 - s_1s_2s_3)s_4 + (-s_1c_2s_3 - s_1s_2c_3)c_4 \tag{34} $$

$$ s_{1z} = -(s_2c_3 + c_2s_3)s_4 + (-s_2s_3 + c_2c_3)c_4 \tag{35} $$

$$ a_{1x} = s_1 \tag{36} $$

$$ a_{1y} = -c_1 \tag{37} $$

$$ a_{1z} = 0 \tag{38} $$

$$ P_{1x} = (c_1c_2c_3 - c_1s_2s_3)c_4L_3 + (-c_1c_2s_3 - c_1s_2c_3)s_4L_3 + (c_1c_2c_3 - c_1s_2s_3)L_2 + c_1c_2L_1 \tag{39} $$

$$ P_{1y} = (s_1c_2c_3 - s_1s_2s_3)c_4L_3 + (-s_1c_2s_3 - s_1s_2c_3)s_4L_3 + (s_1c_2c_3 - s_1s_2s_3)L_2 + s_1c_2L_1 \tag{40} $$

$$ P_{1z} = (s_2c_3 + c_2s_3)c_4L_3 + (-s_2s_3 + c_2c_3)s_4L_3 + (s_2c_3 + c_2s_3)L_2 + s_2L_1 \tag{41} $$

2.2 Inverse Kinematics

The model of the two finger types described in the forward kinematics section can generate a combination of movement with flexions, extensions, abductions, adductions and redundant cases. Figure 2 shows examples of flexion, extension and abductions.

The solution of the inverse kinematics can be derived from geometric methods (J.M. Selig, 2005), such as the relation of triangles. The hand can reproduce positive or negative movements with regard to a reference line for some joints. The movements of the fingers that can be in two different quadrants are: flexion/extension and abduction/adduction. The inverse kinematics is solved in all these cases of movement. In addition, the kinematic behaviour of the kinematic chain depends on the MCP abduction/adduction for the model of the index, middle, ring and little fingers, because this joint can cause situations such as redundancy (Figure 3) if the value of $\theta_{i,MCP\_aa} = 0$. Also, this joint can produce a more complex situation if the adduction is high. The most stable behaviour of the kinematic chain is when there are only abductions.

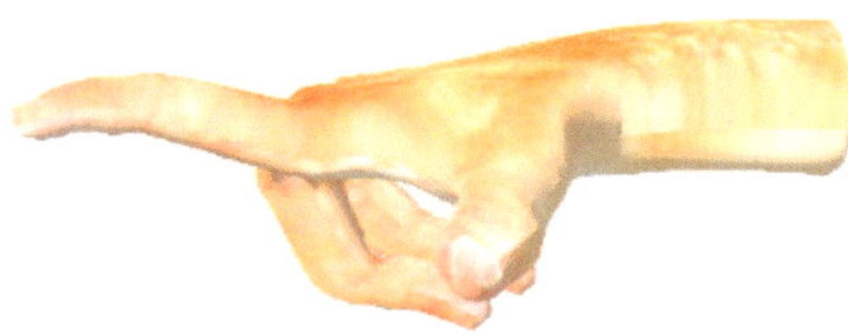

a)

b)

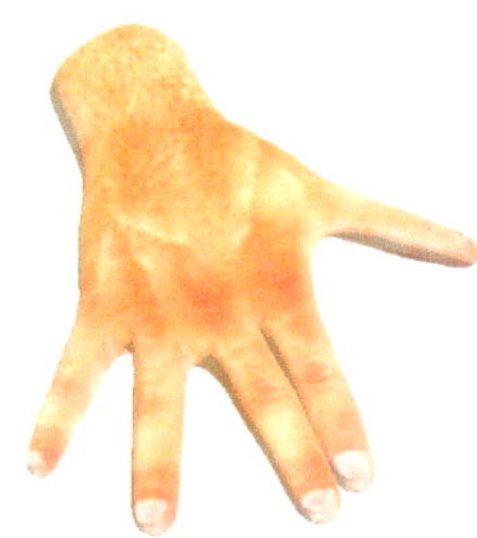

c)

Fig. 2. a) Extension in the MCP joint of the index finger b) Flexion in all joints c) Index and ring fingers show abduction while index and middle finger show abduction in the MCP joint.

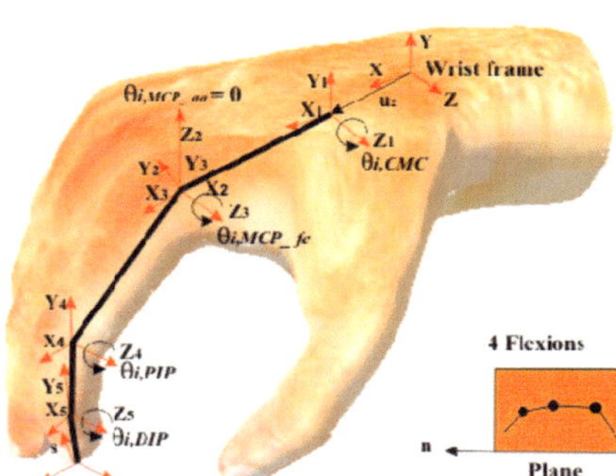

Fig. 3. Example of a redundant case of the index finger model when $\theta_{i,MCP\_aa} = 0$.

When $\theta_{i,MCP\_aa} = 0$, the model of the human hand is a redundant case and several solutions therefore exist. In this type of situation, the inverse kinematics can be solved by means of iterative methods such as the Newton–Raphson method that uses the Jacobean matrix (W.A.

Wolovich & H. Elliot, 1984) and (A. Balestino et al., 1984). To solve this redundant case correctly, constraints have been implemented to solve with a convergent solution.

2.2.1 Inverse kinematics for the index, middle, ring and little fingers

To solve the inverse problem for each finger, the first step to take is to measure the orientation and end position of the fingertip.

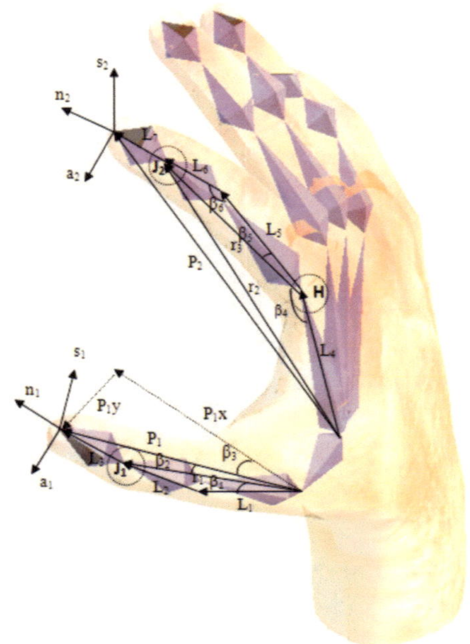

Fig. 4. Vectorial representation of the inverse kinematic model.

With this, it is possible to obtain a homogeneous matrix [n s a P]. Algebraically, for the first degree of freedom of the CMC joint the following is obtained:

$$\theta_{i,CMC} = a\tan\left[\frac{a_{2y}}{a_{2x}}\right] \tag{42}$$

In the same way, the value of the CMC joint was obtained algebraically, the MCP abduction/adduction joint as

$$\theta_{i,MCP\_aa} = a\tan\left[\frac{\dfrac{a_{2x}}{\cos(\theta_{i,CMC})}}{a_{2z}}\right] \tag{43}$$

Dependency exists among the following joints: flexions for MCP_*fe*, PIP and DIP. They are solved by means of geometric methods. With vector P_2 it is possible to obtain a point J_2 with the following expression:

$$J_2 = P_2 - [L_7 * n_2] \tag{44}$$

Another vector H is calculated with the CMC information and length L_4 of the metacarpal link such as:

$$\begin{gathered} H_x = L_4 * \cos(\theta_{i,CMC}) \\ H_y = L_4 * \sin(\theta_{i,CMC}) \\ H_z = 0 \\ H = [H_x \quad H_y \quad H_z]^T \end{gathered} \tag{45}$$

With information J_2 and H it is possible to obtain the vectors u, r_2 and r_3 as shown in Figure 4.

$$u = J_2 - H \tag{46}$$

$$r_2 = norm(J_2) \tag{47}$$

$$r_3 = norm(u) \tag{48}$$

With vectors u, r_2, r_3 and lengths L_4, L_5, L_6 and using the law of cosines the angles β_4, β_5 and β_6 are obtained.

$$\beta_4 = a\cos\left(\frac{L_4^2 + r_3^2 - r_2^2}{2L_4 r_3}\right) \tag{49}$$

$$\beta_5 = a\cos\left(\frac{L_5^2 + r_3^2 - L_6^2}{2L_5 r_3}\right) \tag{50}$$

$$\beta_6 = a\cos\left(\frac{L_6^2 + L_5^2 - r_3^2}{2L_6 L_5}\right) \tag{51}$$

MCP flexion/extension is obtained as:

$$\theta_{i,MCP\_fe} = \pi - \beta_5 - \beta_4 \tag{52}$$

The extension & hyperextension for the MCP joint is obtained as:

$$\theta_{i,MCP\_hyp} = \beta_4 - \pi - \beta_5 \tag{53}$$

The PIP joint is obtained as:

$$\theta_{i,PIP} = \pi - \beta_6 \tag{54}$$

Finally, the DIP joint is obtained by the algebraic method.

$$\begin{aligned} k_1 = {} & (c_2 c_4 c_3 c_1 - c_4 s_1 s_3 - c_2 s_4 c_1 s_3 - s_4 c_3 s_1) n_x + \\ & (c_2 c_4 c_3 s_1 + c_4 c_1 s_3 - c_2 s_4 s_1 s_3 + s_4 c_3 c_1) n_y + \\ & (s_4 s_3 - c_4 c_3) s_2) n_z \end{aligned} \tag{55}$$

$$\begin{aligned} k_2 = {} & (-s_4 c_2 c_1 c_3 + s_4 s_1 s_3 - c_4 c_1 c_2 s_3 - c_4 c_3 s_1) n_x + \\ & (-s_4 c_2 s_1 c_3 - s_4 c_1 s_3 - c_4 s_1 c_2 s_3 + c_4 c_3 c_1) n_y + \\ & (c_4 s_3 + s_4 c_3) s_2) n_z \end{aligned} \tag{56}$$

$$\theta_{i,DIP} = \text{atan2}[k_2, k_1] \tag{57}$$

2.2.2 Inverse kinematics of the thumb

For the thumb the inverse kinematic is obtained by means of algebraic method.

TMC_aa is obtained as:

$$\theta_{T,TMC\_aa} = a\tan\left[\frac{P_{1y}}{P_{1x}}\right] \tag{58}$$

J_1 and vectors r_1, x_1 and x_2 are obtained as:

$$\begin{aligned} J_1 &= P_1 - [L_3 n_1] \\ r_1 &= norm\ (J_1) \\ x_1 &= \frac{r_1 - L_1^2 - L_2^2}{2L_1L_2} \end{aligned} \tag{59}$$

$$x_2 = \sqrt{1 - x_1^2}$$

MCP is obtained as:

$$\theta_{T,MCP} = a \tan 2[x_2, x_1] \tag{60}$$

Auxiliary variables are calculated such as:

$$\begin{aligned} x_3 &= (L_1 + (L_2 x_1)) \\ x_4 &= L_2 x_2 \\ J_{xy} &= \sqrt{J_{1x}^2 + J_{1y}^2} \\ x_5 &= \frac{J_{1z} x_4 + J_{xy} x_3}{x_3^2 + x_4^2} \\ x_6 &= \frac{J_{1z} x_3 + J_{xy} x_4}{x_3^2 + x_4^2} \end{aligned} \tag{61}$$

TMC_*fe* is obtained as:

$$\theta_{T,TMC\_fe} = a \tan 2[x_6, x_5] \tag{62}$$

$$\theta = a \tan 2[n_{1z}, \sqrt{n_{1x}^2 + n_{1y}^2}] \tag{63}$$

Finally, IP is obtained as:

$$\theta_{T,IP} = \theta - \theta_{T,TMC\_fe} - \theta_{T,MCP} \tag{64}$$

3. Main Constraints of Finger Movements

Joint finger movements are limited to a specific range because of static constraints, intra-finger constraints and inter-finger constraints. Intra-finger and inter-finger constraints are often called kinematic constraints, and these are the ones responsible for producing natural movements both statically and kynematically. However, this range of movement is somewhat ambiguous because the range depends on various factors involving human hand biomechanics.

3.1 Inter-finger constraints

This type of constraint refers to some dependency among fingers while they are in motion (J. Lin et al., 2000). The inter-finger constraints were obtained by using the hand model and Cyberglove. These types of constraints are coupled movements among the index, middle, ring and little fingers. The relationship among angles with the middle, ring and little fingers has been measured to represent real movements of the hand model.

Middle, ring and little fingers share common flexor tendons. It implies involuntary movements due to this strong coupling. These cases are defined by:

- Flexion in the ring finger θ_{R,MCP_fe} due to a flexion in the middle finger and no flexions of the index and little fingers. When this unique flexion is generated, the flexion of the ring finger is equal to a percentage of the middle finger flexion, as described in equation (65).

$$\theta_{R,MCP\_fe} \approx \frac{2}{3}\theta_{M,MCP\_fe} \tag{65}$$

- Flexion in the ring finger θ_{R,MCP_fe} is produced when there is flexion solely in the little finger. It is described in equation (66). A small flexion is also produced in the middle when flexion in the little is high.

$$\theta_{R,MCP\_fe} \approx \frac{7}{12}\theta_{L,MCP\_fe} \tag{66}$$

- Flexion of middle and little fingers θ_{M,MCP_fe}, θ_{L,MCP_fe} are produced when there is flexion solely in the ring finger. Involuntary flexions are similar and proportional to the ring flexion as described in equation (67) and (68).

$$\theta_{R,MCP\_fe} - \theta_{M,MCP\_fe} < 60^{\circ} \tag{67}$$

$$\theta_{R,MCP\_fe} - \theta_{L,MCP\_fe} < 50^{\circ} \tag{68}$$

- Ring and little abduction/adductions are similar. In most cases is very difficult to change the abduction of a finger without changes in the other. Equation (69) represents this constraint.

$$\theta_{R,MCP\_aa} \approx \theta_{L,MCP\_aa} \tag{69}$$

- Finally, another involuntary movement appears in the middle flexion when a unique flexion is done in the index finger. This relation is described in equation (70).

$$\theta_{M,MCP\_fe} \approx \frac{1}{5}\theta_{I,MCP\_fe} \tag{70}$$

3.2 Intra-finger constraints

Intra-finger constraints refer to the dependency of internal joints. New intra-finger have been obtained according to the hand gesture. Gestures have been classified as prismatic and circular regardless of power or precision grasping (M.R. Cutkosky, 1989). Some experiments have been carried out in order to obtain these intra-finger relations.

Circular grasping implies a strong relation among finger joints. The most accepted intra-finger constraint is:

$$\theta_{DIP} \approx \frac{2}{3}\theta_{PIP} \tag{71}$$

It has been analyzed by several researchers, such as (C.S. Fahn & H. Sun, 2005) and has been efficiently checked in our experiments.

Equations (72) and (73) show two constraints used for index, middle, ring and little fingers in circular grasping.

$$\theta_{i,DIP} \approx \frac{2}{3}\theta_{i,PIP} \tag{72}$$

$$\theta_{i,MCP\_fe} \approx \frac{4}{3}\theta_{i,PIP} \tag{73}$$

The following equations show intra-finger constraints used for circular grasping:

$$\theta_{T,TMC\_fe} \approx \frac{11}{10}\theta_{T,MCP} \tag{74}$$

$$\theta_{T,MCP} \approx \frac{4}{5}\theta_{T,IP} \tag{75}$$

$$\theta_{I,CMC} \approx \theta_{M,CMC} \tag{76}$$

$$\theta_{M,CMC} \approx \frac{1}{2}\theta_{R,CMC} \tag{77}$$

$$\theta_{M,MCP\_aa} \approx \frac{1}{5}\theta_{I,MCP\_aa} \tag{78}$$

$$\theta_{R,CMC} \approx \frac{2}{3}\theta_{L,CMC} \tag{79}$$

$$\theta_{R,MCP\_aa} \approx \frac{1}{2}\theta_{L,MCP\_aa} \tag{80}$$

$$\theta_{R,DIP} \approx \frac{1}{2}(\theta_{M,DIP} + \theta_{L,DIP}) \tag{81}$$

Also, equations (76), (77), (78), (79) and (80) are used for prismatic trajectories. Equations (82) and (83) show two constraints used for index, middle, ring and little fingers in prismatic grasping.

$$\theta_{i,MCP\_fe} \approx \frac{3}{2}\theta_{i,PIP} \tag{82}$$

$$\theta_{i,PIP} \approx 2\theta_{i,DIP} \tag{83}$$

Finally, the following equations show intra-finger constraints for prismatic grasping.

$$\theta_{T,TMC\_fe} \approx \frac{10}{11}\theta_{T,MCP} \tag{84}$$

$$\theta_{T,MCP} \approx \frac{6}{5}\theta_{T,IP} \tag{85}$$

4. Simplified Human Hand Models

This section describes simplified human hand models that properly represent the kinematic behaviour of the human hand in accordance with the precision and application required.
The human hand model of 24 DoF is used as a basis for comparison among simplified hand models with fewer degrees of freedom than the 24 DoF of the hand model described in section 2. Kinematic constraints are used in order to obtain simplified hand models, which allow for reducing the number of independent variables or joints in the original model. In other words, with few independent variables is possible to reconstruct a gesture of 24 degrees of freedom with an acceptable error with respect to the original gesture of 24 degrees of freedom not reconstructed. In previous works, two simplified hand models with 6 and 9 DoF have been evaluated in (S. Cobos et al., 2008a) and (S. Cobos et al. 2008b). Simplified human hand models are obtained using dependent and independent variables, theses dependent variables or dependent joints are calculated using kinematic constraints such as are showed in equations (65) to (85).
Table 3 shows simplified hand models from 1 to 24 degrees of freedom, and the independent variables used by each simplified hand model. Models 1 to 6 DoF are appropriate for circular power grasps. To control a gesture with one degree of freedom has been demonstrated previously in a robotic hand e.g. The Tuat/Karlsruhe Hand (N. Fukaya et al., 2000) is designed with 24 joints and 1 DoF controlled, this type of device is able to do circular power grasps. In this category the models are capable of performing power grasps with security and stability in the grip without achieving a great precision and skill in handling for precision grasps. Greater precision and dexterity is derived from 9 degrees of freedom, thus is possible to carry out precision grasps. Simplified hand models of 9 to 13 DoF are more precise for both types of grasp: precision and power grasps. Finally, a higher level of realism and sensitivity is achieved with models from 15 to 24 DoF. Table 3, the model from 1 to 9 DoF shows all independent variables that are used. From the model with 10 to 24 DoF uses the same variables from the previous model plus an additional variable e.g. the model with 10 DoF uses the same variables of the model with 9 DoF plus an additional variable.
The reduction of elements from 13 to 1 DoF leads to increasingly rely on interpolations and constraints associated with an increased error in the grip trajectory when the dependent variables are obtained. Somehow, this technique depends on optimizing the functionality of a particular inter-finger or intra-finger constraint.

Simplified hands (Num. DoF)	Independent Joints								
1	$\theta_{I,DIP}$								
2	$\theta_{I,DIP}$	θ_{T,TMC_aa}							
3	$\theta_{I,DIP}$	θ_{T,TMC_aa}	$\theta_{T,IP}$						
4	$\theta_{I,DIP}$	θ_{T,TMC_aa}	$\theta_{T,IP}$	θ_{I,MCP_aa}					
5	$\theta_{I,DIP}$	θ_{T,TMC_aa}	$\theta_{T,IP}$	θ_{I,MCP_aa}	θ_{L,MCP_aa}				
6	$\theta_{I,DIP}$	θ_{T,TMC_aa}	$\theta_{T,IP}$	θ_{I,MCP_aa}	θ_{L,MCP_aa}	$\theta_{L,CMC}$			
7	$\theta_{I,DIP}$	θ_{T,TMC_aa}	$\theta_{T,IP}$	θ_{I,MCP_aa}	θ_{L,MCP_aa}	$\theta_{L,CMC}$	$\theta_{R,DIP}$		
8	$\theta_{I,DIP}$	θ_{T,TMC_aa}	$\theta_{T,IP}$	θ_{I,MCP_aa}	θ_{L,MCP_aa}	$\theta_{L,CMC}$	$\theta_{R,DIP}$	$\theta_{L,DIP}$	
9	$\theta_{I,DIP}$	θ_{T,TMC_aa}	$\theta_{T,IP}$	θ_{I,MCP_aa}	θ_{L,MCP_aa}	$\theta_{L,CMC}$	$\theta_{R,DIP}$	$\theta_{L,DIP}$	$\theta_{M,DIP}$
10	Simplified hand with 9 DoF + $\theta_{I,PIP}$								
11	Simplified hand with 10 DoF + $\theta_{R,PIP}$								
12	Simplified hand with 11 DoF + $\theta_{L,PIP}$								
13	Simplified hand with 12 DoF + $\theta_{M,PIP}$								
14	Simplified hand with 13 DoF + $\theta_{T,MCP}$								
15	Simplified hand with 14 DoF + $\theta_{I,MCP}$								
16	Simplified hand with 15 DoF + $\theta_{L,MCP}$								
17	Simplified hand with 16 DoF + $\theta_{R,MCP}$								
18	Simplified hand with 17 DoF + $\theta_{M,MCP}$								
19	Simplified hand with 18 DoF + θ_{R,MCP_aa}								
20	Simplified hand with 19 DoF + θ_{T,TMC_fe}								
21	Simplified hand with 20 DoF + $\theta_{R,CMC}$								
22	Simplified hand with 21 DoF + $\theta_{M,CMC}$								
23	Simplified hand with 22 DoF + $\theta_{I,CMC}$								
24	Simplified hand with 23 DoF + θ_{M,MCP_aa} (Original human hand model)								

Table 3. Simplified hand models reconstructed with kinematic constraints.

Many manipulations involve similar movements among fingers, e.g., a gesture done with the information of five fingers can be simplified by using only the information provided for three fingers, in this case these three fingers can be thumb, index and ring, creating the same

movement for the middle finger through the information of the index and the little finger through the information of the ring finger. The simplified hand models should be used depending on the relation between the number of degrees of freedom and the allowed error in the application.
The degree of dexterity that can be achieved depends largely on the largest number of independent variables having thumb and index finger inside the SHH. The abduction of the thumb and index fingers are very important because at least one degree of freedom from these fingers are considered as independent variables in all the simplified hand models, thus the flexion of the DIP joint of the index finger is included in all the simplified hand models. The abduction/adduction of the thumb TMC joint is important because with a flexion of the IP joint can produce the opposition of the thumb with the other fingers. It's from SHH with 4 DoF where the abduction of the MCP joint takes importance because allows better positioning of the finger, not SHH with 3 DoF that contain just a flexion in the index finger. In summary the universal joints of the thumb and index fingers are very important for obtaining simplified hand models because of the information they provide.

4.1 Error of simplified hand models

The original posture of the 24 DoF is considered the "ideal posture" or "ideal trajectory" to determine the final position of the fingertip with the use of the forward kinematics; the same forward kinematics is used to obtain the final position with the reconstructed vectors. The relative error (δ_i) is obtained with the original position (op_i) and the reconstructed position (rp_i) such as:

$$\delta_i = \frac{|rp_i - op_i|}{|rp_i|} \times 100\ \% \tag{86}$$

The trajectory is conformed of n positions; thus the average error for each finger is:

$$\overline{\delta_i} = \frac{\sum_{k=1}^{n} \delta_i}{n} \tag{87}$$

Where n is the number of positions and i = Thumb, Index, Middle, Ring, Little.
Finally, the reconstruction error is calculated using the following expression.

$$error = \frac{\sum_{k=1}^{n} \overline{\delta_i}}{5} + \Delta c \tag{88}$$

Where Δc is a parameter of calibration this can vary 1-2% among users according to their hand size.

5. Conclusion

In this chapter, the forward and inverse kinematic models of the human hand with 24 DoF have been presented in detail. Thanks to the kinematic model and with the use of the cyberglove several kinematic constraints were developed for consequently obtained simplified human hand models. This simplified human hand models are useful for diverse applications that require security, stability, sensitivity, dexterity, realism or velocity in calculate kinematic parameters.
These simplified human hand models represent a significant reduction in the number of independent variables needed to describe a hand gesture, taking into account the opportunity to perform a specific manipulation with fewer elements to control for applications where control with many degrees of freedom is complex or computationally expensive. Finally, the human hand model with 24 DoF serves for applications that requires a greater realism, sensitivity in handling or description of a human hand gesture.

6. References

I. A. Kapandji. (1970). *The Physiology of the Joints*, volume 1. E&S Livingstone, Edinburgh and London, 2 editions.

I. A. Kapandji. (1981) *The hand. Biomechanics of the thumb.* In R. Tubiana (Ed.), pp. 404 - 422, Philadelphia: W. B. Saunders.

K.S. Salisbury and B.Roth. (1983). Kinematics and force analysis of articulated mechanical hands, *Journal of Mechanims, Transmissions and Actuation in Design*, 105, pp. 35 - 41.

S.C. Jacobsen, E. K. Iversen, D. F. Knutti, R. T. Johnson and K. B. Biggers. (1986). Design of the Utah/MIT Dexterous Hand, *In Proc. IEEE International conference on robotics and automation*, pp. 1520 - 1532.

T. Okada. (1982). Computer control of multijointed finger system for precise object handling, *IEEE Transactions on systems, Man, and Cybernetics, Vol. Smc-12, No. 3*, pp. 289 - 299

G. Bekey, R. Tomovic, and I. Zeljkovic. (1990), Control architecture for the Belgrade/usc hand, in T. I. S.T. Venkataraman (ed.), *Dexterous Robot Hands*, Springer-Verlag.

C. Melchiorri and G. Vassura. (1992). Mechanical and control features of the UB Hand Version II, *IEEE-RSJ Int. Conf. on intelligent Robots and Systems*, IROS'92, Raleigh, NC,1192

C. Melchiorri and G. Vassura. (1993). Mechanical and Control Issues for Integration of an Arm-Hand Robotic System, in *Experimental Robotics II, the 2nd Int. Symposium* Raja Chatile and Gerd Hirzinger Eds., Springer Verlag.

J. Butterfass, G. Hirzinger, S. Knoch and H. Liu. (1998). DLR's Multisensory articulated Part I: Hard- and Software Architecture, *in Proc. IEEE International Conference on Robotics and Automation*, pp. 2081 - 2086.

J. Butterfass, M. Grebenstein, H. Liu and G. Hirzinger. (2001). DLR-Hand II: Next Generation of a Dextrous Robot Hand, *Proc. IEEE Int. Conf. on Robotics and Automation*, Seoul, Korea, pp. 109 - 114.

Y. K. Lee and I. Shimoyama. (1999). A skeletal Framework Artificial Hand Actuated by Pneumatic Artificial Muscles, *IEEE Int. Conf. On Robotics & Automation*, Detroit, Michigan, pp. 926 - 931.

W.T. Townsend. (2000). MCB - Industrial robot feature article-Barrett Hand grasper, in *Industrial Robot: An International Journal* Vol.27 No. 3 pp.181 - 188.

C.S. Lovchik and M.A. Diftler. (1999). The Robonaut Hand: a Dexterous Robot Hand for Space, *Proc. IEEE International Conference on Robotics and Automation*, pp. 907 - 912.

S. Schulz, C. Pylatiuk and G. Bretthauer. (2001). A new ultralight anthropomorphic hand, *Proc. IEEE Int. Conf. on Robotics and Automation*, Seoul, Korea, vol.3, pp. 2437 - 2441.

H. Kawasaki, H. Shimomura and Y. Shimizu. (2001). Educational-industrial complex development of an anthropomorphic robot hand 'Gifu hand, *Advanced Robotics*, Vol. 15, No. 3, pp. 357 - 363.

Shadow Robot Company, "The Shadow Dextrous Hand." http://www.shadow.org.uk/

F. Röthling, R. Haschke, J. J. Steil, and H. Ritter. (2007). Platform Portable Anthropomorphic Grasping with the Bielefeld 20-DOF Shadow and 9-DOF TUM Hand. *In Proc. IEEE/RSJ International Conference on Intelligent Robots and Systems*, pp. 2951 - 2956.

H. Yokoi, A. Hernandez , R. Katoh, W. Yu, I. Watanabe, and M. Maruishi. (2004). Mutual Adaptation in a Prosthetics Application. In F. Iida et al. (Eds.): *Embodied Artificial Intelligence*, LNAI 3139, Springer Verlag, pp. 146 - 159.

H. Liu, P. Meusel, G. Hirzinger, M. Jin, Y. Liu, and Z. Xie. (2008). The Modular Multisensory DLR-HIT-Hand: Hardware and Software Architecture, *IEEE/ASME Transactions on mechatronics*, Vol. 13, No. 4, pp. 461 - 469.

Immersion corporation webpage. http://www.immersion.com/

M.W. Spong, S. Hutchinson, M. Vidyasagar. (2006). *Robot Modeling and control*. John Whiley & sons.

J.M. Selig. (2005). Geometric fundamentals of robotics. *Monographs in computer science. Springer*, pp. 85 - 112.

W. A. Wolovich and H. Elliot. (1984). A computational technique for inverse kinematics. *In Proc. 23rd IEEE Conference on Decision and Control*, pp. 1359 - 1363.

A. Balestrino, G. De Maria, and L. Sciavicco. (1984). Robust Control of robotic manipulators, *In Proc. of the 9th IFAC World Congress*, pp. 2435 - 2440.

J. Lin, Y. Wu, and T.S. Huang. (2000). Modeling the Constraints of Human Hand Motion, *IEEE Human Motion Proceedings*, pp. 121 - 126.

M.R. Cutkosky. (1989). On grasp choice, grasp models, and the design of hands for manufacturing tasks. *IEEE trans. Robotics and automation*, pp. 269 - 279.

C.S. Fahn and H. Sun. (2005). Development of a Data Glove with Reducing Sensors Based on Magnetic Induction. *IEEE Transactions on Industrial Electronics*, vol. 52, No.2, pp. 585 -594

S. Cobos, M. Ferre, M.A. Sanchéz-Urán, J. Ortego and C. Peña. (2008a). Efficient Human Hand Kinematics for manipulation Task. *IEEE/RSJ International conference on intelligent Robots and Systems*, pp. 2246 - 2250.

S. Cobos, M. Ferre, M.A. Sanchéz-Urán, J. Ortego. (2008b). Simplified Hand configuration for object manipulation. *In Proc. Eurohaptics 2008*, pp. 730 - 735.

N. Fukaya, S. Toyama, T. Asfour and R. Dillmann. (2000). Design of the TUAT/Karlsruhe Humanoid Hand. *In Proc. IEEE/RSJ International Conference on Intelligent Robots and Systems*, pp. 1754 - 1759.

An Impact Motion Generation Support Software

Teppei Tsujita, Atsushi Konno, Yuki Nomura,
Shunsuke Komizunai, Yasar Ayaz and Masaru Uchiyama
Tohoku University
Japan

1. Introduction

When a robot applies force statically on a target object, magnitude of the force is limited by the maximum force or torque of the actuators. In order to exert a large force on the target beyond this limitation, it is effective to apply impulsive force. We describe the motions that perform tasks by applying impulsive force as "impact motion." There are difficult problems introduced by impacts between a robot and a target.
Uchiyama proposed a control algorithm constitution method and dynamic control modes for performing a nailing task by a 3 DOF manipulator (Uchiyama, 1975). Zheng and Hemami discussed mathematical modelling of a robot that collides with the environment (Zheng & Hemami, 1985). Asada and Ogawa proposed the *virtual mass* for analyzing dynamic behaviour of a manipulator arm and its end effecter that interacts with the environment (Asada & Ogawa, 1987). Around the same time, Khatib and Burdick proposed the effective mass (Khatib & Burdick, 1986). Walker investigated the effect of different configurations of kinematically redundant arms with impact force at their end effectors during contact (Walker, 1994). These works mentioned above used rigid robotic manipulators fixed on the ground.
Yoshida and Sashida investigated impact dynamics in free-floating multibody systems in space (Yoshida & Sashida, 1993). Lew discussed about contact force control of a long-reach flexible micro/macro manipulator (Lew, 1997). These studies focused on trying to minimize the impulsive force since the force causes fatal problems in a space robot or a flexible arm.
A few attempts on tasks applying impulsive force by a humanoid robot have been reported in recent years. Arisumi et al. discussed a motion generation method for dynamic lifting by a humanoid robot based on a planar model (Arisumi et al., 2007). The strategy for lifting is based on centre of percussion for maintaining stability.
The main goal of our research is to develop a scheme to generate an optimal humanoid robot's impact motion for a given task considering multibody dynamics. To effectively generate impact motion, impact motion generation software is developed as the first step of the impact motion research.
The developed impact motion generation support software visualizes not only a designed motion but also an experimented motion. Force and torque measured in experiments are visualized on the experimented motion. The visualized ZMP (Zero-Moment Point) (Vukobratović et al., 1990), GCoM (Ground Projection of the Center of Mass), force, moment

and difference between the designed motion and the experimental result will help a motion designer improve the designed impact motion. Nailing task is taken as examples of an impact motion. A motion of a humanoid robot to drive a nail into a wooden block with a hammer is designed using the developed software. Furthermore, the software reveals a situation at the impact. The details of the software and the nailing experiment performed by the humanoid robot are presented.

2. Features of the Impact Motion Generation Support Software

In order to generate whole body motion, many motion design software are developed. Yamane and Nakamura developed an interface for creating whole body motions for human and animal characters without reference motion (Yamane & Nakamura, 2002). By dragging a link to an arbitrary position, a whole body motion can be generated intuitively. Nakaoka et al. developed a software platform for motion composition and generated robot performances of traditional folk dances based on human performances (Nakaoka et al., 2004). These software are mainly for generation natural motions which are slimier to human motions. Therefore, it is significant to design postures. However, contact velocity is significant for an impact motion in addition to postures. Thus, the developed software can design a posture and joint velocities. The details of the software are described in this section.

2.1 System Configuration

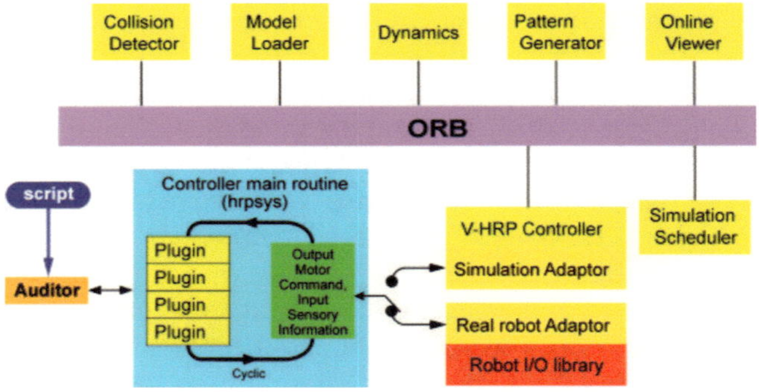

Fig. 1. Control system software of HRP-2 with OpenHRP (the figure is quoted from http://www.generalrobotix.com/product/openhrp/products.htm)

The impact motion generation support software is designed for a humanoid robot HRP-2 (Kaneko et al., 2004) and a humanoid robotics platform OpenHRP (Kanehiro et al., 2004). HRP-2 has 30 DOF (Degrees Of Freedom). The structure of the hands of HRP-2 used in this work is modified from that of the original HRP-2 to realize a natural swing down arm motion. The detail of the hand is described in Section 3.1. The control system software *hrpsys* and its GUI client software *Auditor* are supplied and supported by General Robotics, Inc.
As shown in Figure 1, *hrpsys* is shared by HRP-2 and a dynamics simulation server. Therefore, users are able to alternate between a real robot and a simulation transparently. The impact motion generation software can be used as an add-on application of the control

system software. Figure 2 (a) shows the relationship between the generation software and the control system software.

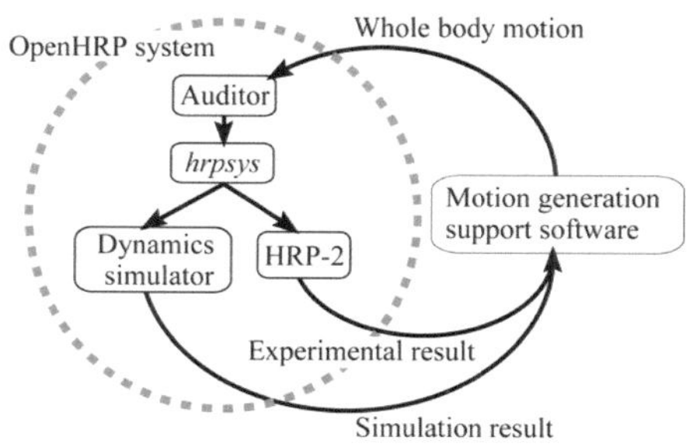

(a) Communication diagram for OpenHRP

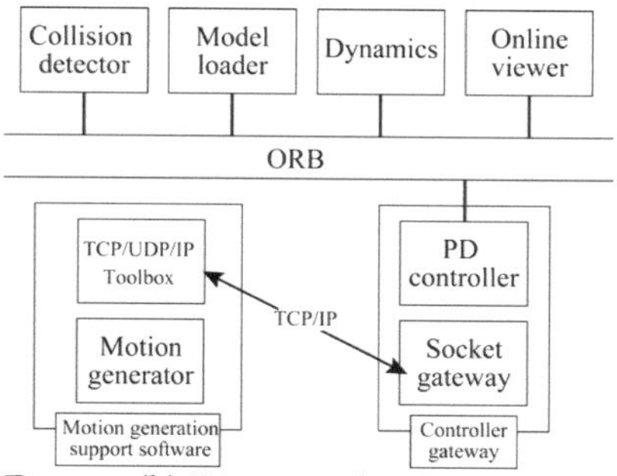

(b) Diagram for OpenHRP3

Fig. 2. A relationship between the developed software and the control system software.

The developed software consists of two main functions as follows.

- Designing the impact motion in a heuristic way
- Supporting the impact motion analysis visually

The motion design function generates a whole body motion of HRP-2. The generated motion is sent to *hrpsys* via *Auditor* for a real robot experiment or simulation. In order to analyze the motion, the results, e.g., force/torque sensor data at the wrists and ankles and measured joint angles, of the experiment or simulation are sent back to the developed software. The software is developed by using the technical computing language MATLAB R2007b (The MathWorks, Inc.). The details of these functions are described in Section 2.2 and 2.3.

2.2 Motion Design

In order to design an impact motion, the impact motion generation support software computes forward kinematics, inverse kinematics, position of the centre of mass, velocity of all links, momentum and angular momentum, GCoM (Ground Projection of the Center of Mass), ZMP (Zero-Moment Point) and a boundary of the possible region of ZMP are also computed as a measure of postural stability of the robot. The developed software is derivative of the Kajita's MATLAB toolbox for a humanoid robot (available from http://www.ohmsha.co.jp/data/link/4-274-20058-2/) and the MATLAB toolbox for space and mobile robots SpaceDyn (Yoshida, 1999) is referred for writing its code. The link parameters, i.e., lengths of the each link, rotation axes of the each joint, inertia matrices, masses and positions of the centre of the mass, are obtained from a VRML (Virtual Reality Modelling Language) format file of HRP-2 supplied by General Robotics, Inc.

Motion is designed by deciding initial and final points of the end effecter or joint angles. These descriptions are written by MALTAB script code. A designed motion is previewed with a GUI viewer. Figure 3 shows an overview of the developed software. In order to draw a HRP-2 model in a motion previewer window and obtain the parameters of the VRML format file, Virtual Reality Toolbox (The MathWorks, Inc.) is used. Virtual Reality Toolbox is a plug-in software of MATLAB that supplies functions to draw virtual reality graphics, enabling MATLAB to control the 3D model. The designed motion is saved as a motion sequence of joints in text format. Real robot experiments or dynamics simulations are executed by loading the file on Auditor. In addition, in order to evaluate a designed motion in the full-featured dynamics simulation environment quickly, the developed software can send the joint trajectories to OpenHRP3 (Nakaoka et al., 2007) directly via TCP/IP socket communication as shown in Figure 2 (b). Simulation results, i.e., joint torques, force/moment and ZMP, are stored in MATLAB workspace directly. To handle the socket communication between the developed software and OpenHRP3, TCP/UDP/IP Toolbox (available from http://www.mathworks.com/matlabcentral/fileexchange/345) is used.

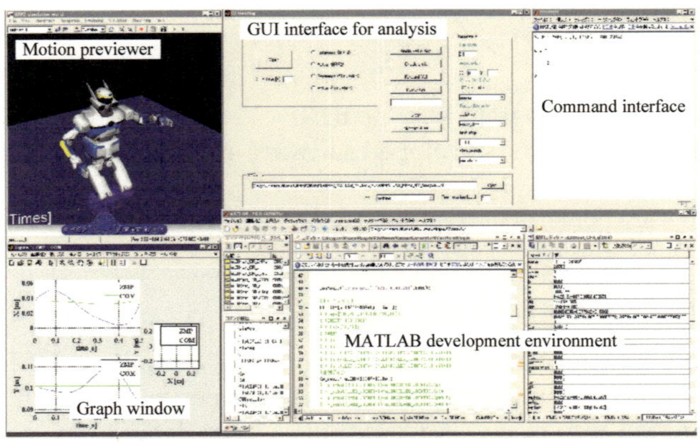

Fig. 3. An overview of the developed impact motion generation support software.

2.3 Analysis Support

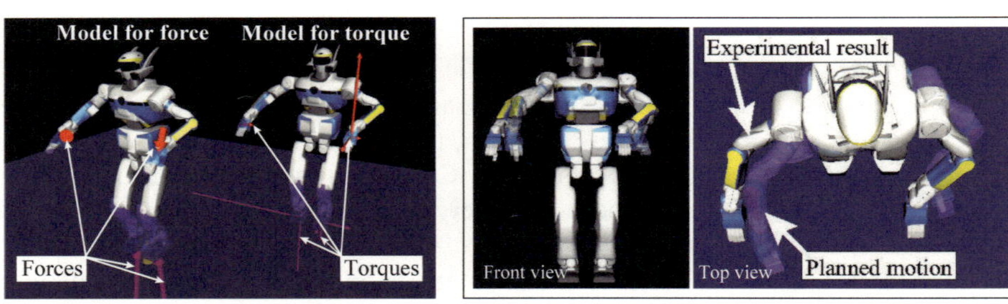

(a) Force/torque analysis (b) Motion analysis

Fig. 4. Examples of force/torque and motion analysis.

It is difficult to extract significant value from enormous amount of experiment/simulation data. For instance, HRP-2 has 30 optical encoders, four force/torque sensors, a gyro sensor and an acceleration sensor (Kaneko et al., 2004). In order to clearly show what happens in an experimentation/simulation, the generation software visualizes the force/torque data, ZMP, GCoM and difference between the designed motion and the experimental result. The resultant sensor data are loaded from a log file recorded by a logging software of HRP-2 or the dynamics simulator using a GUI (Graphical User Interface) control panel.

Figure 4 (a) shows force and torque displayed on CG models of HRP-2. In the figure, the arrows indicate forces and torques measured by each force/torque sensor. In order to avoid confusing the force arrows with the torque ones, two HRP-2 models are displayed in the same viewer. The left model is for displaying forces, the right model is for torques. The direction and length of the arrow displayed on the left model indicates the direction and magnitude of the applied force, respectively. The pointed tops of the arrows are the position of the force sensors mounted on the wrists and the ankles. The torques are expressed by the tree orthogonal arrows. The length of the arrow indicates the magnitude of the torque around the axis. The direction of the arrow indicates rotation direction using the right-handed screw definition.

Playback speed of the reproduced motion can be variable. Slow motion is useful to understand what happens to the robot during impact phase. Due to the dynamic property of the robot, the robot may not track the designed trajectory precisely. Furthermore, end effecter position is geometrically constrained at a point of collision with its environment. In order to understand the difference between the designed motion and the resultant motion intuitively, the impact motion generation support software visualizes the two motions. Figure 4 (b) shows the motion analysis viewer. In the viewer, the transparent and solid models are displayed to compare two motions. The transparent and solid models correspond to the designed motion and the resultant motion, respectively.

2.4 An example of motion generation

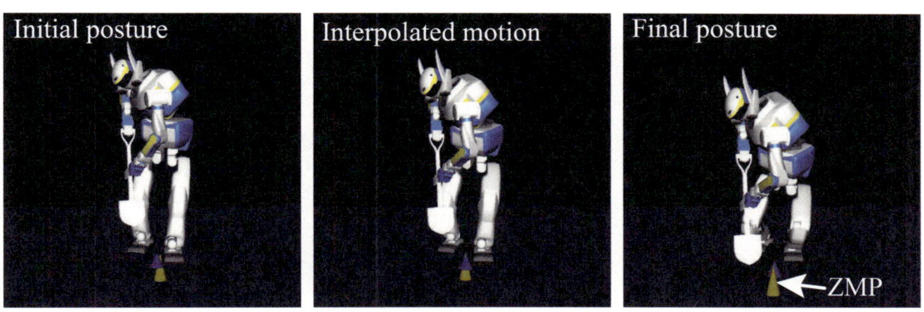

(a) GUI viewer

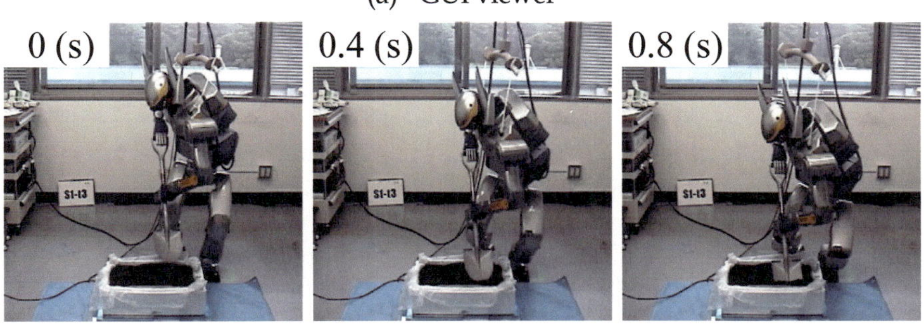

(b) Experiment

Fig. 5. HRP-2 digs a shovel blade into the dirt.

Figure 5 shows an example of motion designed by using the developed software. The motion is designed as follows.

1. The initial position and orientation of the shovel and the waist are decided. The developed software computes the joint angles of the arms and the legs by solving inverse kinematics.
2. The chest joint angles are decided and a whole posture is generated by combining the arms, legs and chest joint angles.
3. The final posture is designed in the same manner.
4. The joint angles are interpolated considering the kinematics closures.

The motion can be improved without real robot experiments since ZMP position is displayed in the GUI viewer.

3. Preliminary Nailing Motion Experiment

In this section, a nailing task is taken as a case study. A nailing task is one of valuable tasks in housework, construction works, and so on. In order to evaluate the effect of reaction force and understand collision situation, the external forces and torques are gauged in the nailing task experiment.

3.1 Experimental Setup

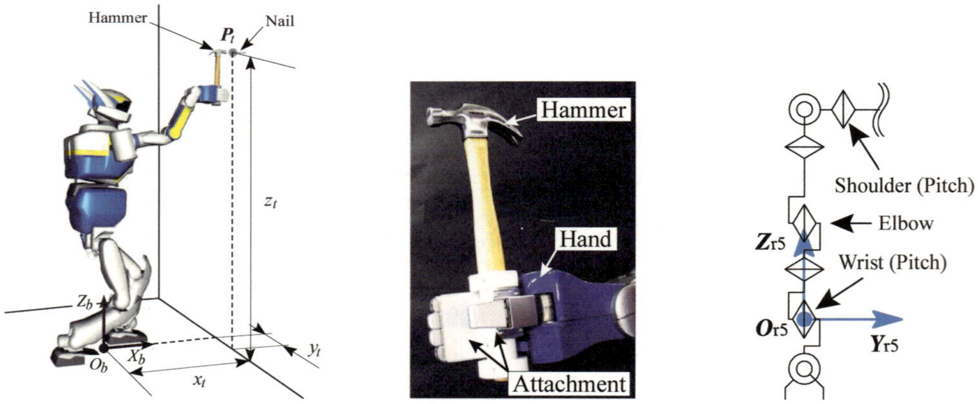

(a) Experimented nailing task (b) HRP-2 grasps a hammer (c) Coordinate notation

Fig. 6. HRP-2 grasps a hammer by its right hand and brings down the hammer to a nail.

The robot is placed to face a wall and to drive a nail over its head as shown in Figure 6 (a). The weight of the hammer is 0.44 kg. HRP-2 grasps the hammer as shown in Figure 6 (b). Figure 7 (a) shows an experimental setup for the nailing task. A wooden block is mounted on the base at a height of about 1.3 m. The wooden block is made with 10 balsa plates whose thickness is 5 mm. A nail is driven into the wooden block. As shown in Figure 7 (b), the dimensions of the nail are 2.5 mm diameter and 45 mm long. The head of the nail is 6 mm diameter.

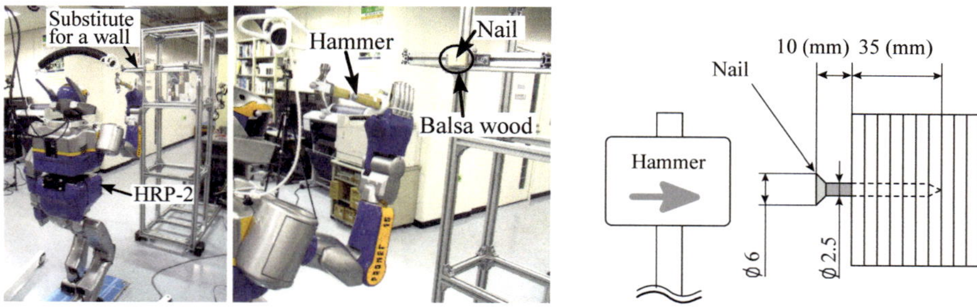

(a) An experimental setup (b) Condition of a nail and a wood block
Fig. 7. A target wooden block and a nail driven by HRP-2.

3.2 Nailing Task Motion Generation

In the preliminary motion generating method, two coordinate frames are defined:

Σ_b : A reference frame fixed on the floor

$(O_b,\ X_b,\ Y_b,\ Z_b)$,

Σ_{r5} :A frame fixed on the right wrist

$(O_{r5},\ X_{r5},\ Y_{r5},\ Z_{r5})$.

The reference frame Σ_b is defined as illustrated in Figure 6 (a). The forward and upward direction of the robot are defined as X_b and Z_b, respectively. Y_b is defined following the right-hand rule. The wrist frame Σ_{r5} is defined as illustrated in Figure 6 (c). In order to swing the hammer, pitch joints at the shoulder, elbow, and wrist are synchronously moved.

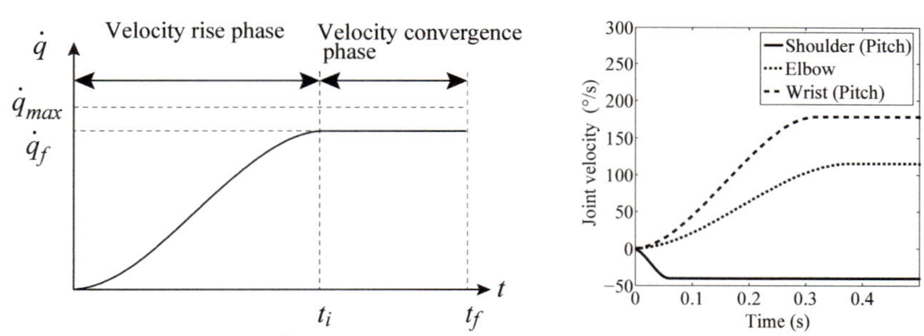

(a) Design scheme of joint velocities (b) Designed joint velocities
Fig. 8. Design of the shoulder, elbow and wrist joint angles.

Joint trajectories are designed in velocity domain considering the velocity limit as follows.

1. The joint velocity $\dot{q}_f$ at the impact, the displacement of the angle from initial angle q_s to final angle q_f and the travelling time t_f are given. The joint velocity $\dot{q}_s$ and acceleration $\ddot{q}_s$ at the initial position are also given.
2. As shown in Figure 8 (a), the joint velocity reference is divided into velocity rise phase and velocity convergence phase.
3. The end of the rise phase t_i is derived to satisfy the displacement of the angle.

4. The joint trajectory is obtained by integrating the angle velocity reference.

In step 1, $\dot{q}_f$ is set slightly under the maximum velocity $\dot{q}_{max}$. $\dot{\boldsymbol{q}}_f$ is a 6×1 vector of joint angle velocities of the right arm at the impact. $\dot{\boldsymbol{q}}_f$ is derived as follows.

$$\dot{\boldsymbol{q}}_f = \boldsymbol{J}_{rarm}^{-1} \begin{bmatrix} \boldsymbol{v}_f \\ \boldsymbol{\omega}_f \end{bmatrix}, \tag{1}$$

where $\boldsymbol{J}_{rarm}$, $\boldsymbol{v}_f$ and $\boldsymbol{\omega}_f$ are respectively jacobian matrix of the right arm, velocity and rotational velocity of O_{r5} with respect to Σ_b. $\dot{q}_s$ and $\ddot{q}_s$ are usually set to be zero.

In step 2, since the joint velocity becomes constant before the impact, the joint acceleration $\ddot{q}_f$ at the impact is zero and the joint velocity reaches $\dot{q}_f$ at t_i. During the velocity rise phase, the joint velocity is interpolated with cubic interpolation.

In step 3, t_i is derived so as to fulfil the following equation:

$$\int_0^{t_i} (a_0 + a_1 t + a_2 t^2 + a_3 t^3) dt + (t_f - t_i)\dot{q}_f = q_f - q_s, \tag{2}$$

where the coefficient a_0 and a_1 are zero, since the boundary condition of cubic interpolation must satisfy $\dot{q}_s = \ddot{q}_s = 0$. The coefficient a_2 and a_3 are functions of the time t_i and the velocity $\dot{q}_f$ as expressed by:

$$a_2 = \frac{3}{t_i^2}\dot{q}_f, \quad a_3 = -\frac{2}{t_i^3}\dot{q}_f. \tag{3}$$

Solving (2), a solution is obtained for t_i as follows:

$$t_i = \frac{2\{t_f \dot{q}_f - (q_f - q_s)\}}{t_f}. \tag{4}$$

In step 4, the joint angle q_t at the time t is obtained by integrating the angle velocity reference. If $t < t_i$, q_t is derived as follows:

$$q_t = \int_0^{t} \left(\frac{3}{t_i^2}\dot{q}_f t^2 - \frac{2}{t_i^3}\dot{q}_f t^3 \right) dt. \tag{5}$$

If $t \geq t_i$, q_t is derived as follows:

$$q_t = \int_0^{t_i} \left(\frac{3}{t_i^2}\dot{q}_f t^2 - \frac{2}{t_i^3}\dot{q}_f t^3 \right) dt + (t - t_i)\dot{q}_f. \tag{6}$$

3.3 Experiment

The point of driving a nail P_t is set at x_t = 0.53 m, y_t = 0.23 m and z_t = 1.5 m with respect to the reference frame Σ_b (Figure 6 (a)). The initial position of the point O_{r5} is set at x_s = 0.06 m, y_s = 0.23 m and z_s = 1.4 m with respect to the reference frame Σ_{r5}. The initial and final pose of Y_b are respectively set at −170 ° and −90 ° around the Y_b axis with respect to Σ_b. Velocity v_f and joint velocity w_f are set at $\begin{bmatrix} 0.3 & 0 & 0 \end{bmatrix}^T$ m/s and $\begin{bmatrix} 0 & 254 & 0 \end{bmatrix}^T$ °/s, respectively. The travelling time t_f is 0.5 s. From (1), the joint velocity $\dot{\boldsymbol{q}}_f$ is derived. Under this condition, the time t is obtained from (4). Consequently the joint trajectories are obtained from (5) and (6). Figure 8 (b) shows the designed joint velocities of the three joints. The speed of the hammer head in X_b direction is about 1.33 m/s at the impact. To swing the hammer back to the initial position, swing up motion is designed too. The swing up motion moves the hammer head from impact point to initial position. It is designed assuming that the velocity of the hammer is zero after collision and the duration of impact is instantaneous. The nail is driven into a depth of 35 mm in advance by a human operator. Then, HRP-2 drives the remaining 10 mm into the balsa block by the above-mentioned motion as shown in Figure 9. By repeating the same motion 13 times, HRP-2 completes driving the nail.

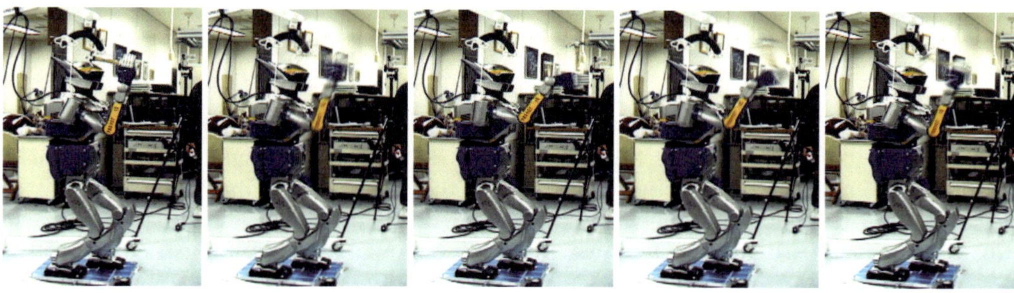

(a) 0 s (b) 0.33 s (c) 0.66 s (d) 0.99 s (e) 1.32 s

Fig. 9. HRP-2 drives a nail into a balsa wood block.

4. Analysis of Results and Problems

4.1 Motion

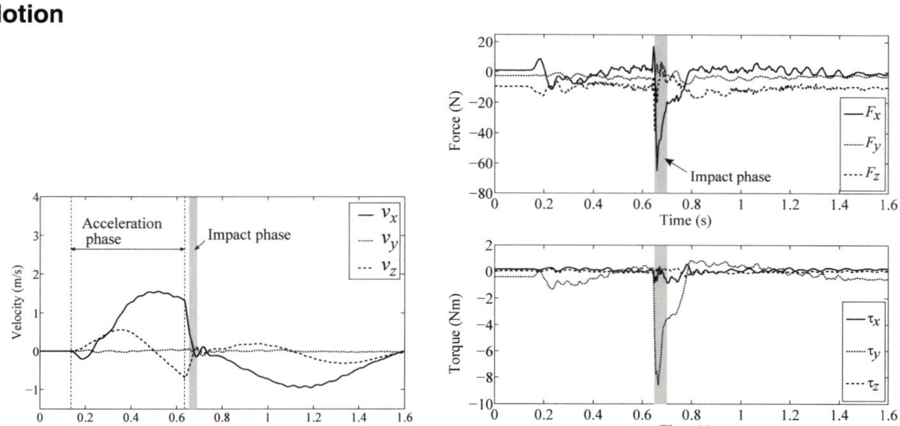

(a) Velocity of the hammer head (b) Force and torque of the right wrist

Fig. 10 Actual velocity of the hammer head computed by the joint angle velocities and force/torque measured by a sensor mounted on the right wrist.

Figure 10 (a) represents the velocity of the hammer head. The velocity is calculated by the measured joint angles and the dimensions of the hammer. At the impact (0.655 s), the hammer collides with a nail at 0.49 m/s in X_b direction. The speed is lower than expected.

The hammer trajectory is designed to collide with the nail 0.5 s after the beginning of the motion (about 0.14 s) and the speed is expected to be 1.33 m/s. This result suggests that the nail is not set in the correct place, the dimensions of the hammer are inaccurate, or the stabilizer slightly changes the trajectory of the hammer head.

4.2 Reaction Force

Figure 10 (b) shows the force and torque measured by the force/torque sensor mounted on the right wrist with respect to Σ_b. As shown in Figure 10 (b), peak of force in the negative direction of X_b is seen from 0.655 s to 0.700 s. The phase is considered as the impact phase. The duration of the impact is about 0.045 s.

Figure 11 shows the force and torque before and after contact. Before contact, large force can be seen in negative direction of Z_b. At the impact, a large force is produced in the negative direction of X_b. 5 ms after impact, peak force is measured. Experimental data show dynamic change of reaction force. Since the pulling up motion is designed assuming that the velocity of the hammer is zero after collision and the duration of impact is quite small, the acceleration is discontinuous between the swing down and up motions. Moreover, the impact happens later than expected. Therefore, it seems that the large force seen at 0.650 s is caused by inertial force. The force in X_b direction rapidly increases from 0.655 s to 0.660 s. The nail point compresses the balsa block during this phase and the balsa wood has elasticity, thus, the force increases gradually.

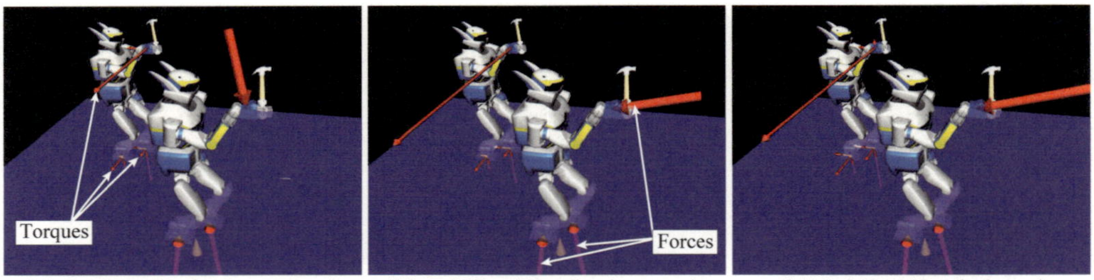

(a) 0.650 s (Before contact) (b) 0.655 s (At the impact) (c) 0.660 s (Peak force)

Fig. 11 Visualization of force and torque.

4.3 Stability

Figure 12 (a) shows ZMP and support polygon. The solid line, dotted line and gray box represent ZMP, support polygon and footprints, respectively. The ZMP has enough margins to the edge of the support polygon. Therefore it can be concluded that the motion is stable.

Figure 12 (b) shows trajectories of ZMP. The external force affects the ZMP. In this case, since large force in the negative direction of X_b and large torque around Y_b axis are measured, the position moves to the negative direction of X_b. However, during the impact phase, significant change of the position is not observed. After about 0.07 s of the collision, peak reaction force in X_b direction is measured (0.660 s), and the ZMP in X_b direction rapidly moves from about 0.05 m to 0.01 m. This change is thought to be the effect of the reaction force. To confirm the effect of the reaction force on stability, ZMP when the hammer does not collide with the nail are also shown in Figure 12 (b). The change of ZMP in X_b is observed after about 0.07 s of the expected moment of impact. However, the direction is opposite. This change is believed to be effect of inertial force, since the velocity of the hammer changes rapidly around the expected moment of impact. Since the deflection of ZMP at about 0.730 s depends on whether the hammer head collides with the nail or not, there is a time delay in relationships between reaction force and the stability and it is estimated to be about 0.07 s.

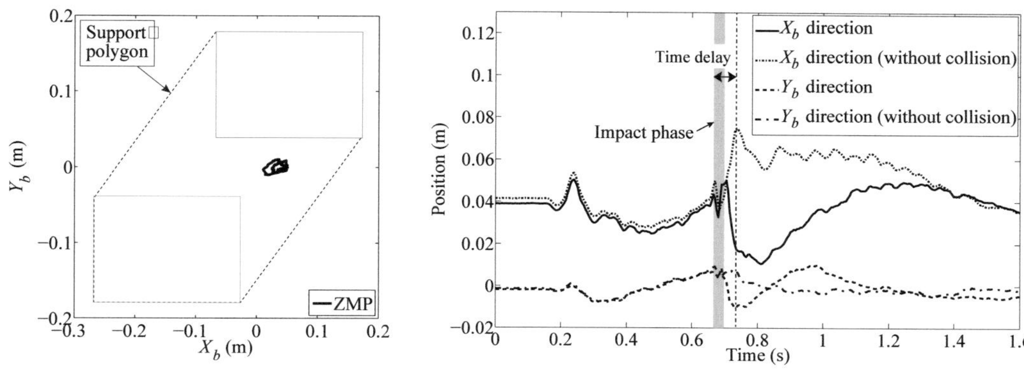

(a) ZMP and a support polygon (b) ZMP with and without collision
Fig. 12. Behaviour of ZMP during impact phase.

5. Conclusions

The detail of impact motion generation support software is presented in this paper. The developed software supports impact motion design with OpenHRP or OpenHRP3. A preliminary impact motion experiment is performed by a humanoid robot and the analyses of its result are presented. The analysis reveals that the designed motion is not robust against error in the position of the nail since the timing of pulling up the hammer is defined in the designed motion in advance. Therefore, a robust feedback control method for driving a nail is required.

The motions which are described in this paper are designed heuristically. However, the developed software can communicate with OpenHRP3 and use MATLAB toolboxes. As a next step, an impact motion will be optimized by using these features.

6. Acknowledgement

This research was supported by NEDO Industrial Technology Research Grant Program (project ID: 05A30703a) and JSPS Grant-in-Aid for JSPS Fellows (20•6273).

7. References

Arisumi, H.; Chardonnet, J.-R.; Kheddar, A. & Yokoi, K. (2007). Dynamic Lifting Motion of Humanoid Robots, *Proc. of Int. Conf. on Robotics and Automation*, pp. 2661-2667, ISBN 1-4244-0602-1, Roma, Italy, Apr., 2007, IEEE.

Asada, H. & Ogawa, K. (1987). On the dynamic analysis of a manipulator and its end effector interacting with the environment, *Proc. of IEEE Int. Conf. on Robotics and Automation*, pp. 751–756, ISBN 0-8186-0787-4, NC, USA, Mar.–Apr., 1987, IEEE.

Kanehiro, F.; Hirukawa, H. & Kajita, S. (2004). OpenHRP: Open architecture humanoid robotics platform, *Int. J. of Robotics Research*, Vol. 23, No. 2, pp. 155–165, ISSN 0278-3649.

Kaneko, K.; Kanehiro, F.; Kajita, S.; Hirukawa, H.; Kawasaki, T.; Hirata, M.; Akachi, K. & Isozumi, T. (2004). Humanoid robot HRP-2, *Proc. of IEEE Int. Conf. on Robotics and Automation*, pp. 1083–1090, ISBN 0-7803-8232-3, CA, USA, Apr., 2004, IEEE..

Khatib, O. & Burdick, J. (1986). Motion and force control of robot manipulators, *Proc. of IEEE Int. Conf. on Robotics and Automation*, pp. 1381–1386, ISBN 0–8186–0695–9, CA, USA, Apr., 1986, IEEE.

Nakaoka, S.; Nakazawa, A. & Ikeuchi, K. (2004). An Efficient Method for Composing Whole Body Motions of a Humanoid Robot, *Proc. of Int. Conf. on Virtual Systems and Multimedia*, pp. 1142-1151, ISBN 978-4274906343, Gifu, Japan, Nov., 2004, Ohmsha.

Nakaoka, S.; Hattori, S.; Kanehiro, F.; Kajita, S. & Hirukawa, H. (2007). Constraint-based Dynamics Simulator for Humanoid Robots with Shock Absorbing Mechanisms, *Proc. of IEEE/RSJ Int. Conf. on Intelligent Robots and Systems*, pp. 3641-3647, ISBN 978-1-4244-0912-9, CA, USA, Oct., 2007, IEEE.

Uchiyama, M. (1975). A control algorithm constitution method for artificial arm and dynamic control modes, In : *Biomechanism 3*, Society of Biomechanism Japan, pp. 172-181, University of Tokyo Press, ISBN 978-4-13-060069-9 (in Japanese).

Vukobratović, M.; Borovac, D.; Surla, D. & Stokić, D. (1990) *Biped Locomotion – Dynamics, Stability, Control and Application*, Springer-Verlag, ISBN 978-0387174563.

Walker, I. D. (1994). Impact configurations and measures for kinematically redundant and multiple armed robot systems, *IEEE Trans. on Robotics and Automation*, Vol. 10, No. 5, Oct., 1994, pp. 670–683, ISSN 1042–296X.

Yamane, K. & Nakamura, Y. (2002). Synergetic CG choreography through constraining and deconstraining at will, *Proc. of IEEE Int. Conf. on Robotics and Automation*, pp. 855–862, ISBN 0-7803-7272-7, DC, USA, May, 2002, IEEE.

Yoshida, K. & Sashida, N. (1993). Modeling of impact dynamics and impulse minimization for space robots, *Proc. of IEEE/RSJ Int. Conf. on Intelligent Robots and Systems*, pp. 2064–2069, ISBN 0-7803-0823-9, Yokohama, Japan, Jul., 1993, IEEE.

Yoshida, K. (1999). The SpaceDyn: a MATLAB toolbox for space and mobile robots, *Proc. of IEEE/RSJ Int. Conf. on Intelligent Robots and Systems*, pp. 1633–1638, ISBN 0-7803-5184-3, Kyongju, South Korea, Oct., 1999, IEEE.

Lew, J. Y. (1997). Contact control of flexible micro/macro-manipulators, *Proc. of IEEE Int. Conf. on Robotics and Automation*, pp. 2850–2855, ISBN 0-7803-3612-7, NM, USA, Apr., 1997, IEEE.

Zheng, Y. F. & Hemami, H. (1985). Mathematical modeling of a robot collision with its environment, *J. of Robotic Systems*, Vol. 2, No. 3, 1985, pp. 289-307, ISSN 0741-2223.

12

Peltier-Based Freeze-Thaw Connector for Waterborne Self-Assembly Systems*

Shuhei Miyashita, Flurin Casanova, Max Lungarella and Rolf Pfeifer
Artificial Intelligence Laboratory, University of Zurich
Switzerland

1. Introduction

Manufacturing technologies and industries heavily rely on robots. For macroscopic objects industrial robots are not only economical but are also reliable, fast, and accurate. Such robots, however, hit a barrier - entailing lower yields and higher fabrication costs - as the assembled objects become too small and complex. One potential solution to this problem is to exploit processes of self-assembly, that is, processes in which the interaction of pre-existing components leads to organized structures without human intervention. Such components could be, for instance, identical mechanical units (modules). Self-assembly is of crucial importance in the biological realm at all scales, e.g. for the formation of the protein shells of viruses and for protein folding.

By taking inspiration from nature, many inroads have already been made to realize self-assembly systems. For instance, the possibility of using self-assembly for the fabrication of structures from a given set components (potentially of nano- or micrometer scale) has been suggested by (Winfree *et al.* 1998; Rothemund 2006; Whitesides *et al.* 2001; Bowden *et al.* 2001; Mao *et al.* 2002; Boncheva *et al.* 2003). In the related field of self-reconfigurable systems, research effort has been devoted to realizing robots that can rearrange the connectivity of their structural units and create new topologies to accomplish a task (Mondada *et al.* 2005). Special attention has been paid to the design and construction of basic building blocks of a typically small repertoire, binding or docking interfaces allowing transfer of mechanical forces and moments, electrical power, and the sharing of information between the modules (Fukuda & Kawauch 1990; Chirikjian 1994; Murata *et al.* 1994; Murata *et al.* 1998; Murata *et al.* 1999; Yim 1994; Kotay *et al.* 1998; Rus & Vona 2001; Castano *et al.* 2002; Jorgensen *et al.* 2004; Zykov *et al.* 2005; Detweiler *et al.* 2006).

Most modular systems are "deterministically self-reconfigurable" implying that the exact location of the unit is known all the time. That is, the units are moved or directly manipulated into their target locations through deliberate active motion. In contrast to such systems, self-assembly systems are "stochastically self-reconfigurable" implying that (1) there are uncertainties in the knowledge of the modules' location (the location is known exactly only when the unit

* Parts of the material in this chapter previously appeared in; S. Miyashita, F. Casanova, M. Lungarella, and R. Pfeifer (2008), "Peltier-Based Freeze-Thaw Connector for Waterborne Self-Assembly Systems", Proc. IEEE Int. Conf. on Intelligent Robots and Systems, pp. 1325-1330.

docks to the main structure); and (2) the modules have only limited (or no) computational (deliberative) abilities. To date, a few self-reconfigurable modular robots relying on stochastic self-assembly have been built (White *et al.* 2004; White *et al.* 2005; Griffith *et al.* 2005; Shimizu *et al.* 2005; Bishop *et al.* 2005; Napp *et al.* 2006; Miyashita *et al.* 2008). Although in all these systems the units interact asynchronously and concurrently, a certain amount of state-based control is still required for the modules to move, communicate, and dock. Such docking/undocking is one of the main challenges towards the realization of self-assembly system (other two challenges are how to actuate the modules, so that they can move, and how to supply power to them).

In this chapter, we address the docking/undocking challenge by presenting the design and construction of a novel kind of connection mechanism. The connector works by freezing water close to it so that when another module is in its neighbourhood, the two modules stick to each other. We also show how this connector can be embedded in a stochastic modular robot. In the following Section 2, we provide a brief review of available connection mechanisms with a special emphasis on the ones used in the field of modular robotics. Then, in Section 3, we describe our Peltier-based freezing-thawing connection mechanism and validate its functioning as a connector. In Section 4, we describe the experiments of proposed connector by embedding it into a group of modular robots. This is followed by a discussion (Section 5), some pointers to future work and a brief conclusion (Section 6).

2. Connection mechanism

For a modular robot, the ability to attach to and detach from another module or to parts of the environment is of fundamental importance. With the connection mechanism described in this chapter we tackle the following problems that arise especially at smaller scales (<1 cm): (1) The actuation that is necessary for mechanical connectors is not easy to scale down. (2) The connection strength has to be sufficiently strong to fulfil the robot's purpose. (3) The precise alignment of the connector is crucial for a successful binding for some connection types. (4) The electromechanical complexity of the connector has to be small enough to allow for mass fabrication.

Table 1 lists various popular connection types for modular robots. In what follows, we will review some of them.

type	sub-type	main problem	reference
surface tension	-	weak bond / controllability	Boncheva 2003, Rothemund 2000, Hosokawa *et al.* 2005, Bowden *et al.* 1997, Chengde Mao *et al.* 2002
permanent magnets	-	detachment	Hosokawa *et al.* 1994
	+ electrical actuator	heavy weight	Bishop *et al.* 2005
	+ SMA spring	duration to detach	Murata *et al.* 1999
electrical magnets	-	strength-to-weight ratio	Murata 1994
	+ electrical actuator	heavy weight	Kotay *et al.* 1998, Zykov *et al.* 2005, Bhat *et al.* 2006
mechanical (hook, latch,	-	actuation / alignment	Penrose 1959

lock)	+ electrical actuator	heavy weight	Yim 1994, Rus *et al.* 2001, Jorgensen *et al.* 2004, Murata *et al.* 1998, Terada *et al.* 2004
	+ electrical magnets		Griffith 2005
	+ SMA		Fukuda *et al.* 1988, Castano *et al.* 2002
Velcro	-	detachment	
	+ pneumatic actuator	energy consumption	Shimizu *et al.* 2005
	+ electrical actuator	heavy weight	Moeckel *et al.* 2005
Peltier	-	heat dissipation	Walle *et al.* 2007

Table 1. Classes of connectors (μm - cm scale)

The exploitation of surface tension through the use of hydrophilic and hydrophobic materials provides a binding mechanism for modular robots in a fluidic environment that is often used for research on cm-scale stochastic self-assembly systems (Bowden *et al.* 2001; Mao *et al.* 2002; Boncheva *et al.* 2003; Hosokawa *et al.* 1996). The connection strength is weak compared to other mechanisms but is adequate because the modules are lightweight. Additional properties make this mechanism useful: (1) no power has to be provided for attachment and detachment; (2) the alignment is done by the connecting force itself; and (3) the connection mechanism is easy to produce.

Permanent magnets are a second type of popular connection mechanism for modular robots. They have many useful properties: (1) they do not require any power for binding; (2) the relatively strong attractive force eases the alignment problem; and (3) they are rather straightforward to manufacture. However, because their attractive force is constantly active, a repelling force is necessary to revoke the connection. Some robots use a mechanism to push modules away from each other until the attraction force has no more effect, for example with a Shape Memory Alloy (SMA) (Murata *et al.* 1999). Others rotate the permanent magnets so that they repel each other (Bishop *et al.* 2005). Permanent magnets are useful at the cm scale though their attractive force decreases with third power with respect to the size.

A third type of popular connection mechanism is an electro-magnet. They allow for selective connections and are simple to fabricate and implement. However, they need to be constantly powered to ensure the connection, and their strength-to-weight ratio decreases with size. It follows that for the use on a scale smaller than the cm scale they are not applicable.

Mechanical connectors such as latches lock and key, as well as hooking mechanisms provide a high connection strength. The docking and undocking is usually driven by electrical motors. However, they are not a viable solution at small scales because of the high demands on the precision of the required alignment. Furthermore, it is a difficult engineering task to build and actuate small and robust mechanical systems. Velcro has the advantage that the connection mechanism itself does not have to be actuated (Shimizu *et al.* 2005; Moeckel *et al.* 2005). A repelling force has to be provided only for the detachment (through an actuator). A further advantage of Velcro is that it does not have to be aligned precisely to connect and also works at small scales.

3. Peltier-based freeze-thaw connector for lightweight self-assembly robots

The size of 1 cm is a critical size for self-assembly systems. For objects in water around that size, viscosity is as important as inertia; for such objects, the Reynolds number (the ratio of inertial forces and viscous forces) is ≈ 1. It follows that objects smaller than this size are affected more by viscous forces whereas larger objects are affected more by inertial forces. In order to make a step towards smaller and more lightweight systems, we describe a novel connection mechanism: the water close to the docking interface of one module is frozen to ice building a local bridge to another module.

3.1 The Peltier connector

The core of our connector is the Peltier heat pump: a double-faced cooling/heating device that can transfer thermal energy from one side of the device to the other, with consumption of electrical energy. We used the Peltier device (Quick-OHM Kupper & Co.GmbH) to freeze (and thaw) the water between two modules and thus realize binding (and unbinding) between modules. The polarity of the current applied to the device defines which side is cooled down or heated up. The device consists of different types of semiconducting materials that are connected in series to take advantage of the so-called thermoelectric or Peltier effect. This effect is the direct conversion of an electric voltage into a temperature difference and vice versa, and allows the element to work as a heat pump. Peltier devices are available in various sizes. For our purpose we used an 8 × 8 mm element that weighs 0.8 g (Fig. 1). Theoretically, the Peltier heat pump can induce a temperature difference of up to 72 $°C$ while consuming approximatively 2.60 W.

One particular advantage of this type of inter-module connection mechanism is due to the absence of mechanical parts which makes it scalable. The fact that the connector is devoid of moving parts makes it also intrinsically less prone to failures. One disadvantage is that in order to sustain the connections, energy has to be supplied permanently to the heat pump. For the detachment process, however, there is no need to supply energy because the ice melts when it is not cooled down; moreover, the flow of heat from the hot surface supports the thawing process speeding it up.

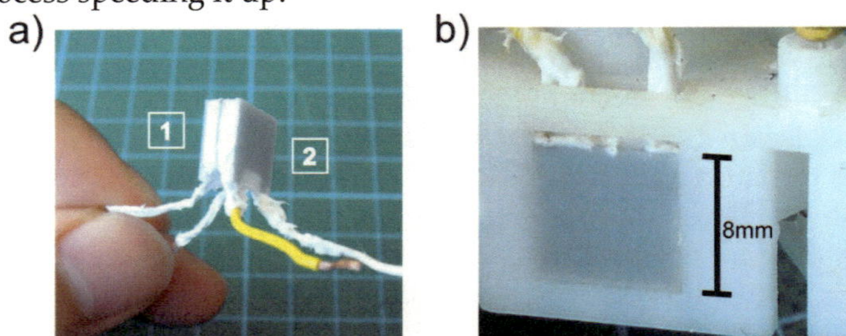

Fig. 1. Illustration of Peltier elements. a) Frozen Peltier elements (8 × 8 mm) sticking together. b) A Peltier element embedded into a module.

3.2 Feasibility study

To test whether Peltier devices can be used to build a connector, we conducted two experiments. First, we tested the connector's functionality for different water temperatures.

Second, we measured how much force was required to separate two modules (once they had connected to each other) also for different water temperatures.
Table 2 shows the conditions for which a connection was realized as a function of the voltage V applied to the Peltier elements and the time T necessary to achieve a connection. On average it took about one minute to establish a connection. For low temperatures and high voltages the two Peltier devices bound quickly to each other, for higher temperatures, however, the required time increased (e.g. at room temperature for V = 2 V, it took T = 3 min).
Although the time for the connection takes more time than for most other types of interconnection, the duration and the energy required for freezing decreases with the square of the size of the element's active area (one of the smallest commercially available Peltier elements have an area of approximately 1 mm^2, and we assume that a reduction in size can solve the heat dissipation problem which comes with this device).
We measured the bond strength for several water temperatures by applying a force perpendicular to the binding side until the connection broke (Fig. 2). A voltage of V = 2 V

water temperature 0 °C

V\T	30s	60s	120s
0.5V	✕	✓	✓
1.0V	✓	✓	✓
2.0V	✓	✓	✓

water temperature 5 °C

V\T	30s	60s	120s
0.5V	✕	✕	✕
1.0V	✕	✕	✓
2.0V	✓	✓	✓

water temperature 10 °C

V\T	30s	60s	120s
0.5V	✕	✕	✕
1.0V	✕	✕	✓
2.0V	✓	✓	✓

water temperature 15 °C

V\T	30s	60s	120s
0.5V	✕	✕	✕
1.0V	✕	✕	✕
2.0V	✓	✓	✓

Table 2. Experimental results of establishing a connection under different conditions (Voltage [V], Time [seconds]).

was applied to two Peltier elements for 60 seconds to bond, and the two elements were separated by hand from each other. On average, the connector withstood a pulling force of 17.56 N (the standard deviation was $\sigma = 2.03$). Considering the weight cumulated of the two Peltier elements (1.6 g), it becomes clear that the bond strength/weight ratio is higher than for other known connection mechanisms (Table 1). Moreover, we observe that the bonding force remains roughly constant for different water temperatures. This is because the volume of ice that the two Peltier elements built up stayed relatively constant irrespective of the water temperature. A further advantage is that in order to connect, the two Peltier elements do not have to be aligned precisely. This property allows the system to be not only strong and scalable, but also robust to misalignments.

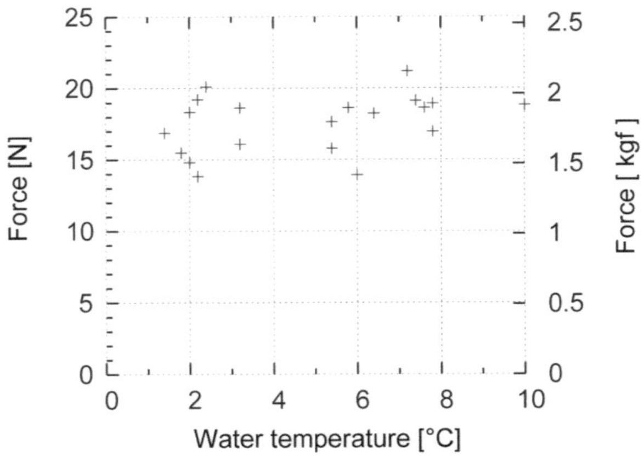

Fig. 2. Temperature - bonding strength comparison. X-axis: water temperature [°C], Y-axis: force to detach [N, kgf]

4. Implementation of the Peltier connectors to Self-assembly robots

Followed by the feasibility of the connector idea (described in Section 3), we embedded the Peltier-based connector into two types of stochastic modular robots.

4.1 Kite-shaped model

The experimental setup was composed of a power supply, a metallic ceiling, a water tank, and six modules immersed halfway in water (Fig. 3). Each module consisted of a kite-shaped wedge made of durable plastic (acrylnitrile butadene styrene; ABS) spanning angles of 60 degrees and 30 degrees. The modules (H: 13 mm, L: 30 mm) contained a permanent magnet oriented orthogonally to their main axis to attract or repel other modules (Fig. 4 a). A vibration motor was used to endow the modules with a minimal locomotive ability which allowed the modules to move randomly around vaguely reminiscent of Brownian motion. Rather than using batteries, electricity was supplied to the modules through a pantograph that drew current from a metal ceiling. This solution not only led to lightweight modules (m = 6.0 g), but it ensured that all modules received approximately the same amount of energy in a particular experiment. When an electrical potential was applied to the metallic ceiling plate, current flowed through the pantograph to the vibration motor returning to ground via the electrodes (platinum) immersed in the water (8 % concentration of electrolyte (salt) was added to the water to make it conductive). To speed up the connection between two modules, the water in which the modules moved was cooled down to approximately 3 °C (due to the concentration of salt this was slightly higher than the freezing temperature). Two diodes were used to switch the direction of the current. Current flowed either through the Peltier element or the vibration motor depending on the direction of the voltage applied to the system (Fig. 3 c, switch).

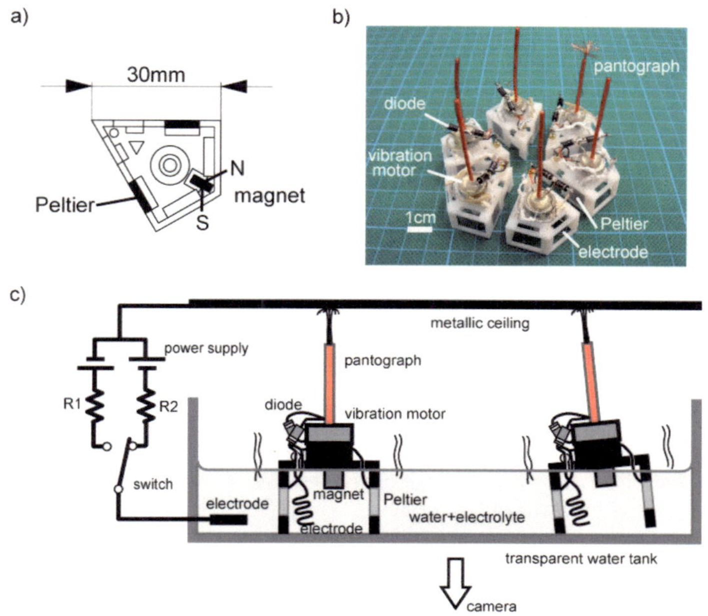

Fig. 3. Experimental setup. a) Schematic illustration of a module (bottom view). b) Picture of 6 modules. c) Experimental setup with 2 modules.

We first carried out experiments to test the reliability of the connector and to investigate the reconfigurability of our self-assembly system. The result is shown in Fig. 4. In the beginning of the experiment the modules were placed in the arena (Fig. 4 a) and arranged by hand to form a hexagonal shape (Fig. 4 b). Voltage was applied via the metallic ceiling (Fig. 4 c). After one minute, all six modules were connected to each other forming one unit (Fig. 4 d). We then flipped the polarity of the current supplied through the pantograph. The Peltier connectors stopped cooling and the vibration motors started to vibrate causing a disassembly of the hexagonal shape into 6 separate modules (Fig. 4 e). As a result of the vibrations of the motor, the modules moved around in the arena where they eventually got magnetically attracted by another module and started to form triangles (Fig. 4 f, g). The experiment was considered completed when the six modules had formed two triangles (Fig. 4 h). We conducted the experiment several times. For sufficiently long waiting times T, we always observed two different ways of convergence to the final states: one is in Fig. 4 h (two 3-clusters), the other is three 2-clusters (not on the picture, yield problem (Hosokawa *et al.* 1994)).

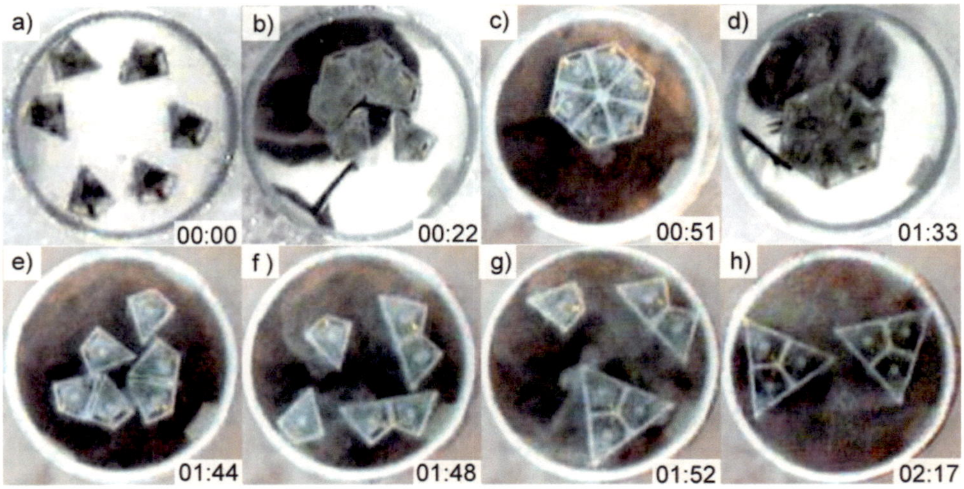

Fig. 4. Snapshots of the experiment.

4.2 Hinge-connected chain model

The outcome of the experiments described in Section 4.1 led to the question of how to take our modular system to another level of operation. To answer this question, we added permanent magnets to each module. Because with many modules the outcome of this change was not easy to anticipate, we built a new type of module. Each module was physically linked through hinge joints to two modules forming a chain (Fig. 5, m = 12.8 g). By taking inspiration from protein folding, we expected a drastic reduction of dimensionality of the search space. The main advantage of this implementation is that it avoids a crucial problem: the increased number of magnets generates also undesired configurations. Note that the positions of all the other magnets were replaced and rearranged. Only the center module (red colored, Fig. 5b) had a large magnet oriented perpendicularly to the symmetry axis. The other small cylindrical magnets were oriented vertically. "S"outh poles attracting "N"orth poles and vice versa. As in the modules described in the previous section, diodes were used to direct the current flow. Depending on the polarity of the applied voltage, current only flowed either through the Peltier elements (12 V) or through the vibration motors (10 V).

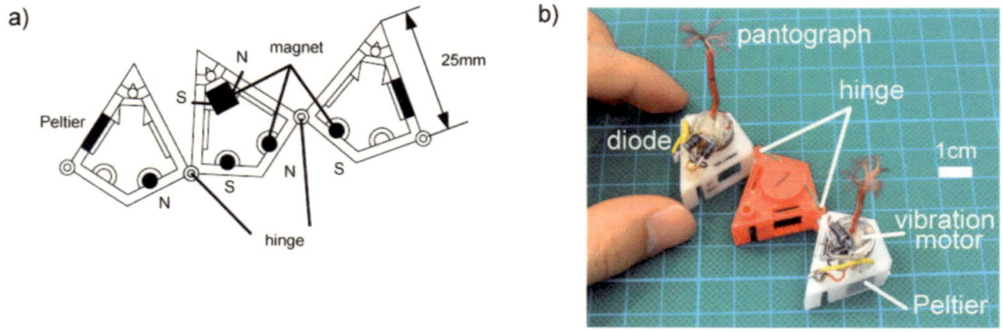

Fig. 5. Chain model. a) Schematic illustration of a chain. b) Picture of a chain. Three modules are connected by hinges.

Snapshots from a representative experiment are shown in Fig. 6. At the beginning of the experiment, we arranged by hand two chains of 3 modules each to form a hexagonal shape (Fig. 6 b). Voltage was applied to the system so that the Peltier elements were powered (Fig. 6 c). After one minute, an ice layer built up between the modules causing them to attach to each other yielding one single piece (Fig. 6 d). We then inverted the polarity of the applied voltage and let the current flow to the vibration motors (Fig. 6 e). The ice melted and the modules altered their configuration guided by the magnetic forces (Fig. 6 f, g). The transformation was completed in a minute, and two magnetically connected triangle chains were obtained (Fig. 6 h). The success rate of the reconfiguration just described was not as high as expected. We suspect that this low yield rate was mainly due to a design problem: the position of the large magnet was too close to the edge of the module. Therefore a rather strong movement of the modules was required to induce a disassembly of the initial hexagonal configuration.
The main implication of the two experiments described in this section 4 is that the restriction of the geometric constraint of modules, in other words, the "dimensionality reduction" of the reconfiguration problems enables the system to transit to a different level of functionality, bringing the modules to magnetically connected triangular clusters while avoiding undesirable formations (yield problem).

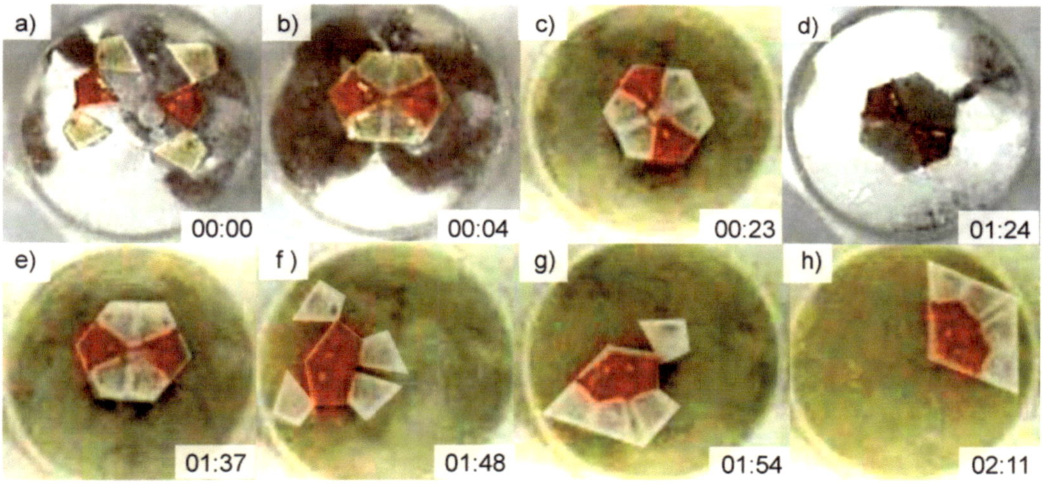

Fig. 6. Snapshots of the experiment with chain modules.

5. Discussion

An important goal of the growing field of self-assembly is the development of a better formal understanding of the specific mechanisms and general principles underlying it. It is clear that the discovery of principles which hold at all scales will require substantial input from various fields. At the molecular scale, biological systems are one of the examples that achieve robust self-assembly system through an intricate web of well ordered reactions. Attention must be paid to the fact that all components are passively interacting even if it looks as if they are actively reacting (Leiman *et al.* 2003). In order to achieve highly

autonomous self-assembly modules, the realization of a sufficient number of degrees of freedom that are controllable is necessary for such small scale autonomous-distributed systems.
In particular, because of the difficulty in including different types of attractive forces within the same system, realizing a new kind of connection mechanism endows the module with a better means of reacting in the environment. In this sense, the idea of a connector exploiting the thermoelectric effect may open yet another possibility for the state of the art of self-assembly systems.

6. Conclusion

We presented a novel type of connection mechanism for small-scale modular robotic systems. The mechanism exploits the thermoelectric effect to cool down the temperature and freeze water close to the modules and induce a strong bonding between modules. To test the connector, we embedded it into water based modular robots. The results obtained in this research demonstrate the utility of the proposed connection mechanism for lightweight self-assembling systems, and open a door towards more resilient self-assembly system at smaller scales.

7.Acknowledgment

The authors thank Nathan Labhart and Peter Eggenberger Hotz for helpful suggestions. This research was supported by the Swiss National Science Foundation project #200021-105634/1.

8. References

Winfree, E., Liu, F., Wenzler, L.A., Seeman, N.C. (1998). Design and selfassembly of two-dimensional dna crystals. Nature 394, pp. 539-544

Rothemund, P.W.K. (2006). Folding DNA to create nanoscale shapes and patterns. Nature 440 (7082), pp. 297-302

Whitesides, G.M., Grzybowski, B. (2001). Self-assembly at all scales. Science 295, pp. 2418-2421

Bowden, N., Weck, M., Choi, I.S., Whitesides, G.M. (2001). Molecule-mimetic chemistry and mesoscale self-assembly. Accounts of Chemical Research 34, pp. 231-238

Mao, C., Thalladi, V.R., Wolfe, D.B., Whitesides, S., Whitesides, G.M. (2002). Dissections: Self-assembled aggregates that spontaneously reconfigure their structures when their environment changes. J. Am. Chem. Soc 124 (49), pp. 14508-14509

Boncheva, M., Ferrigno, R., Bruzewicz, D.A., Whitesides, G.M. (2003). Plasticity in self-assembly: Templating generates functionally different circuits from a single precursor. Angew. Chem. Int. Ed. 42, pp. 3368-3371

Mondada, F., Gambardella, L.M., Floreano, D., Nolfi, S., Deneubourg, J.L., Dorigo., M. (2005). The cooperation of swarm-bots: Physical interactions in collective robotics. IEEE Robotics & Automation Magazine 12 (2), pp. 21-28

Fukuda, T., Kawauch, Y. (1990). Cellular robotic system (cebot) as one of the realizations of self-organizing intelligent universal manipulator. In: Proc. Int. Conf. on Robotics and Automation, pp. 662-667

Chirikjian, G.S. (1994). Kinematics of a metamorphic robotic system. In: Proc. Int. Conf. on Robotics and Automation, pp. 449-455

Murata, S., Kurokawa, H., Kokaji, S. (1994). Self-assembling machine. In: Proc. Int. Conf. on Robotics and Automation, pp. 441-448

Murata, S., Kurokawa, H., Yoshida, E., Tomita, K., Kokaji, S. (1998). A 3-D selfreconfigurable structure. In: Proc. Int. Conf. on Robotics and Automation, pp. 432-439

Murata, S., Tomita, K., Yoshida, E., Kurokawa, H., Kokaji, S. (1999). Selfreconfigurable robot. In: Proc. Int. Conf. on Intelligent Autonomous Systems, pp. 911-917

Yim, M. (1994). New locomotion gaits. In: Proc. Int. Conf. on Robotics and Automation. Volume 3, pp. 2508-2514

Kotay, K., Rus, D., Vona, M., McGray, C. (1998). The self-reconfiguring robotic molecule. In: Proc. Int. Conf. on Intelligent Robots and Systems. Volume 1, pp. 424-431

Rus, D., Vona, M. (2001). Crystalline robots: Self-reconfiguration with compressible unit modules. Autonomous Robots 10 (1), pp. 107-124

Castano, A., Behar, A., Will, P.M. (2002). The conro modules for reconfigurable robots. IEEE/ASME Trans. on Mechatronics 7 (4), pp. 403-409

Jorgensen, M.W., Ostergaard, E.H., Lund, H.H. (2004). Modular atron: Modules for a self-reconfigurable robot. In: Proc. Int. Conf. on Intelligent Robots and Systems. Volume 2, pp. 2068-2073

Zykov, V., Mutilinaios, E., Adams, B., Lipson, H. (2005). Self-reproducing machines. Nature 435 (7039), pp. 163-164

Detweiler, C., Vona, M., Kotay, K., Rus, D. (2006). Hierarchical control for selfassembling mobile trusses with passive and active links. In: Proc. Int. Conf. on Robotics and Automation, pp. 1483-1490

White, P., Kopanski, K., Lipson, H. (2004). Stochastic self-reconfigurable cellular robotics. In: Proc. Int. Conf. on Robotics and Automation. Volume 3, pp. 2888-2893

White, P., Zykov, V., Bongard, J., Lipson, H. (2005). Three dimensional stochastic reconfiguration of modular robots. In: Proc. Int. Conf. on Robotics Science and Systems, pp. 161-168

Griffith, S., Goldwater, D., Jacobson, J. (2005). Robotics: Self-replication from random parts. Nature 437, pp. 636

Shimizu, M., Ishiguro, A. (2005). A modular robot that exploits a spontaneous connectivity control mechanism. In: Proc. Int. Conf. on Robotics and Automation, pp. 2658-2663

Bishop, J., Burden, S., Klavins, E., Kreisberg, R., Malone, W., Napp, N., Nguyen, T. (2005). Programmable parts: A demonstration of the grammatical approach to self-organization. In: Proc. Int. Conf. on Intelligent Robots and Systems, pp. 3684-3691

Napp, N., Burden, S., Klavins, E. (2006). The statistical dynamics of programmed self-assembly. In: Proc. Int. Conf. on Robotics and Automation, pp. 1469-1476

Miyashita, S., Kessler, M., Lungarella, M. (2008). How morphology affects selfassembly in a stochastic modular robot. In: Proc. Int. Conf. on Robotics and Automation, pp. 3533-3538

Hosokawa, K., Shimoyama, I., Miura, H. (1996). 2-d micro-self-assembly using the surface tension of water. Sensors and Actuators A 57, pp. 117-125

Hosokawa, K., Shimoyama, I., Miura, H. (1994). Dynamics of self-assembling systems: Analogy with chemical kinetics. Artificial Life 1 (4), pp. 413-427

Bhat, P., Kuffner, J., Goldstein, S., Srinivasa, S. (2006). Hierarchical motion planning for self-reconfigurable modular robots. In: Proc. Int. Conf. on Intelligent Robots and Systems, pp. 4108-4115

Penrose, L.S. (1959). Self-reproducing. Sci. Amer. 200-6, pp. 105-114

Terada, Y., Murata, S. (2004). Automatic assembly system for a large-scale modular structure. In: Proc. Int. Conf. on Intelligent Robots and Systems. Volume 3, pp. 2349-2355

Moeckel, R., Jaquier, C., Drapel, K., Upegui, A., Ijspeert, A. (2005). Yamor and bluemove - an autonomous modular robot with bluetooth interface for exploring adaptive locomotion. In: Proc. Int. Conf. on Climbing and Walking Robots, pp. 685-692

Walle, B.L., Gauthier, M., Chaillet, N. (2007). Dynamic modelling of a submerged freeze microgripper using a thermal network. In: Proc. Int. Conf. Advanced Intelligent Mechatronics.

Leiman, P.G., Kanamaru, S., Mesyanzhinov, V.V., Arisaka, F., Rossmann, M.G. (2003). Structure and morphogenesis of bacteriophage t4. Cellular and Molecular Life Sciences 60, pp. 2356-2370

13

Adhesion Forces Reduction by Oscillation and Its Application to Micro Manipulation

Tetsuyou Watanabe† and ZhongWei Jiang‡
†*Kanazawa University, Japan*
‡*Yamaguchi University, Japan*

1. Introduction

In a micro range, it is a key issue how to cope with adhesion which arises between an object and tools (see Fig. 1). To resolve the problem, many methods have been developed, including the adhesion-type micro end-effector (Arai et al., 1996 a) (Arai et al., 1996 b), the vacuum gripping tool (Zesch et al., 1997), releasing by slopping and oscillating an end-effector (Rollot et al., 2002) (Haliyo et al., 2002) (Haliyo et al., 2003), releasing by control of electrostatic force (Saito et al. 2003), and releasing strategy based on environment information (Saito et al., 2001). However, there are still unresolved problems such as the difficulty of object motion control after the release, the difficulty of practical execution of the method (Saito et al., 2001) (since it is almost impossible to acquire all required information).

In this chapter, we propose a new strategy to cope with adhesion forces and develop an automatic micro manipulation system. By minutely oscillating an end-effector, bringing it near to an object on a substrate (table) and contacting it with the object, the adhesion force between the end-effector and the object becomes small comparing with the adhesion force between the substrate (table) and the object (see Fig. 2). Then, it is easy to remove the end-effector from the object while the object adheres to the substrate. Hereafter, ***we call this (phenomenon) relaxation of adhesion force. If the object is removed from the end-effector at any time, we say the adhesion force is relaxed enough. If it is not and the object sometimes adheres to the end-effector, we say the adhesion force is not relaxed enough***. Using this strategy, it is expected to accurately manipulate micro object like macro manipulation.

But, this strategy is not always available. When the pushing force applied to the object by end-effector is large, the effect of oscillation decreases, and then the adhesion force is not reduced enough. Then, we propose a method to check the adhesion state; whether the adhesion force is relaxed enough or not. We measure the oscillation of the end-effector, and apply FFT to it. If the lower mode frequencies (than the frequency of the input oscillation) are excited, the adhesion force is reduced enough. Otherwise, the adhesion force is not reduced enough.

However, this method is available in only limited situations if using the laser displacement meter to measure the oscillation, because 1) the end-effector must be located at the specific

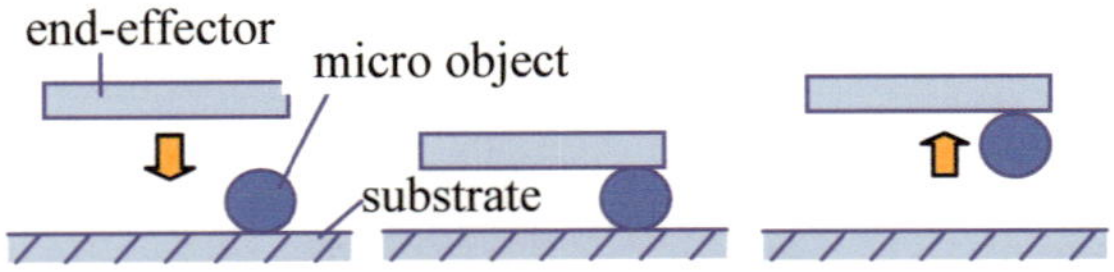

Fig. 1. Micro object adhered to end-effector

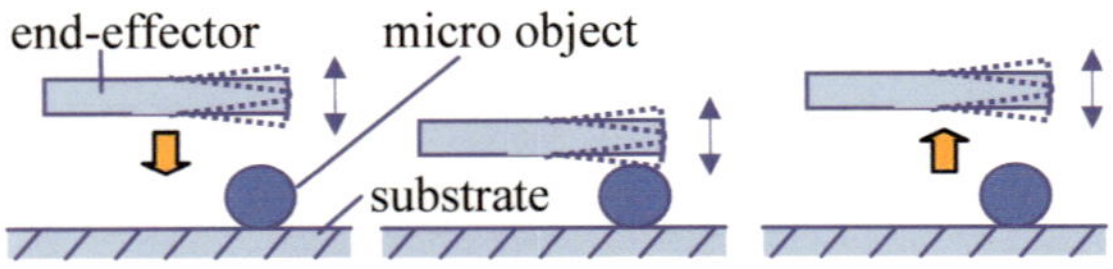

Fig. 2. Relaxation of adhesion force

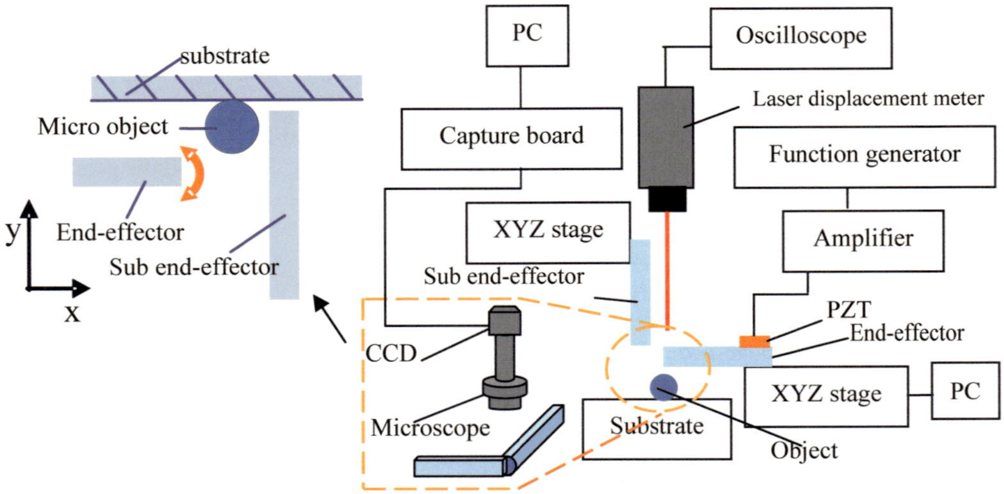

Fig. 3. Target system (Overview of experimental set up)

point where laser displacement meter can measure oscillation; 2) the adhesion state can not be checked if something blocks the light/laser or the target leaves the measuring point. So, it is hard to apply this method to micro manipulation directly. Then, we propose a method to check the adhesion state by vision. The oscillation of end-effector can not be perfectly caught by camera. Instead, the blur of the oscillation appears in the captured image. The amount of the blur is associated with the amplitude of the oscillation. Then, we develop a method to estimate the amplitude of oscillation by the blur. Subsequently, we focus that lower mode oscillation has large amplitude comparing with oscillation with higher mode frequency or no-resonance frequency. Utilizing this findings, we develop a method to detect lower mode frequencies by a blur. The adhesion state can be checked by checking whether lower mode frequencies are excited or not. Then, based on this findings and the method for detecting lower mode frequencies, we propose a method to check the adhesion state by vision. This method can be applied to any areas in the captured image and can be used all the time during manipulation. Its computational load is low. Since vision sensor is usually used in micro manipulation and any other sensors are not needed, the total system is very simple and low cost. Lastly, applying this method to micro manipulation, we develop an automatic control system for micro manipulation.

2. Target system

Fig. 3 shows the target system, which consists of manipulation part, image-capturing part, end-effector-oscillating part, and displacement-measuring part. For the simplicity, we assume that: (1) the manipulation is done in a planner space and a gravity force doesn't work, (2) the object is a sphere, (3) the end-effector, the sub end-effector and the substrate are made of a same material, (4) the end-effector, the sub end-effector and the substrate are grounded for preventing an extra charge at the initial state.

The manipulation part consists of end-effector, sub end-effector, micro object, and substrate. The end-effector and the sub end-effector are cantilever beams made of copper in size of 3x40x0.3[mm]. The beams are rolled copper, and any surface treatments such as grinding are not conducted. Young's modulus of copper is 1.02×10^{11} [N/m^2], its Poisson's ratio is 0.35, and its density is 8900 [kg/m^3]. On the end-effector, the PZT (piezocell) (Fuji ceramics, Z0.2T50x50x50S-W C6) of 3x3x0.2 [mm] is bonded at the position of 1 [mm] from the clamped end for oscillating the end-effector. The surface of substrate is a copper cut bonded on an aluminum board. The end-effector and the sub end-effector are attached on XYZ stage (Surugaseiki, PMZG413) which can be controlled by PC. The object is a glass sphere (Union, unibeads) with a radius of 100 or 200 [10^{-6}m] and a copper sphere with a radius of 150 [10^{-6}m]. Young's modulus of glass is 7.05×10^{10} [N/m^2], its Poisson's ratio is 0.17, and its density is 2500 [kg/m^3].

The image-capturing part consists of Video-microscope (Surugaseiki, VMU-V) with objective lens (Mitsutoyo S72M-5), CCD camera (Lumenera, LU135), and PC. The overview of the manipulation is captured by the CCD camera through the microscope and sent to PC. We use maximum illumination of light source (Schott MegaLight100-ROHS) whose maximum illumination is 24000[Lx] at the 100[mm] from the tip of the lighting system.

The end-effector is oscillated by oscillating the PZT by a function generator (NF, DF1906) through a power amplifier. The power amplifier is handmade circuit and its amplification ratio is set to 3.2.

The tip motion of the end-effector is measured by laser displacement meter (Sony VL10). The measured data is sent to PC through oscilloscope (Yokogawa DL1700).

3. Adhesion force relaxation and adhesion state estimation by laser displacement meter

By minutely oscillating the end-effector, bringing it near to an object on a substrate and contacting it with the object, the adhesion force between the end-effector and the object becomes small comparing with the adhesion force between the substrate and the object (see Fig. 2). This is thought to be mainly due to a hitting (impulse) effect and smaller time of

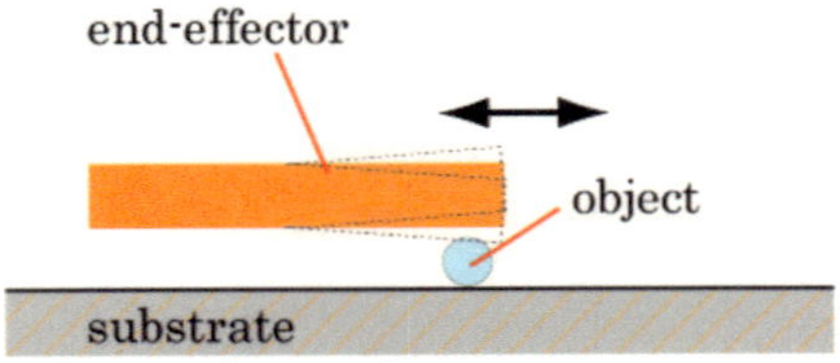

Fig. 4. Experiment to show adhesion force relaxation

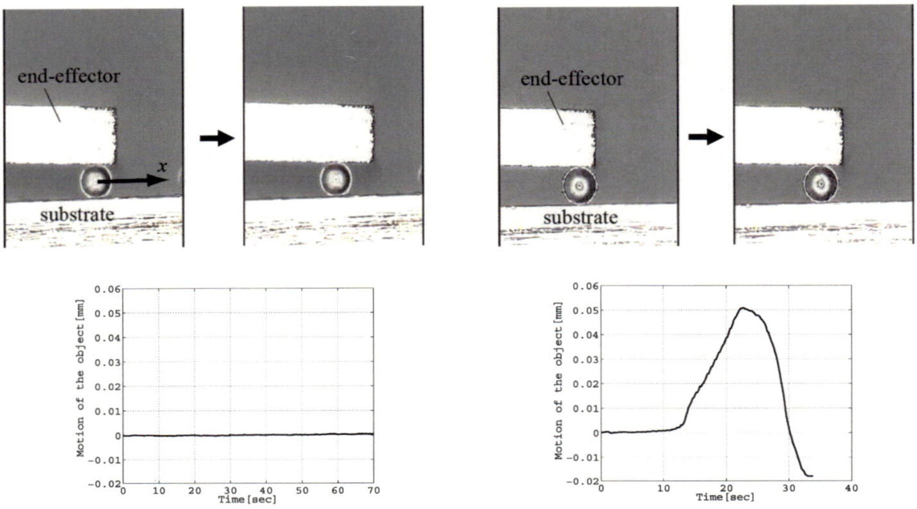

(a) when oscillating the end-effector

(b) when not oscillating the end-effector

Fig. 5. Overview of the experiment (upper figures) and the motion of the object center (lower figures)

contact between the object and the end-effector. Then, it is easy to remove the end-effector from the object while the object adheres to the substrate. Here we show simple experiment to show the effect as shown in Fig. 4. We bring the end-effector near to the object on the substrate, and contact it with the object. Subsequently, we move the end-effector in the left and right directions (of this page). We perform the experiments when the end-effector is oscillated and when it is not oscillated. We observe the motion of the object center. The object is a glass sphere. The input voltage to PZT is sine wave with the amplitude of 10[V]. Its frequency is 4th mode resonance frequency. Fig. 5 shows the result. The vertical axis denotes the position of the object center while horizontal axis denotes time. It can be seen that the end-effector slides on the object when oscillating the end-effector while the object rotates when not oscillating the end-effector. It means that end-effector oscillation can relax the adhesion force.

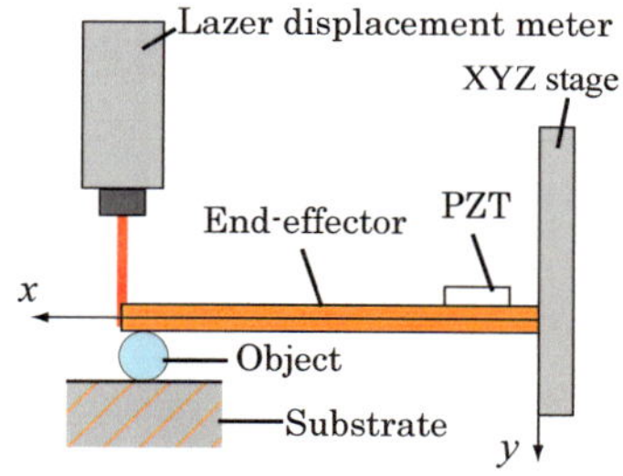

Fig. 6. Experimental set up for adhesion state check and its coordinate frame

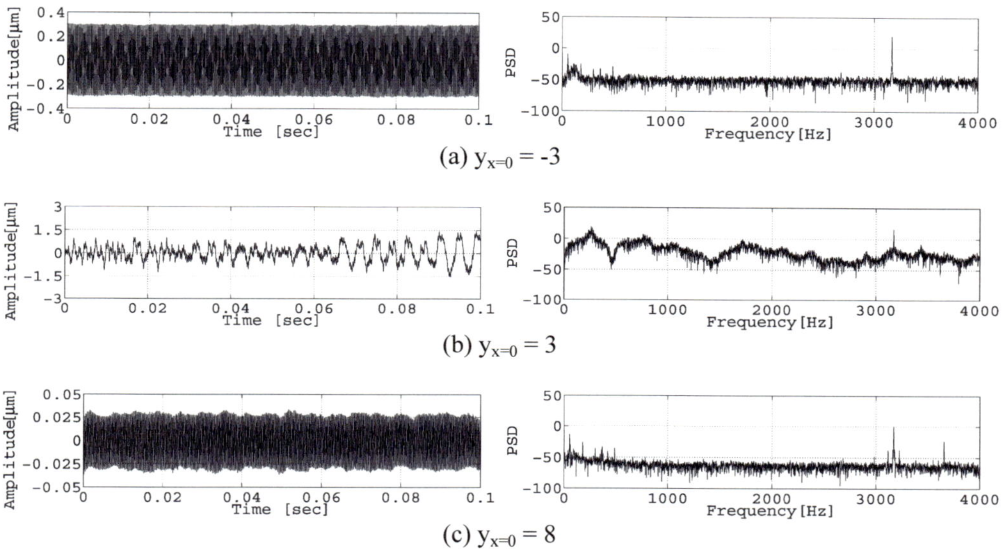

Fig. 7. Tip displacement (left side) and power spectrum density (right side) when pushing the glass sphere by oscillated end-effector (5V)

3.1 adhesion state check by laser displacement meter

This method is not always available. If the pushing force applied to the object is large, oscillation effect decreases, and the adhesion force is not relaxed enough. Therefore, adhesion state has to be checked. Here, we propose a method for the check. Fig. 6 shows the experimental set up. We set y direction so that y can be orthogonal to the long side of the end-effector as shown in Fig. 6. We move the oscillated end-effector by moving the clamped end by XYZ stage, along y positive direction with the step of 1 [μm] from $y(x=0)=-3$ to y(x=0)=8 [μm]. Let y(x=0) when the end-effector firstly contacts with the object be 0. At the initial state (y(x=0)=-3), the end-effector does not contact with the object. At y(x=0)=0, the end-effector contacts with the object. At y(x=0)≥0, the end-effector pushes the object. y(x=0) corresponds to the magnitude of the pushing force. We measure the oscillation of the end-effector by the laser displacement meter. The input signal for the oscillation is sine curve whose amplitude is 5 [V], and whose frequency is 4th mode resonance frequency (this mode is selected so that enough large kinetic energy can be got while the amplitude can be small enough not to disturb the manipulation). The object is a

y[μm]	Frequency of adhesion / number of trial times
0~7	0/10
8	2/10

Table 1. Frequency of adhesion to the end-effector when removing the end-effector from the substrate.

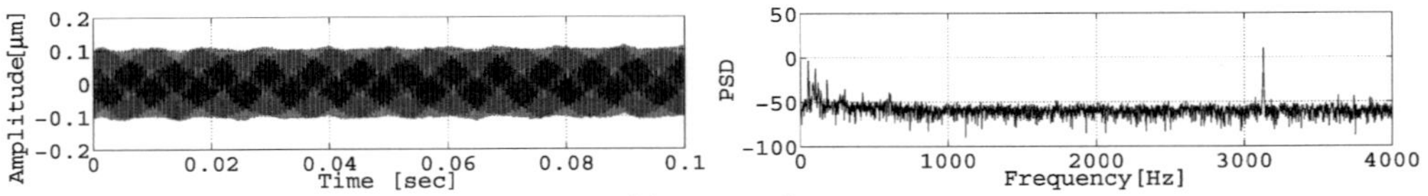

(a) $y_{x=0} = -3$

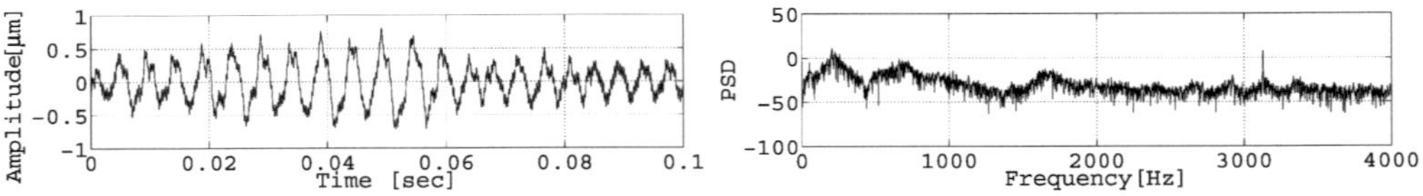

(b) $y_{x=0} = 0$

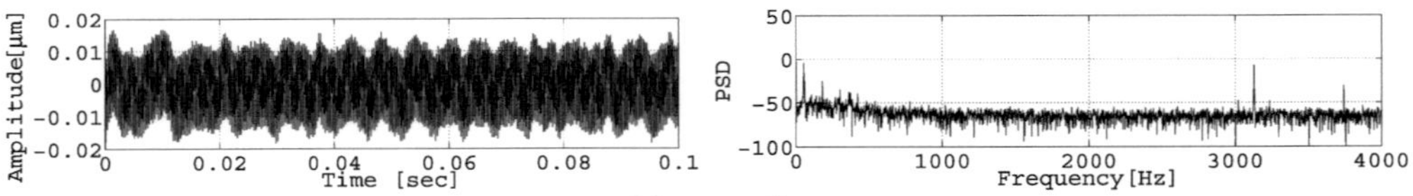

(c) $y_{x=0} = 3$

Fig. 8. Tip displacement (left side) and power spectrum density (right side) when pushing the glass sphere by oscillated end-effector (2.5V)

glass sphere with radius of 200 [μm]. The left figures of Fig. 7 show the tip displacement, and the right figures show the power spectrum density obtained by applying FFT to the measured tip displacement. The horizontal axis denotes time and the vertical axis denotes the amplitude at the left figures while the horizontal axis denotes the frequency and the vertical axis denotes the power spectrum density at the right figures. At $y(x=0)=-3$ [μm] (before contact), only inputted 4th mode frequency was observed as shown in Fig. 7 (a). At $y(x=0)=0\sim7$ [μm], the amplitude is larger than at $y(x=0)=-3$ [μm], and not only inputted 4th mode frequency but also lower mode frequencies were observed. Here, we show the case at $y(x=0)=3$ [μm] as a representative of the results (see Fig. 7 (b)). At $y(x=0)=8$ [μm], the amplitude is smaller than the other cases, and lower mode frequencies were not observed. At every case, we perform the experiment in which the end-effector is moved along y negative direction (removed from the substrate) 10 times. Table 1 shows the result. At $y(x=0)=0\sim7$ [μm], the object did not adhere to the end-effector at any time. Then, the adhesion force is thought to be relaxed enough. On the other hand, at $y(x=0)=8$ [μm], the object adhered to the end-effector twice. Then, the adhesion force is thought to be not relaxed enough due to smaller amplitude of the oscillation. It indicates that we can estimate whether adhesion force is relaxed enough or not by checking the excitation of the lower mode frequencies.

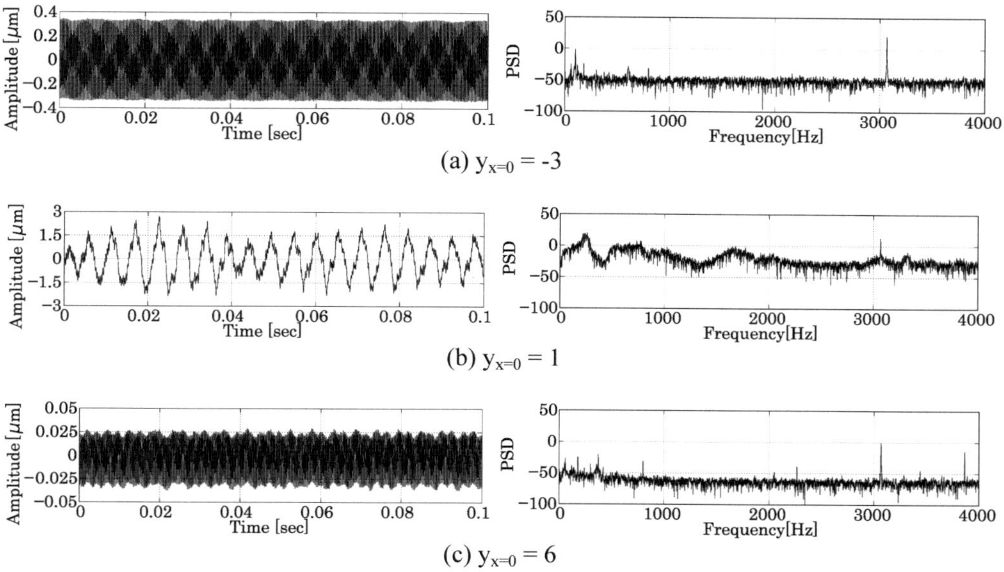

Fig. 9. Tip displacement (left side) and power spectrum density (right side) when pushing the copper sphere by oscillated end-effector (5V)

Next, to investigate the effect of amplitude of input voltage to PZT, we change the amplitude from 5 [V] to 2.5 [V] and perform the same experiment. Fig. 8 shows the results. In addition, to investigate the effect of material of the object, we change the object from glass sphere to copper sphere with a radius of 150 [μm] and perform the same experiment. Fig. 9 shows the results. From Fig. 7, 8 and 9, it can be seen that our approach is available at any case. But, the amplitude of input voltage and the material of the object affect the available range of the proposed method (how much we can push the object, relaxing adhesion force). It can be seen that in order to get larger available range, the oscillation with larger energy (larger amplitude of input voltage) should be applied. Since surface energy of copper is 2 [J/m^2] while surface energy of glass is 0.08 [J/m^2] (Israelachvili, 1996), the adhesion force for copper sphere is larger than that for glass sphere. It is thought to be the reason why the available range for copper sphere is smaller than that for glass sphere.

4. Adhesion state estimation by vision

As mentioned the above, the proposed method to check adhesion state is available in only limited situations due to the use of laser displacement meter: 1) The end-effector must be located at the specific point where laser displacement meter can measure oscillation; 2) the adhesion state can not be checked if something blocks the light/laser or the target leaves the measuring point. So, it is hard to apply this method to micro manipulation directly. Concerning these problems, here we present a method to check the adhesion state by vision.

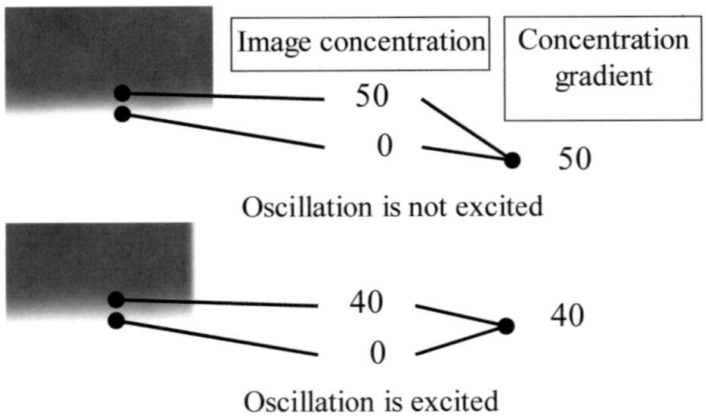

Fig. 10. Image concentration gradient

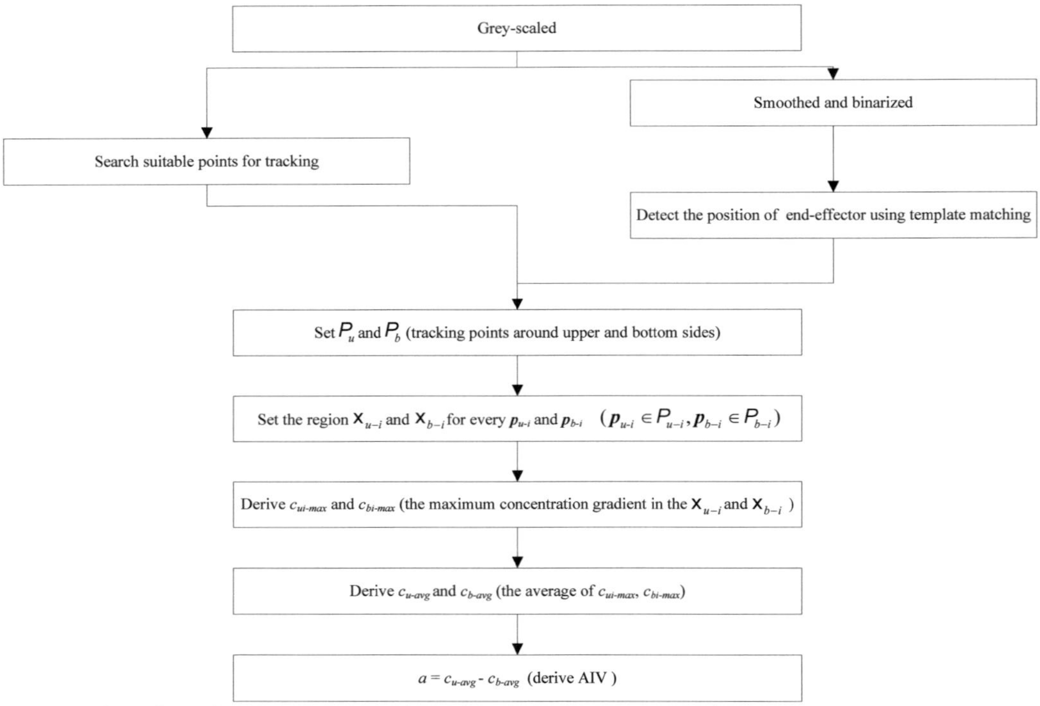

Fig. 11. Flowchart for deriving AIV (amplitude indicating value)

4.1 Oscillation amplitude estimation by vision

Firstly, we develop a method to estimate amplitude of oscillation using only vision information. Oscillation is too fast to be perfectly caught by camera. However, a blur resulted from the oscillation appears in the captured image. Then, we try to estimate the amplitude of the oscillation using the blur. When oscillation is not excited, the image concentration gradient (see Fig. 10) around the edge of the end-effector is large. On the other

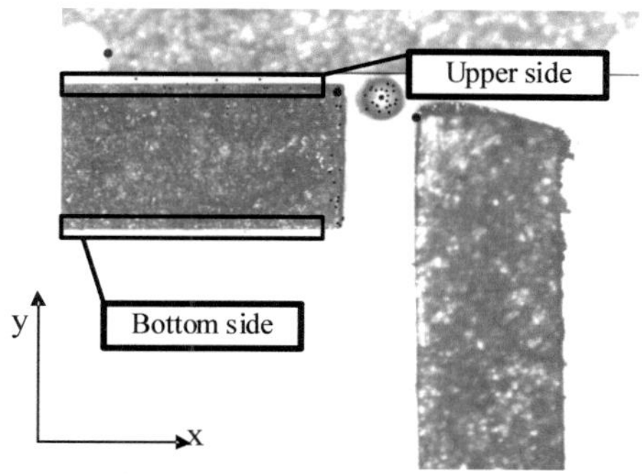

Fig. 12. Upper and bottom sides of the end-effector

hand, when oscillation is excited, it is small due to a blur. Using the concentration gradient, we estimate the amplitude of the oscillation. We call the estimated amplitude AIV (amplitude indicating value). The procedure for deriving AIV is shown in Fig. 11. First, the captured image is grey-scaled. Next, it is smoothed and binarized. Previously, we prepare the tip area image of the end-effector as a template image. By template matching technique which finds the part in the binarized image which matches the template image, we detect the position of end-effector. On the other hand, we search suitable points for tracking, $\boldsymbol{p}$, around the edge of the end-effector in the firstly grey-scaled image. Here, we pick up $\boldsymbol{p}$'s located around the upper side (area of 10 pixel (about 10μm) from the upper edge), and let P_u be a set of the picked up points (see Fig. 12). Similarly, we pick up $\boldsymbol{p}$'s located around the bottom side and let P_b be a set of the picked up points. Also, let $\boldsymbol{p}_{k\text{-}i}$ be ith $\boldsymbol{p}$ contained in P_k ($k \in \{u, b\}$), let n be the number of $\boldsymbol{p}_{u\text{-}i} \in P_u$, and let m be the number of $\boldsymbol{p}_{b\text{-}i} \in P_b$. We calculate $l_{k\text{-}i}$ which is the length between $\boldsymbol{p}_{k\text{-}i}$ and its nearest side/edge. Then, we derive the maximum value of $l_{k\text{-}i}$:

$$l_{\max} = \max_{i,k} l_{k-i} \tag{1}$$

Using $l_{\max}$, we define the following region $X_{k\text{-}i}$ for $\boldsymbol{p}_{k\text{-}i}$:

$$X_{k-i} = \left\{ (x,y) \middle| \begin{matrix} p_{k\text{-}ix} - 0.6 * l_{max} \le x \le p_{k\text{-}ix} + 0.6 * l_{max} \\ p_{k\text{-}iy} - 1.4 * l_{max} \le y \le p_{k\text{-}iy} + 1.4 * l_{max} \end{matrix} \right\} \tag{2}$$

where $p_{k\text{-}ix}$ and $p_{k\text{-}iy}$ are, respectively, x and y components of $\boldsymbol{p}_{k\text{-}i}$. Here, 0.6 and 1.4 are set by trial and error so that the nearest side/edge can be contained in $X_{k\text{-}i}$ and the variation of the maximum image concentration gradient in the $X_{k\text{-}i}$ can be small. Let $c_{ki\text{-}max}$ be the maximum concentration gradient in the $X_{k\text{-}i}$. We compute $c_{ki\text{-}max}$ for every $\boldsymbol{p}_{k\text{-}i}$ ($X_{k\text{-}i}$) and derive its average value $c_{k\text{-}avg}$:

$$\begin{aligned} c_{u-avg} &= (\textstyle\sum_{i=1}^{n} c_{ui-\max})/n \\ c_{b-avg} &= (\textstyle\sum_{i=1}^{m} c_{bi-\max})/m \end{aligned} \tag{3}$$

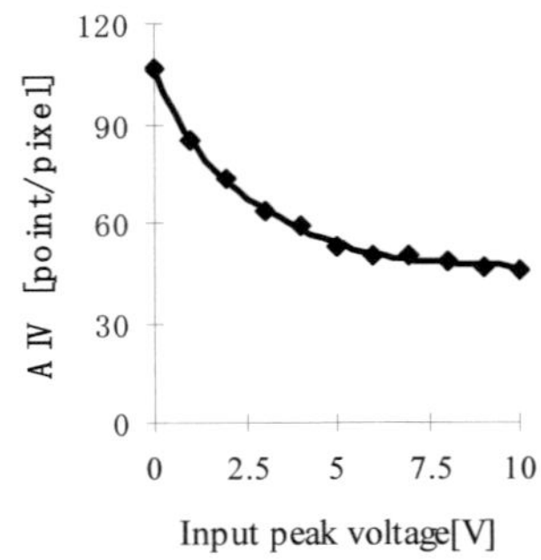

Fig. 13. Relation between the input peak voltage and AIV

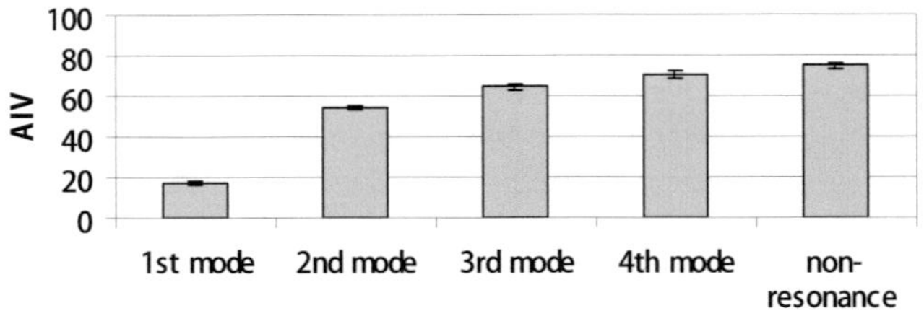

Fig. 14. AIV for several mode frequencies

From (3), we define the amplitude indicating value (AIV) a as follows:

$$a = C_{u-avg} - C_{b-avg} \tag{4}$$

Here, we confirm the efficiency of AIV by experiment. We oscillate the cantilevered end-effector freely. The input voltage for PZT is square wave whose peak to peak is from 0 to 0-10 [V], whose duty ratio is 50[%] , and whose frequency is 1st mode frequency 0.18[kHz]. AIV is computed by the program written by C++ language using OPEN CV library.

Fig. 13 shows the result. The horizontal axis expresses the input peak voltage, and the vertical axis expresses the computed AIV. Note that the input peak voltage indicates amplitude of oscillation since the voltage is proportional to the amplitude. By applying regression analysis, the relation is expressed by v=85$a^{-0.27}$ where v denotes the input peak voltage. From the result, it can be seen that the amplitude of oscillation can be estimated by AIV at 0≤v≤ 5 [V]. On the other hand, it is hard to estimate the amplitude at v ≥ 6 [V], although it can be detected that the oscillation has larger amplitude than a certain constant value (for example, the amplitude at v=5 [V]).

4.2 Discrimination between higher and lower mode frequencies by AIV

When bringing the oscillated end-effector with high mode frequency close to the object on the substrate and lower mode frequencies are excited, the adhesion force between the end-

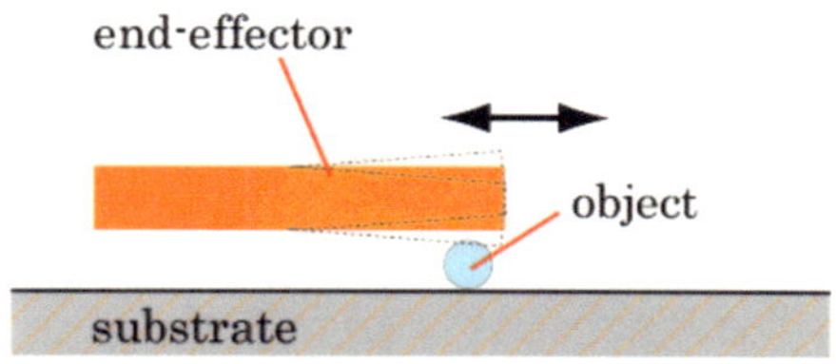

Fig. 15. Overview of the experiment for checking whether adhesion state can be estimated by AIV

effector and the object is reduced. Then, if discriminating between higher and lower mode frequencies by AIV, we can detect the adhesion state by AIV. Here, we investigate whether or not higher and lower mode frequencies can be discriminated by AIV by experiment.

We oscillate the cantilevered end-effector freely. The input voltage for PZT is square wave whose peak to peak is from 0 to 24 [V], whose duty ratio is 50[%], and whose frequency is 1st - 4th mode frequency. For the comparison, we also select non-resonance frequency of 2 [kHz].

Fig. 14 shows the results. From Fig. 14, it can be seen that the higher the frequency mode is, the larger AIV is. It is thought to come from that the higher frequency mode is, the smaller the amplitude is. The difference between AIV's for 1st mode and the other modes (including non-resonance frequency) is very large, and then the 1st mode oscillation can be easily detected. The 2nd mode oscillation can also be discriminated from the other higher mode oscillations by checking the difference of AIV. On the other hand, the discriminations between 3rd and 4th modes and between 4th mode and non-resonance frequencies are not easy. If setting the threshold for the discrimination is 3, we can discriminate 3rd and 4th modes frequencies, and 4th mode and non-resonance frequencies. If setting it is over 5, we can discriminate neither. In short, we can detect lower mode frequencies by AIV.

4.3 Detection of adhesion state by AIV

Base on the previous subsection results, we investigate whether adhesion state can be estimated by AIV.

We take the following way (see Fig. 15). First, we bring the oscillated end-effector close to the object and contact the end-effector with the object. Next, we move the end-effector in the left and right directions (of this page). If adhesion force is reduced enough, the end-effector slides on the object while the object is at stationary state. If it is not reduced enough, the object rotates. In the case when the end-effector slides on the object, we increase the pushing force applied to the object by moving the end-effector along y positive direction (refer to y direction in Fig. 6)..., and move the end-effector in the left and right directions again. This procedure is repeated until the object rotates. The input voltage for the oscillation is square wave whose peak to peak is from 0 to 24 [V], whose duty ratio is 50[%], and whose frequency is 4th mode frequency. The experience was done 5 times.

The results are shown in Fig. 16. Free means the end-effector is oscillated without contacting with the object. Note that in this experiment, the value of AIV when adhesion force is reduced enough changes with the change of the pushing force. Then, AIV in that case is

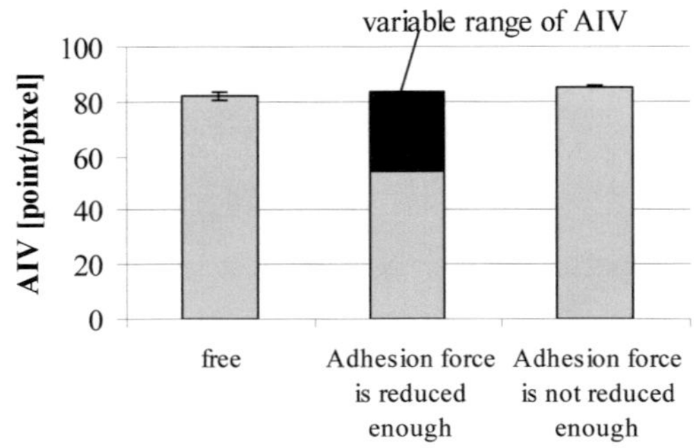

Fig. 16. AIV's when adhesion force is reduced enough and not reduced enough (AIV is shown by range because AIV when adhesion force is reduced enough changes with the change of the pushing force applied to the object)

shown by range. From Fig. 16, it can be seen that AIV when adhesion force is reduced enough is smaller than or equal to AIV for free state. The maximum difference is about 20. It is the reason why the lower mode frequencies (than the frequency of the inputted oscillation) are excited. On the other hand, AIV when adhesion force is not reduced enough is larger than AIV for free state. It is the reason why if the pushing force applied to the object is large, the energy of oscillation decreases and then the amplitude of the oscillation becomes smaller than that in the free state (refer to Fig. 7).

Note that AIV for 4th mode in Fig. 14 is different from AIV for free state in Fig. 16. It is due to the large difference of the end-effector position. The illumination or light intensity differs from place to place. Then, if the end-effector position changes largely, AIV also changes. Therefore, in order to check adhesion state, we use the difference between AIV when the end-effector is freely oscillated around the target point and AIV when the end-effector contacts with the object. Note also that there is the case when AIV when adhesion force is reduced enough is almost same as AIV for free state. It is thought that the lower mode frequencies are excited but their amplitude is small, and then AIV is large. In such a case, it is difficult to estimate adhesion state: whether adhesion force is reduced enough or not. However, a precious control of the pushing force applied to the object does not need in the target operation. Therefore, we only have to control the end-effector so that the difference between AIV's for free case and the case when the end-effector contacts with the object can be included in the appropriately defined range. Then, we can keep adhesion force reduced enough, while pushing the object with enough large force.

5. Automatic micro manipulation system

Using the developed method for estimating the adhesion state by vision, we develop a system which automatically pick and place a micro object. We use the experimental set up described at section 2 (see Fig. 3).

We present a procedure for picking operation in Fig. 17 (refer to the real movement shown in Fig. 19). First, using template matching technique, we find the tip positions of the end-

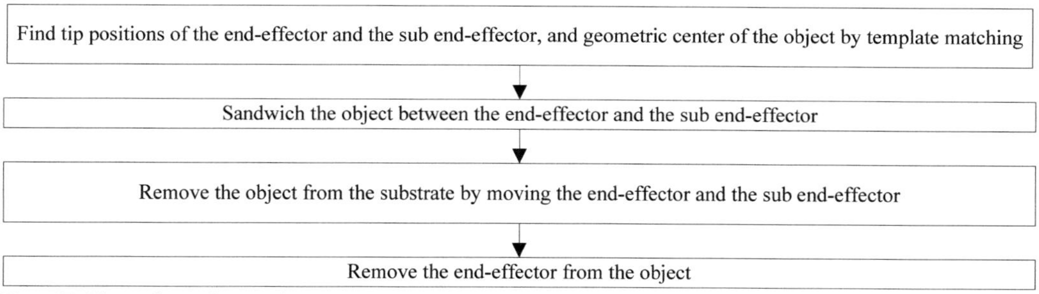

Fig. 17. Flowchart for picking operation

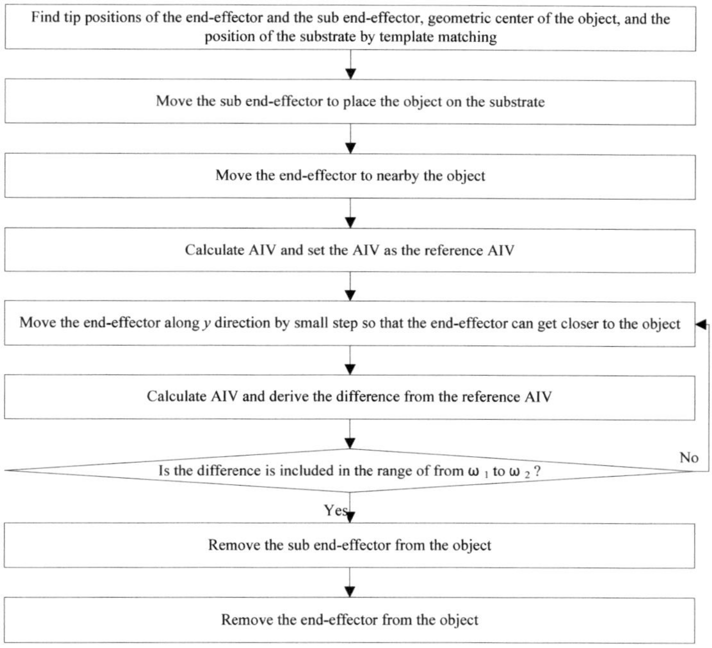

Fig. 18. Flowchart for placing operation

effector and the sub end-effector, and the geometric center of the object. The end-effector is oscillated. Using the position information, we sandwich the object between them. Next, we remove the object from the substrate by moving (controlling) the end-effector and the sub end-effector. Subsequently, we remove the end-effector from the object. In this case, the oscillation of the end-effector can reduce the adhesion force between the end-effector and the object since the oscillation in not only the bending but also the longitudinal directions of the end-effector is excited. We check the difference of AIV and judge the control of end-effector (adhesion force) is not needed, and then the checking and controlling procedures are not included in the flowchart shown in Fig. 17.
Next, we present a procedure for placing operation in Fig. 18. First, using template matching technique, we find the tip positions of the end-effector and the sub end-effector,

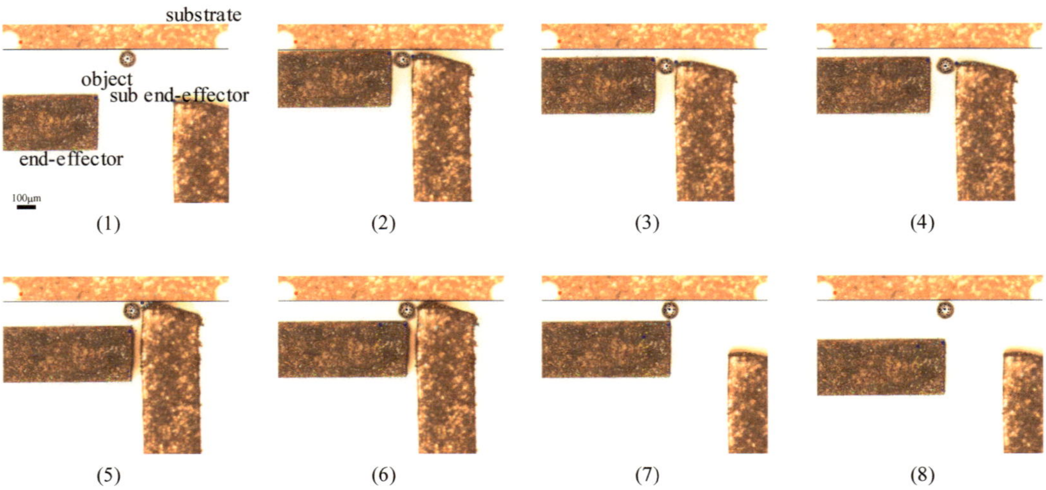

Fig. 19. Overview of the experiment for automatic micro manipulation for pick and place operation

the geometric center of the object, and the position of the substrate. Using the information, we place the object on the substrate by moving (controlling) the sub end-effector. Next, we move the end-effector to nearby the object so that the end-effector can contact with the object and push it if moving the end-effector along y positive direction (refer to y direction in Fig. 3). The end-effector is oscillated. Here, we calculate AIV and set it as the reference AIV. Next, we move the end-effector along y positive direction by small step so that the end-effector can get closer to the object. We calculate AIV and derive the difference from the reference AIV. If the difference is not included in the range of from ω_1 to ω_2, we move the end-effector along y direction by small step, and calculate AIV and the difference from the reference AIV again. If the difference is included in the range, we judge that the end-effector push the object by enough large force, reducing the adhesion force enough. In this case, we stop moving the end-effector and remove the sub end-effector from the object. Subsequently, we remove the end-effector from the object, and finish the operation.

The input voltage for PZT is square wave whose peak to peak is from 0 to 24 [V], whose duty ratio is 50[%], and whose frequency is 4th mode frequency. The threshold value ω_1 and ω_2 are, respectively, set to 5 and 10 by try and error so that the difference from the reference AIV can be detected, while the end-effector can apply enough large pushing force to the object.

The outline of the result is shown in Fig. 19. The number denotes the order of the time line. We did this operation 5 times, and all operations were successfully done. These results indicates the validity of our approach.

6. Conclusion

In this chapter, we proposed a novel method for reducing adhesion force by oscillation. By minutely oscillating the end-effector, bringing it near to an object on a substrate (table) and contacting it with the object, the adhesion force between the end-effector and the object becomes small comparing with the adhesion force between the substrate (table) and the object. Then, it is easy to remove the end-effector from the object while the object adheres to the substrate. We showed the available range of this method can be checked by checking the excitation of lower mode frequencies (than the inputted oscillation frequency), and controlled by oscillation energy. But, the method for the check is available in only limited situations due to the use of laser displacement meter: 1) The end-effector must be located at the specific point where laser displacement meter can measure oscillation; 2) the adhesion state can not be checked if something blocks the light/laser or the target leaves the measuring point. It is hard to apply this method to micro manipulation directly. Then, we developed a method to check adhesion state, using only vision information. Firstly, we developed a method to estimate the amplitude of the oscillation, using the blur resulted from the oscillation in the captured image. We call the estimated amplitude AIV (amplitude indicating value). Subsequently, we showed lower mode oscillations can be detected by AIV. Then, we developed a method for checking adhesion state (whether the adhesion force is reduced enough or not) by AIV. Based on the checking method, we developed a automatic micro manipulation system for pick and place operation. The validity of our method was shown by experiments.

7. References

Arai, F.; Andou, D. & Fukuda, T. (1996 a). Micro Manipulation Based on Physical Phenomena in Micro World (1st Report, The Reduction Method of Van Der Waals Force). *Trans. of the Japan Society of Mech. Eng., Series C*, vol.62, no.603, pp. 4286-4293.

Arai, F.; Andou, D.; Nonoda, Y. & Fukuda, T. (1996 b). Micro Manipulation Based on Physical Phenomena in Micro World(Principle and Prototype Experiments of Adhesion type Micro Endeffector). *Trans. of the Japan society of Mech. Eng., series C*, Vol.62, No.604, pp.4630-4635.

Fearing, R. S. (1995). Survey of sticking effects for micro-parts. *Proceeding of the IEEE/RSJ International Conference on Intelligent Robots and Systems*, pp. 212-217.

Haliyo, D. S.; Rollot, Y. & Regnier, (2002). S. Manipulation of micro-objects using adhesion forces and dynamical effects *Proc. of the IEEE Int. Conf. on Robotics and Automation*, pp.1949-1954.

Haliyo, D. S. & Regnier, S. (2003). Advanced applications using mad, the adhesion based dynamic micro-manipulatior. *Proc. of the IEEE/ASME Int. Conf. on Advanced Intelligent Mechatronics*, pp. 880-885.

Israelachvili, J. N. (1996) *Intermolecular and Surface Forces (Colloid Science)*, Academic Pr.

Lucas, B. D. & Kanade, T. (1981). An lteretive Image Registration Technique with an Application to Stereo Vision. *Proc. 7th Int. Conf. on Artificial Intelligence*, pp.674-679.

Rollot, Y.; Regnier, S. & Guinot, J. (2002). Dynamical model for the micromanipulation by adhesion : Experimental validations for determined conditions. *J. of Micromechatronics*, Vol.1, No.4, pp.273-297.

Saito, S.; Miyazaki, H. T.; Sato, T.; Takahashi, K. & Onzawa, T. (2001). Analysis of micro-object operation based on the dynamics considering the adhesion under an sem. *Proc. of the IEEE/RSJ Int. Conf. on Intelligent Robots and Systems*, pp. 1349-1357.

Saito, S. ; Himeno, H. & Takahashi, K. (2003). Electrostatic detachment of an adhering particle from a micromanipulated probe. *J. of Applied Physics*, Vol. 93, No. 4, pp. 2219-2224.

Zesch, W. ; Brunner, M. & Weber, A. (1997). Vacuum tool for handling micro objects with a nano-robot. *Proc. of the IEEE Int. Con. on Robotics and Automation*, pp.1761-1766.

OpenCV Reference manual, Available: http://opencv.jp/opencv-1.0.0_org/docs/index.htm.

14

Passivity based control of hydraulic linear arms using natural Casimir functions

Satoru Sakai
Chiba University
Japan

1. INTRODUCTION

This chapter discusses a modeling and passivity based control of hydraulic arms which are robotic, that is, have asymmetric cylinders. Hydraulic arms are very important components in field robotics, such as construction, agriculture, rescue , demining robotics and so on since hydraulic arms are superior to electric arms with respect to the power to weight ratio and also can keep joint forces even when the energy source (the hydraulic pump) does not work.
In many cases of electric arms, the driving system (or the actuator dynamics) is simple and almost static, for example, the input torque (or velocity) is just proportional to the control input. On the other hand, in many cases of hydraulic arms, the driving system is complex and consists of compressible fluid systems, that is, *nonlinear* dynamical systems with *unknown (or hard-to-be identified) parameters*. To solve these problems, this chapter gives some results about modeling and control of hydraulic arms by applying and developing port-Hamiltonian systems and control theory.

Port-Hamiltonian systems van der Schaft (2000) are generalization of Hamiltonian systems in classical mechanics but can model many systems such as electro-mechanical systems, mechanical systems with nonholonomic constraints Maschke & van der Schaft (1994), distributed systems. The first important property of port-Hamiltonian systems is that the interconnection of port-Hamiltonian systems gives again another port-Hamiltonian system. That is, it is easy to treat more complex systems consisting of these finite systems and the infinite systems such as the flexible beams Macchelli & Melchiorri (2005). The second important property of port-Hamiltonian systems is passivity and some passivity based control methods, originally from the chapter Takegaki & Arimoto (1981), were developed, such as, Energy-Casimir methods van der Schaft (2000) , the generalized canonical transformations Fujimoto & Sugie (2001), IDA-PBC Ortega & Garcia-Canseco (2004) and IPC approach Sakai & Stramigioli (2007); Stramigioli et al. (1998) and so on. These methods can give nonlinear robust controllers, and not only stabilization, but also tracking and dynamic output feedback stabilization Sakai & Fujimoto (2005) are achieved already.

For hydraulic arms, some nonliner robust (or adaptive) controllers were already proposed Bonchis et al. (2001); Mazenc & Richard (n.d.); Yao et al. (2000); Zhu & Piedboeuf (n.d.). However, in these approaches, the closed-loop systems are not port-Hamiltonian systems any more, even when the controlled systems can be described as the port-Hamiltonian systems.

That is, it is difficult to treat and extend the results in these approaches to more complex systems such as the high-degree of freedom (hydraulic) mechanisms and the dynamic energy sources.
Recently, modeling and control of several fluid systems are discussed in port-Hamiltonian form. For example, Ramkrishna et al. discuss infinite dimensional canal systems in three dimensional space Ramkrishna & van. der. Schaft (2006) . These fluid systems have free-surface and are incompressible, that is, differ from compressible fluid systems in hydraulic arms. Riccardo et al. discuss the modeling of hydraulic arms and show some experimental results Riccardo et al. (2006). Gernot et al. discuss the control of hydraulic arms Gernot & Schlacher (2005). However, these approaches are based on the standard procedure using only the first or the second properties above. Apart from these approaches, we discuss the modeling and control of hydraulic arms based on a new (the third important) structural property, that is, not only passivity but also Casimir functions are used in the modeling and control as a new structural property. In our approach, the bulk modulus is not identified at all.

This chapter gives a new passivity based control of hydraulic arms based on a new model using natural Casimir functions. In Section II, we refer port-Hamiltonian systems and their properties. In Section III, we propose two stabilization methods, a new dynamic asymptotic stabilization method and a new partial stabilization method. In Section IV, we give a new model of hydraulic arms using Casimir functions. At the same time, a very fundamental state of hydraulic arms is discovered. In Section V, the proposed two stabilization methods are applied to the new model of hydraulic arms and a new passivity based control of hydraulic arms are proposed. In Section VI, the validity of our methods are confirmed by numerical simulation and finally we conclude this chapter in Section VII.
In this chapter, I_n is $n \times n$ identity matrix, $\mathbb{R}^{m \times n}$ is the space of the m rows and n columns real matrix. Class-$\mathcal{K}$ functions are strictly increasing functions which have the origin Khalil (1996).

2. Port-Hamiltonian systems

2.1 Port-Hamiltonian systems

A port-Hamiltonian system with a Hamiltonian $H(x) \in \mathbb{R}$ is a system described by

$$\left\{ \begin{array}{rcl} \dot{x} & = & J(x)\dfrac{\partial H(x)}{\partial x}^{\mathrm{T}} + g(x)u \\ y & = & g(x)^{\mathrm{T}}\dfrac{\partial H(x)}{\partial x}^{\mathrm{T}} \end{array} \right. \tag{1}$$

with $u, y \in \mathbb{R}^m$, $x \in \mathbb{R}^n$ and a skew symmetric matrix $J(x)$, i.e. $-J(x) = J(x)^{\mathrm{T}}$ holds [1]. Hamiltonian H is bounded from below. The following first property is known Maschke & van der Schaft (1992).

Lemma 1 van der Schaft (2000) *Consider the port-Hamiltonian system (1). Suppose the Hamiltonian $H(x)$ satisfies $H(x) \geq H(0) = 0$. Then the input-output mapping $u \mapsto y$ of the system is passive with respect to the storage function H, and the feedback*

$$u = -D(x)\,y \tag{2}$$

[1] J was replaced by a negative semidefinite matrix in order to describe dissipative elements van der Schaft (2000). The results in this chapter can be directly applicable to this modification, Nevertheless we concentrate on a skew-symmetric matrix J in this chapter for simplicity.

with a matrix $D(x) \geq \varepsilon I > 0 \in \mathbb{R}^{m\times m}$ renders $(u,y) \to 0$. Furthermore if $H(x)$ is positive definite and if the system is zero-state detectable, then the feedback (2) renders the origin asymptotically stable.
The zero-state detectability and the positive definiteness of the Hamiltonian assumed in Lemma 1 do not always hold for general port-Hamiltonian systems. In such a case, the generalized canonical transformation is useful.

2.2 Casimir functions

One of other properties of port-Hamiltonian systems are the existence of Casimir functions. Casimir functions (with respect to J) are defined as solutions of the following PDE,

$$\frac{\partial C(x)}{\partial x} J(x) \equiv 0. \tag{3}$$

Not only Hamiltonian, but also Casimir functions are the special first integrals, that is,

$$\dot{C} \equiv 0 \tag{4}$$

holds for any Hamiltonian $H(x)$ when $u = 0$. Unlike Hamiltonian, Casimir functions are not bounded from below nor upper in general and do not alway exist for port-Hamiltonian systems (but exist in some important cases in robotics).
Note that we do not treat Casimir functions for closed-loop systems but treat "Casimir functions for controlled systems (plants)". The former Casimir functions are artificial and discussed in controller design phase such as the Energy-Casimir method van der Schaft (2000). The latter Casimir functions are discussed in (at least) modeling phase and referred to as "natural Casimir functions" to avoid confusions in this chapter.

3. Dynamic and partial stabilization for port-Hamiltonian systems

3.1 Dynamic asymptotic stabilization for port-Hamiltonian systems

In this subsection, we give a dynamic stabilization method for port-Hamiltonian systems. As we refereed in the previous section, port-Hamiltonian systems can be stabilized by the static stabilizers. However, a dynamic stabilizers are also useful and will be applied to hydraulic arms later.

Theorem 1 *Consider the following (mechanical) port-Hamiltonian systems*

$$\Sigma_m^{pH}: \begin{cases} \begin{bmatrix} \dot{q} \\ \dot{p} \end{bmatrix} = \begin{bmatrix} 0 & I \\ -I & 0 \end{bmatrix} \begin{bmatrix} \frac{\partial H_m}{\partial q}^T \\ \frac{\partial H_m}{\partial p}^T \end{bmatrix} + \begin{bmatrix} 0 \\ G \end{bmatrix} u \\ y = G^T \frac{\partial H_m}{\partial p}^T \end{cases} \tag{5}$$

where $q, p \in \mathbb{R}^n$ are the (generalized) position and momentum, G is nonsingular matrix and the Hamiltonian $H_m = (1/2)(p^T M(q)^{-1} p) + U(q)$ with $M = M^T > 0$ and $U(q) \geq U(0) = 0$. Then, the following dynamic controller

$$\Sigma_{dd}: \begin{cases} \dot{r} = P(x,r)^T G^{-T} y - D(x,r) \frac{\partial H_r}{\partial r}^T \\ u = -P(x,r) \frac{\partial H_r}{\partial r}^T \end{cases} \tag{6}$$

makes the set $\Omega_0 = \{(q,p)|y = u = 0\}$ *asymptotically stable, where* $x = (q,p)^T \in \mathbb{R}^{2n}$, $r \in \mathbb{R}^r$ $H_r = (1/2)r^T R(r) r$ *with* $R = R^T > 0$, $P^T \in \mathbb{R}^{r\times m}$ *and* $D = D^T > 0$.

Proof of Theorem 1

From Equations (5) and (6), the closed-loop system is

$$\Sigma_{cl}^{pH} : \left\{ \begin{bmatrix} \dot{q} \\ \dot{p} \\ \dot{r} \end{bmatrix} = \begin{bmatrix} 0 & I & 0 \\ -I & 0 & -P(x,r) \\ 0 & P(x,r)^T & -D(x,r) \end{bmatrix} \begin{bmatrix} \frac{\partial(H_m+H_r)}{\partial q}^T \\ \frac{\partial(H_m+H_r)}{\partial p}^T \\ \frac{\partial(H_m+H_r)}{\partial r}^T \end{bmatrix} \right. \tag{7}$$

and (zero-input) port-Hamiltonian systems with a new Hamiltonian $H_m + H_r$ and a dissipation. Since the time derivative of $H + H_r$ (along the trajectory) is

$$\dot{H} + \dot{H}_r = -\frac{\partial H_r}{\partial r} D \frac{\partial H_r}{\partial r}^T \leq 0 \tag{8}$$

and $H + H_r$ is bounded from below,

$$\frac{\partial H_r}{\partial r}^T \to 0 \tag{9}$$

as $t \to \infty$, that is, the set $\{u = 0\}$ is asymptotically stable from Equation (6).

At the same time, Equation (9) implies $r \to 0$, and thus $\dot{r} \to 0$ because H_r is a class-$\mathcal{K}$ function with respect to $\|r\|$. Since we have Equation (6) and

$$P(x)^T \frac{\partial(H_m + H_r)}{\partial p}^T = 0 \Rightarrow \frac{\partial H_m}{\partial p}^T = 0 \tag{10}$$

due to the condition $m \geq r$, the set $\{y = 0\}$ is asymptotically stable from Equation (5). In all, the set $\{y = u = 0\}$ is asymptotically stable. (Q.E.D.)

Note that the proposed dynamic asymptotic stabilization is a generalization of our result Sakai & Fujimoto (2005). The origin of the closed-loop system (7) is asymptotically stable if and only if the origin of a closed-loop system from Equations (5) and (2) is asymptotically stable.

It is easy for readers to extend the mechanical port-Hamiltonian system Σ_m^{pH} in Theorem 1 to more general port-Hamiltonian systems even though Theorem 1 will be applied directly later.

3.2 Partial stabilization for port-Hamiltonian systems with Casimir function

Advantages of port-Hamiltonian systems are from their structural properties, such as passivity, which do not exist in general nonlinear systems. In this subsection, we focus a new structural property of a special port-Hamiltonian systems, that is, port-Hamiltonian systems with Casimir functions. As we referred in Section II, Casimir functions do not exist in general port-Hamiltonian systems.

One of the new advantages of this special port-Hamitonian systems are presented in the following theorem.

Theorem 2 *Consider the following port-Hamiltonian systems with Casimir functions* $C(x)$

$$\Sigma_c^{pH} : \left\{ \begin{array}{rcl} \begin{bmatrix} \dot{q} \\ \dot{p} \end{bmatrix} & = & J(q,p) \begin{bmatrix} \frac{\partial H}{\partial q}^T \\ \frac{\partial H}{\partial p}^T \end{bmatrix} + \begin{bmatrix} 0 \\ G \end{bmatrix} u \\ y & = & G^T \frac{\partial H}{\partial p}^T \end{array} \right. \tag{11}$$

where $q, p \in \mathbb{R}^n$ are the (generalized) position and momentum, G is nonsingular matrix and the Hamiltonian $H \geq H(0) = 0$. Suppose there exists a coordinate transformation $x \mapsto \phi(x) = (x_r, C)^T$ such that

$$H(\phi) = H_r(x_r) + H_c(C), \tag{12}$$

where $u, y \in \mathbb{R}^m$, $x_r \in \mathbb{R}^r$, $r \leq \text{rank}(J)$ and H_r is bounded from below.
Then the feedback

$$u = -D(x)\, y_r \tag{13}$$

with $D = D^T > 0$ and

$$y_r = [\,0 \;\; G^T]\, \frac{\partial \phi}{\partial x}^T \begin{bmatrix} I_r \\ 0 \end{bmatrix} \frac{\partial H_r}{\partial x_r}^T \tag{14}$$

makes the set $\{y_r = 0\}$ asymptotically stable.

Proof of Theorem 2

From the existence of Casimir functions $C(x)$, there exists a coordinate transformation which converts the system (11) into

$$\begin{cases} \begin{bmatrix} \dot{x_r} \\ \dot{C} \end{bmatrix} = \begin{bmatrix} J_r(x_r, C) & 0 \\ 0 & 0 \end{bmatrix} \begin{bmatrix} \frac{\partial H(x_r,C)}{\partial x_r}^T \\ \frac{\partial H(x_r,C)}{\partial C}^T \end{bmatrix} + \frac{\partial \phi}{\partial x} \begin{bmatrix} 0 \\ G \end{bmatrix} u \\ y = [0\; G^T] \frac{\partial \phi}{\partial x}^T \frac{\partial H}{\partial \phi}^T \end{cases} \tag{15}$$

with the skew-symmetrix matrix J_r.
The time derivative (along the trajectory) of the function H_r is given as

$$\begin{aligned} \dot{H_r} &= \frac{\partial H_r}{\partial x_r}\dot{x}_r + \frac{\partial H_r}{\partial C}\dot{C} \\ &= \frac{\partial H_r}{\partial x_r}\left[J_r \frac{\partial H(x_r, C)}{\partial x_r}^T + [I_r\; 0^{r\times 2n-r}] \frac{\partial \phi}{\partial x} \begin{bmatrix} 0 \\ G \end{bmatrix} u \right] \\ &= \frac{\partial H_r}{\partial x_r}\left[J_r \frac{\partial H_r}{\partial x_r}^T + [I_r\; 0] \frac{\partial \phi}{\partial x} \begin{bmatrix} 0 \\ G \end{bmatrix} u \right] \\ &= \frac{\partial H_r}{\partial x_r} [I_r\; 0] \frac{\partial \phi}{\partial x} \begin{bmatrix} 0 \\ G \end{bmatrix} u \end{aligned} \tag{16}$$

due to the special Hamiltonian structure $H = H_r(x_r) + H_c(C)$.
This means that the system with input u and output y_r is passive (lossless) with respect to the storage function H_r, that is,

$$\dot{H_r} = y_r^T u \tag{17}$$

holds. Finally the controller (13) makes the set $\{y_r = 0\}$ asymptotically stable since

$$\dot{H_r} = -y_r^T D y_r \leq 0 \tag{18}$$

and H_r is bounded from below. (Q.E.D.)

Note that the above output y_r is different from the usual output y of port-Hamiltonian function and a new output based on the structural properties of special port-Hamiltonian systems, that is, port-Hamiltonian systems with Casimir functions. Furthermore, not all states, but only the partial state x_r is stabilized in Theorem 2 because H_r is the function of x_r.
By combination of the above two proposed stabilization methods, that is, the dynamic stabilization method and the partial stabilization method, robotic hydraulic arms will be controlled later based on the modeling in the next section.

4. Modeling of hydraulic arms using natural Casimir functions

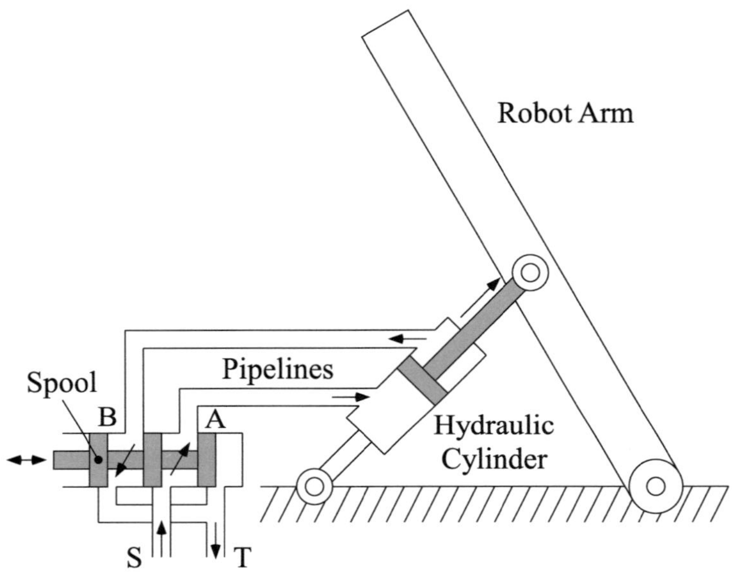

Fig. 1. Hydraulic arms

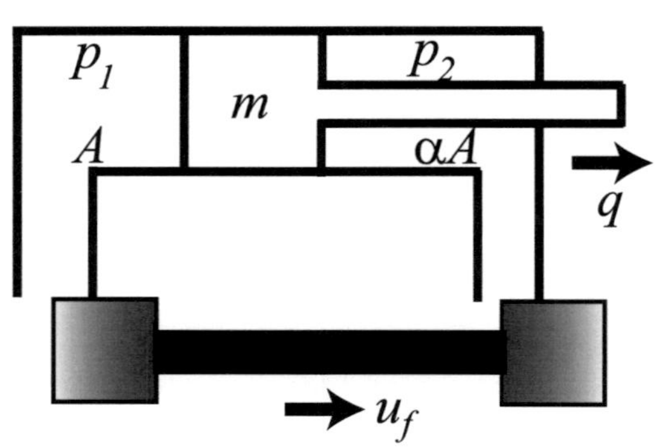

Fig. 2. The equivalent model of hydraulic arms

In this section, modeling of robotic hydraulic arm (with asymmetric cylinders as shown in Fig. 1) is discussed in port-Hamiltonian framework. This section discusses an empirical model in Merrit (1967) and Jelali & Kroll (2002), which is equivalent to the model in Fig. 2. The spool in Fig. 2 has only two slots while the original one in Fig. 1 has three slots. The empirical model is not infinite dimensional but finite dimensional model, and takes input not as the driving force of the spool, but as the spool displacement. This finite dimensional model with spool displacement input is based some assumptions but already used in many theoretical and experimental works Bonchis et al. (2001); Mazenc & Richard (n.d.); Yao et al. (2000); Zhu & Piedboeuf (n.d.).

4.1 Generalized continuous law

A continuous law (in integral form) for compressible fluid is

$$\frac{V}{E}\dot{p}_i = -\dot{V} + Q_i^{in} - Q_i^{out} \tag{19}$$

where p_i is the pressure of chamber i, Q_i^{in} is the flow into chamber i, Q_i^{out} is the flow from chamber i, E is the bulk modulus and V is the fluid volume. See Jelali & Kroll (2002) for this modeling assumptions in detail.

As the port-Hamiltonian systems are generalization of classical Hamiltonian system (the energy conservation law), the following system is given as a generalization of the continuous law (19)

$$\Sigma_f : \begin{cases} \begin{bmatrix} \dot{x}_{f1} \\ \dot{x}_{f2} \end{bmatrix} = \begin{bmatrix} -A \\ \alpha A \end{bmatrix} u_o + \begin{bmatrix} g_{p1} \\ g_{p2} \end{bmatrix} u_f \\ y_f = [-A \;\; \alpha A] \dfrac{\partial H_f}{\partial x_f}^T \end{cases} \tag{20}$$

where

$$H_f = \frac{E}{2V}(x_{f1}^2 + x_{f2}^2) \tag{21}$$

$x_f = (x_{f1}, x_{f2})^T$, $x_{fi} = (V/E)p_i$, u_o is the cylinder velocity and $g_{pi} = g_{pi}(x_{fi})$ are the flow velocity from Bernoulli's equation (omitted from its uniqueness), u_f is the spool input displacement, A and αA are the square in the chambers. Since the cylinder is asymmetric, $0 < \alpha < 1$ holds. The state is not pressure and different from that in the previous model Gernot & Schlacher (2005).

Now the mechanical system Σ_m^{pH} and the fluid system Σ_f are interconnected by the following

$$\begin{cases} u_o = y \\ u = -y_f \end{cases} \tag{22}$$

because u_o is the cylinder velocity and y_f is the driving force of the cylinder. Then the interconnected system is given as

$$\Sigma_{fm}^{pH} : \begin{cases} \begin{bmatrix} \dot{q} \\ \dot{p} \\ \dot{x}_{f1} \\ \dot{x}_{f2} \end{bmatrix} = \begin{bmatrix} 0 & I & 0 & 0 \\ -I & 0 & GA & -\alpha GA \\ 0 & -AG^T & 0 & 0 \\ 0 & \alpha AG^T & 0 & 0 \end{bmatrix} \begin{bmatrix} \dfrac{\partial H_{fm}}{\partial q}^T \\ \dfrac{\partial H_{fm}}{\partial p}^T \\ \dfrac{\partial H_{fm}}{\partial x_{f1}}^T \\ \dfrac{\partial H_{fm}}{\partial x_{f2}}^T \end{bmatrix} + \begin{bmatrix} 0 \\ G_f \end{bmatrix} u_f \\ y_{f1} = G_f^T \dfrac{\partial H_{fm}}{\partial x_f}^T \end{cases} \tag{23}$$

where $H_{fm} = H_f + H_m$ and $G_f = [g_{p1}\ g_{p2}]^T$. Σ_{fm}^{pH} is easily confirmed to be port-Hamiltonian systems since J matrix part is again skew symmetric.

It is interesting that the interconnected system of the fluid system Σ_f and the mechanical system Σ_m^{pH} is again a port-Hamiltonian system even though the fluid system Σ_f is not port-Hamiltonian system. By the generalized continuous law, we can take not only the energy conservation but also the mass conservation into account simultaneously. This situation is different from that in fluid mechanics where only the energy conservation is taken in Navier-Stokes equations and only the mass conservation is taken in the continuous law.

4.2 Casimir functions

In this section, we give the most important result on the modeling of hydraulic arms.
Lemma 2 *Consider the fluid-mechanical port-Hamiltonian systems Σ_{fm}. Then there exists a Casimir function*

$$C_f = \frac{1}{\sqrt{1+\alpha^2}}\left(\alpha x_{f1} + x_{f2}\right). \tag{24}$$

Proof of Lemma 2.
By a direct calculation, it is confirmed that C_f satisfies the PDE (3), that is, $\dot{C}_f \equiv 0$ holds for any Hamiltonian H_{fm} at zero-input. (Q.E.D.)

Theorem 3 *Consider the fluid-mechanical port-Hamiltonian systems Σ_{fm}. Then there exists a coordinate transformation ϕ such that the transformed systems satisfy the condition (12).*
Proof of Theorem 3.
Consider the following coordinate transformation,

$$\begin{bmatrix} q \\ p \\ x_{fr} \\ C_f \end{bmatrix} = \begin{bmatrix} I_{2n} & 0 & 0 \\ 0 & \frac{1}{\sqrt{1+\alpha^2}} & \frac{-\alpha}{\sqrt{1+\alpha^2}} \\ 0 & \frac{\alpha}{\sqrt{1+\alpha^2}} & \frac{-1}{\sqrt{1+\alpha^2}} \end{bmatrix} \begin{bmatrix} q \\ p \\ x_{f1} \\ x_{f2} \end{bmatrix} \tag{25}$$

where C_f is already given in Lemma 2. It is calculated that the system (23) are transformed to a new port-Hamiltonian systems which satisfies the conditions (12) because we have $H = H_r + (1/2)C_f^T C_f$ where H_r is given in the following J_r-part dynamics (a reduced dynamics) in (15)

$$\left\{ \begin{array}{l} \begin{bmatrix} \dot{q} \\ \dot{p} \\ \dot{x_{fr}} \end{bmatrix} = \begin{bmatrix} 0 & I & 0 \\ -I & 0 & \sqrt{1+\alpha^2}A \\ 0 & -\sqrt{1+\alpha^2}A & 0 \end{bmatrix} \begin{bmatrix} \frac{\partial H_r}{\partial q}^T \\ \frac{\partial H_r}{\partial p}^T \\ \frac{\partial H_r}{\partial x_{fr}}^T \end{bmatrix} + \begin{bmatrix} 0 \\ g_r \end{bmatrix} u_f \\ \\ y_r = g_r^T \frac{\partial H_r}{\partial x_{fr}}^T \end{array} \right. \tag{26}$$

where

$$H_r = H_m + \frac{E}{2V} x_{fr}^2 \tag{27}$$

and g_r is omitted because of its uniqueness. (Q.E.D.)

5. Control of hydraulic arms using natural Casimir functions

In this section, we give a new stabilization method based on the previous results in this chapter. The proposed controller in this section can stabilize only mechanical part, even if the parameters of the fluid systems, the bulk modules E, is unknown.

Lemma 3 *Consider the fluid-mechanical systems Σ_{fm}^{ph} and suppose that $U(q)$ is the positive definite function. Then the feedback*

$$u_f = -D(x) g_r^T \frac{\partial H_r}{\partial x_{fr}}^T \tag{28}$$

with a matrix $D(x) \geq \varepsilon I > 0 \in \mathbb{R}^{m \times m}$ makes the set $\{(q,p) = 0\}$ asymptotically stable.

Proof of Lemma 3.

First, from Theorem 2 and Theorem 3, the set

$$\{y_r = g_r^T \frac{\partial H_r}{\partial x_{fr}}^T = 0\} \tag{29}$$

is asymptotically stable. Second, since the closed-loop system of (26) and (28) is equivalent to the system (7) in Theorem 1, the feedback (28) makes the set

$$\{y = \frac{\partial H_m}{\partial p}^T = 0\} \tag{30}$$

asymptotically stable. This implies that the set $\{(q,p) = 0\}$ is asymptotically stabilized because the zero-state detectability holds due to the positive definiteness of $U(q)$. (Q.E.D.)

Note that the above stabilization is achieved without the exact value of E, that is, the parameter identification is not required since all inequalities hold for any $E > 0$. Furthermore, the closed-loop system keeps the structure of port-Hamiltonian systems. That is, many port-Hamiltonian techniques (re-design, learning) can be applied on this close-loop systems.

Remark 1 *Lemma 3 is easily extend to the case of the following (mechanical) port-Hamiltonian systems with dissipation (frictional effect)*

$$\Sigma_{mr}^{pH} : \begin{cases} \begin{bmatrix} \dot{q} \\ \dot{p} \end{bmatrix} = \begin{bmatrix} 0 & I \\ -I & -R(q,p) \end{bmatrix} \begin{bmatrix} \frac{\partial H_m}{\partial q}^T \\ \frac{\partial H_m}{\partial p}^T \end{bmatrix} + \begin{bmatrix} 0 \\ G \end{bmatrix} u \\ y = G^T \frac{\partial H_m}{\partial p}^T \end{cases} \tag{31}$$

where $R(x) = R(x)^T > 0$. (See the footnote in Section II.)

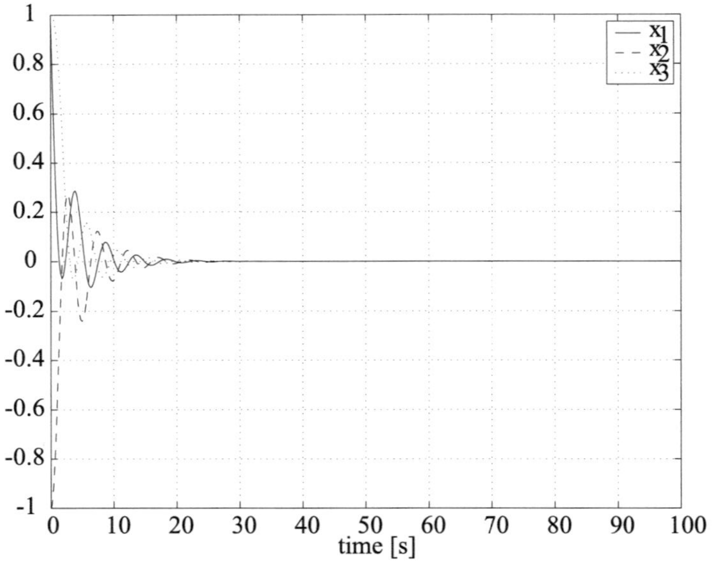

Fig. 3. Time responses of the linear mechanical-spring system

6. Numerical simulations

In this section, we confirm the validity of our methods by numerical simulation. Fig. 3 shows the time response of a standard (the simplest) linear mechanical-spiring SISO system with the stabilizer in Theorem 1. This SISO system does not have any damping at all. All states of the closed-loop system $(x_i, i = 1, 2, 3)$ convert to the origin smoothly and the validity of Theorem 1 is confirmed.

Figures 4-5 show the results by the stabilizer in Lemma 3. Initial states are $(1, -1, 1, 1)$ and all parameters are normalized as 1. Fig. 4 shows the time responses of the state at $D = 1/2$. Only the state of mechanical systems (q, p) converts to the origin smoothly. This implies that the validity of the partial stabilization methods in Theorem 2. The settling time is about 20 s. Fig. 5 shows the time responses of the state at $D = 1/5$. In this case, only the state of mechanical systems (q, p) also converts to the origin smoothly. However the settling time is about 12s even though the gain D is lower than $1/2$. This implies that the linearized system around the origin has not only poles but also zeros and the gain-tuning guideline Sakai & Fujimoto (2005) will be useful. (See Sakai & Fujimoto (2005).) The states converts to the origin. In all, the validity of our methods are confirmed.

7. Conclusions

This chapter gives a new modeling and control of hydraulic linear arms. First, we propose two stabilization methods a dynamic asymptotic stabilization method and a partial stabilization method. Second, we give a new model of hydraulic arms using Casimir functions. Third, the proposed two stabilization methods are applied to this new model and a new passivity based control are proposed. Finally, the validity of our methods are confirmed by simulations even though we did not identify the bulk modulus E at all. In future work, not only the friction but also the gravity compensation (which are important in the field applications) will be achieved in the similar approach.

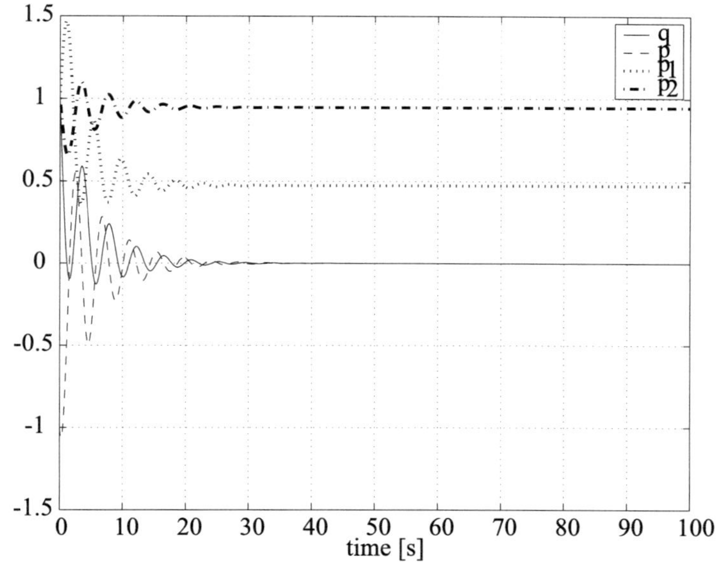

Fig. 4. Time response of all states ($D = 1/2$)

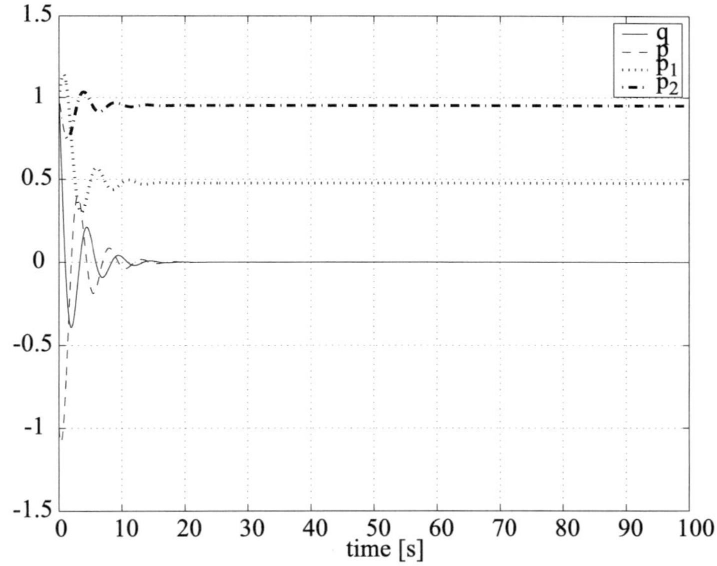

Fig. 5. Time response of all states ($D = 1/5$)

8. References

Bonchis, A., Corke, P., D.C.Rye & Ha, Q. (2001). Variable structure methods in hydraulic servo systems control, *Automatica* **37**(1): 589–595.

Fujimoto, K. & Sugie, T. (2001). Canonical transformation and stabilization of generalized hamiltonian systems, *Systems & Control Letters* **42**(3): 217–227.

Gernot, G. & Schlacher, K. (2005). Energy-based nonlinear control of hydraulically actuated mechanical systems, *Proc. of the 44th Conference on Decision and Control*, pp. 7520–7525.

Jelali, M. & Kroll, A. (2002). *Hydraulic Servo Systems*, Springer.

Khalil, H. K. (1996). *Nonlinear Systems*, third edn, Macmillan Publishing Company, New York.

Macchelli, A. & Melchiorri, C. (2005). Control by interconnection of mixed port hamiltonian systems, *IEEE Trans. Automatic Control* **50**(11): 1839–1844.

Maschke, B. M. J. & van der Schaft, A. J. (1992). Port-controlled Hamiltonian systems: modeling origins and system-theoretic properties, *IFAC Symp. Nonlinear Control Systems*, pp. 282–288.

Maschke, B. M. J. & van der Schaft, A. J. (1994). A Hamiltonian approach to stabilization of nonholonomic mechanical systems, *Proc. 33rd IEEE Conf. on Decision and Control*, pp. 2950–2954.

Mazenc, F. & Richard, E. (n.d.). Stabilization of hydraulic systems using passive property, *Systems and Control Letters* **44**.

Merrit, H. (1967). *Hydraulic Control Systems*, John Willey & Sons.

Ortega, R. & Garcia-Canseco, E. (2004). Interconection and damping assignment passivity-based control: A survey, *European Journal of Control* pp. 1–27.

Ramkrishna, P. & van. der. Schaft, A. (2006). A port-hamiltonian approach to modeling and interconnections of canal systems, *Network modeling and control of physical systems*, p. WeA08.

Riccardo, M., Roberto, Z. & Paolo, F. (2006). Dynamic model of an electro-hydraulic three point hitch, *Proc. of the 2006 Amecican Control Conference*, pp. 1868–1873.

Sakai, S. & Fujimoto, K. (2005). Dynamic output feedback stabilization of a class of nonholonomic hamiltonian systems, *Proc. IFAC World Congress 2005*, pp. 1967–1970.

Sakai, S. & Stramigioli, S. (2007). Port-hamiltonian approaches to motion generations for mechanical systems, *Proc. of IEEE Conference on Robotics and Automation*, pp. 69–74.

Stramigioli, S., Maschke, B. M. J. & van der Schaft, A. J. (1998). Passive output feedback and port interconnection, *Proc. 4th IFAC Symp. Nonlinear Control Systems*, pp. 613–618.

Takegaki, M. & Arimoto, S. (1981). A new feedback method for dynamic control of manipulators, *Trans. ASME, J. Dyn. Syst., Meas., Control* **103**: 119–125.

van der Schaft, A. J. (2000). *L_2-Gain and Passivity Techniques in Nonlinear Control*, Springer-Verlag, London.

Yao, B., Bu, F., Reedy, J. & Chiu, G. T.-C. (2000). Adaptive robust motion control of single rod hydraulic actuators: Thorey and experiment, *IEEE/ASME Transactions of Mechatoronics* **5**(1): 79–91.

Zhu, W. & Piedboeuf, J. (n.d.). Adaptive output force tracking control of hydraulic cylinders with applications to robot manipulators, *Transactions of the ASME Dynamic systems, Measurement and Control* **127**.

15

The Formation Stability of a Multi-Robotic Formation Control System

Chih-Fu Chang and Li-chen Fu
National Taiwan University
Taiwan, R.O.C.

1. Introduction

The recent trend in the design of a robot system towards to increase the capibility of a single robot for the increasing demand in different applications at home, industry or military(Murry 2007). However, it greatly increases not only the unit production cost but also the design complexity. One of the best solution for this problem is certainly identified as using many of the low cost robots in terms of the cooperative manner. Accordingly, the formation system may constitute a more effective achievement than a single robot in regard to the union of the different functions in the different Wheeled Mobile Robots(WMRs) of a team such that the cooperative system tends to the better signal resolution and the higher performance of the information system *etc.* Sometimes we could also suffer the task which is very diffcult to be achieved with a single robot, *i.e.*, move a large size object from one place to another one or deal with the highly flexible tasks.

The Multi-Robot Formation System (MRFS) is generally defined as a system contains a collection of the robotic subsystems which is able to be cooperation/competition with each other. The interconnected structure of the MRFS represents the physical or nonphysical interconnected relationship between any robots based on the specific robot in the formation team. Usually, the interconnected structure of the MRFS can be regarded as a topological structure with the communication system(Fax and Murray 2004; Olfati-Saber and Murray 2004). In other words, the topological structure is the foundation of the information system of the MRFS that is used to exchange their state information for the centralized coordinator(Matinez, Cortes et al. 2007). From geometrical persepective, the topological structure governs the shape stability or equivalently the graph rigidity(Lin, Francis et al. 2005) of the MRFS.

Reviewing with the information aspect, the interconnected structure is naturally constructed with respect to the physical limitation of the sensors or the communication power of the WMRs, *i.e.*, the resolution of the range sensor or the communication range *etc.* Sometimes, the interconnected structure may be assigned autonomously by the task manager who receives the global information of the systems and manages the performance of the task. The MRFS could surely be explained through the hierarchical system, ref. (Singh 1977). We emphasize that the MRFS is topologically constructed by the two hierarchy with low(subsystem) and high(interconnected system) level in this research. In most of the applications, the hierarchical system is diffeomorphic from low to high level or vice versa, see (Pappas, Lafferriere et al. 2000). However, in our research, it is complex beyond that: for the proposed hierarchical system, if we define an injective map: $f_M : \mathbb{R}^m \to \mathbb{R}^n$, it is said that f_M is a smooth map or, furthermore we said that the interconnected system is locally submersion to the subsystems of the MRFS.

The field of the nonholonomic system has been extensively studied since it is founded from Euler and Lagrange's research result in the rolling without sliding mechanical system on a plane. Recently, the related control issues in the nonholonomic system are still an active research area either in the mathematics or in the engineering field, see (BLOCH, DRAKUNOV et al. 2000; Monforte 2002; Fernandez and Bloch 2008). Without loss of generality, the subject of the subsystem in this research can be practically restricted to a Wheeled Mobile Robot (WMR) so that the subsystem belongs to the nonholonomic system which statisfies both no sliding and no slipping constraints. More precisely, the motion of the nonholonomic system is physically subject to the nonholonomic constraints. Hence, we say that the MRFS is again the nonholonomic system so that the unreachable region locally varied with the instant of the velocity of the WMRs in a team. With this aspect, the research in (Brokett 1983) has indicated that there doesn't exist smooth feedback controller for a single WMR. Fortunately, the later works in (Koiller 1992; Murry and Sastry 1993) proposed that the periodic solution can be adopted for the control of the nonholonomic system. For instance, the asymptotically stable of the WMR is guaranteed.

Therefore, we consider the ways of relating the interconnected stability(or so-called the graph rigidity) to the stability of the interconnected nonholonomic subsystems. More prcisely, the task of the MRFS is generally classified as formation maintain and the formation switch. The task of the formation maintain is simply to maintain the formation shape when moves from the start point to the end point so that the control goal is to stabilize the formation variables and the subsystem variables simetinuously by the given trajectory, see (Chang and Fu 2008). Several elegant researches have been proposed in terms of the issue. The disctribute formation control architecture with respect to the consensus problem of the communication system is experimentally implemented and valited on the neighbour to neighbour information exchange in (Ren and Sorensen 2008). The control strategy with the input constraint in associated with the leader-follower control architecture of the MRFS has been obtained (Consolinia, Morbidib et al. 2008). A game theoretic modelling approach for the MRFS is provided by (Harmati and Skrzypczyk 2008). (Kaminka, Schechter-Glick et al. 2008) has been proposed the sensor based MRFS to achieve the minimal cost sensing system design. In addition, for overcomimg the trade off in the control goal between formation system and subsystems, a differential game approach has been used to model such the problem and a differential game based controller has been derived in (Kaminka, Schechter-Glick et al. 2008). Moreover, the construction for the MRFS can be additionally identified as virtual structure approach, leader-follower approach and the behaviour approach.

Also. the task of the formation switch is naturally defined that the swithing topology of the MRFS is physically performed while executing the task. Comparing to the proceed problem in the formation maintain, the switching stability has to be additionally cared particullarly in the nonholonomic system. Namely, it is lead to the topological structure switch of the formation system such that two main issues have to be concerned: the structure stability of the MRFS and the subsystem stability under the impluse response. Additionally, the impulse response of the nonholonomic system may lead to relase the nonholonomic constrains due to the desired state of the subsystem may locally stay on the unreachable region. The decentralized receding horizen controllers has been proposed in (Keviczky, Borrelli et al. 2008) that reside on each vehicle to achieve coordination among team members. The control of the MRFS which considered the changing formation has been studied in (Desai, Ostrowski et al. 2001). (Das, Fierro et al. 2002) has developed a framework for cooperative control of the MRFS that has been applied in the vision-based formation system. We do not have a study to the overall problem in the MRFS but we focus the problem of the formation stability analysis and the nonholonomic multi-robotic formation control design with respect to the capacity of the switching formation topology on-line which extends from the previous work(Chang and Fu 2008). This chapter is organized as follows: the general modelling of the MRFS is presented in Section 2. Additionally, the interconnected stability and formation control design is examined in Section 3. In Sections 4, the simulation are presented. Finally, in Section 5, the conclusions are made.

2. General Modelling of the MRFS

Consider a MRFS with the formation state : $z = \left\{ z_{ij} \in \mathbb{R}^2 \middle| i \neq j; 1 \leq i, j \leq n \right\}$. The MRFS is composed by n WMRs whose state can be described as a vector matrix: $q = [q_1 \ \cdots \ q_n] \in \mathbb{R}^{3n \times 2}$. More precisely, the ith WMR with the jth interconnection in a MRFS implies $z_{ij} \neq \phi$ if the ith WMR connects to the jth WMR. Suppose that the desired interconnected structure and a desired formation state $z_d = \left\{ z_{dij} \in \mathbb{R}^2 \middle| i \neq j; 1 \leq i, j \leq n \right\}$ are given and a virtual formation center $q_c \in \mathbb{R}^2$ which locates inside the closed region of the formation shape is chosen. The kinematics of the WMR can be generally regarded as an driftless affine control system: $\dot{q}_i = \sum_{j=1}^{k} g_j(q_i) u_j$ with u_j being the control; k being a constant number of the control variables.

Now the virtual center of the MRFS moves along a desired trajectory $c_d(t)$ so that the MRFS is driven from an initial state $q_{c0} = c_d(t_0)$ to a final state $q_{cf} = c_d(t_f)$ where t_0 and t_f denote the initial and final time respectively in addition to maintain the given formation shape derived from the desired formation state simultaneously. It is obviously that the set of the desired state $\{q_{d1}, q_{d2}, \cdots, q_{dn}\}$ of the WMRs in the formation team could be obtained with the well-known $c_d(t)$ and z_d, *i.e.*, $q_{di} = c_d + f(z_d)$ where f denotes a differentiable function. For all positive $\varepsilon_z, \varepsilon_q$, there exists positive $\delta_z, \delta_\varepsilon$ such that $\sup B_{zi}(\varepsilon_z) = r_{zi}$ and $\sup B_{qi}(\varepsilon_q) = r_{qi}$ where $B_{zi}(\varepsilon_z)$ and $B_{qi}(\varepsilon_q)$ denote small enough balls ; r_{zi} and r_{qi} are the radius of the balls respectively;. Now we set $\varepsilon_z = \min(\delta_z, r_i)$, and the following definitions can be made:

Definition 2.1(Interconnection stable): Consider a nonholonomic MRFS with its interconnected structure. Initially, we set $\sum_j \left\| z_{ij}(t_0) - z_{dij}(t_0) \right\| \leq \delta_z$. If there exists the following condition: $\lim_{t\in[t_0,t_f]}\sum_j\left\|z_{ij}(t)-z_{dij}(t)\right\|\le\varepsilon_z$ for all i, the MRFS is said to be *interconnection stable.*

Definition 2.2(Formation system stable): Let z_{ij} be continuous in t. The equilibrium point $z_{ij}(0)=0$ and $q_i(0)=0$ information variable and individual variable respectively for all i,j is

- *formation system stable*: Definition 2.1 holds and if there exists $\sum_j\left\|z_{ij}(t_0)-z_{dij}(t_0)\right\|\le\delta_z$; $\left\|q_i(t_0)-q_{di}(t_0)\right\|\le\delta_q$ then $\lim_{t\in[t_0,t_f]}\left\|q_i(t)-q_{di}(t)\right\|\le\varepsilon_q$, for all i;
- *asymptotically formation system stable*: Definition 2.1 holds and if there exists $\sum_j\left\|z_{ij}(t_0)-z_{dij}(t_0)\right\|\le\delta_z$; $\left\|q_i(t_0)-q_{di}(t_0)\right\|\le\delta_q$ then $\lim_{t\to\infty}\left\|q_i(t)-q_{di}(t)\right\|\to 0$, for all i;
- *formation system unstable*: if it is not formation system stable.

According to Definition 2.2, if the MRFS has the formation system stable, one of the necessary condition is that the interconnection sable has to be held. On the contrary, the interconnection stable cannot be the necessary condition for the formation system stable. In other words, the interconnection stability is clearly defined as the sufficient condition for achieving the formation stable. The formation system stability, no doubt, is thus based on the interconnection stable and the subsystem stable simeltineously. In addition, we have proved that if the Definition 2.2 is commitment, then the final state of the WMRs in the MRFS will be reached: $q_{df}=c_d(t_f)$, in section IV.

Remark 2.3: Considering the Definition 2.2, the following condition yields:

- if there exists $\lim_{t\in[t_0,t_f]}\left\|q_i(t)-q_{di}(t)\right\|\le\varepsilon_q$ then $\lim_{t\in[t_0,t_f]}\sum_j\left\|z_{ij}(t)-z_{dij}(t)\right\|\le\varepsilon_z$;
- if there exists $\lim_{t\to\infty}\left\|q_i(t)-q_{di}(t)\right\|\to 0$ then $\lim_{t\to\infty}\sum_j\left\|z_{ij}(t)-z_{dij}(t)\right\|\to 0$.

Thus, the formation system stable can be guaranteed by evaluating the convergence property of the individual states while performing the full state formation tracking.
As we know, the formation variables: the relative length and the relative heading angle, is abstracted from a collection of the states of nonholonomic WMRs. Also, the formation states can be written by general functions:

$$z_{ij}=\begin{bmatrix}l_{ij}\\ \varphi_{ij}\end{bmatrix}=\begin{bmatrix}f_{pij}\left(q_{pi},q_{pj}\right)\\ f_{ij}\left(q_{pi},q_{pj},q_{\theta i},q_{\theta j}\right)\end{bmatrix}\in Q$$

with $q_i=\begin{bmatrix}q_{pi} & q_{\theta i}\end{bmatrix}^T\in N_i$ and $q_j=\begin{bmatrix}q_{pj} & q_{\theta j}\end{bmatrix}^T\in N_j$ where $Q\in\mathbb{R}^m, N_i\in\mathbb{R}^k$ and $N_j\in\mathbb{R}^k$ denote the compact and differentiable manifolds.
Suppose the desired formation states are given and the formation system satisfies the condition of interconnection stable such that the solution of the individual states may not unique. For example, $\left(q_{pi},q_{pj}\right)=f_{pij}^{-1}\left(l_{ij}\right)$ and $\left(q_{pi},q_{pj},q_{\theta i},q_{\theta j}\right)=f_{ij}^{-1}\left(\varphi_{ij}\right)$, there are two equations but more than two unknown variables in both of the equations. Figure 1 shows the illustrated scenario with three WMRs in the MRFS.

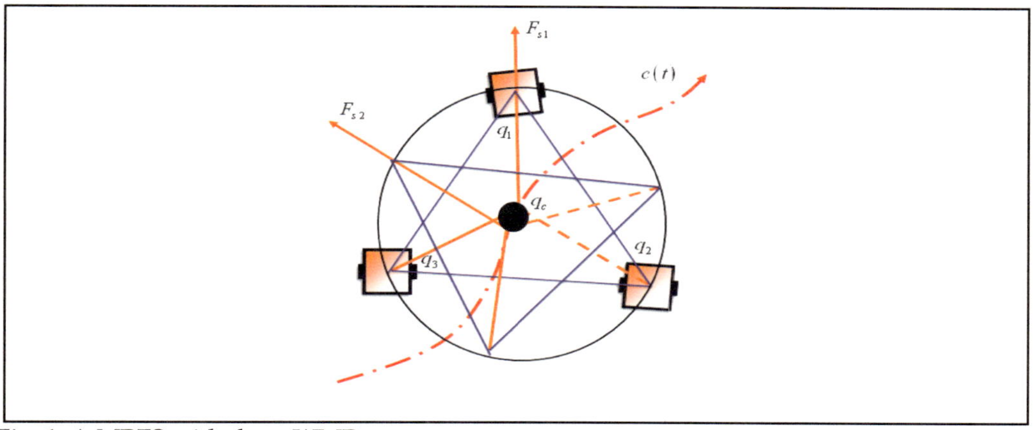

Fig. 1. A MRFS with three WMRs.

In Figure 1, the interconnected structures: F_{s1} and F_{s2}, are both the solutions. If the additional nonholonomic constraints in each of the WMRs are called the nonholonomy, the design challenge of the MRFCS immediately arises that there may be infinite solutions or conversely no solutions. Thus we can conclude that the conditions of the solution depends on the nonholonomy. We can further explain that the nonholonomic constraint always forbids locally to reach some of the neighborhood of the WMR so that the nonholonomic system with redundent nonholonomy or holonomy equations(ususally the total equation number is over or equal to the dimension of the system) may not have physical solution.

Now we set oriented direction of the MRFS from q_c to q_1 tangent to the desired path $c(t)$, see Figure 1. With respect to the interconnection stability and the subsystem stability, Definition 2.2 shall be further modified.

Definition 2.4: Let z_{ij} be piecewise continuous in t. The equilibrium point $z_{ij}(0)=0$ and $q_i(0)=0$ in formation variable and individual variable respectively for all i,j is

- *formation system stable*: Definition 2.1 holds and if there exist $\sum_j \|z_{ij}(t_0)-z_{dij}(t_0)\| \le \delta_z$ and $\|q_i(t_0)-q_{di}(t_0)\| \le \delta_q$ then $\lim_{t\in[t_0,t_f]} \sum_j \|z_{ij}(t)-z_{dij}(t)\| \le \varepsilon_z$ and $\lim_{t\in[t_0,t_f]} \|q_i(t)-q_{di}(t)\| \le \varepsilon$, for all i;
- *asymptotically formation system stable*: Definition 2.1 holds and if there exist $\sum_j \|z_{ij}(t_0)-z_{dij}(t_0)\| \le \delta_z$ and $\|q_i(t_0)-q_{di}(t_0)\| \le \delta_q$ then $\lim_{t\in[t_0,t_f]} \|q_i(t)-q_{di}(t)\| \le \varepsilon_q$ and $\lim_{t\to\infty} \|q_i(t)-q_{di}(t)\| \to 0$, for all i;
- *formation system unstable*: if it is not formation system stable.

No doubt, Definition 2.4 is more rigorous than Definiton 2.2 particularly it can be put on the condition after releasing the constraints on the formation state. So far, we got two unsolved problems in the design of the MRFS: first, the the uniqueness of the solution; second, the subsystem stability with respect to the interconnection stability.

For the first point, coneptually, the key step is how to select the adequate stable interconnected structure which corresponds to the number of the additional constraints.

Actually, this idea is simple but it is much complex than we expect in the design process resulted by the nonholonomic system of the WMR. As we know, the choice of the state of the MRFS can be either the relative length or the relative angle or even mix both of them and they are all capable to be the abstractive variables which are abstracted from the states of the nonholonomic subsystems. There also exists the nonlinear transfomation between the position and the oriented angle of the WMR so that, in the MRFS, the relative length couples the relative angle or vice versa. We, therefore, usually select one of them as the abstractive variables for simplifing the design complexity. With this aspect, if the minimal interconnected structure of the MRFS is performed, the process is the way regarded as to release some redundent abstracted equations. In this research, for this issue, we have proposed the minimal relization with respect to the stable interconnected structure in the controller design of the MRFS.
The second issue requires more detail study on the nonholonomic system. The nonholonomic constraints are assumed to be strictly satisfied in this research for applying the kinematics of the WMR. Hence, the output of the control velocity and the angular velocity is limited for avoiding to generate the large torque of the WMR. It immediately implies us that the unreachable region of the nonholonomic system is locally restricted by the limited torque. In real application of the MRFS, the desired state is usually given in the abstracted space. When we switch the interconnected topology, following the Remark 2.3, the nonholonomic subsystem may not be stable if $\lim_{t\to\infty}\sum_j \left\| z_{ij}(t) - z_{dij}(t) \right\| \to 0$. In this research, the Lyapunov based approach is proposed for dealing with this design issue.

3. Interconnected Stability and Formation Control Design

Formally, considering the nonholonomic constraints in a differential type WMR, the kinematics is able to be written by

$$\dot{q}_i = S_i u_i \tag{1}$$

where $q_i = \begin{bmatrix} q_{pi} & q_{\theta i} \end{bmatrix}^T \in \mathbb{R}^3$ denote the state of the WMR; $S_i = \begin{bmatrix} 0 & 0 & 1 \\ \cos q_{\theta i} & \sin q_{\theta i} & 0 \end{bmatrix}^T$ denotes the distribution; $u_i = \begin{bmatrix} v_i & w_i \end{bmatrix}^T \in \mathbb{R}^2$ denotes the control input. The formation state between two WMRs is distinctly defined as

$$z_{ij} = \begin{bmatrix} l_{ij} \\ \varphi_{ij} \end{bmatrix} \triangleq \begin{bmatrix} \left\| q_{pj} - q_{pi} \right\| \\ \left\| q_{\theta j} - q_{\theta i} \right\| \end{bmatrix} \tag{2}$$

In contrast to the relative formulation with two WMRs, the formation state to the ith WMR with respect to all j^{th} connection without regarding with the interconnection structure is simply defined as the sum of the relative state:

$$z_i = \sum_j z_{ij} = \begin{bmatrix} \left\| q_{p1} - q_{pi} \right\| \\ \left\| q_{\theta 1} - q_{\theta i} \right\| \end{bmatrix} + \cdots + \begin{bmatrix} \left\| q_{pn} - q_{pi} \right\| \\ \left\| q_{\theta n} - q_{\theta i} \right\| \end{bmatrix} \tag{3}$$

and if $i = j$, $z_{ij} = 0$. Taking partial derivative to Eq. (3), we have the following equation:

$$\dot{z}_i = \begin{bmatrix} \sum_j \dot{l}_{ij} \\ \sum_j \dot{\varphi}_{ij} \end{bmatrix} = \sum_j \left(\frac{\partial z_i}{\partial q_i} + \frac{\partial z_i}{\partial q_j} \right) \tag{4}$$

For a MRFS, the neighbours of the ith WMR is noted as $q_j \sim q_i$ which corresponds to the interconnected structure and can be equivalently interpreted as an adjacency matrix. The adjacency matrix(Chung 1949) (or so-called interconnection matrix), A_G , is represented as a binary matrix which is one-one maps from the interconnected structure to the elements of the matrix, *i.e.*, q_j acts on q_i if the element in i^{th} row and j^{th} column of the matrix equals "1", $A_G(i,j)=1$ but if $i=j$, $A_G(i,j)=0$. It is the fact that all of the connections of the ith WMR to the neighbour ones are a set: $a_{ij} = \{A_G(i,j) | 1 \le j \le n\}$ where i and j denotes the ith raw and j^{th} column in the adjacency matrix. Therefore Eq. (4) could be naturally rewritten as

$$\dot{z}_i = \sum_j \frac{a_{ij}}{l_{ij}} \begin{bmatrix} q_{pij}^T I_2 \dot{q}_{pij} \\ q_{pij}^T J_2 \dot{q}_{pij} \end{bmatrix} = \sum_j \begin{bmatrix} q_{pij}^T \Omega_{ij}^I \dot{q}_{pij} \\ q_{pij}^T \Omega_{ij}^J \dot{q}_{pij} \end{bmatrix} \tag{5}$$

with $q_{ij} = \begin{bmatrix} q_{pij} & q_{\theta ij} \end{bmatrix}^T \in \mathbb{R}^3$; $\Omega_{ij}^I = \frac{a_{ij} I_2}{l_{ij}}$; $\Omega_{ij}^J = \frac{a_{ij} J_2}{l_{ij}}$; $I_2 = \begin{bmatrix} 1 & 0 \\ 0 & 1 \end{bmatrix}$; $J_2 = \begin{bmatrix} 0 & -1 \\ 1 & 0 \end{bmatrix}$.

Now we summarize the result to the general formation dynamics form Eq. (1) and Eq. (5):

$$\left.\begin{aligned} \dot{z}_1 &= \sum_j a_{1j} \left(\frac{\partial z_1}{\partial q_i} + \frac{\partial z_1}{\partial q_j} \right) \\ &\vdots \\ \dot{z}_n &= \sum_j a_{nj} \left(\frac{\partial z_n}{\partial q_i} + \frac{\partial z_n}{\partial q_j} \right) \end{aligned}\right\} 2n \tag{6}$$

$$\left.\begin{aligned} \dot{q}_1 &= S_1 u_1 \\ &\vdots \\ \dot{q}_n &= S_n u_n \end{aligned}\right\} 3n \tag{7}$$

There are totally $5n$ equations in Eq. (6-7). Obviously, a number of $3n$ physical variables need to be solved so that we can freely choose $2n$ equations as a constraints, for example, minimizing Eq.(7) subject to Eq.(6) or minimizing the position subject to the heading angle of each WMRs and Eq.(6) and so forth. However, regarding with the interconnected structure, two problems yield: first, how to determine the minimal stable interconnected structure; second, how to guarantee the existence of the solution. For the first question, the following lemma will help us to make such a design:

Lemma 3.1: Considering the MRFS with a selective interconnection structure with totally p connections, the stable minimal connection number of p is $2n-3$.

The proof follows the rigidity condition of the two dimensional graph, see (Laman 1970).

Now we begin with the second question for the existence of the MRFS. The existence of the solution is somehow linked to the subsystem stability if the designed nonholonomic control can derive the WMR to the admissible region within the control time. In other words, the existence of the solution is in the sense that there locally exist the reachable states of the

nonholonomic subsystem such that the WMR moves within the reachable region such that the sufficient condition of the subsystem stability is achieved.
Moreover, the coupling effect of the states in the WMR has to be considered. The state equation in Eq. (1) can be generally rewritten as

$$\begin{aligned} \dot{q}_{pi} &= f_{pi}(q_{\theta i})v_i \\ \dot{q}_{\theta i} &= w_i \end{aligned} \tag{8}$$

where $f_{pi}:\mathbb{R}\to\mathbb{R}^2$ denotes a continuous and differentiable function; v_i and w_i denote the velocity and angular velocity respectively. Eq. (8) clearly represents the coupled effect between q_{pi} and $q_{\theta i}$ in the nonholonomic system. It may be safety to assume that the velocity is a constant in the practical control design, the position and oriented angle can be derived by the assigned angular velocity simultaneously due to non-invloutive characteristic from Frobenious Thorem(Abraham and Marsden 1967). Conversely, if we set the angular velocity as a constant, the WMR is restricted to move along a line for the constrained oriented angle in the abstracted space. (BLOC and CROUC 1998) has indicated the general design rule of the nonholonomic control design which is stated in the following Remark:

Remark 3.2: Consider the nonholonomic system in Eq. (8). The system stability holds if the controller is designed for the WMR whose convergence rate of $q_{\theta i}$ is always faster than the one of q_{pi}.

Remark 3.2, for the MRFS, implies us that the subsystem stability is able to be designed by choosing the interconnected structure with respect to the relative length which is the function of q_{pi}. Through the way, another variable $q_{\theta i}$ is set free and is configurable. Therefore, the MRFS will be stable if the controller of the MRFS is carefully designed for satisfying Remark 3.2. Hence the formation dynamics for the ith WMR in Eq.(5) could be further reduced:

$$\dot{z}_i = \sum_j \frac{1}{l_{ij}} a_{ij}\left(q_{pij}^T I_2 \dot{q}_{pij}\right) = \sum_j q_{pij}^T \Omega_{ij}^I \dot{q}_{pij} \tag{9}$$

Rearranging the equation, the canonical form of the MRFS is further obtained with Eq. (6):

$$\begin{cases} \dot{z}_i = \sum_j q_{pij}^T \Omega_{ij}^I \left(f_{pj}(q_{\theta j})v_j - f_{pi}(q_{\theta i})v_i\right) \\ \dot{q}_{\theta 1} = w_1 \\ \vdots \\ \dot{q}_{\theta n} = w_n \end{cases} \tag{10}$$

Corollary 3.3: Consider the formation dynamics in Eq. (10), the state flow of the MRFS is equivalent to the state flow of the nonholonomic WMR. It can generally be written as the following formula:

$$\begin{aligned}
\dot{z}_i &= \sum_j f_i^1\left(a_{ij}, z_{ij}, q_{pi}, q_{pj}, q_{\theta j}\right) + \left(\sum_j f_i^2\left(a_{ij}, z_{ij}, q_{pi}, q_{pj}, q_{\theta i}\right)\right) v_i \\
\dot{q}_{\theta 1} &= w_1 \\
&\vdots \\
\dot{q}_{\theta n} &= w_n
\end{aligned} \tag{11}$$

Figure 2. shows the nonholonomic hierarchical structure in the nonholonomic formation dynamics in Eq. (11).

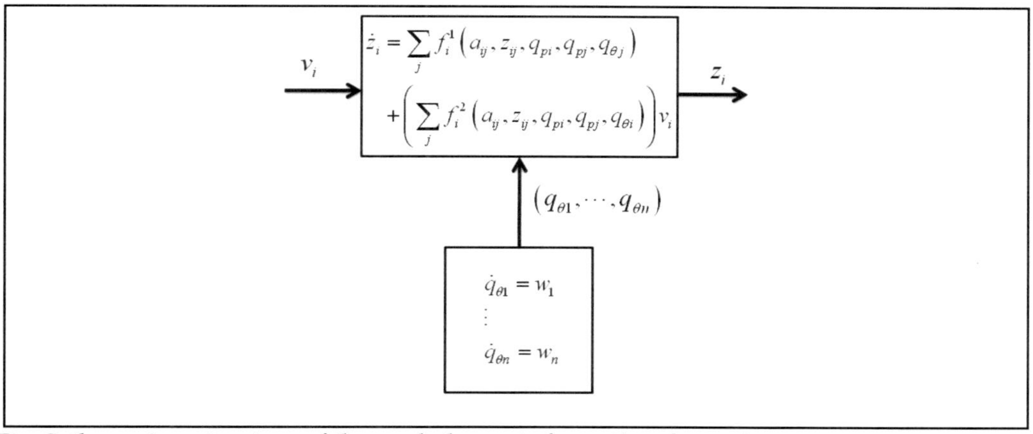

Fig. 2. the system structure of the nonholonomic formation dynamics.

Remark 3.4: Considering the MRFS, the interconnection matrix can be regarded as a linear operator of the formation dynamics.

For the Remark 3.4, an immediately result can be observed in Eq. (10). Hence, once the interconnected structure of the MRFS changes on-line so as to the interconnection matrix, the formation shape is able to be dynamically modified by applying the operator with the refreshed interconnection matrix. It is helpful in the implementation of the MRFS.

Now we shall prove the following statement: the interconnection stable is hold if and only if all of the eigenvalues of the interconnection matrix is positive. Purposely, the Lyapunov approach is adopted for minimizing the energy generated from the individual WMRs and the formation system. We select the Lyapunov function: $L_i = \frac{1}{2} a_{ii} q_i^T q_i$, in each of the subsystem. This leads into the convergence rate of the heading angle of the WMR could be under our control. For helping the judgement, we also define the interconnection Lyapunov function: $L_{ij} = \sum_{j:j\neq i} \frac{1}{2} a_{ij} z_{ij}^T z_{ij}$. Following these definitions, the formation Lyapunov function L_i^F can be simply split into two parts: the individual Lyapunov function of the ith WMR and the interconnection Lyapunov functions of the jth WMR which acts on the ith WMR:

$$L_i^F = L_i + \sum_j L_{ij} \tag{12}$$

In Eq. (12), L_i is generated from the ith subsystem and $\sum_j L_{ij}$ is produced by the interconnection of the MRFS for the ith subsystem. In the component form, it is able to be written as

$$L_i^F = \frac{1}{2}\begin{bmatrix} q_{pi} & q_{\theta i} \end{bmatrix} P_i \begin{bmatrix} q_{pi} \\ q_{\theta i} \end{bmatrix} + \frac{1}{2} A_{Gi} \begin{bmatrix} z_1^T z_1 & \cdots & z_n^T z_n \end{bmatrix}^T \quad (13)$$

where $P_i \in \mathbb{R}^{3\times 3}$ denotes the positive diagonal matrix of the ith WMR; A_{Gi} denotes the ith raw of the interconnection matrix. Hence the necessary condition for the asymptotically formation stable is established via the following theorem:

Theorem 3.5: Considering the MRFS described in Eq. (11), the system, follows Definition 2.2, is said to be asymptotically interconnection stable.

Proof. Using Eq. (9), the time derivative of the Eq. (12) can be written as:

$$\sum_j \dot{L}_{ij} = \frac{1}{2}\left(\sum_j \dot{q}_{pij}^T \Omega_{ij}^I q_{pij} + q_{pij}^T \Omega_{ij}^I \dot{q}_{pij} \right) = \frac{1}{2}\left(\sum_j q_{pij}^T \left(F_i \Omega_{ij}^I + \Omega_{ij}^I F_i \right) q_{pij} \right) \quad (14)$$

where $F_i = \partial f_{pi} / \partial q_i$ denotes a linearized matrix from the nonlinear function f_{pi} in Eq. (8). In order to state the stability condition on the MRFS, the Lyapunov function can be reproduced by Eq. (14) from single WMR to all WMRs in a formation team. Thus we reformulate the result in Eq. (14) in associated with a matrix formula:

$$F\Omega^I + \Omega^I F = -Q \quad (15)$$

where Q_i are positive matrix. According to the Lyapunov stability theorem, if Ω^I and Q_i are positive definite, then the MRFS in Eq. (11) is asymptotically stable. Q. E. D.

So far, the analysis result of the interconnection stability reveals us that the sufficient condition of the formation stable satisfies not only the existence of the positive definite interconnection matrix but also the subsystem stable by the Definition 2.4. Namely, if the formation stable holds, the necessary condition is that the interconnection matrix has to be positive definite. Note that the formation dynamics can be identified without driving the formation dynamics via Theorem 3.5. Practically, let us now consider the design of the control of the MRFS. The Lyapunov function in Eq. (12) can be further taken the partial derivative:

$$\begin{aligned} \dot{L}_i^F &= \frac{\partial L_i}{\partial q_i} + \sum_j \frac{\partial L_{ij}}{\partial q_j} \\ &= \sum_j f_i^1\left(a_{ij}, z_{ij}, q_{pi}, q_{pj}, q_{\theta j}\right) + \left(q_{pi} S_i + \sum_j f_i^2\left(a_{ij}, z_{ij}, q_{pi}, q_{pj}, q_{\theta i}\right) \right) v_i + q_{\theta i} w_i \end{aligned} \quad (16)$$

Therefore, the formation control can be chosen by the following Theorem:

Theorem 3.6: Considering the MRFS follows Eq. (11), if the velocity and angular velocity is chosen by:

$$v_i = \begin{cases} \left(\sum_j f_i^1\left(a_{ij}, z_{ij}, q_{pi}, q_{pj}, q_{\theta j}\right) - K_{pi} L_i^F \right) \Big/ \left(q_{pi} S_i + \sum_j f_i^2\left(a_{ij}, z_{ij}, q_{pi}, q_{pj}, q_{\theta i}\right) \right); \\ 0 \quad \text{if} \left(q_{pi} S_i + \sum_j f_i^2\left(a_{ij}, z_{ij}, q_{pi}, q_{pj}, q_{\theta i}\right) \right) = 0; \end{cases} \tag{17}$$

$$w_i = -K_{\theta i} q_{\theta i}.$$

then the MRFS is exponentially stable where $K_{pi} \geq K_{\theta i} \geq 0$ denote the constant real number.

Proof: After taking the controller in Eq. (17) into Eq. (16), the Lyapunov equation is obtained:

$$\dot{L}_i^F = -K_{pi} L_i^F - \left(K_{pi} - K_{\theta i}\right) q_{\theta i}^2 \leq -K_{pi} L_i^F \tag{18}$$

Consequently, the system is exponentially stable.

Remark 3.7 According to Theorem 3.6, the controller is capable of switching the interconnection structure in real-time by modifying the parameter: a_{ij} .

Finally, the proposed formation stability theories and control design process in this section can be regarded as a useful tool.

4. Simulation

In this section, a simulation is performed for demonstrating the performance of the proposed nonholonomic multi-robotic formation control with respect to the formation stability. Figure 3 shows the simulation scenario with four WMRs in the MRFS. The team begins with the triangular shape and moves along a curve to the target with a square shape that shall change the interconnected structure on the middle way of the motion curve drawn as the solid line in Figure 3. Observing the interconnected structures, they satisfy the rigid condition which implies the interconnection stable of the MRFS in Lemma 3.1 so that the interconnection stability is promised by Definition 2.2.

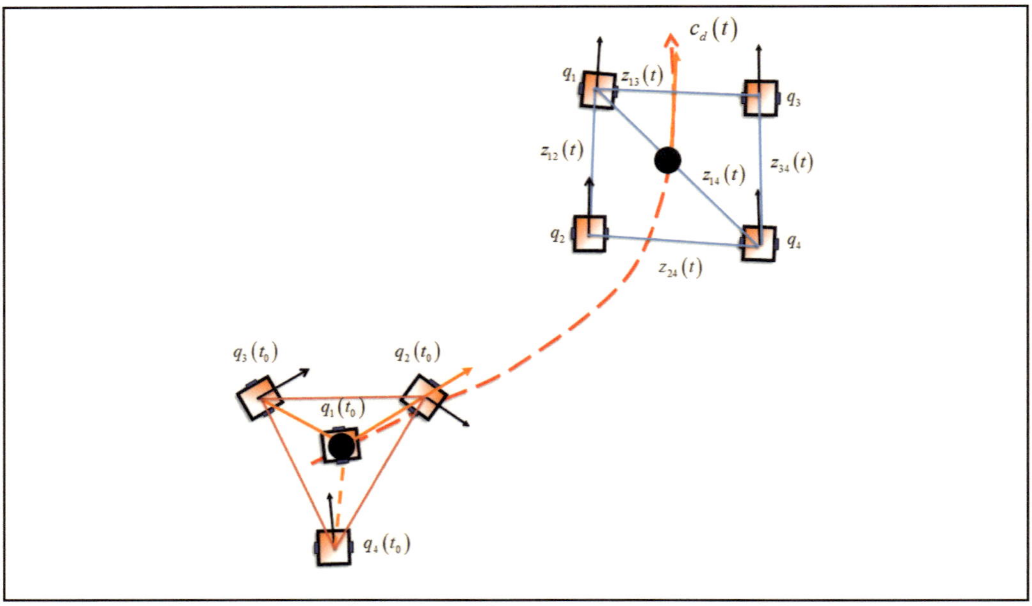

Fig. 3. the simulation scenario: from triangular to square structure of the MRFS.

In this simulation, we suppose that each of the WMRs is able to know the states from rest of the WMRs within the control time. Also, the physical configurations for the simulation are listed: the desired relative length is $l_{12} = l_{13} = l_{14} = 5(m)$; $l_{23} = l_{34} = l_{24} = 5\sqrt{3}(m)$ and the initial relative length is $l_{12} = l_{13} = l_{14} = 4(m)$; $l_{23} = l_{34} = l_{24} = 4\sqrt{3}(m)$ in the triangular shape and $l_{12} = l_{24} = l_{34} = l_{13} = 5(m)$; $l_{14} = 5\sqrt{2}(m)$ in the squire shape respectively. Considering the configuration of the single WMR, the initial oriented angles of the WMRs set to zero. The radius of the active wheels are $0.3(m)$ and the length of the axis of the active wheels is $0.5(m)$. Practically, the control time is set to $0.01(\sec)$ in each of the WMRs.

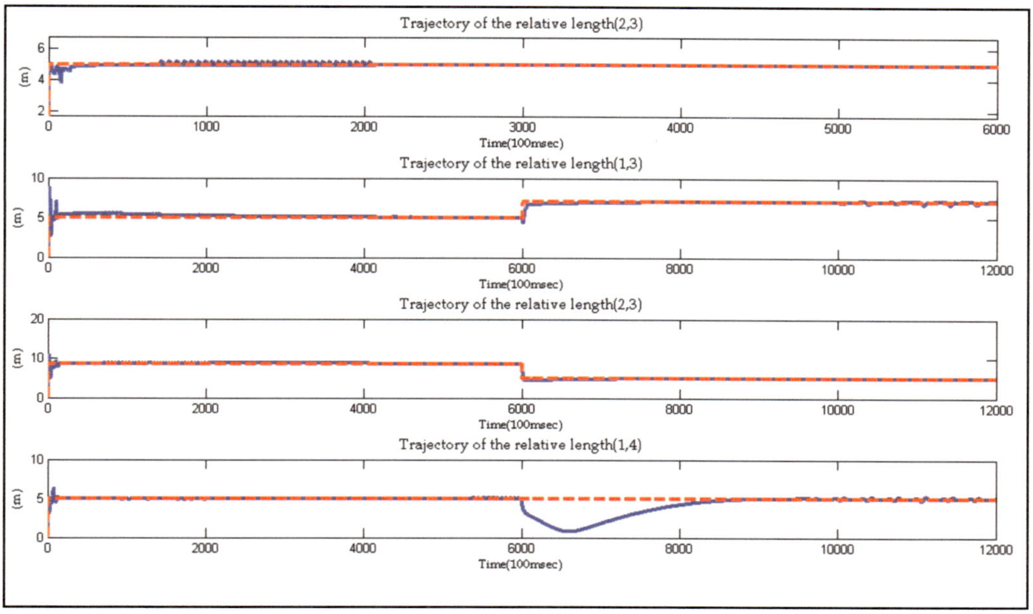

Fig. 4. The trajectory error of the relative length: $l_{23};l_{13};l_{23};l_{14}$.

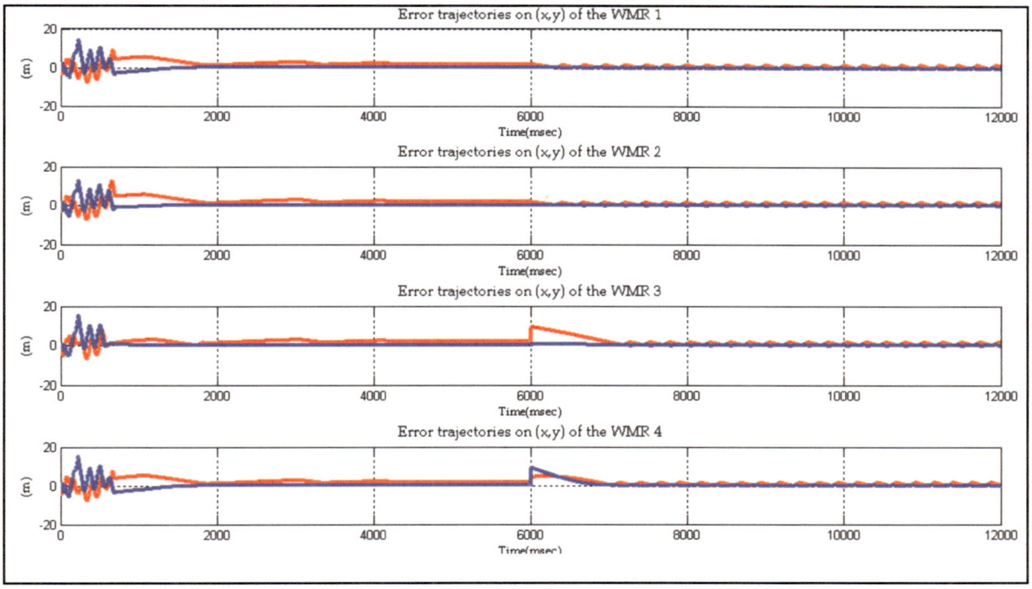

Fig. 5. The error trajectories on the X(red)-Y(blue) Plane from WMR 1-4.

The simulation results are drawn in Figure 4-5 where Figure 4 describes the relative lengths of the WMRs in the MRFS; Figure 5 draws the tracking error of the WMRs respectively. The diagrams indicate that the there exists impulse responses on each of the states of the subsystems when the interconnected structure is changed. In our proposed design, the subsystem stability can easily be handled.

5. Conclusion

The research reveal several important results: first, the formation stability could be hierarchically decoupled with the interconnection stability and the subsystem stability; second, the general framework of the MRFS with respect to the nonholonomic subsystems is obtained; third, the practical exponentially stable formation control is derived with respect to the minimal interconnection structure of the MRFS that can guarantee the subsystem stability. Clearly, our study provides a framework for designing and studying the modelling and the control problem in the nonholonomic MRFS. Finally, the simulation result shows the control performance so that the approach can be practically used in the switching interconnected structure of the MRFS on-line without adjusting any control parameters.

6. References

Abraham, R. and J. E. Marsden (1967). Foundations of mechanics. New York, W. A. Benjamin Inc.

BLOCH, A. M. and P. E. CROUC (1998). "NEWTON'S LAW AND INTEGRABILITY OF NONHOLONOMIC SYSTEMS." SIAM JOURNAL OF CONTROL OPTIMIZATION **36**(6): 2020-2039.

BLOCH, A. M., S. V. DRAKUNOV, et al. (2000). "STABILIZATION OF NONHOLONOMIC SYSTEMS USING ISOSPECTRAL FLOWS." SIAM JOURNAL OF CONTROL OPTIMIZATION **38**(3): 855–874.

Brokett, R. W. (1983). "Asymptotic Stability and Feedback Stabilization." Differential Geometric Control Theory: 181-191.

Chang, C.-F. and L.-C. Fu (2008). A Formation Control Framework Based on Lyapunov Approach. IEEE IROS, Nice, France.

Chung, F. R. K. (1949). Spectral Graph Theory, American Mathematical Society.

Consolinia, L., F. Morbidib, et al. (2008). "Leader–follower formation control of nonholonomic mobile robots with input constraints." Automatica.

Das, A. K., R. Fierro, et al. (2002). "A Vision-Based Formation Control Framework." IEEE TRANSACTIONS ON ROBOTICS AND AUTOMATION **18**(5): 813-825.

Desai, J. P., J. P. Ostrowski, et al. (2001). "Modeling and Control of Formation of Nonholonomic Mobile Robots." IEEE TRANSACTIONS ON ROBOTICS AND AUTOMATION **17**(6).

Fax, J. A. and R. M. Murray (2004). "Information Flow and Cooperative Control of Vehicle Formations." IEEE TRANSACTIONS ON AUTOMATIC CONTROL **49**(9).

Fernandez, O. E. and A. M. Bloch (2008). "Equivalence of The Dynamics of Nonholonomic And Variational Nonholonomic Systems For Certain Initial Data." Journal of physics A: Mathematical and Theoretical **41**: 1-20.

Harmati, I. and K. Skrzypczyk (2008). "Robot team coordination for target tracking using fuzzy logic controller in game theoretic framework." Robotics and Automated System.

Kaminka, G. A., R. Schechter-Glick, et al. (2008). "Using Sensor Morphology for Multirobot Formations." IEEE TRANSACTIONS ON ROBOTICS AND AUTOMATION **24**(2): 271-282.

Keviczky, T., F. Borrelli, et al. (2008). "Decentralized Receding Horizon Control and Coordination of Autonomous Vehicle Formations." IEEE TRANSACTIONS ON CONTROL SYSTEMS TECHNOLOGY, **16**(1): 19-33.

Koiller, J. (1992). "Reduction of Some Classical Nonholonomic Systems with Symmetry." Archive for rational mechanics and analysis **118**(2): 113-148.

Laman, G. (1970). "On Graphs and Rigidity of Plane Skeletal Structures." Journal of Engineering Mathematics **4**(4): 331-341.

Lin, Z., B. Francis, et al. (2005). "Necessary and Sufficient Graphical Conditions for Formation Control of Unicycles." IEEE TRANSACTIONS ON AUTOMATIC CONTROL **50**(1): 121-127.

Matinez, S., J. Cortes, et al. (2007). Motion Coordination with Distributed Information. IEEE Control System Magazine: 75-88.

Monforte, J. C. (2002). Geometric, Control and Numerical Aspects of Nonholonomic Systems. New Yourk, Springer Verlag.

Murry, R. M. (2007). "Recent Research in Cooperative Control of Multi-Vehicle System." Journal of Dynamics **129**: 571-583.

Murry, R. M. and S. S. Sastry (1993). "Nonholonomic Motion Planning: Steering Using Sinusoids." IEEE Transaction on Automatic Control **38**(5): 700-716.

Olfati-Saber, R. and R. M. Murray (2004). "Consensus Problems in Networks of Agents With Switching Topology and Time Delay." IEEE TRANSACTIONS ON AUTOMATIC CONTROL **49**(9): 1520-1533.

Pappas, G. J., G. Lafferriere, et al. (2000). "Hierarchically Consistent Control Systems." IEEE Transactions on Automatic Control **45**(6): 1144-1159.

Ren, W. and N. Sorensen (2008). "Distributed coordination architecture for multi-robot formation control." Robotics and Automated System **56**: 324-333.

Singh, M. G. (1977). Dynamical Hierarchical Control. New York, North-Holland.

Estimation of User's Request for Attentive Deskwork Support System

Yusuke Tamura, Masao Sugi, Tamio Arai and Jun Ota
The University of Tokyo Japan

1. Introduction

Since the late 1990s, several studies have been conducted on intelligent systems that support daily life in the home or office environments (Sato et al., 1996; Pentland, 1996; Brooks, 1997). In daily life, people spend a significant amount of time at desks to operate computers, read and write documents and books, eat, and assemble objects, among other activities. Therefore it can be said that supporting deskwork by intelligent systems is of extreme importance. Many kinds of intelligent systems have been proposed to provide desktop support. In particular, augmented desk interface systems have been eagerly studied. DigitalDesk is one of the earliest augmented desk interface systems (Wellner, 1993). It requires a CCD camera and a video projector to integrate physical paper documents and electronic documents. Koike et al. proposed EnhancedDesk, which uses an infrared camera instead of a CCD camera to improve sensitivity to changes in lighting conditions and a complex background (Koike et al., 2001). In addition, Leibe et al. proposed one called Perceptive Workbench, which requires both a CCD and an infrared camera (Leibe et al., 2000), and Rekimoto proposed SmartSkin, which is based on capacitive sensing without cameras (Rekimoto, 2002).
Raghavan et al. proposed a system that requires a head-mounted display to show how to assemble products (Raghavan et al., 1999). These systems have been limited to show some information to the user. Ishii & Ullmer proposed an idea referred to as "tangible bits (Ishii & Ullmer, 1997)," which seeks to realize a seamless interface among humans, digital information, and the physical environment by using manipulable objects. Based on this idea, they proposed metaDESK (Ullmer & Ishii, 1997).
Pangaro et al. proposed a system called Actuated Workbench (Pangaro et al., 2002), and Noma et al. proposed one called Proactive Desk (Noma et al., 2004). Both systems convey only information to the user through movement of physical objects. They do not support the user from physical aspects.
On the other hand, especially in rehabilitation robotics, several studies have been conducted on supporting humans working at desks from a physical aspect (Harwin et al., 1995; Dallaway et al., 1995). Dallaway & Jackson proposed RAID (Robot for Assisting the Integration of Disabled people) workstation (Dallaway & Jackson, 1994). In RAID, a user selects an object through a GUI, and a manipulator carries it to the user. Ishii et al. proposed a meal-assistance robot for disabled individuals (Ishii et al., 1995). The system user points a laser attached to his head to operate a manipulator. Topping proposed a system, Handy 1,

which assists severely disabled people with tasks such as eating, drinking, washing, and shaving (Topping, 2002). In these systems, every time a user wants to be supported, the user is required to consciously and explicitly instruct their intention to the systems. Such systems are not really helpful.

Moreover, a few studies have focused on the physical act of passing an object from a human to a manipulator, or vice versa (Kajikawa et al., 1995; Agah & Tanie, 1997). These studies focused on the realization of human-like motion of the manipulators. When a user needs to be supported, on the other hand, the systems are required to support the user as fast as possible. The studies did not consider the requirement.

In this study, we propose a robotic deskwork support system that delivers objects properly and quickly to a user who is working at a desk. The intended applications of the proposed system are assembly, repair, simple experiment, etc. In such applications, the system often cannot know a sequence of used objects by workers in advance. To achieve the objectives, the system fulfils two primary functions: It estimates the user7 s intention, and it delivers objects to the user.

Intelligent systems are used by ordinary people; therefore, it is important that the systems be intuitive and simple to use. One of the most intuitive ways to control such systems is using gestures, especially pointing (Bolt, 1980; Cipolla & Hollinghurst, 1996; Mori et al., 1998; Sato & Sakane, 2000; Tamura et al., 2004; Sugiyama et al., 2005). Although pointing is intuitive, it is bothersome for a user to explicitly instruct the systems every time he/she wants to get objects. Furthermore, as pointing direction can be determined only when the user's hand and finger remain stationary, the recognition process takes long time. In the approach proposed here, the system estimates a user's intention inherent in his action without explicit instructions. In fact, the system 1) detects a user's act of reaching, 2) predicts the target object required by the user by measuring continuous movement of his body parts, especially hands and eyes, and finally 3) delivers the object to a user (Figure 1).

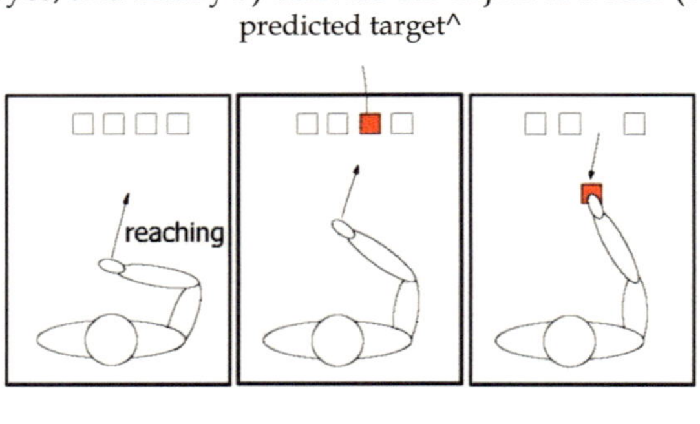

Fig. 1. Concept image of the proposed system

In this chapter, the first two items, involving detection and prediction, are mainly described and discussed.

For the third problem, it is unreasonable to use manipulators for carrying objects. Using manipulators for delivering objects has the following difficulties:

- Weight capacities of manipulators are generally low for their size.
- As manipulators move three-dimensionally, there is a tremendous danger in their high-speed movements.

- Because of the large size of manipulators, many manipulators cannot be operated simultaneously at a desk. Therefore, a manipulator can deliver a target object only after it grasps the object.

As a result, a system using manipulators cannot quickly and safely support a user. Moreover, small wheeled mobile robots present problems relative to speed and accuracy of movement.

One solution for the quick and accurate delivery of multiple objects to a user is to use movable trays driven with a Sawyer-type 2-DOF stepping motor (Sawyer, 1969). The motors are small and have high speed, positioning accuracy, and thrust.

The movable tray has high weight capacity, and moves only on a desk plane. Furthermore, because multiple trays can be placed simultaneously on a desk, multiple objects can be loaded on the trays. Therefore, a system using the movable trays can quickly and safely support a user.

In this chapter, we assume that our deskwork support system uses such movable trays and objects are loaded onto the trays. Assumed size of each tray is 130 x 135 x 25 (mm). In this study, we assume a normal size desk for the system. The width of a normal desk is at most 1200 (mm). According to this, the number of trays lined up in one row sideways is less than nine. In order to quickly deliver objects, a straight route is preferable for each tray. Even if the arrangement of the trays is schemed, the possible number of trays on a desk will be at most ten. We also assume that the distance between the trays and a user is greater than the user's reach. This assumption is for not obstructing a user's work.

In order to quickly deliver objects to a user, the trays are required not only to move fast but also to start early. Considering the speed of the user's hand and the movable trays, the preparation time for carrying objects (detection of the user's reach and prediction of the target object) should be less than a half of an average duration of reaching movements. According to a preliminary experiment, the average duration is about 0.8 (s) without any help. Therefore, the preparation time should be less than 0.4 (s).

In section 2, an algorithm used to detect reaching movement of a user is presented. A method used to predict a target object among multiple objects is described in section 3. In section 4, experiments for verifying the proposed method are described and discussed. In the experiments, the movable trays are not used. Experiments using the movable trays are presented in section 5. We conclude this chapter and refer to the future research in section 6.

2. Detection of human reaching movements

To deliver an object to a user, it is necessary that the system determine whether the user is performing an unrelated task or reaching for the object in question. When an individual reaches for an object, his hand and eyes move almost simultaneously toward the object. It has been reported that saccadic eye movement occurs before the onset of a reaching movement (Prablanc et al., 1979; Biguer et al., 1982; Abrams et al., 1990) and the saccade is followed about 100 (ms) later by a hand movement (Prablanc et al., 1979). In this study, therefore, a user's hand movements are measured to detect his reaching movements. When individuals perform tasks at desks, their hand movements are limited to a specific area, and their hands turn around frequently. When reaching for objects, on the other hand, individuals move their hands toward the outside of the working area at a high speed. The trajectories of hand movements are known to be relatively straight and smooth (Morasso, 1981). In addition to these characteristics of hand movements, eyes move toward a target object to localize the position of the object for guiding hand movements (Abrams et al., 1990). Based on the facts reported above, in this study, the deskwork support system interprets a hand movement as a reaching movement if the following conditions are satisfied:

- The speed of a hand movement is rapid,
- The trajectory of a hand movement is relatively smooth and straight, and
- The directions of the gaze and hand (see Figure 2) are close, and the hand and gaze point are far from the head position.

We define a hand movement as the trajectory of the center of a user's hand. To measure hand movements, we use a color CCD camera attached to a ceiling. The RGB video data is first converted to the hue, saturation, and value (HSV) space. These values are then thresholded to acquire binarized hand images. After that, we apply a morphological erosion operator to the obtained hand region until it becomes smaller than a predetermined threshold value, and the center of a user's hand is given as the resulting region's center of mass. This procedure makes the hand's center insensitive to changes of the shape of the hand image due to a closing or opening motion of the hand (Oka et al., 2002). A tracking system that requires no physical contact is used to measure head and eye movements.
The parameters are defined in Figure 2.

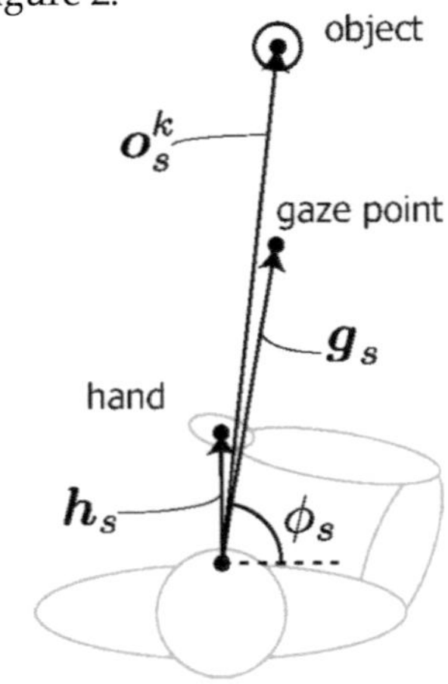

Fig. 2. Definition of parameters

Head and eye positions are measured three-dimensionally. However, in what follows, all positions are projected within a desk plane and are considered to be two-dimensional. Thus, all vectors are also two-dimensional. h_s is a vector from the user's head to user's hand at time s; g_s is a vector from the user's head to a gaze point; and a vector from the user's head to object k is denoted by O_s. In this study, v_s, the speed of a hand movement, is defined using the following equation:

$$v_s = \frac{\|h_s\| - \|h_{s-1}\|}{\Delta t} \tag{1}$$

where At is the sampling time of the camera.
To enable the system to determine whether a hand movement is a reach or some unrelated movement is difficult. Failures to detect the target movement can be eliminated by

integrating multiple criteria. Therefore, probabilities are established for three criteria, speed of hand movement, curvature of hand trajectory, and the relationship between the hand position and gaze point, which are used to detect the act of reaching.

2.1 Speed of a hand movement

The speed of a hand movement during reaching is much greater than that when performing tasks that occur close to the trunk of the body.
Therefore, we assume that the faster the relative speed of a user7 s hand to his head is, the higher the probability that the hand movement is an act of reaching will be. Here, we adopt a function whose output ranges between 0 and 1 and increases monotonically with its input as a probability function. Following this policy, we define R_v, the estimated probability from a hand speed at time s, as the following equation:

$$R_V = \frac{1}{1+\exp\{-\alpha(v_s-\beta)\}}, \tag{2}$$

where a and ft are parameters representing the motion characteristics of each user.

2.2 Curvature of a hand trajectory

In this study, the curvature of a user's hand trajectory is used as a criterion to indicate straightness and smoothness.
We regard the curvature of the circle passing through points h_{s-2}, h_{s-1}, and h_s as the curvature of the hand trajectory at time s (Figure 3).

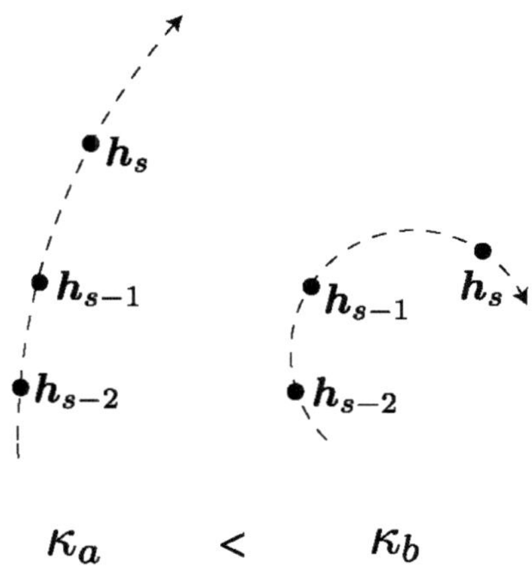

Fig. 3. Definition of the curvature of a hand trajectory at time

κ_s is the curvature of the hand trajectory at time s calculated by the following equation:

$$K_s = \frac{2|h_{s-1}\times h_{s-2} + h_{s-1} + h_{s-2}\times h_s|}{\sqrt{\|h_{s-1}-h_{s-2}\|\|h_s-h_{s-1}\|\|h_{s-2}-h_2\|}} \tag{3}$$

As reported earlier, reaching movements are generally straight and smooth. Therefore, the smaller the curvature of the hand trajectory, the greater the probability that the movement is an act of reaching. Based on this, we define ^s, the estimated probability from a hand trajectory at time *s*, with the following equation:

$$R_s = \gamma^{-K_s} \tag{4}$$

where y is a parameter representing the motion characteristics of each user.

2.3 Relationship between the hand position and gaze point

When an individual reaches for an object, he first locates the object and then reaches for it. To map the location of the target object, a saccadic eye movement occurs about 100 (ms) before the reaching motion begins (Prablanc et al., 1979), as reported above. Because the trajectories of reaching movements are relatively straight (Morasso, 1981), the gaze direction and the direction from head to hand are supposed to be almost equal during the act of reaching. Furthermore, in the act of reaching, the hand position and gaze point are farther from an individual's body (Figure 4-a) than during other unrelated tasks (Figure 4-b).

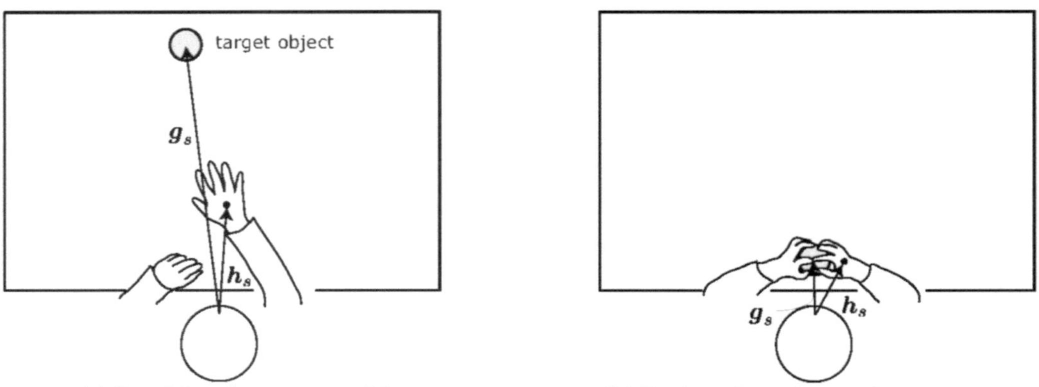

(a) Reaching for a target object (b) Performing other tasks

Fig. 4. Relationship between hand position and gaze

Based on these facts, we use the inner products of h_s and g_s at time *s* to detect acts of reaching.

$$I_s = h_S \bullet g_S \tag{5}$$

The large values of I_s suggest that the directions of the hand and gaze are close and the hand position and gaze point are far from the person's head. R_I, the estimated probability from the relationship between the hand position and gaze point at time s, is defined as follows:

$$R_I = \frac{1}{1 + \exp\{-\delta(I_S - \zeta)\}}, \tag{6}$$

where ó and Z are parameters representing the motion characteristics of each user.

2.4 Parameter determination

Because individuals differ in the motion characteristics of their hands and eyes, the five parameters reported above $(\alpha, \beta, \gamma, \delta, and\ \zeta)$ are required to determine a specific individual's characteristics.

In this study, we determine the parameters through the following sequence. In this chapter, we take two parameters of R_v $(\alpha$ and $\beta)$ as an example.

1. A hand movement while a user is performing some tasks and occasionally reaches for objects is measured.
2. Acquired velocity data v_s are discretized into several ranges, and times that a hand movement in a certain range of velocity is a reaching movement and those that the movement is not a reaching movement are counted respectively.
3. $p\left(reaching \middle| v_{B_i} \le v_s \prec v_{B_{i+1}}\right)$, the likelihood that a hand movement in a certain

   range of velocity is a reaching movement, is plotted as Figure 5, where v_{B_i} is the boundary value between range *i-1* and range i. Here, because we assume that the system cannot determine a prior probability p(reaching), we normalize the observed data.
4. A curve represented as (2) is fitted to the plotted data points by the Levenberg-Marquardt method, which is a non-linear least-squares method, to acquire *a* and *i5*. Here, we adopt the inverse of the probabilities

   $p\left(reaching \middle| v_{B_i} \le v_s \prec v_{B_{i+1}}\right)$ as weight factors for fitting. An example of fitting

   R_v curve

   to the observed data points is shown in Figure 5.

The other parameters are determined in the same way.

2.5 Detection of reaching movements

The proposed system detects a hand movement as a reaching movement when *R,* the integrated probability at time *s,* exceeds the predetermined threshold value. *R* is defined as follows:

$$R = R_v R_s R_i \tag{7}$$

To reduce missed detections, a large threshold value *r]* was used to indicate the distance between the user's head and hand, which acts as a safety net. Even if *R* does not exceed the predetermined threshold value, the hand movement is detected as an act of reaching when $||\ \mathbf{h}_s\ ||$ is larger than *n* . *n* is a sufficiently large value which is used to prevent missed detections, and it is empirically set to 400 (mm).

Moreover, *R* is set to 0 for a given length of time after detection, where the length of time is empirically set to 1.0 (s). This rule is used to prevent false detections.

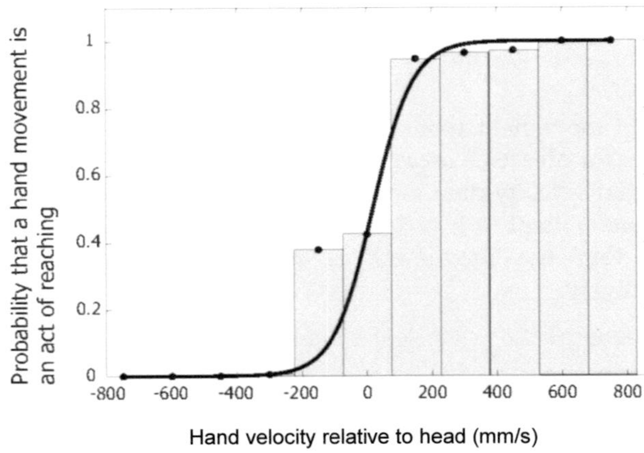

Fig. 5. Fitting a curve of the estimated Rv to the observed probabilities that a hand movement is a reaching movement

3. Prediction of the target object among multiple objects

People usually use multiple objects at a desk, and the sequence in which the objects are used is not predetermined. To deliver a required object to a user, a system must be able to correctly interpret a user's act of reaching at an early stage. It is also necessary that the system predict the required object as soon as possible.

Accurately predicting the user's hand trajectory seems to offer the right way to predict the target object. Studies of mathematical models for human arm trajectory planning have attracted considerable attention; such models include the minimum jerk model (Flash & Hogan, 1985), the minimum torque change model (Uno et al., 1989), and the minimum variance model (Harris & Wolpert, 1998). As these models do not consider human trunk movements and some of them require musculoskeletal parameters that are not easily acquired, it is difficult to apply them here.

In this study, knowledge of precise trajectories is not necessary; however it is necessary to identify the target object. There have been several researches on prediction of the target icon based on movements of a cursor in graphical user interfaces (Balakrishnan, 2004; Asano et al., 2005), however it cannot be applied to our situation because movements of hands and cursors are different. The following two assumptions are made to predict the target object.

- The approach rate of human hand to the target object is higher than to any other objects in the presence of multiple objects, and
- When an individual reaches for an object, his gaze directions are distributed around the direction of the object.

Based on these assumptions, certainty values from hand movements and eye movements are calculated and integrated probabilistically for each object.

3.1 Inference from hand movements

The certainty that object O^k is a target object given the trajectory of the user's hand after starting a reaching movement $H_s=[ho, h_1, \ldots, h_s\}$ is defined as follows:

$$p\left(O^* = O^k \middle| H_s\right) = \frac{f\left(o_s^k\right)}{\sum_{i=1}^{N} f\left(o_s^i\right)} \tag{8}$$

where O* represents the target object and o_s^i is a vector from the user's head to the object O^i ('=1, 2, N) at time s. Furthermore, $f\,(o_s^k)$ is calculated with the following equations:

$$f\left(o_s^k\right) = \begin{cases} g\left(o_s^k\right)^2 & \left(g\left(o_s^k\right) < 0\right) \\ 0 & \left(g\left(o_s^k\right) \geq 0\right) \end{cases}, \tag{9}$$

$$g\left(o_s^k\right) = \sum_{j=1}^{s} \frac{\left\|o_j^k - h_j\right\| - \left\|o_{j-1}^k - h_{j-1}\right\|}{\left\|o_0^k - h_0\right\|}. \tag{10}$$

Equation (10) is transformed into the following equation:

$$g\left(o_s^k\right) = \frac{\left\|o_s^k - h_s\right\| - \left\|o_0^k - h_0\right\|}{\left\|o_0^k - h_0\right\|} \tag{11}$$

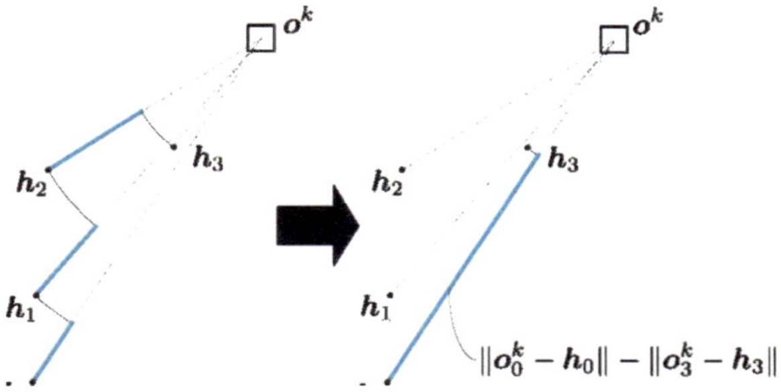

Fig. 6. Prediction of the target object on the basis of the ratio of the reduction in the distance between the hand and the object

The above equation yields the ratio of the reduction in the distance between the hand and the object O^k from the time the reaching movement is detected (Figure 6).

3.2 Inference from eye movements

In this study, we assume that the user's gaze direction $\arg(g_s)$ follows a wrapped normal distribution (Gumbel et al., 1953), which is for circular data, with mean $\arg(o_s)$, the direction of the target object. Thus, we can represent the conditional probability density function for $\arg(g_s)$ = fas given target $O^*=O^k$ as the following equation:

$$p\left(\arg\left(g_s\right) = \varnothing_s \middle| O^* = O^k\right) = \frac{1}{\sqrt{2\pi\sigma}} \left\{ -\frac{\left(\varnothing_s - \arg\left(o_s^k\right) + 2\pi i\right)^2}{2\sigma^2} \right\}, \tag{12}$$

where *a* is estimated using previous data for each user.

3.3 Integration of information from hand and eye movements

The system integrates the probabilities from the user's hand and eye movements based on Bayes' rule and predicts O, the target object, using the following equation:

$$\begin{aligned}\hat{o}^* &= \arg\max p\left(O^* = O^k \middle| H_s, \arg(g_s) = \varnothing_s\right) \\ &= \arg\max_{O^k} p\left(O^* = O^k \middle| H_s\right) p\left(O^* = O^k \middle| \arg(g_s) = \varnothing_s\right).\end{aligned} \tag{13}$$

4. Evaluation of the methods

To examine the usefulness of the methods, experiments were conducted. Generally, human motion in experiments is not reproducible. If the experiments are conducted in different conditions, therefore, it is impossible to fairly compare the methods. To tackle this problem, in this section, we conducted the estimation using three methods for the same human motion to fairly compare the methods. In the next section, we demonstrated the usefulness of the proposed method in the system that the movable trays can move.

4.1 Subjects

A total of 11 volunteers (10 males and 1 female, aged 21-42 years old) participated in the experiments. All subjects were right-handed, and three of them wore eyeglasses.

4.2 Experimental apparatus

An overhead digital color CCD camera (VCC-8350CL, CIS) measured the subjects' hand movements two-dimensionally.

For image processing, we used a Windows PC (Intel Xeon 3.0GHz Dual) with an image-processing board (GINGA digital CL-2, Linx) and image-processing software (HALCON7.0, MVTec).

To measure the subjects' head and eye movements, a head and eye tracking system (faceLAB4.2, Seeing Machines) that requires no physical contact and a Windows PC (Intel Pentium4 3.8GHz) were used. The frame rates for the measurement of the hand and eye movements are 30 (fps) and 60 (fps), respectively. Acquired three-dimensional data were projected to a desk plane and transformed into two-dimensional data.

4.3 Experimental procedure

In the experiments, subjects assembled a plastic model of a car from five types of subassemblies (Figure 7) five times, and the movements of each subject's hand, head, and eyes were recorded.

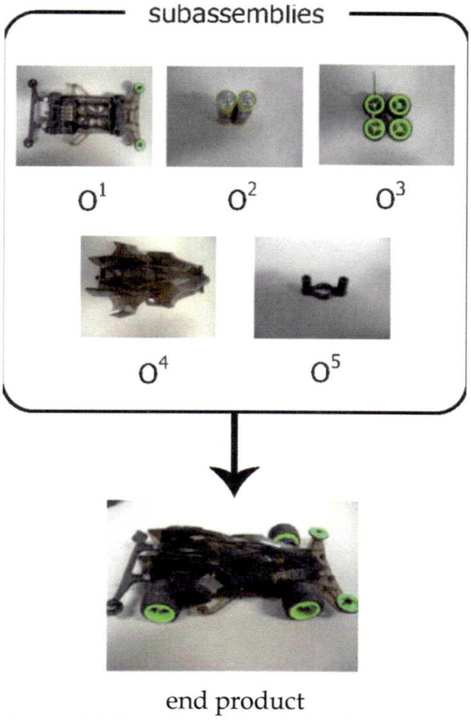

Fig. 7. Assembly of a plastic model from five types of subassemblies

The subjects reached with their dominant hand (right hand) for the subassemblies. In what follows, "hand" indicates the dominant hand (right hand). The subjects were asked to grasp only one type of subassembly at a time.
The arrangement of a subject, a desk, and the subassemblies are shown in Figure 8.
During the experiments, the subjects were asked to sit at the desk and assembled a plastic model from five types of subassemblies in no particular order.

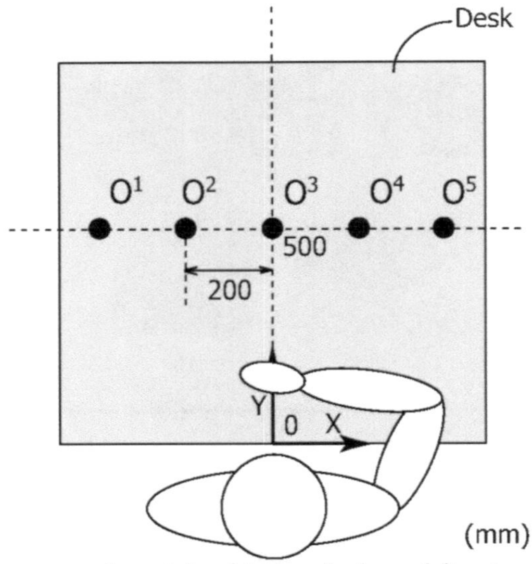

Fig. 8. Arrangement of an experimental subject, a desk, and five types of subassemblies (O^1, O^2, ... , O^5). The initial position of a subject's head is around (0, -200) (mm).

4.4 Experimental results

An example of the observed trajectories of a hand, head, and gaze point is shown in Figure 9.

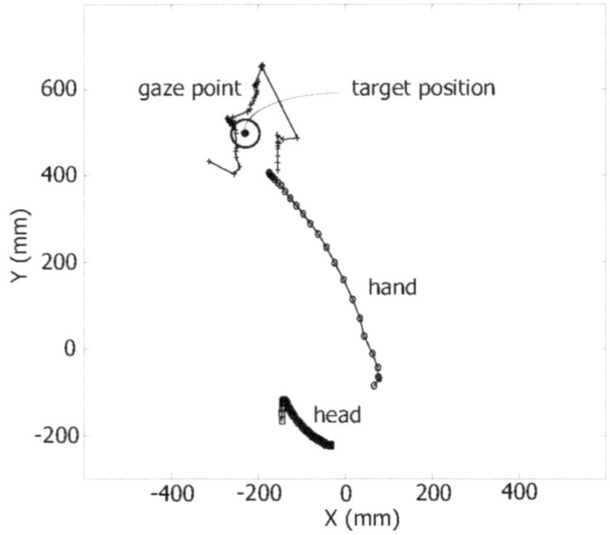

Fig. 9. Example of the observed trajectories of hand, head, and gaze point while a subject reached for O^2

The observed trajectories of the hand are relatively straight and smooth as presented in the literature (Morasso, 1981), and the gaze points are distributed around the target object. As evident in Figure 9, gaze directions are distributed around the target direction. In the proposed estimation method, detection and prediction processes are independent each other. Thus we independently examined the usefulness of the detection method and prediction one.

Detection of reaching movements

We compared the detection performances of the proposed method with those of MD and MS. MD represents a detection method that uses only a threshold value of distance. MS represents a detection method uses only a threshold value of hand speed. In MD, we set a threshold value for the distance between the user's head and hand. When the distance $||\mathbf{h}_s||$ exceeds the predetermined threshold value, the hand movement is detected as a reaching movement. In MS, we set a threshold value for speed of a hand movement. When the speed exceeds the predetermined threshold value, the hand movement is detected as a reaching movement.

We used two metrics, required detection time and detection accuracy, to evaluate the methods. Because there is a tradeoff between false detections and missed detections, we defined the detection accuracy using a following equation:

$$DA = \frac{N_{correct}}{N_{correct} + N_{false} + _{missed}}, \tag{14}$$

where Ncorrect, Nfalse, and Nmissed mean the number of correct detections, false detections, and missed detections, respectively. The threshold values of MD and MS were determined for each experimental subject to maximize DA. The required time for detection and calculated values of DA are shown in Table 1 and Table 2, respectively.

Subject	MD	MS	Proposed method
CR	0.363	0.187	0.173
GJ	0.348	0.126	0.196
HH	0.488	0.215	0.296
KO	0.407	0.215	0.297
MI	0.385	0.123	0.314
MK	0.462	0.230	0.234
MT	0.555	0.295	0.369
OJ	0.475	0.296	0.312
SH	0.264	0.097	0.185
SM	0.519	0.184	0.230
YY	0.374	0.258	0.212
Average	0.422	0.202	0.256

Table 1. Comparison of the required time (s) to detect an act of reaching

The average required time to detect an act of reaching is 0.256 (s). This is about 0.16 (s) shorter than that of MD and 0.05 (s) longer than that of MS. A total of five acts of reaching went undetected when the proposed method applied. The reason for the occurrence of the missed detections is that false detections occurred just before the act of reaching and R was set to 0 as a result of the false detections.

Subject	MD				MS				Proposed method			
	C	M	F	DA	C	M	F	DA	C	M	F	DA
CR	23	2	19	0.52	22	3	43	0.32	25	0	2	0.93
GJ	24	4	14	0.57	23	5	15	0.54	28	0	4	0.88
HH	26	2	21	0.53	21	7	41	0.30	28	0	1	0.97
KO	29	8	7	0.66	20	17	37	0.27	37	0	2	0.95
MI	22	4	19	0.49	16	10	28	0.30	25	1	10	0.69
MK	38	2	13	0.72	33	7	76	0.28	40	0	7	0.85
MT	45	0	8	0.85	36	9	63	0.33	45	0	4	0.92
OJ	27	3	39	0.39	22	8	68	0.22	29	1	8	0.76
SH	37	1	7	0.82	32	6	24	0.52	37	1	1	0.95
SM	44	1	32	0.57	38	7	62	0.36	43	2	7	0.83
YY	25	0	15	0.63	15	10	63	0.17	25	0	0	1.0
Average	30.9	2.5	17.6	0.61	25.3	8.1	47.3	0.31	32.9	0.45	4.2	0.88

Table 2. Comparison of the detection accuracy (C: Correct detection, M: Missed detection, F: False detection)

However, the number of missed detections of the proposed method was less than 20% of MD and less than 6% of MS. Furthermore, the number of false detections of the proposed method was less than 25% of MD and less than 10% of MS. As a consequence, the average DA of the proposed method was much larger than that of MD and MS.

According to the results, MD is inferior to the proposed method in both accuracy and required time. Although MS can detect an act of reaching a little sooner than the proposed method, its accuracy of detection is much lower than the proposed method. Therefore, the proposed detection method is more useful than MD and MS.

Prediction of target objects

We compared the prediction performances of the proposed method with those of a method using only hand movements (MH) and one using only eye movements (ME). Figure 9 shows the relationship between the timing of the prediction and the rate of correct target predictions.

As shown in Figure 10, the rates of the proposed method and MH increased with time. On the other hand, the rates of ME decreased slightly with time. This could be attributed to the fact that the subjects sometimes did not see the target object in the final stage of the reaching movements.

Differences in the prediction performances of the three methods were tested using the paired t-test with a significance level of 5%. The difference in the average rates of correct prediction between the proposed method and MH was statistically significant from 0 (s) to 0.1 (s) ($p<0.05$), and the difference between the proposed method and ME was also significant after 0.067 (s) ($p<0.05$).

As reported before, it takes about 0.26 (s) to detect a user's reaching movements. Therefore, it is necessary for the system to predict the target object within 0.14 (s) in order to meet the time requirement described in section 1. At 0.133 (s) after detection, while the rates of correct predictions of MH and ME were 84% and 62%, respectively, the proposed method's rate of correct predictions was 92%.

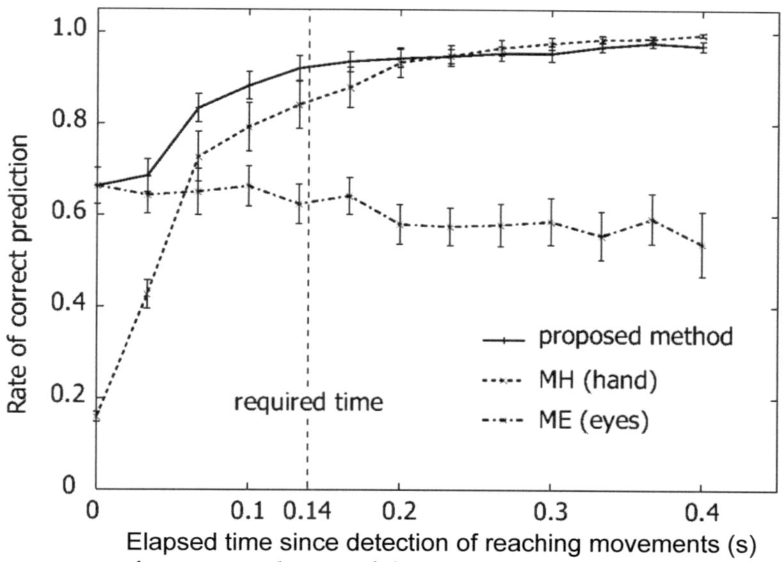

Fig. 10. Average rate of correct prediction of the targets with the standard error of the mean, N=11

The difference in the rates of correct predictions between MH and the proposed method at the time was not significant, but a statistical trend in which the proposed method exceeds MH (p=0.069) was observed.

5. Application possibility to the deskwork support system

To verify the possibility of applying the proposed methods to our deskwork support system, experiments using movable trays were conducted.

5.1 Subjects

A total of 4 male volunteers (aged 24-33 years old) participated in the experiments. All subjects were right-handed, and they had also participated in the experiments stated in section 4.

5.2 Experimental apparatus

Our deskwork support system is shown in Figure 10.

To measure movements of the subjects' hands, heads, and eyes, we used the same apparatus with section 4.

To deliver objects to the subjects, five self-moving trays, which are driven by a Sawyer-type 2-DOF stepping motor, were used. The trays move on a motor *platen* (i.e. iron plate) with its size being 1200 x 1000 (mm). The size of each tray is 130 x 135 x 25 (mm). The trays can move 1000 (mm/s) at maximum for each axis.

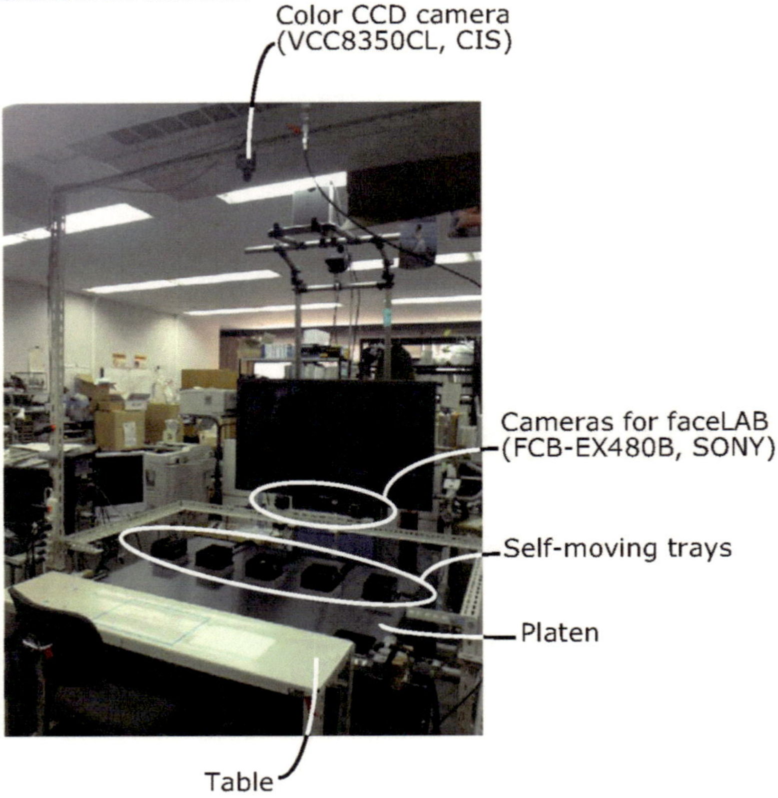

Fig. 11. An overview of the implemented deskwork support system

5.3 Experimental procedure

In the experiments, the subjects assembled a plastic model of a car from five types of subassemblies four times as well as the experiments described in section 4. The subassemblies (Figure 7) were loaded on each tray, and when a subject reaches for an object, the deskwork support system detects an act of reaching, predicts a target object, and delivers the object using a self-moving tray.

In the experiments, we set the duration for prediction to 0.133 (s), and we use the same parameters for the prediction process as in section 4. In other words, the parameters were obtained in the reachable target condition.

The initial arrangement of a subject and the subassemblies on the trays are shown in Figure 12. In this experiment, the subassemblies are located farther from the position of the subjects than the experiment in section 4.

During the experiments, the subjects were asked to sit at the desk and assembled a plastic model in no particular order, and place the end product on tray 3.

5.4 Experimental results

An example of the experimental scene where the proposed method was applied is shown in Figure 12.

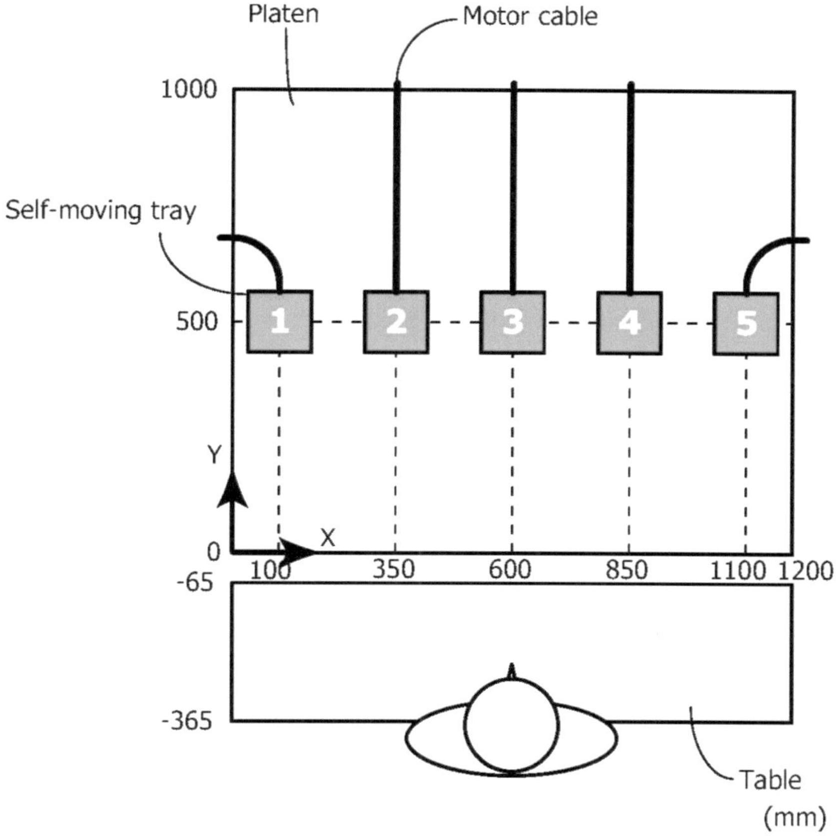

Fig. 12. Arrangement of an experimental subject and five trays (O is on tray i)

The calculated values of DA are shown in Table 3.

Subject	Correct	Missed	False	DA
CR	40	9	0	0.82
MI	40	4	0	0.91
MK	40	4	0	0.91
SM	40	3	0	0.93
Average	40	5	0	0.89

Table 3. Detection accuracy when applied to the deskwork support system

The average value of DA was 0.89. This was almost the same as the result in section 4. The proposed detection method can be applied to the deskwork support system in which the self-moving trays deliver the objects.

We also test the validity of the proposed prediction method. The calculated rates of correct prediction of the targets are shown in Table 4.

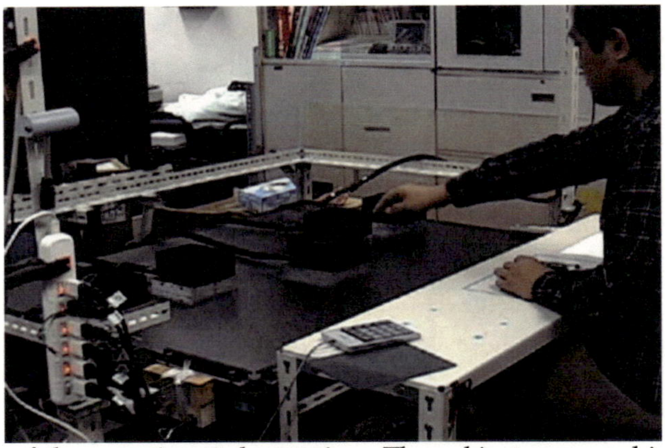

Fig. 13. An example of the experimental overview. The subject was reaching for tray 3, and the tray was moving towards the subject.

Subject	Correct	False	Correct rate
CR	40	2	0.95
MI	40	0	1.0
MK	40	1	0.98
SM	40	2	0.95
Average	40	1.25	0.97

Table 4. Rate of correct prediction of the targets when applied to the deskwork support system

The average rate of correct predictions was 97%. This was higher than the result in section 4. This result confirms that the proposed prediction method can be applied to the deskwork support system.
Based on the above experimental results, the proposed method is useful in the deskwork support system, and the parameters acquired from the reachable target condition can be applied to the unreachable one.

6. Conclusion

We have presented methods to detect an act of reaching among other hand movements and to predict target objects based on measurements of a user's hand and eye movements. In the detection method, we adopted speed, the smoothness and straightness of a user's hand movements, and the relationship between hand and eye movements. The usefulness of the proposed method was experimentally demonstrated.
In the future, an error recovery algorithm should be developed for more reliable deskwork support system.

7. References

Abrams, R.A.; Meyer, D.E. & Kornblum, S. (1990). Eye-hand coordination: Oculomotorcontrol in rapid aimed limb movements. *Journal of Experimental Psychology: Humanperception and Performance,* Vol. 16, No. 2, 248-267

Agah, A. & Tanie, K. (1997). Human interaction with a service robot: Mobile-manipulator handing over an object to a human. *Proceedings of the 1997 IEEE InternationalConference on Robotics and Automation,* pp. 575-580

Asano, T.; Sharlin, E.; Kitamura, Y.; Takashima, K. & Kishino, F. (2005). Predictive interaction using the Delphian desktop. *Proceedings of the 18th Annual ACM Symposium on User Interface Software and Technology,* pp. 133-141

Balakrishnan, R. (2004). "Beating" Fitts' law: Virtual enhancements for pointing facilitation. *International Journal of Human-Computer Studies,* Vol. 61, No. 6, 857-874

Biguer, B.; Jeannerod, M. & Prablanc, C. (1982). The coordination of eye, head, and arm movements during reaching at a single visual target. *Experimental Brain Research,* Vol.46, 301-304

Bolt, R.A. (1980). "Put-that-there": Voice and gesture at the graphics interface. *Proceedings of the 7th Annual Conference on Computer Graphics and Interactive Techniques,* pp. 262-270 Brooks, R.A. (1997). The intelligent room project. *Proceedings of the 2nd International Cognitive Technology Conference,* pp. 271-278

Cipolla, R. & Hollinghurst, N.J. (1996). Human-robot interface by pointing with uncalibrated stereo vision. *Image and Vision Computing,* Vol. 14, 171-178 Dallaway, J.L. & Jackson, R.D. (1994). The user interface for interactive robotic workstations. *Proceedings of the 1994 IEEE/RSJ International Conference on Intelligent Robots and Systems,* pp.1682-1686

Dallaway, J.L.; Jackson, R.D. & Timmers, P.H.A. (1995). Rehabilitation robotics in Europe. *IEEE Transactions on Rehabilitation Engineering,* Vol. 3, No. 1, pp. 35-45 Flash, H. & Hogan, N. (1985). The coordination of arm movements: An experimentally confirmed mathematical model. *Journal ofNeuroscience,* Vol. 5, No. 7, 1688-1703

Gumbel, E.; Greenwood, J. & Durand, D. (1953). The circular normal distribution: Theory and tables. *Journal ofthe American Statistical Association,* Vol. 48, No. 261, 131-152

Harris, C.M. & Wolpert, D.M. (1998). Signal-dependent noise determines motor planning.*Nature,* Vol.394, 780-784

Harwin, W.S.; Rahman, T. & Foulds, R.A. (1995). A review of design issues in rehabilitation robotics with reference to North American research. *IEEE Transactions on Rehabilitation Engineering,* Vol. 3, No. 1, 3-13

Ishii, H. & Ullmer, B. (1997). Tangible bits: Towards seamless interfaces between people, bits, and atoms. *Proceedings of the SIGCHI Conference on Human Factors in Computing Systems,* pp. 234-241

Ishii, S.; Tanaka, S. & Hiramatsu, F. (1995). Meal assistance robot for severely handicapped people. *Proceedings of the IEEE International Conference on Robotics and Automation,* pp.1308-1313

Kajikawa, S.; Okino, T.; Ohba, K. & Inooka, H. (1995). Motion planning for hand-over between human and robot. *Proceedings of the 1995 IEEE/RSJ International Conference on Intelligent Robots and Systems,* pp. 193-199

Koike, H.; Sato, Y. & Kobayashi, Y. (2001). Integrating paper and digital information on EnhancedDesk: A method for realtime finger tracking on an augmented desk system. *ACM Transactions on Computer-Human Interaction,* Vol. 8, No. 4, 307-322

Leibe, B.; Starner, T.; Ribarsky; Wartell, Z.; Krum, D.; Singletary, B. & Hodges, L. (2000). The perceptive workbench: Toward spontaneous and natural interaction in semi immersive virtual environments. *Proceedings of the IEEE Virtual Reality 2000,* pp. 13-20

Morasso, P. (1981). Spatial control of arm movements. *Experimental Brain Research,* Vol.42, 223-227

Mori, T.; Yokokawa, T. & Sato, T. (1998). Recognition of human pointing action based on color extraction and stereo tracking. *Intelligent Autonomous Systems 5,* Kakazu, Y.; Wada, M. & Sato, T. (Eds.), pp. 93-100, IOS Press

Noma, H.; Yoshida, S.; Yanagida, Y. & Tetsutani, N. (2004). The proactive desk: A new haptic display system for a digital desk using a 2-DOF linear induction motor. *Presence,* Vol. 13, No. 2, 146-153

Oka, K.; Sato, Y. & Koike, H. (2002). Real-time fingertip tracking and gesture recognition. *IEEE Computer Graphics and Applications,* Vol. 22, No. 6, 64-71

Pangaro, G.; Aminzade, D.M. & Ishii, H. (2002). The actuated workbench: Computer-controlled actuation in tabletop tangible interfaces. *Proceedings of the 15th Annual ACM Symposium on User Interface Software and Technology,* pp. 181-190

Pentland, A. (1996). Smart rooms. *Scientific American,* Vol. 274, No. 4, 54-62

Prablanc, C.; Echallier, J.F.; Komilis, E. & Jeannerod, M. (1979). Optimal response of eye and hand motor systems in pointing at a visual target - I. Spatio-temporal characteristics of eye and hand movements and their relationships when varying the amount of visual information. *Biological Cybernetics,* Vol. 35, 113-124

Raghavan, V.; Molineros, J. & Sharma, R. (1999). Interactive evaluation of assembly sequences using augmented reality. *IEEE Transactions on Robotics and Automation,* Vol. 15, No. 3, 435-449

Rekimoto, J. (2002). SmartSkin: An infrastructure for freehand manipulation on interactive surfaces. *Proceedings of the SIGCHI Conference on Human Factors in Computing Systems,* pp. 113-120

Sato, S. & Sakane, S. (2000). A human-robot interface using an interactive hand pointer that projects a mark in the real work space. *Proceedings of the 2000 IEEE International Conference on Robotics and Automation,* pp. 589-595

Sato, T.; Nishida, Y. & Mizoguchi, H. (1996). Robotic room: Symbiosis with human through behavior media. *Robotics and Autonomous Systems,* Vol. 18, 185-194 Sawyer, B.A. (1969). Magnetic positioning device. *US patent,* 3,457,482

Sugiyama, O.; Kanda, T.; Imai, M.; Ishiguro, H. & Hagita, N. (2005). Three-layered draw-attention model for humanoid robots with gestures and verbal cues. *Proceedings of the 2005 IEEE/RSJ International Conference on Intelligent Robots and Systems,* pp. 2140-2145

Tamura, Y.; Sugi, M.; Ota, J. & Arai, T. (2004). Deskwork support system based on the estimation of human intentions. *Proceedings of the 13th IEEE International Workshop on Robot and Human Interactive Communication,* pp. 413-418 Topping, M. (2002). An overview of the development of Handy 1, a rehabilitation robot to assist the severely disabled. *Journal of Intelligent and Robotic Systems,* Vol. 34, 253-263

Ullmer, B. & Ishii, H. (1997). The metaDESK: Models and prototypes for tangible user interfaces. *Proceedings of the 10th Annual ACM Symposium on User Interface Software and Technology,* pp. 223-232

Uno, Y.; Kawato, M. & Suzuki, R. (1989). Formation and control of optimal trajectory in human multijoint arm movement. *Biological Cybernetics,* Vol. 61, 89-101 Wellner, P. (1993). Interacting with paper on the DigitalDesk. *Communications of the ACM,* Vol. 36, No. 7, 87-96

17

Adaptive Swarm Formation Control for Hybrid Ground and Aerial Assets

Laura Barnes[1], Richard Garcia[2], MaryAnne Fields[2] and Kimon Valavanis[3]

[1]*Automation and Robotics Research Institute, University of Texas at Arlington Fort Worth, TX USA*

[2]*U.S. Army Research Lab Aberdeen Proving Grounds, MD USA*

[3]*Department of Electrical and Computer Engineering, University of Denver Denver, CO USA*

1. Introduction

The use of Unmanned Aerial Vehicles (UAVs) with Unmanned Ground Vehicles (UGVs) allows for cooperation, coordination, and tight or loose collaboration related to multiple missions. UAVs can provide a global perspective of the surrounding environment, obstacles, and possible threats, broadcasting goals, sub-goals and alterations to the overall mission of the swarm. Further, the deployment of UAVs creates a 3-D sensor network increasing communication capabilities allowing for more complete information about the environment.

UAV-UGV coordination has obvious applicability in military applications due to the line of sight issue. Air vehicles can detect items of interest long before UGVs. Related literature in the area refers to general frameworks and simulation results only. In (Chaimowicz and Kumar 2004; Chaimowicz and Kumar 2004), UGV swarms are coordinated and directed by "shepherd" UAVs. A hierarchy is formed between the UAV and the UGVs. UAVs are responsible for grouping and merging swarms as well as controlling swarm distributions and motion. In (Sukhatme, Montgomery et al. 2001), an architecture is proposed for coordinating an autonomous helicopter and a group of UGVs using decentralized controllers. In (Tanner 2007), an approach is proposed to coordinate groups of ground and aerial vehicles for the purpose of locating a moving target in a given area. This is done by combining decentralized flocking algorithms with navigation functions. Other instances utilizing coordination between air and ground vehicles can be seen in (Elfes, Bergerman et al. 1999; Lacroix, Jung et al. 2001; Stentz, Kelly et al. 2002).

In this work, the problem of controlling and coordinating heterogeneous unmanned systems required to move as a group is addressed. A strategy is proposed to coordinate groups of Unmaned Ground Vehicles with one or more Unmanned Aerial Vehicles (UAVs). UAVs can be utilized in one of two ways: (1) as alpha robots to guide and oversee the UGVs; and (2) as beta robots to surround the UGVs and adapt accordingly. In the first approach, the UAV guides a swarm of UGVs controlling their overall formation. In the second approach, the

UGVs guide the UAVs controlling their formation. The unmanned systems are brought into a formation utilizing artificial potential fields generated from normal and sigmoid functions. These functions control the overall swarm geometry. Nonlinear limiting functions are defined to provide tighter swarm control by modifying and adjusting a set of control variables forcing the swarm to behave according to set constraints. Formations derived are subsets of elliptical curves but can be generalized to any curvilinear shape. The formation control strategy is a hybrid which can be either completely distributed using only local information summing individually calculated weighted vectors for formation keeping and obstacle avoidance. Moreover, a hiearchical approach with leaders and followers can also be utilized to create a tighter formation and coordinate UAVs and UGVs. The proposed strategy is platform and controller independent as the vector generation is not dependent on the specific robot. Previous research reported in (Barnes, Alvis et al. 2006; Barnes, Fields et al. 2007) presents extensive simulation results and field experiments to validate the formation control methodology.
Both approaches are demonstrated in simulation and experimentally. The first approach is demonstrated experimentally with a fully autonomous UAV for coordination and three UGVs. The autonomous UAV take-off, landing and waypoint navigation is controlled via fuzzy logic controllers. The UGVs utilize identical navigation and formation controllers. To demonstrate the second approach in simulation, a swarm of forty UAVs is utilized in a convoy protection mission. As a convoy of UGVs travels, UAVs dynamically and intelligently adapt their formation in order to protect the convoy of vehicles as it moves. Section 2 discusses the swarm formation controller followed by the UAV controllers in Section 3. Results are presented in Section 4 and 5.

2. Swarm Formation Controller

The objective of the formation controller is to attract elements of a swarm into a bounded formation and allow the swarm to stay in that formation as it moves in a mission space. Vector fields and weights are utilized to attract swarm members to the desired surface and keep them distributed about that surface.

2.1 Generation of Formation Surface

At any instant in time, the robots can be visualized as particles moving in a potential field generated from a bivariate normal "hill" that controls the velocity and heading of the swarm members. A bivariate normal function with form given in (1):

$$f(x,y) = e^{-\alpha(x-x_c)^2+\gamma^2(y-y_c)^2} \tag{1}$$

produces an oval/ellipsoid shaped function. Assuming that the current robot location is at (x, y), the center of the function in (1) is represented by (x_c, y_c) in the world reference frame. The 'control' variable γ determines the ratio of the minor axis (y-direction) to the major axis (x-direction) affecting the eccentricity of the swarm. The x and y partial derivatives create the velocity vectors that are used to determine the heading and velocity of each member of the swarm as shown in (2):

$$d_x = -2\alpha f(x,y)(x - x_c)$$
$$d_y = -2\alpha\gamma f(x,y)(y - y_c) \quad (2)$$

The swarm formation shape has both a local reference and a world reference frame. For the swarm to follow a trajectory in the world reference frame, an axis rotation is required. The heading, φ, between the swarm formation's x-axis and the center (x_c, y_c) must be found; the translated and rotated coordinates can be found using (3):

$$x_{rot} = \cos(\phi)(x - x_c) - \sin(\phi)(y - y_c)$$
$$y_{rot} = \sin(\phi)(x - x_c) + \cos(\phi)(y - y_c) \quad (3)$$

The rotated coordinates are then substituted back in to find d_x and d_y.

2.2 Formation Description

By attracting swarm members to a specific elliptical ring R^* shown in Fig. 1. The swarm can be closely associated with the UAV with the (x_c, y_c) denoting its location. For a fixed value of γ, we will refer to the set of points $(x, y) \in \Re^2$ satisfying (4) as the R^* ellipse.

$$R^{*2} = (x - x_c)^2 \gamma^2 (y - y_c)^2 \quad (4)$$

A potential field based controller using a small number of physically relevant weights and vectors v_i is developed to attract the robots to a neighborhood of the R* ellipse. This neighborhood is shown in Fig. 1. The variables R* - ΔR_{in} and R* + ΔR_{out} denote the inside and outside boundaries of the R* neighborhood respectively as shown in Fig 1. The desired vector fields will 'trap' the robots in these bands. Typically, this is a very narrow band of allowable space for the robots with a controllable width of ΔR_{in}+ ΔR_{out}.

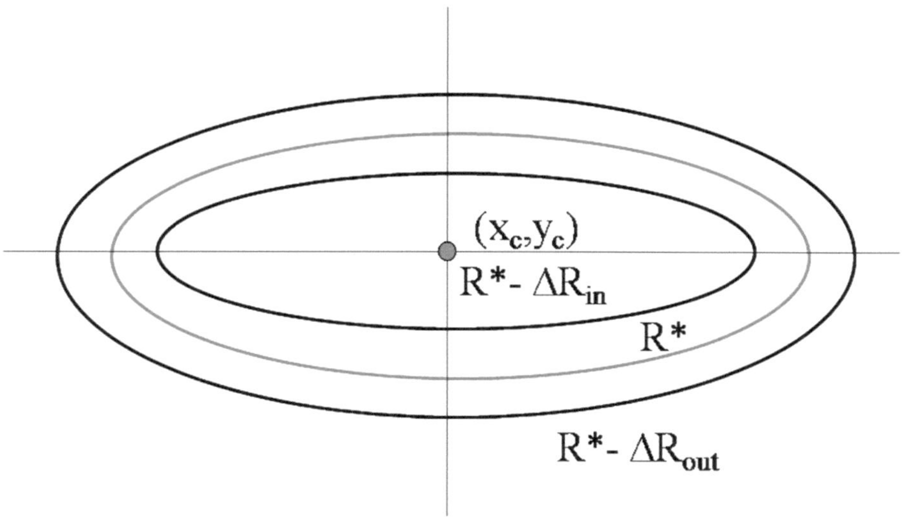

Fig. 1. Elliptical attraction band for the robots

In the defined vector field, robots with position defined as *r*, starting within the *R** ellipse, with:

$$r = \sqrt{(x - x_c)^2 + \gamma^2 (y - y_c)^2} \tag{5}$$

center until they reach the *R** neighborhood. Eventually all the robots will be trapped within the neighborhood given in (6):

$$(R^* - R_{in}) \le r \le (R^* + R_{out}) \tag{6}$$

2.3 Vector Field Generation

In order to generate the desired vector fields to hold the robots inside the R* neighborhood, three fields are needed. The gradient vector field, $G^- = -(d_x, d_y)$ points away from the center. Vector calculus dictates that the gradient vector field, $G^+ = (d_x, d_y)$ points in the direction of greatest increase of the function $f(x,y)$, which is towards the center. The vector fields $(d_x, -d_y)$ and $(-d_x, d_y)$ are perpendicular to the gradient ($G\perp$).
Tighter swarm control is accomplished when restricting the influence of the vector fields to a small region of the *x-y* plane by multiplying each of the fields by a 'limiting function'. This limiting function controls how far from the center the vectors in the field 'die out' or become smaller than some number ε.
In order to create the desired field, the G^- and G^+ fields must be limited to end at the appropriate boundaries. These fields will be limited with *sigmoid* functions. The G^- field should die out at $R^*\text{-}\Delta R_{in}$, and the G^+ field should die out at $R^*+\Delta R_{out}$. The $G\perp$ field will be active only inside the elliptical bands so it will die out at $R^*\text{-}\Delta R_{in}$ and $R^*+\Delta R_{out}$. This field will be limited with a Normal function.

2.4 Limiting Functions

Vector fields 'moving away' from the center (the vectors inside of the ellipse) require a limiting function that approaches zero as the distance from the center is increased; such a limiting function is given in (7):

$$S_{in}(\alpha_{in}, r, R^*_{in}, \Delta R_{in}) = 1 - \frac{1}{1 + e^{\alpha_{in}(r - (R^* - \Delta R_{in})}} \tag{7}$$

Gradient vector fields directed towards the center (those vectors outside of the ellipse) are required to approach zero as the vectors 'move towards' the center; this is achieved using the limiting function in (8):

$$S_{out}(\alpha_{out}, r, R^*, \Delta R_{out}) = 1 - \frac{1}{1 + e^{-\alpha_{out}(r - (R^* + \Delta R_{out})}} \tag{8}$$

Attracting the robot to the R* neighborhood specified in equation (6) is the first step in the construction of the final vector field.

An additional vector field can be used to control the robots once they are in the elliptical band. In this field, the robots need to move along the ellipse in a field perpendicular to the previously described gradient fields. A limiting function accomplishing that is given in (9):

$$N_{\perp}(\alpha_{\perp}, r, R^*) = e^{-\alpha_{\perp}(r-R^*)^2} \tag{9}$$

The plot of the functions S_{in}, S_{out}, and $N_{\perp}$ as a function of r is provided in Fig. 2. S_{out} has its largest influence at points whose distance from the center of the ellipse is small. S_{in} has its greatest influence at points whose distance from the center is large. These functions approach *0* near the R^* band. $N_{\perp}$ is only influential with in the ellipsoid bands.

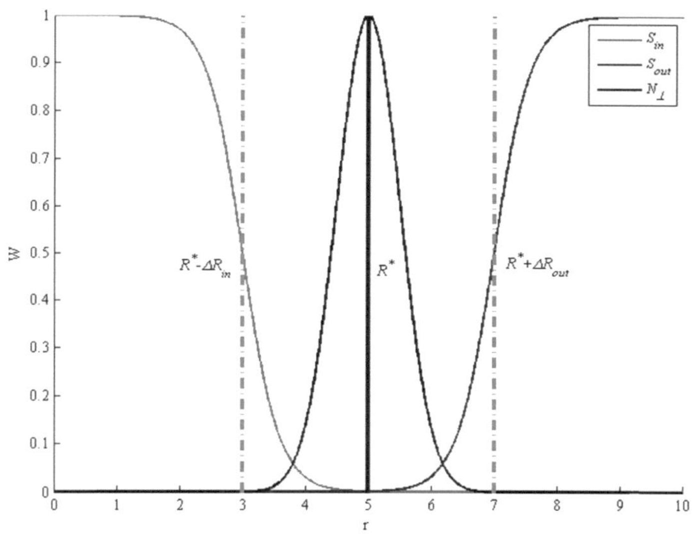

Fig. 2. The weighting functions S_{in}, S_{out}, and $N_{\perp}$ as a function of the weighted distance r defined in (5)

In addition, another multiplier to the perpendicular field must be added so the robots do not circle around the ellipse bands. In order for the perpendicular field to change directions, the field perpendicular to the gradient is multiplied by (10) where y_{rot} is the translated and rotated value of y:

$$SGN(\alpha_{\perp}, y_{rot}) = 1 - 2.0\left(\frac{1}{1+e^{-\alpha_{\perp}(y_{rot})}}\right) \tag{10}$$

Each of the limiting functions in (7) through (10) contains *tuning parameters* that may be used as *vector field control variables*. These functions include one tuning parameter each, which determines how quickly the function approaches zero.

The parameters α_{in}, α_{out}, and $\alpha_{\perp}$ control the slope of $S_{in}(r)$, $S_{out}(r)$, and $N_{\perp}(r)$, respectively, for r in the set $R - \Delta R_{in} < r < R + \Delta R_{out}$.

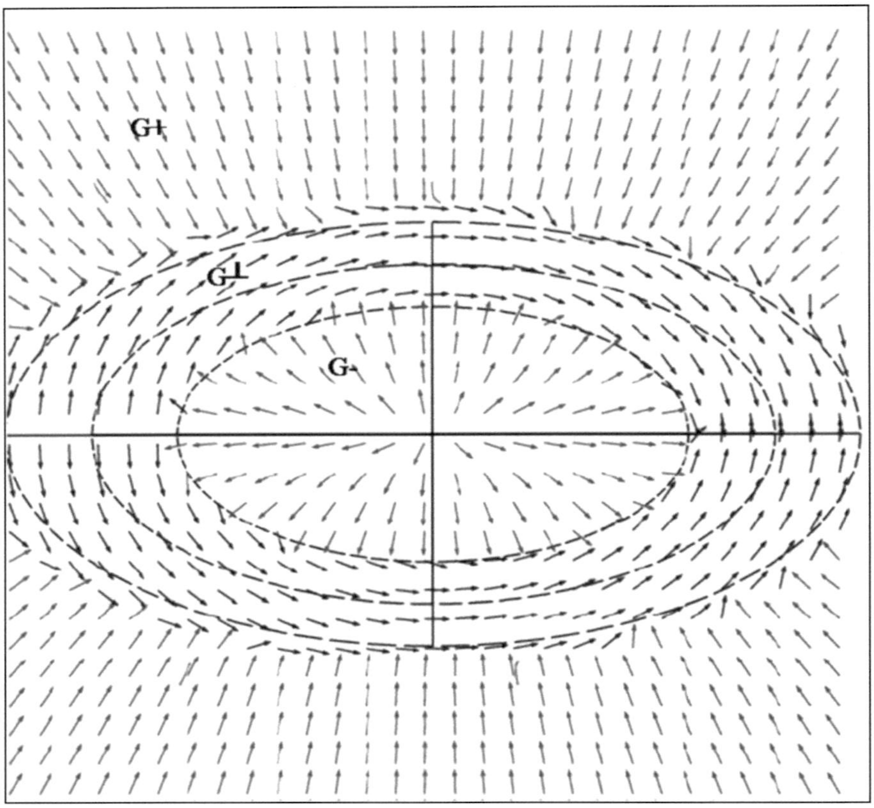

Fig. 3. Final vector field

The value of $S_{in}(R^*)$ can be made arbitrarily small. Let $\varepsilon > 0$ be a small number such that $S_{in}(R^*) = \varepsilon$. Then the value of α_{in} can be determined. The same technique is used in the other limiting functions. The resulting equations are shown in (11) to (13):

$$\alpha_{in} = \frac{1}{\Delta R_{in}} \ln\left(\frac{1-\varepsilon}{\varepsilon}\right) \tag{11}$$

$$\alpha_{out} = \frac{1}{\Delta R_{out}} \ln\left(\frac{1-\varepsilon}{\varepsilon}\right) \tag{12}$$

$$\alpha_{\perp} = \frac{1}{((\Delta R_{out} - \Delta R_{in})/2)^2} \ln(1-\varepsilon) \tag{13}$$

The final vector field is depicted in Fig. 3. Functions S_{out}, $N_{\perp}$ and S_{in} impose additional restrictions and constraints on top of and in addition to the initial swarm function $f(x, y)$. The limiting functions, along with vector fields created by the bivariate normal function, may be summed to create swarm movement in formation as a group. When combined, these equations form the velocity and direction of the swarm movement with respect to the center of the swarm, as shown in:

$$\begin{bmatrix} v_x \\ v_y \end{bmatrix} = (S_{in} - S_{out}) \begin{bmatrix} d_x \\ d_y \end{bmatrix} + SGN * N_{\perp} \begin{bmatrix} d_x \\ d_y \end{bmatrix}_{\perp} \tag{12}$$

2.5 Obstacle Avoidance and Swarm Member Distribution

Vector fields weighted with sigmoid functions may be used for obstacle avoidance as well as controlling member distribution by creating vectors moving away from the center of the obstacle's or other swarm member's location (x_{co}, y_{co}). For the purposes of this work, the concern is formation including distribution of swarm members on the formation. In describing the formation control methodology, it is assumed that the only obstacles are other members of the swarm. The same form of limiting function as S_{in} may be used. Obstacle avoidance between members is accomplished using Equations (13) to (15):

$$r_{avoid} = \sqrt{(x - x_{co})^2 + (y - y_{co})^2} \tag{13}$$

$$S_{avoid}(\alpha_{avoid}, r_{avoid}, \Delta R_{avoid}) = \kappa - \frac{\kappa}{1 + e^{\alpha_{avoid}(\sqrt{r_{avoid}} - \Delta R_{avoid})}} \tag{14}$$

$$\begin{bmatrix} d_{x\_avoid} \\ d_{y\_avoid} \end{bmatrix} = \begin{bmatrix} S_{avoid}(x - x_{co}) \\ S_{avoid}(y - y_{co}) \end{bmatrix} \tag{15}$$

The weight function generated by a single obstacle is a sigmoid with maximum value κ shown in (14). Tuning the parameters κ and a_{avoid} ensures that the collision avoidance field dominates the vector field near an obstacle. Notice that r_{avoid} is similar to r from Equation (5) except that instead of distance from the center, the distance to the swarm member is used. The ΔR_{avoid} parameter denotes the minimum distance from other members. This parameter determines the distribution of swarm members in formation. S_{out} and S_{in} get swarm members to the band, but do not control their distribution.

Avoidance of individual swarm members including their distribution is controlled by the range of influence for the avoidance vector field. The a_{avoid} parameter in (14) controls how quickly vector fields die out near obstacles. As a_{avoid} decreases, the influence range of the avoidance vector field increases. By controlling the a_{avoid} parameter, different types of formations can be made within the elliptical bands.

The a_{avoid} parameter is solved for in the same way as the other sigmoid limiting functions in (12) and (13). The term in (15) is simply summed in (12) to create swarm movement in formation with distribution of swarm members. The ΔR_{avoid} parameter specifies the minimum distance between swarm members. Solving for $S_{avoid}(\Delta R_{avoid})=\varepsilon$ gives:

$$\alpha_{avoid} = \frac{1}{\Delta R_{avoid}} \ln\left(\frac{\kappa - \varepsilon}{\varepsilon}\right) \tag{16}$$

3. UAV Controllers

The individual UAV helicopters are controlled via four distinct fuzzy controllers. These controllers are responsible for four of the five helicopter inputs: roll, pitch, yaw, and collective. The fifth input, throttle, is output as a constant value throughout the helicopter's navigation routines and thus does not utilize a fuzzy controller. It should be noted that throttle control does vary during the startup and shutdown routines. These routines are simply responsible for starting and stopping the motor during the take-off and landing procedures and utilize a linear throttle increase/decrease to transition the throttle between zero and the constant value used during flight.
The four fuzzy controllers utilized on helicopter are designed using Sugeno constant fuzzy logic and a weighted average defuzzification method. All rules for the controllers are based on the 'and' method and use membership products to determine the strength of each rule. Each controller has a single output which ranges from [-1,1] corresponding to the minimum and maximum Pulse Width (PW) for that particular control respectively. The calculation of the PW from the controller output is done using:

$$O_{Pi} = \begin{cases} (Max_{P_i} - N_{P_i}) * \alpha & for\ \alpha \geq 0 \\ (N_{P_i} - Min_{P_i}) * \alpha & for\ \alpha < 0 \end{cases} \tag{17}$$

where Max_{P_i} is the maximum PW value for servo 'i', Min_{P_i} is its minimum PW value, N_{P_i} is its neutral PW value, O_{P_i} is the calculated PW for servo 'i', and α is the controller output. N_{P_i} is the approximate PW value of the vehicle in a level hover and is taken from the radio after the vehicle has been properly setup and trimmed by an expert pilot. Note that the helicopter utilizes a three point swashplate which requires cyclic and collective mixing for vehicle control. The method for cyclic and collective mixing is described in great detail in (Garcia 2008).
It should be noted that the control design assumes that the effects of the tail are negligible with respect to roll and pitch control. This assumption is validated by the controller's non-aggressive flight control design, the use of a heading hold gyro, and the minimal tail surface area which creates very little side slip drag. Under this assumption the only real difference between roll and pitch control is the axis of control. As such, the roll and pitch controllers have exactly the same rules with exactly the same outputs and weights. The only difference is the axis used for input and the axis to which the output is applied.

3.1 Roll / Pitch Controller

The roll / pitch controller each utilize four inputs, positional error, velocity, orientation angle, and acceleration, each of which is in the local coordinate frame. The positional error, velocity, and orientation inputs each utilize five membership functions with the acceleration input utilizing three membership functions. The fuzzy rules were designed to provide a complete set of rules given the inputs, thus the roll and pitch controllers each contain 375 rules, available in (Garcia 2008).

Roll / Pitch, or lateral / longitudinal, control methodology was designed around a hovering technique. The controllers simply attempt to hover at a desired location. Transitions between waypoints are simply an attempt to minimize position error in the hovering location. The fuzzy rule base was designed to first determine a desired input and then compare that to the actual input.

Due to the controllers being designed around a hover routine, the desired positional error is always zero. This value is then compared to the input value. The difference between these values is used to calculate a desired velocity which is consequently compared to the input velocity. The difference between the desired velocity and actual velocity is used to determine a desired orientation. The comparison of the desired orientation is compared to the actual orientation which is utilized to calculate a desired angular rate. This angular rate is then adjusted based on the acceleration input. For example, if the acceleration is currently too high the desired angular rate is decreased. The calcuations referenced here are never truely calculated but describe the mentatiliy used to create the specific fuzzy rules.

3.2 Collective Controller

The collective controller utilizes three inputs, positional error, velocity, and acceleration, each of which is in the local coordinate frame. The positional error and velocity inputs each utilize five membership functions with the acceleration input utilizing three membership functions. The fuzzy rules were designed to provide a complete set of rules given the inputs, thus the roll and pitch controllers each contain 75 rules.

The collective control methodology, like the roll / pitch control methodology, was designed around a hovering technique. As such the desired positional error is always zero. This value is then compared to the input positional error. The difference between these values is used to calculate a desired vertical velocity which is consequently compared to the input velocity. The difference is then used to determine a desired acceleration. The difference between the desired and actual acceleration is then used to calculate the control output.

3.3 Yaw Controller

The yaw controller utilizes a single input: heading error. The heading error utilizes five membership functions with a single rule dedicated to each. Yaw control is simply based on holding a desired heading. Due to the UAV's use of a heading hold gyro, common on all RC based helicopters, control is calculated by determining a desired angular rate. This rate is calculated by determining the difference between the current heading and the desired heading. The desired rate is then obtained and maintained by the gyro.

The UAV control methodology provides controllers that can be easily modified for desired speeds and orientations. Note that the link between the desired angular rate for roll, pitch, and yaw as well as the vertical acceleration for collective and controller output was hand tuned. Further details into the aspects of the UAV controllers, hardware and software design, as well as experimentation can be seen in (Garcia 2008).

4. Application to Convoy Protection Utilizing a UAV Swarm

In order to demonstrate the proposed approach, it will be applied to the convoy protection problem. Suppose that a swarm of UAVs needs to accompany a convoy of vehicles, surrounding them in a particular formation. In the general case, the convoy can be enclosed in some geometric shape, defined loosely by dimensions, direction of travel, and the center of mass as shown in Fig. 4. The length of the convoy along the axis of travel is *2A*. The width of the convoy with respect to the axis of travel is *2B*.

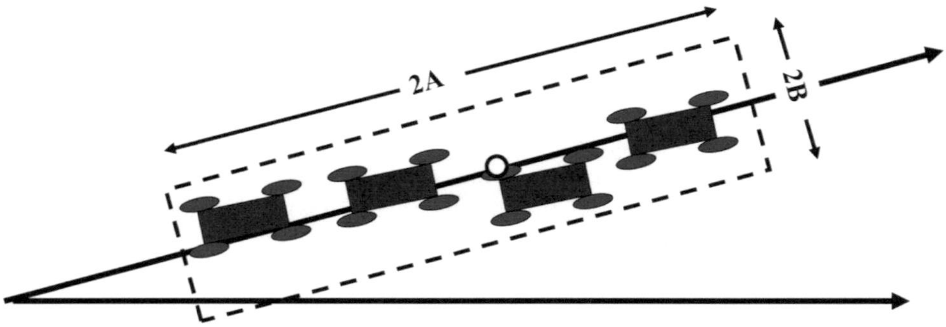

Fig. 4. Convoy description

A field needs to be designed to attract swarm members to surround the convoy in a designated formation. The swarm members need to be close enough to the convoy to offer protection, but far enough to allow the convoy to move safely. The formation controller described in Section 2 is utilized. Assume that the positions of each of the convoy vehicles are known and that the centroid of the convoy is (x_c, y_c). It is possible to enclose the convoy within a sequence of concentric ellipses with center (x_c, y_c). Fig. 5 depicts three elliptical rings with center (x_c, y_c), semi-major axis A, and semi-minor axis B, surrounding a convoy of vehicles.

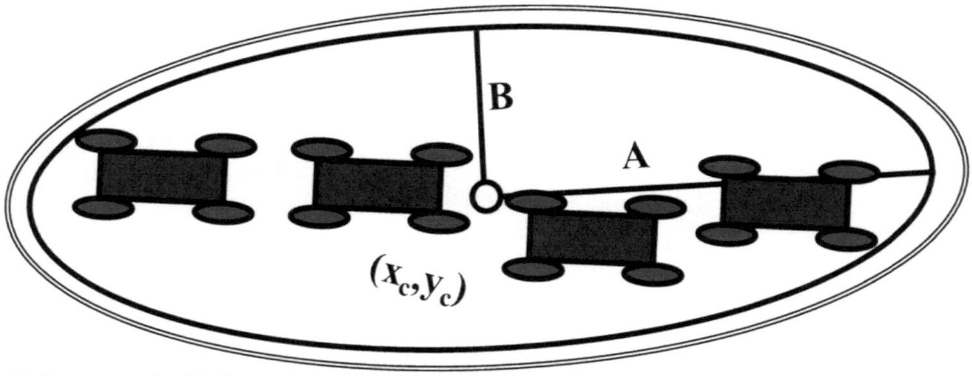

Fig. 5. Convoy of vehicles surrounded by concentric ellipses

Unreal Tournament is utilized to simulate the real world problem of convoy protection. In this simulation, a convoy of three vehicles is given a set of waypoints on a road and a swarm of forty UAVs is utilized to surround this convoy as it travels. The formation dynamically changes as the convoy travels along the road.

The shape of the elliptical formation is determined by the information provided by the convoy of vehicles that are traveling on the road. They send the swarm parameters describing an ellipse enclosing the convoy – the parameters are the center of the ellipse, the orientation, and the length of the major and minor axes. As the convoy turns the corner, the convoy trucks bunch-up causing the ellipse to become circular. In turn, the swarm redistributes as their elliptical ring becomes circular. This illustrates that proposed approach can easily adapt in differing circumstances.

Fig. 6 shows the swarm formation around the convoy at different time slices. The line is the convoy's path of travel and the darkened circles represent the convoy vehicles. Note that the formation widens and narrows when necessary. This can be noted when the convoy goes around the turn in the road. Fig. 7 shows the parameter values changing over time. Fig. 8 depicts a screenshot from the simulation environment.

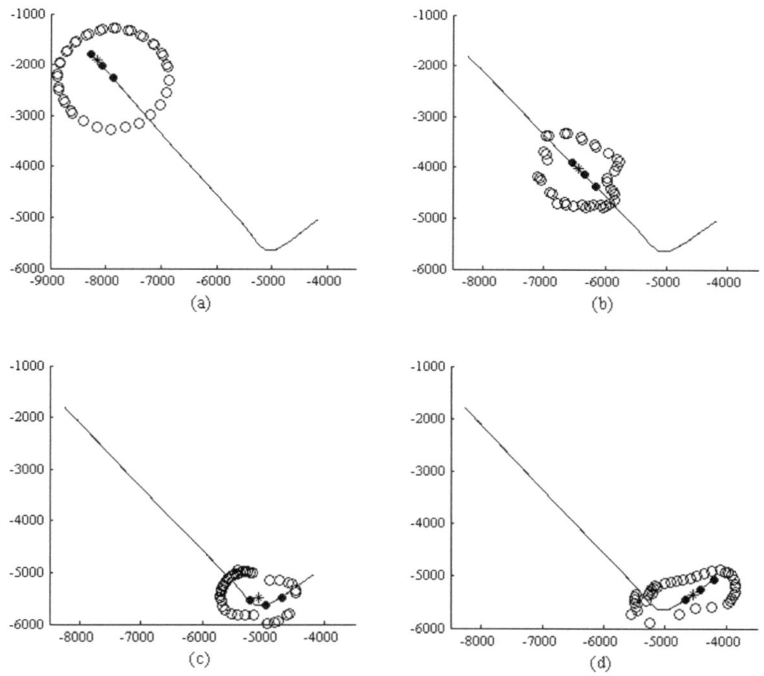

Fig. 6. Snapshot of a swarm of forty robots travelling and surrounding a convoy of vehicles in formation (a) t_1=*1*. (b) t_2=*135*. (c) t_3=*225*. (d) t_4=*260*

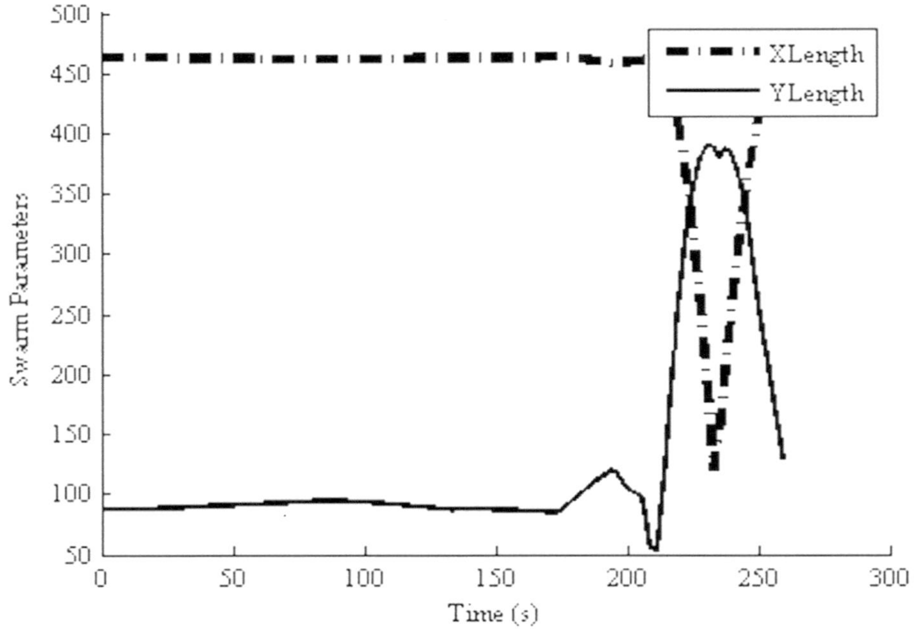

Fig. 7. Swarm formation parameters changing as convoy travels on road network

Fig. 8. Convoy protection utilizing a notional UAV swarm

5. UAV-UGV Coordination

In order to describe the coordination between the UAV and the UGV swarm, consider that a swarm of robots needs to accompany an aerial vehicle by surrounding it in a particular formation. A field needs to be designed to attract the swarm members to surround the UAV in a designated formation.

The centroid of the formation is (x_c, y_c) or the location of the UAV in two dimensions. The UAV is surrounded with a sequence of concentric ellipses with the center (x_c, y_c). Fig. 9 depicts the envisioned framework. The formation of the ground robots is described by a series of ellipsoids with center (x_c, y_c), semi-major axis $2A$, and semi-minor axis $2B$, surrounding the UAV.

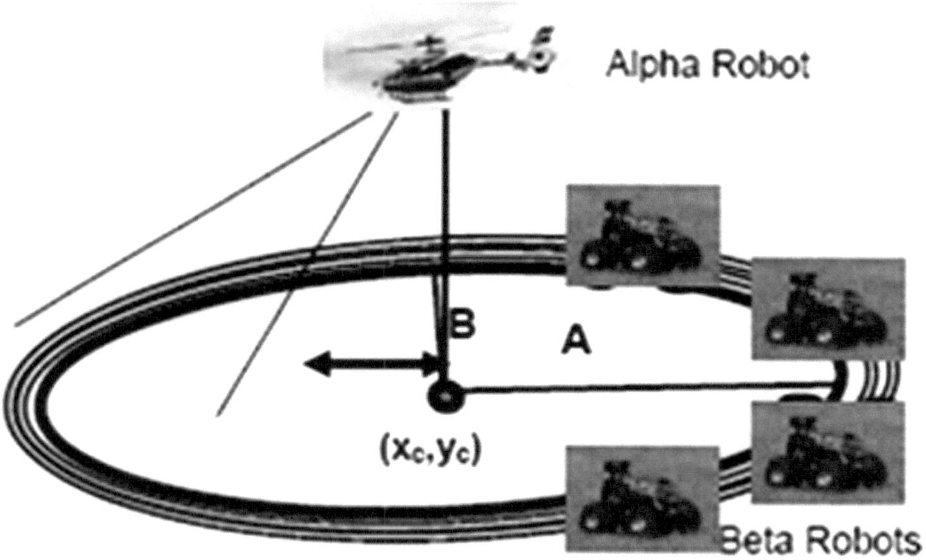

Fig. 9. Framework for UAV-UGV swarm coordination

To validate this work, experiments were performed utilizing an autonomous helicopter as the alpha robot and three custom built RC-trucks as the UGV swarm. The UAV is a Maxi Joker II and the UGVs are Traxxas Emaxx, RC-cars. Both the UAV and UGVs are equipped with a custom built computer system. The UGVs are Ackerman steered and each is equipped with an inertial measurement unit (IMU) and global positioning system (GPS). The UAV is equipped with GPS, IMU, and laser sensors (Garcia and Valavanis 2009). A simple broadcast communication model is used for information relay and exchange between UGVs and UAVs. Fig. 10 depicts the robots and helicopter utilized in these experiments.

In these experiments three UGV vehicles travel in formation surrounding the helicopter. The helicopter, or the alpha robot, acts as the formation center (x_c, y_c). The three UGVs (beta robots), surround the UAV and attempt to stay a minimum specified distance away from one another. Table I shows the control parameters used for this experiment. All units are in meters.

Fig. 11 details the UGV paths with respect to time for a single experiment. The black dotted line represents the path of the UAV. Throughout the mission the UGVs avoid each other while maintaining formation with respect to the UAV's position. Fig. 12 details each of the beta swarm member's distance from the UAV (formation center) over time. The large spikes are due to communication interference which was typical in the experiment field.

Fig. 10. UAV and UGV robots

Parameter	R*	γ	κ	ΔR_{in}	ΔR_{out}	ΔR_{avoid}	ε
Value	7	0.5	1.0	3	4	5	0.0001

Table 1. Formation parameters

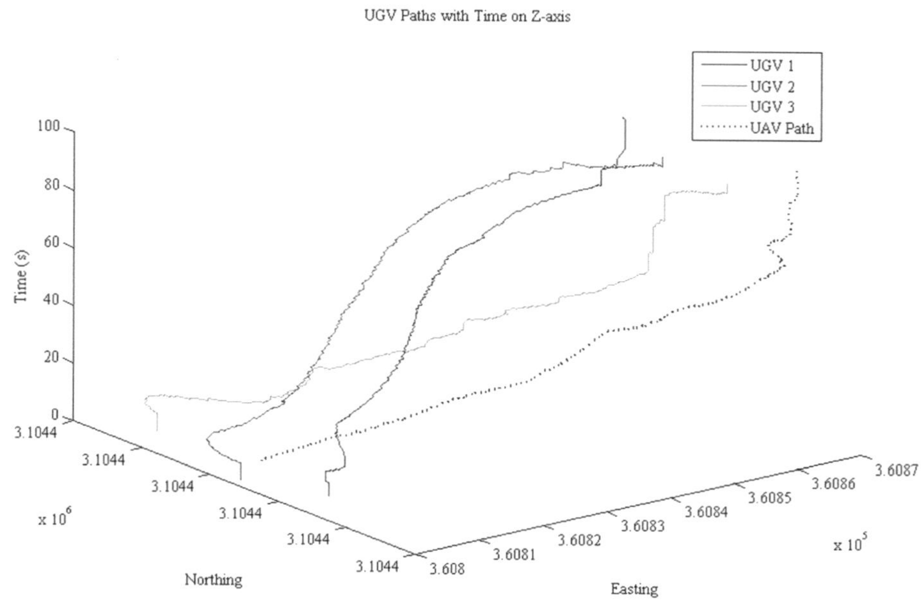

Fig. 11. Robot paths with respect to center (x_c, y_c) with time on z-axis

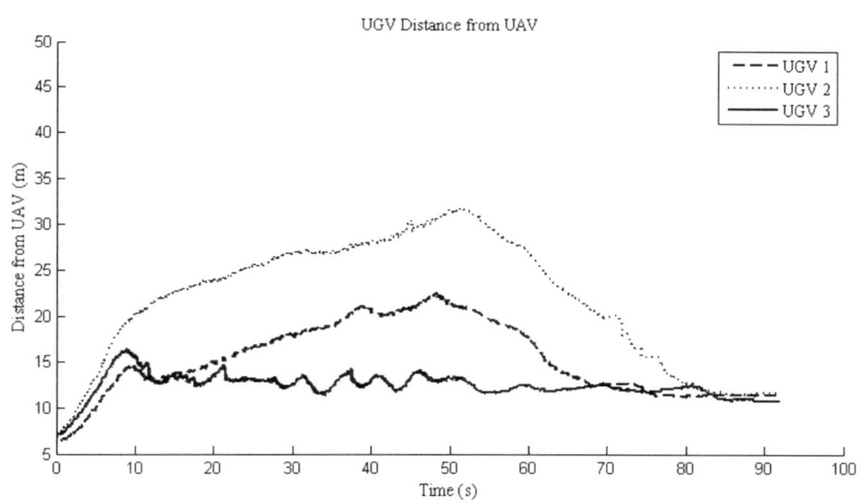

Fig. 12. UGV Distance from UAV

6. Conclusions

In this work, a methodology for control and coordination of UAVs and UGVs has been presented. UAVs and UGVs were integrated into a single team and were able to adapt their formation accordingly. Potential field functions together with limiting functions can be successfully utilized to control UGV and UAV swarm formation, obstacle avoidance and the overall swarm movement. A single UAV was also successfully used to pull the UGV swarm into formation. These formations can move as a unit, adapt to non-uniform surfaces and change dynamically. In the future, this approach will be expanded to swarms of heterogeneous UAVs utilizing manned and unmanned vehicles as well as swarm adaptation based on environmental parameters. In addition, the variety of formations will be expanded.

7. References

Barnes, L., W. Alvis, et al. (2006). Heterogeneous Swarm Formation Control Using Bivariate Normal Functions to Generate Potential Fields. IEEE Workshop on Distributed Intelligent Systems: Collective Intelligence and Its Applications.

Barnes, L., M. A. Fields, et al. (2007). Unmanned Ground Vehicle Swarm Formation Control Using Potential Fields. 15th Mediterranean Conference on Control and Automation.

Chaimowicz, L. and V. Kumar (2004). Aerial Shepherds: Coordination among UAVs and Swarm Robots. International Symposium on Distributed Autonomous Robotic Systems.

Chaimowicz, L. and V. Kumar (2004). A Framework for the Scalable Control of Swarms of Unmanned Ground Vehicles with Unmanned Aerial Vehicles. Proceedings of the International Conference on Robotics and Remote Systems for Hazardous Environments.

Elfes, A., M. Bergerman, et al. (1999). Air-ground robotic ensembles for cooperative applications: concepts and preliminary results. *International Conference on Field and Service Robotics*.

Garcia, R. D. (2008). Designing an Autonomous Helicopter Testbed: From Conception through Implementation. Computer Science and Engineering. Tampa, University of South Florida.

Garcia, R. D. and K. P. Valavanis (2009). "The Implementation of an Autonomous Helicopter Testbed." J. Intell. Robotics Syst. 54(1-3): 423-454.

Lacroix, S., I.-K. Jung, et al. (2001). Towards cooperative air/ground robotics: issues related to environment modeling. *International Conference on Advanced Robotics*.

Stentz, A., A. Kelly, et al. (2002). Integrated air/ground vehicle system for semi-autonomous off-road navigation. *AUVSI Unmanned Systems Symposium*.

Sukhatme, G., J. Montgomery, et al. (2001). Experiments with aerial-ground robots. Robot Teams: From Diversity to Polymorphism, AK Peters.

Tanner, H. G. (2007). Switched UAV-UGV Cooperation Scheme for Target Detection. IEEE International Conference on Robotics and Automation.

Intelligent Robot Systems based on PDA for Home Automation Systems in Ubiquitous

In-Kyu Sa*, Ho Seok Ahn**, Yun Seok Ahn***,
Seon-Kyu Sa**** and Jin Young Choi**
Samsung Electronics Co., Seoul National University**,*
*MyongJi University***, Mokwon University*****
Republic of Korea

1. Introduction

Koreans occasionally introduce their country with phrases like, 'The best internet penetration in the world', or 'internet power'. A huge infra-structure for the internet has been built in recent years. The internet is becoming a general tool for anyone to exploit anytime and anywhere. Additionally, concern about the silver industry, which is related to the life of elderly people, has increased continuously by extending the average life span. In this respect, application of robots also has increased concern about advantages of autonomous robots such as convenience of robots or help from autonomous robots. This increase in concern has caused companies to launch products containing built-in intelligent environments and many research institutes have increased studies on home automation projects.

In this book chapter, we address autonomous systems by designing fusion systems including intelligent mobility and home automation. We created home oriented robots which can be used in a real home environment and developed user friendly external design of robots to enhance user convenience. We feel we have solved some difficulties of the real home environment by fusing PDA based systems and home servers. Wireless Lan attached PDAs can serve as small size and stable wireless environments. By sending states of robots using a home server, our system has the unique feature of remote control combined with 3D home monitoring. We have implemented a system that can adapt a mobile robot to home automation systems and confirm our proposal with experiments using a PBMoRo (PDA Based Mobile Robot).

2. Home Automation system using PDA based Intelligent Robot System (HAuPIRS)

2.1 Limitations and features of the home automation system

There are lots of home automation oriented researches being done together with the spreading of the internet and further development of communication technology. Innovative companies are making an effort to improve their business image through mass media. The main feature of these systems is centralization that roots on home servers and uses external devices to control home appliances.

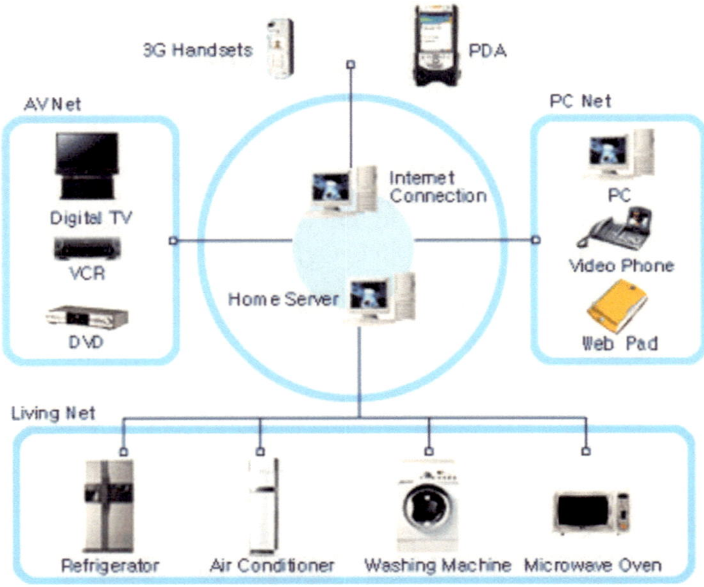

Fig. 1. Home Network concept.

Advantages	Remote control
	Security support
	Serving convince environment
Disadvantages	Consideration of system before construction
	Difficulty of adding and fixing extra home appliances
	Rising cost in proportion to size of home.
	Limitation of autonomous environment
	Passive system
	Simple UI based on text or 2D graphic
	Prerequisite learning for usage of system

Table 1. Advantages and disadvantages of home automation system.

Fig. 1 illustrated home automation system the intelligent home network & security structure (JongWhan Kim, 2002). Although it can serve a high level of security and ubiquitous functions, clients are able to connect anywhere and anytime. In Fig. 1, though they consider designing of home automation system should be done before construction, it is hard to add extra home appliances and fix them. In addition, it may increase building cost in proportion to the size of home. Another limitation exists as a view of system users. People who use this system should move to the right place in order to operate the system and learn about installation documents. Table 1 notes advantages and disadvantages of home automation system.

2.2 Limitations and features of intelligent robot system

Universities, companies and institutes studying robotics have changed their patterns by looking for huge markets such as welfare for elder or physically challenged people rather than factory industrial robots (Sebastian Thrun, 1999). Vacuum cleaning robot markets are already gradually forming (Vacuum cleaner robot from LG Electronics).

Service robot issues are a hot topic of the robotics filed (Ho Seok Ahn, 2008). Although there are many service robots, entertainment robots, escort robots, and medical robots, the common goal of these robots is to help people and increase convenience by actively giving information (Kuk-Jin Yoon, and In-So Kweon, 2001). Another feature of these robots is they are equipped with a PC based processing device, display system, and actuators.

Fig. 2 (Kyung-Sang Bukdo, 2004) shows an exhibition escort robot, another branch of intelligent robots. These robots have a PC based system that is easy to develop and expand. Thus, it has an efficient performance to process complex calculations and smart intelligence. The PC, however, should be running on the robot and it consumes a great amount of power because of the larger body size. Therefore, it is not suitable for home environments because the robot demands too much space and is too heavy (Koide Y., Kanda T., Sumi Y., Kogure K. & Ishiguro H. 2004).

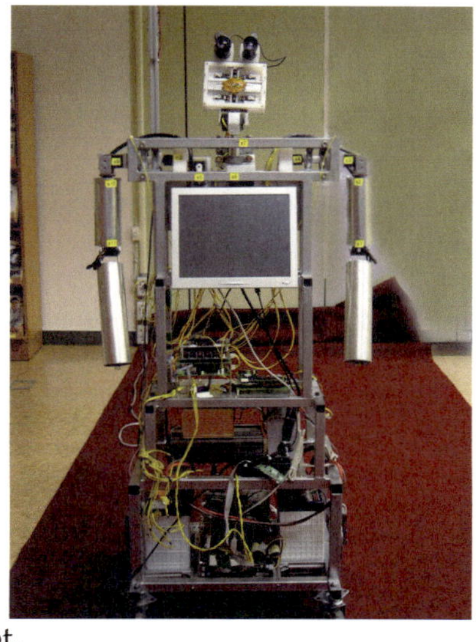

Fig. 2. Exhibition escort robot.

Some companies tried to combine home automation and intelligent systems in Fig. 3 (Most-itech). Despite the good idea to make home robot, they failed to form a market due to the limitation that relied on performance of the mobile phone (NEC personal robot PAPERO, Palm Pilot Robot kit from CMU). The boundary of incorporation of intelligent robot and home automation is displayed in Table 2.

Fig. 3. Home network robot of Most-itech.

Advantages	Convenient developing environment.
	Good expandability of developed modules.
	High performance.
Disadvantages	Difficulty of usage due to heavy shape.
	Much power consumption.
	High cost.
	Difficulty of building home automation.

Table 2. Critical points of combination of intelligent robot and home automation.

2.3 Home Automation system using PDA based Intelligent Robot System (HAuPIRS)

In this book chapter, we propose a system that overcomes the limitation of current difficulty; a fusion system combining home automation systems and intelligent robots. We focus on designing compact size of the robot in order to exploit the home environment and expand controllable area by adapting mobility and activity of the robot. Also, we discuss easy installation and repair of home appliances by RF communication.

To achieve our goal, we use PDA based environments rather than PCs. This gives us a lot of advantages. We could save power consumption by adapting to PDA and create a smaller shape for the body. It also allows a lower cost and higher operation efficiency. The fundamental differences of the HAuPIRS are that they specify a home environment, are easy to use and allow expandability.

Fig. 4 illustrates the system architecture of the HAuPIRS. The HAuPIRS are made up of 3 levels: the executive level, cognitive level and user level (Arkin R. C, 1987, 1998, 1999). The user level enables clients to connect a home server and send commands. Users can move the home robot and control home appliances. Client programs return current status to the home. The cognitive level is a pre-processing stage before execution of commands. It calculates command signals utilizing the information stored in the server, added to the robot input data. On this level, it accompanies artificial intelligent processing; SLAM (Simultaneous Localization and Mapping), image processing. We can not only add third party algorithms, but also recognition technology to the cognitive level. The executive level executes commands and returns sensor data obtained from cameras and ultrasonic range finders to the cognitive level (Brooks R, 1986).

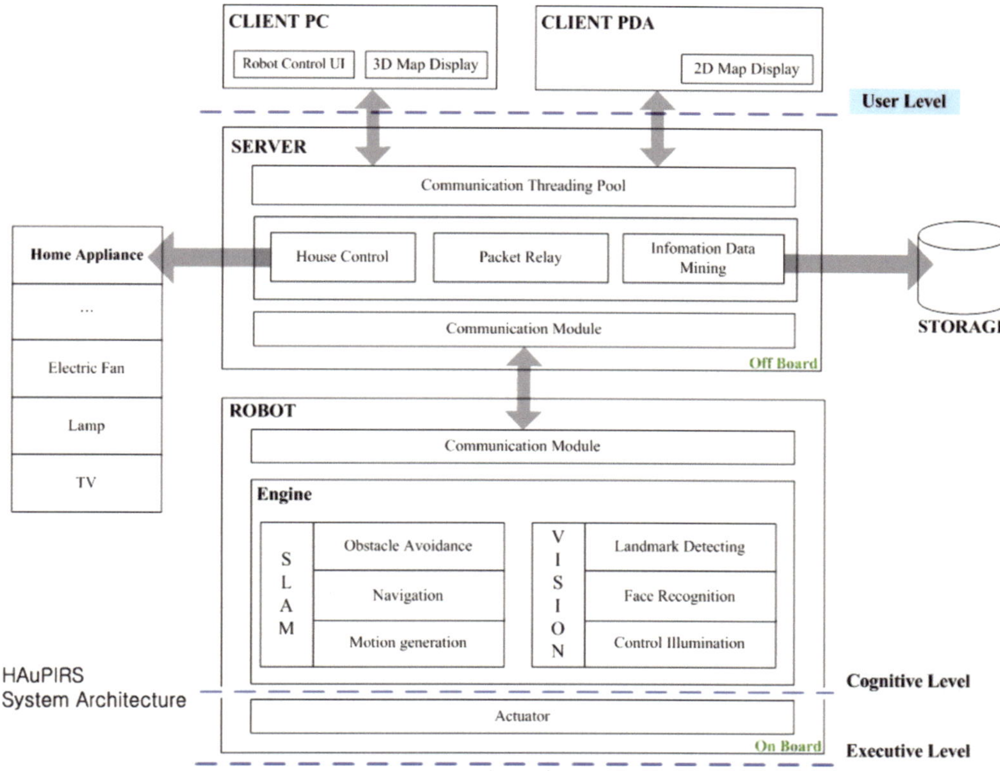

Fig. 4. Home Automation system using PDA based Intelligent Robot System (HAuPIRS).

Specifications of the HAuPIRS are shown in Table 3. HAuPIRS need a diversity of technologies to run the home automation system with the intelligent robot. In order to establish network environment, wide spread internet penetration, wireless technology, image streaming technology, and reliable packet command technology are highly demanded. Additionally, motor and sensor control technology, vision and voice processing technology, path planning, object avoidance, localization for freely navigation of robot, and well-organized scheduling technology are needed in aspects of the robot.

Main system	PDA (iPaq 6360). Home server equipped.
Power	Lithium-Polymer light weight notebook battery,. Guarantee 3 operating hours.
Size & Weight	Suitable size for home environment. Light weight.
Cost	Less expensive than PC. One camera and multi range finders.
Interface	Face detection and voice recognition technology support. Support additional devices; PC, PDA and mobile phone. Client System using 3D Display allows reliability for users. Intuitive user interface support.
Automation	Support mobility and active automation. Enable far distance watching by remote control.
Security	Secure home instead of human being. When it comes to crime, it can make evidence and identity of person.
System expandability	Easy expansion by altering robot instead of home. Easy to plug in or plug-out modules on home appliances.
Design	User friendly robot design. Prevent a wound by smooth design.

Table 3. Specifications of the Home Automation system using PDA based Intelligent Robot System (HAuPIRS).

3. PBMoRo System

In this book chapter, we implement the HAuPIRS Control Architecture system in order to evaluate its performances and refer it as the PBMoRo System (PDA Based Mobile Robot System). It is possible to add intelligence to the PBMoRo System and to recognize the module by adjusting the HAuPIRS. It has also been tested under the home automation conditions for house specified robots.

3.1 Structure of PBMoRo system

The PBMoRo System is consists of three parts, shown in Fig. 5. The first part of the system is a robot system based on the PDA. The second part is a server system that possesses all the data. Finally, a client system plays the role of interacting with the human through display devices and input devices.

To begin, the Robot system rooted on PDA can be divided into a PDA aspect and actuator aspect. The PDA aspect acquires images from cameras and communicates with the server using an Ethernet interface. The actuator aspect executes commands obtained from the

server. The server system plays an important role in spreading usage by supporting the PDA and acting as a home server.
The client system is made up of two systems as well, a PC oriented application and a PDA oriented application, and can run independently. In addition, this system sends commands locally or remotely to operate the PBMoRo system. Users are able to check the current status of the system as desired.

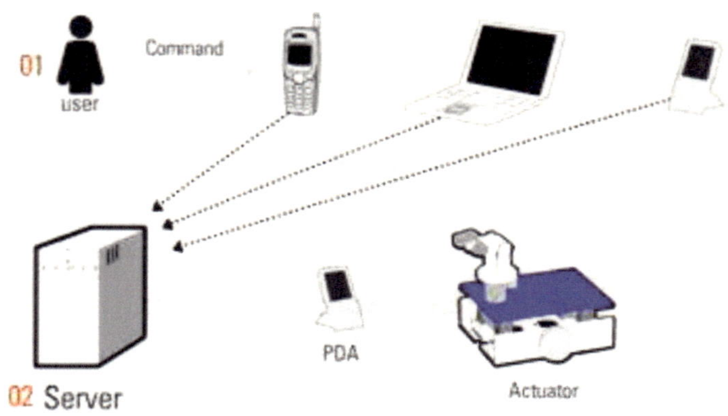

Fig. 5. Architecture of the PBMoRo System using the HAuPIRS.

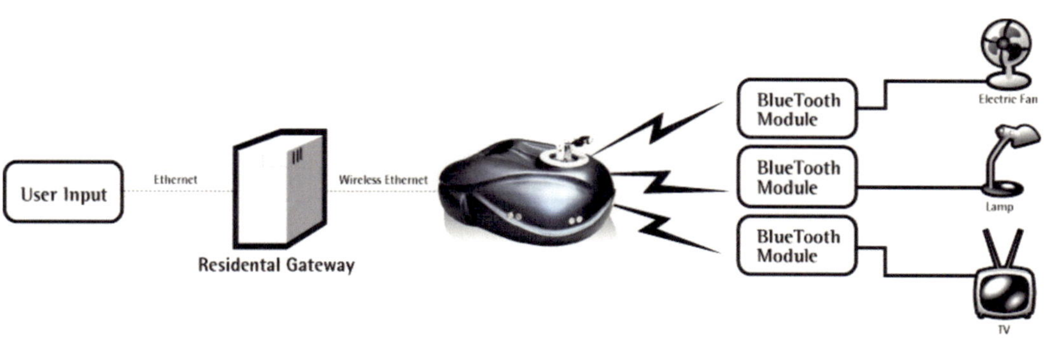

Fig. 6. Architecture of the PBMoRo System using the HAuPIRS with home appliances.

Fig. 6 denotes the architecture of the PBMoRo System using the HAuPIRS with home appliances. The main part of a home network server which plays a significant role in residential gateways sending commands to intelligent robots through a wireless network interface. The robot approaches home appliances in order to manage these with Bluetooth interface. Fig. 7 illustrates the control architecture of the PBMoRo system. Table 4 displays specifications of devices we employed.

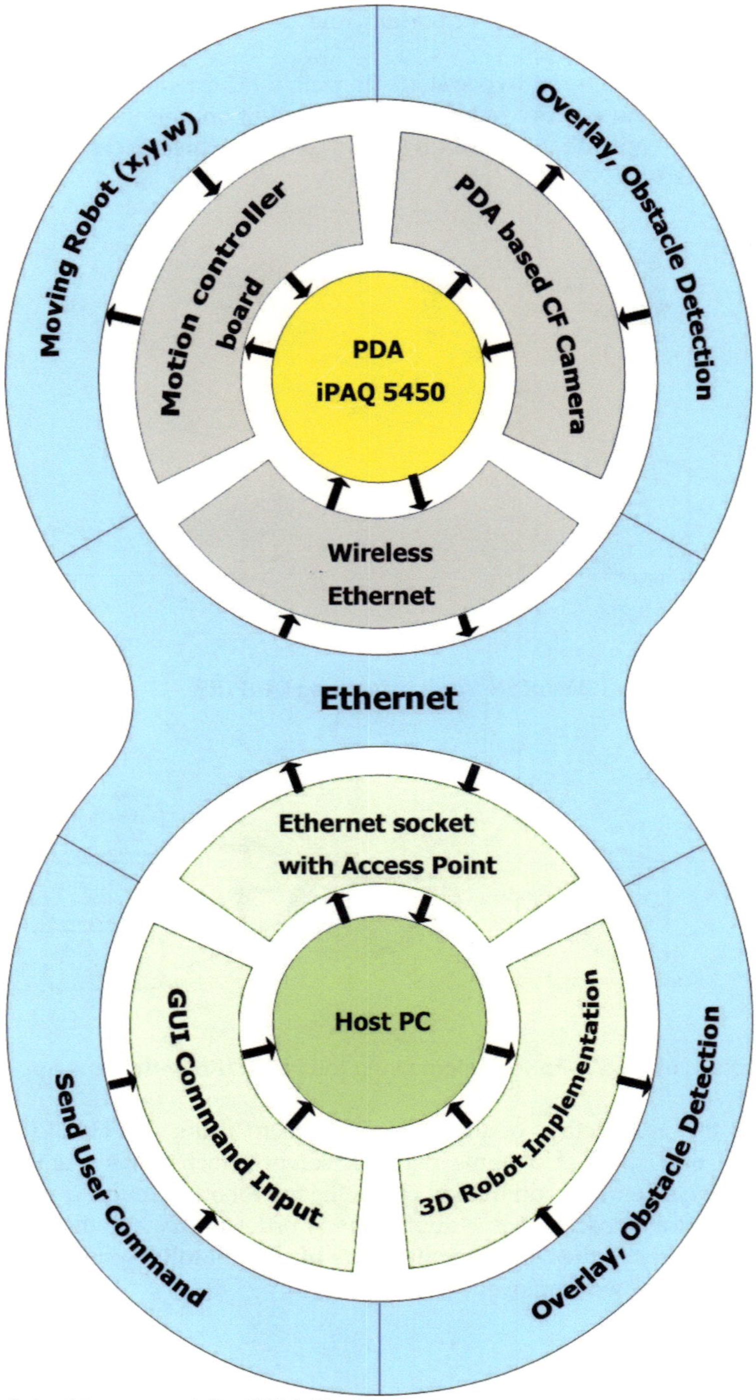

Fig. 7. Control Architecture of the PBMoRo System.

Dimensions	300 x 400 x 300 mm (w x d x h)
Weight	8 kg
PDA	HP iPAQ h5450
Body	Aluminum , acrylic
Motor	Actuators : DC motor x 2 , Pan-Tilt : DC motor x 2
Camera	CF COMS Camera (1.3M pixel, 30 Frame/sec)
Sensors	Ultra Sonic x 5 , CO2 Sensor x 1 , VOC Sensor x 1
Power	Notebook battery x 2 (6000mAh, 15V, 800g)

Table 4. Features of PBMoRo System.

3.2 Robot system of PBMoRo system

As we mentioned above, the robot system consists of two parts. One part is in charge of acquiring image frames, talking to the server, and sending motion commands. The other, the actuator, generates robot motion by commands. Fig. 8 denotes appearances of the robot we implemented.

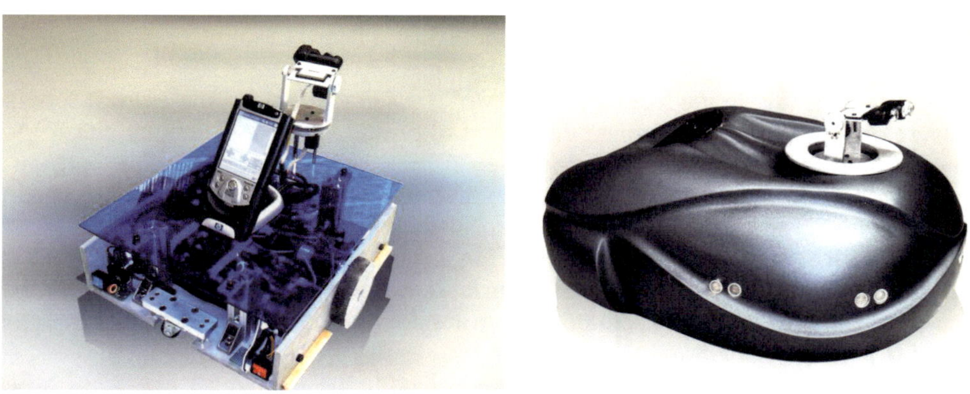

Fig. 8. Shape of the PBMoRo (Left image) before considering robot design, (Right image) after adapting robot design.

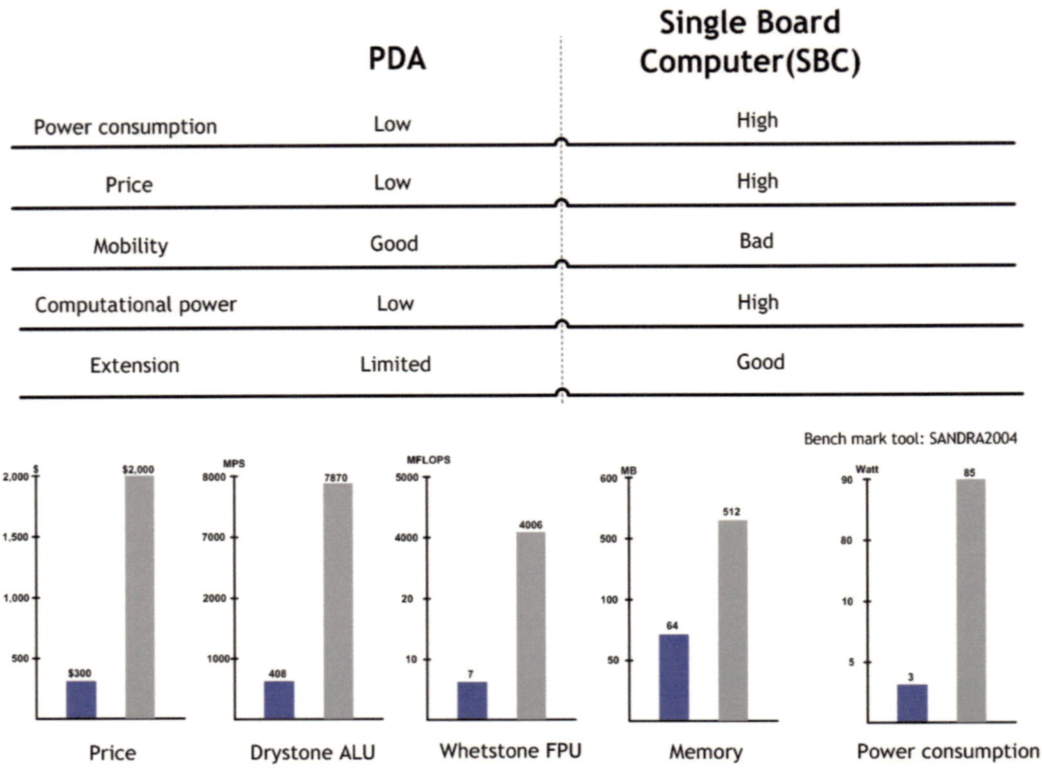

Fig. 9. Performances comparison between PDA and Single Board Computer.

As we see in Fig. 9, due to the fact that the PBMoRo system is rooted on a PDA, it shows a very severe experimental environment compared to primary PC based robots. The PDA is 10 times cheaper than an SBC and it consumes 30 times less energy. One disadvantage is that processing performance is far weaker than an SBC. To improve this aspect we need to optimize the general algorithms for the PBMoRo system.

Fig. 10 shows a conceptual diagram of our PDA program. At first, PDA investigates the current state of the robot through ultrasonic sensor data and motor encoder data. After that, it sends commands to the robot actuator using the data received from motion generators and target position information.

In order to approach the goal, we should control both the left wheel and right wheel velocities. Consequently, we need mathematical equations to evaluate the angle between robot and goal position. Fig. 11 shows the standard angular coordination (SAC) what we used in this robot system. θ_G denotes the goal angle related SAC and means robot angle oriented from SAC. We can say equation (1) in terms of above statement.

$$\theta_G - \theta_R = \theta_{GR}. \tag{1}$$

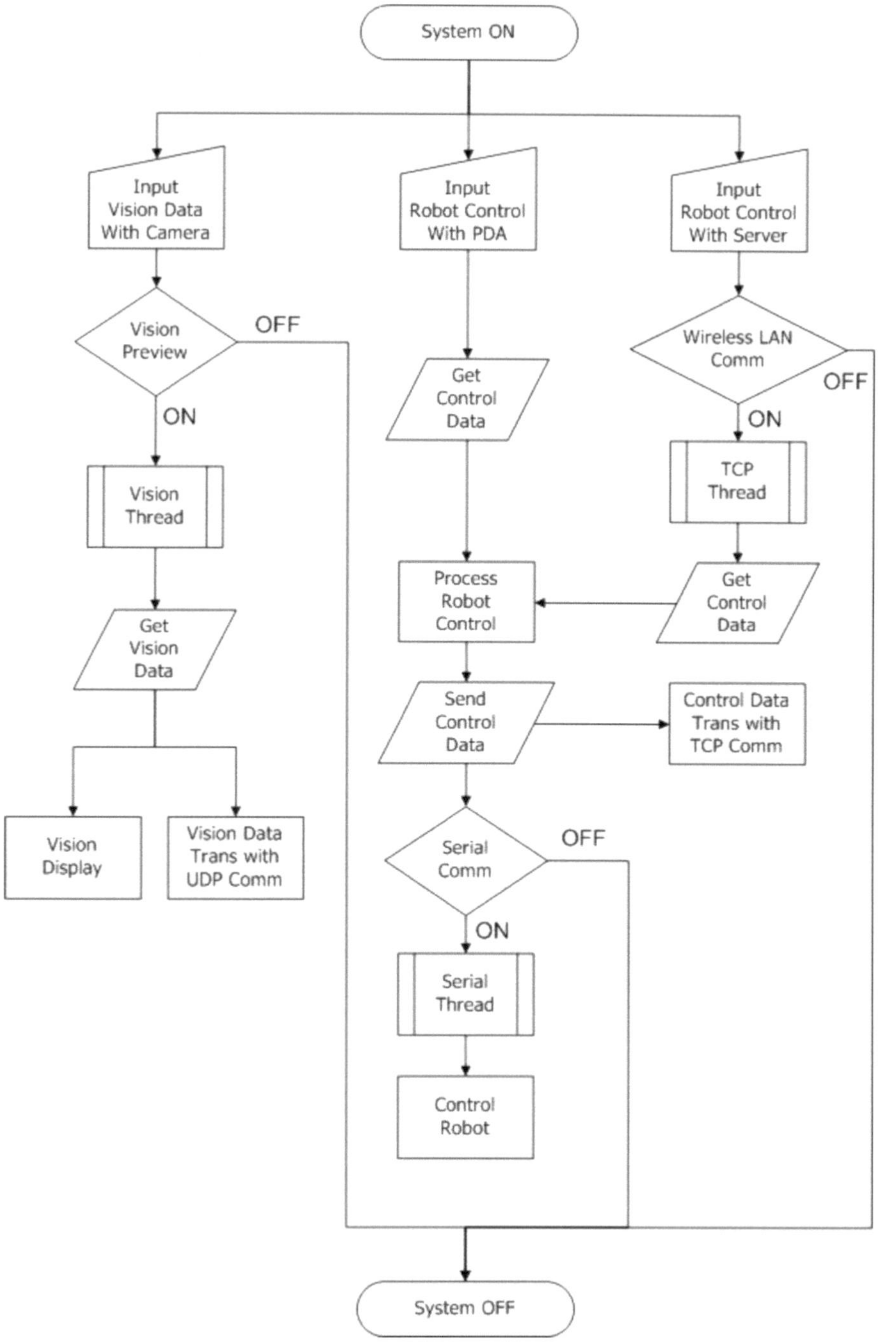

Fig. 10. Flow diagram of the PDA program.

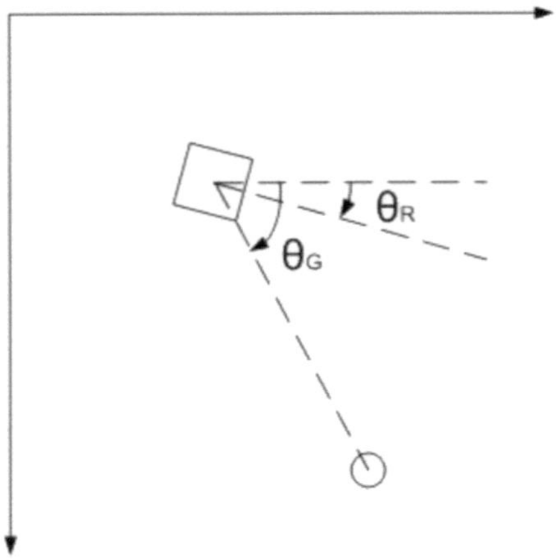

Fig. 11. Standard angular coordination in the PBMoRo System.

To find out the current robot position, many researchers have used odometry information from motor encoder and landmarks. Odometry based dead reckoning method employs encoder data to obtain the current robot speed. By implementing this method with our system, we can estimate current robot position by accumulating movement per sample time. First, we are able to calculate distance and velocity of the two wheels by encoder data during a sample time.

$$V_L = \frac{\delta_L}{T}, \tag{2}$$

$$V_R = \frac{\delta_R}{T}, \tag{3}$$

where, δ_L is movement of left wheel measured from the motor encoder per sample time, δ_R is movement of right wheel measured from the motor encoder per sample time, T is sample time, V_L is velocity of left wheel, and V_R is velocity of left wheel.
We can address linear velocity and angular velocity of robot from equation (2) and (3).

$$\delta_\theta = \frac{\delta_R - \delta_L}{\omega}, \tag{4}$$

$$V_C = \frac{V_R + V_L}{2} = \frac{\delta_R + \delta_L}{2T}, \tag{5}$$

$$V_\omega = \frac{\delta_\theta}{T} = \frac{\delta_R - \delta_L}{T \cdot \omega}, \tag{6}$$

where V_C is linear velocity of robot, V_ω is angular velocity of robot, and δ_θ is angular movement of robot during sample time.

Through equation (4), (5), and (6), we could estimate present robot position and orientation as shown in equation (7), (8), and (9).

$$^{\omega}O_X(k+1) = {}^{\omega}O_X(k) + \frac{\delta_R+\delta_L}{2} \cdot \cos\left({}^{\omega}O_\theta(k) + \frac{\delta_\theta}{2}\right), \tag{7}$$

$$^{\omega}O_Y(k+1) = {}^{\omega}O_Y(k) + \frac{\delta_R+\delta_L}{2} \cdot \sin\left({}^{\omega}O_\theta(k) + \frac{\delta_\theta}{2}\right), \tag{8}$$

$$^{\omega}O_\theta(k+1) = {}^{\omega}O_\theta(k) + \delta_\theta. \tag{9}$$

Although users can control the PBMoRo System using a remote joystick, it is also able to operate automatically. If people input their desired goal position, the robot can navigate in a home environment freely by itself. To accomplish these kinds of jobs, a path planner should make the shortest path and avoid obstacles while a motion generator establishes suitable velocities for the two wheels. However, if the robot faces obstacles which don't exist on the map, collision and slippage problems occur simultaneously. In this case, the robot and home appliances might be damaged and systems could break out. Therefore, the robot should be able to avoid obstacles that are not on the map.

In these respects, we suggest an obstacle avoidance algorithm for the PBMoRo System. The angular coordination can be divided into 4 sectors depending on the angle of the robot as illustrated in Fig. 12.

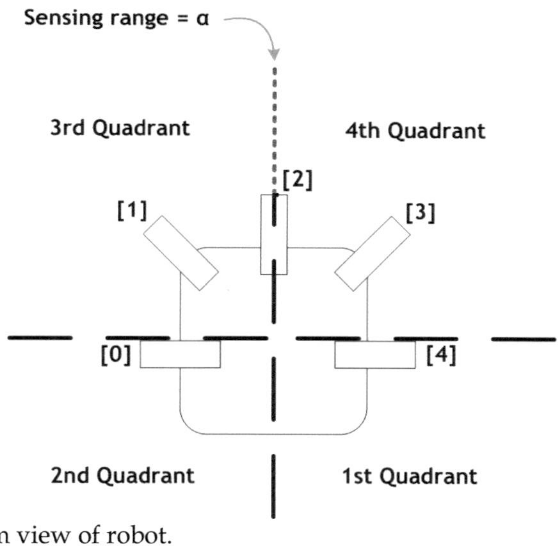

Fig. 12. Quadrant from view of robot.

If there are some obstacles near the robot, ultrasonic range sensors can detect them and transfer their data to a PDA (Beom H. R. and Cho H. S, 2000). The PDA recognizes locations of obstacles and makes a safe path to prevent collision. These processes are shown in Fig. 13.

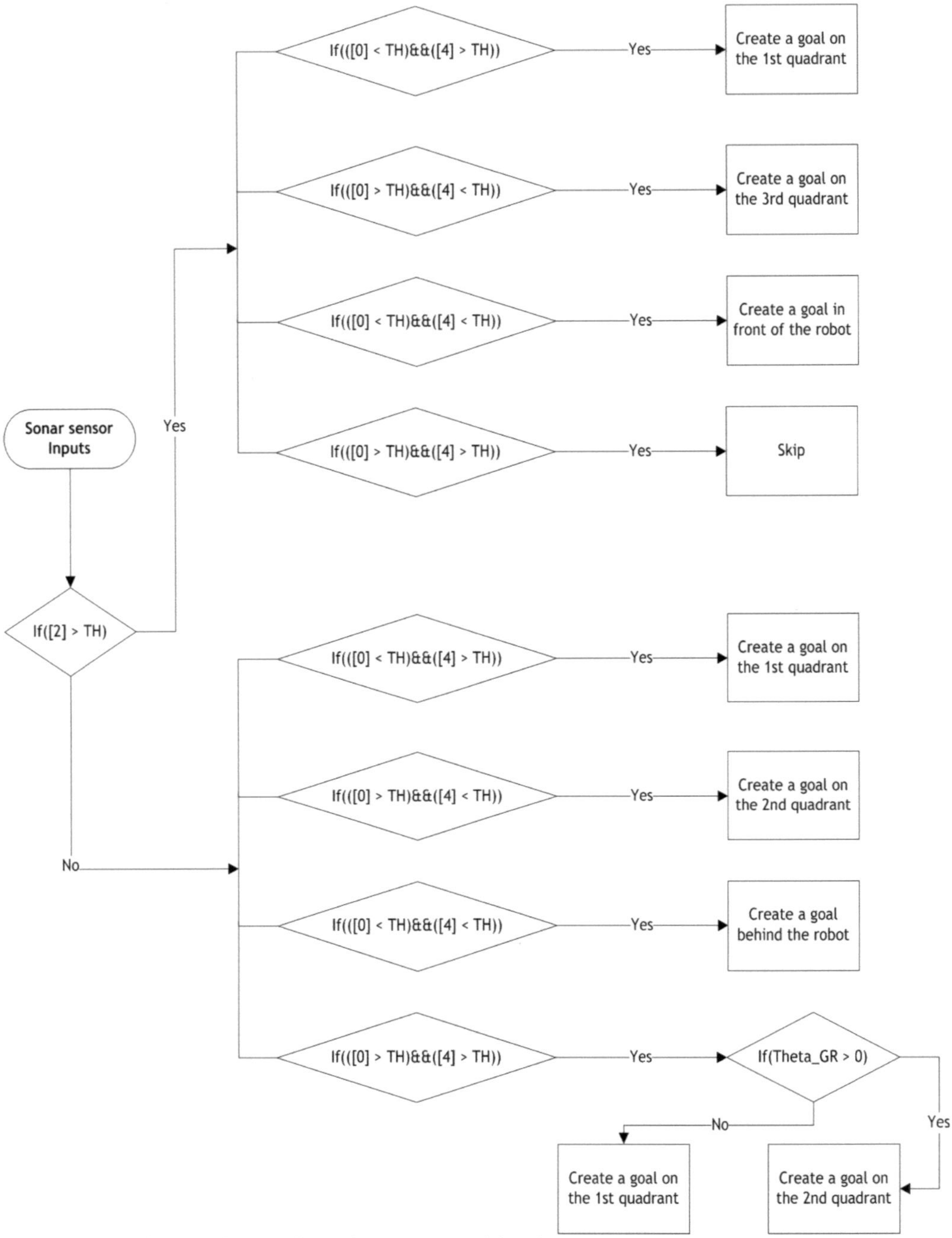

Fig. 13. Obstacle avoidance algorithm proposed by the PBMoRo System.

Because the PDA has a lower performance than the SBC, we exploited half of the ultrasonic sensor data, even filed and odd filed. By using this method, we could make an obstacle detection algorithm in real time. Fig. 13 illustrates even filed data processing using a zero index, two index and four index sensors.
Due to developing map building algorithms, overloading the PDA, we skipped a searching procedure surrounding the robot. Instead of map building on the PDA, the PBMoRo system updates map data into the server system and relies on current sensing data when non-existent obstacles appear in the way of the robot (Roland Siegwart, 2007). This method reduces errors and power consumption because it limits unexpected motion of the robot. In addition, it improves total system performance because it relieves unnecessary procedures. It needs to pay attention to implement intelligence and active sensibilities of the robot.
Fig. 14 shows map building algorithms running on a server system and position where sensors are arranged. In the PBMoRo Robot System, there are 5 ultrasonic sensors equipped. Each of them has a 45° gap between each other and has from 0 to 4 indexes on clockwise. As the illustration on the left of Fig. 14 shows, the front of the robot is designed at 0°, the left semi sphere has a range of from 0 to -179° and opposite side has range of from 0 to 180°.
$\theta_{[n]}$ denotes an angle of nth sensors, each sensor possesses -90°, -45°, 0°, 45°, 90° separately. $\alpha_{[n]}$ denotes a length of nth sensors. If we suppose that θ_R is an angle of the robot, we can address the location of obstacles as shown in equations (10) and (11).

$$X_O = X_R + \alpha_{[n]} \cdot \cos\left(\theta_R + \theta_{[n]}\right), \tag{10}$$

$$Y_O = Y_R + \alpha_{[n]} \cdot \sin\left(\theta_R + \theta_{[n]}\right). \tag{11}$$

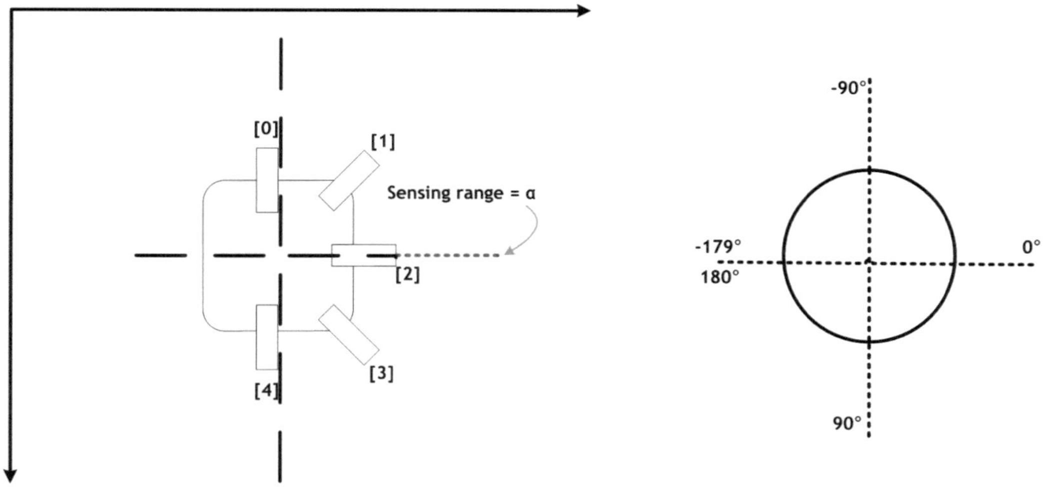

Fig. 14. Positions and angles of ultrasonic sensors.

Although the actuator part consists of 3 parts such as a sensor, moving and pan/tilt, the PDA only has one RS232 port. Thus, we should design a new data passage structure to communicate between PDA and actuator part. We developed multi sensors architecture as illustrated in Fig. 15.

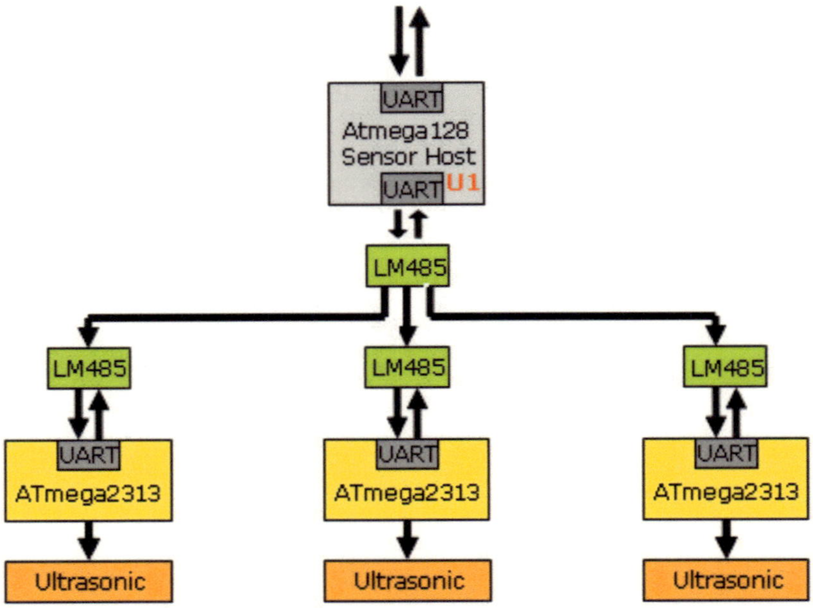

Fig. 15. The PBMoRo RS485 communication architecture.

In this book chapter, we developed an RS485 communication for the purpose of expanding the number of UART ports. Each slave device has a front end microchip to enable interaction with a master device.

3.3 Server system of PBMoRo system

The server system plays the role as an intermediary connecting client system and robot system according to the HAuPIRS. Thus, both robot system and client system should connect to the server system and have an important effect on the design of the entire system. Video streaming and robot information should pass through the server system due to a limit of PDA performances such as computational power and memory boundaries. In addition, we should consider the conditions of multi-connection and security when a number of user clients try to use the PBMoRo system. The dataflow diagram of our server system is shown in Fig. 16. Call flow structure of PDA-Server network system is illustrated in Fig. 17.
We adapted a UDP protocol when transferring video data and TCP protocol exploited the transaction of control signals. This is largely because UDP can reduce the load and TCP can guarantee high reliability.

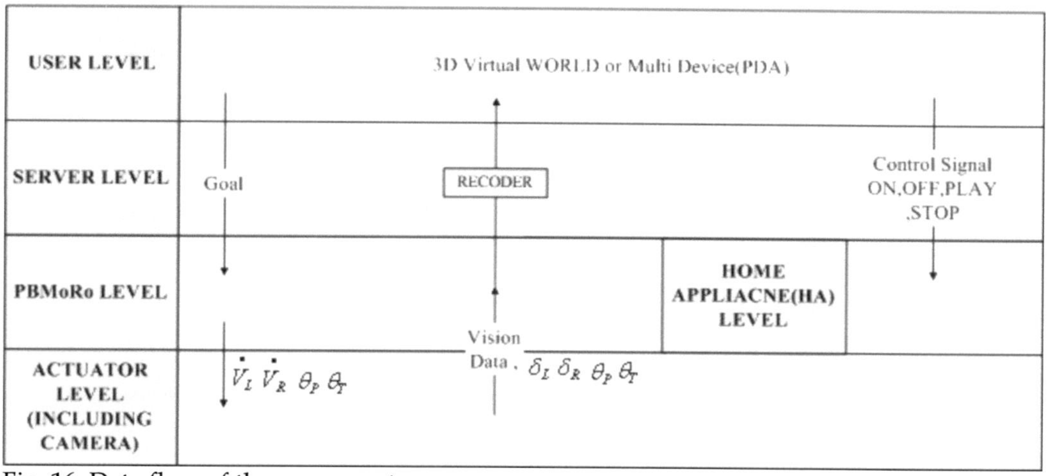

Fig. 16. Data flow of the server system.

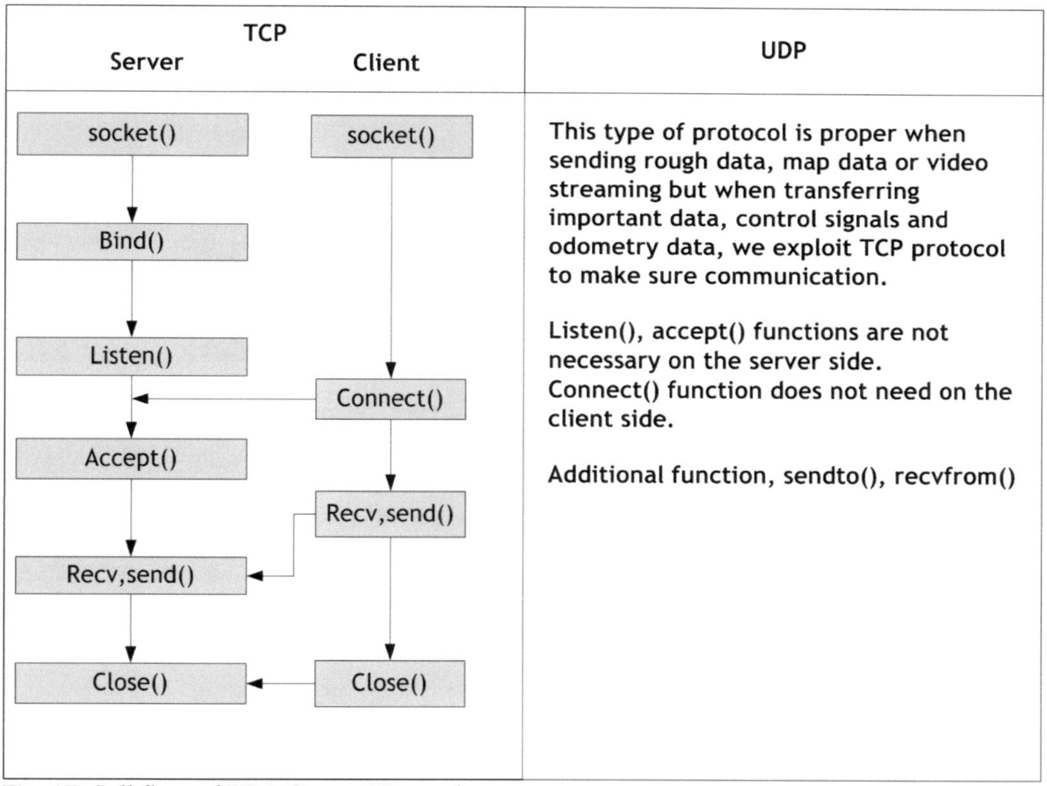

Fig. 17. Call flow of PDA-Server Network system.

3.4 Client system of PBMoRo system

The client system consists of PC client programs and PDA client programs. PC client programs need more details and performances than the latter. We implemented this PC program using a 3D technology to give a better sense of reality. Additionally, designing of the system based on the HAuPIRS enables us to attach extra mobile devices.

4. Experimental Results

4.1 Tracking

The PBMoRo system detects a moving object and tracks it using security processes. When an object moves in security mode of the PBMoRo system, it follows the moving object keeping a suitable distance. The PBMoRo system calls to the owner and police using this function. Fig. 18 shows the experimental results of moving object tracking. In this experiment, the PBMoRo system followed the red glove keeping a suitable distance. We have proven that the PBMoRo system is capable of detecting and tracking a moving object.

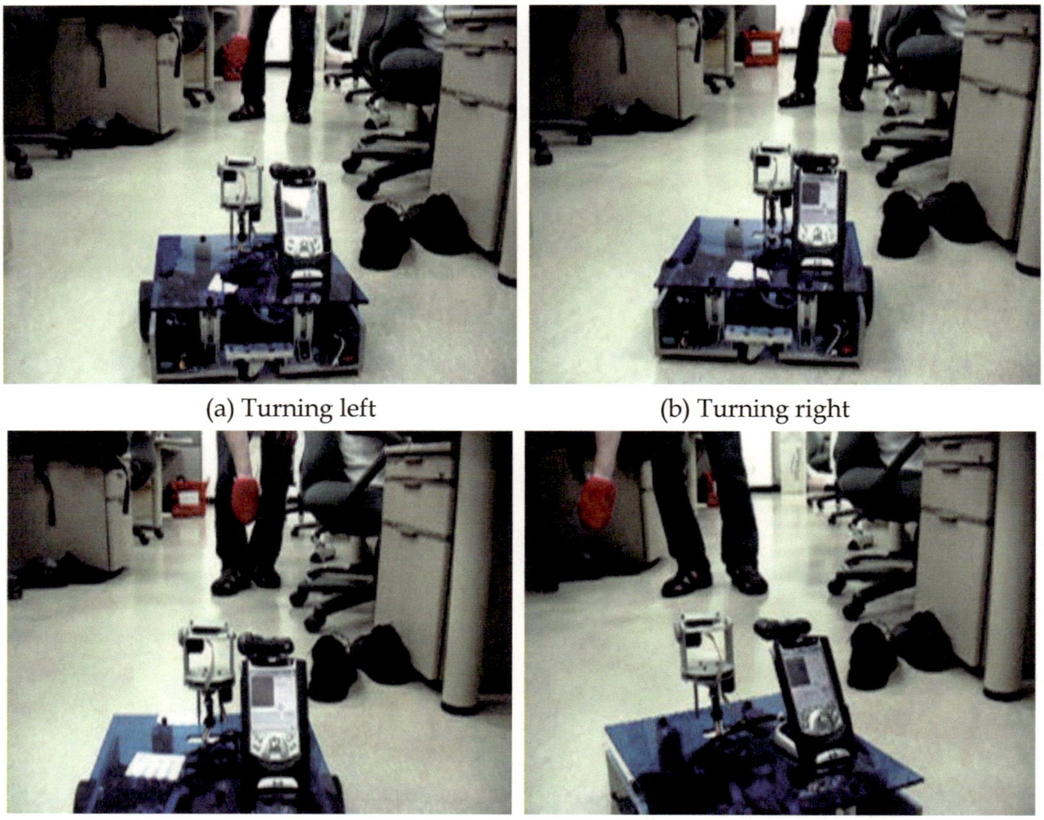

(a) Turning left (b) Turning right

(c) Approach (d) Approach by turning left

Fig. 18. The experimental results of moving object tracking.

4.2 Path planning and localization

This experiment is for remote control and monitoring of the PBMoRo system. When a user sets the goal position of the robot, the robot should move to that position automatically. It should be possible for users to monitor the situation and the states of the robot from a location outside of the home. In this case, the experiment is important because the robot should synchronize the real-world position with the position of cyber-world.

Fig. 19 shows the experimental results of path planning and localization. In this experiment, the PBMoRo system made the path and moved to the goal position. In this process, the PBMoRo system negotiated the present position with the 3D monitoring program. We have verified that the PBMoRo system synchronizes the position of real-world and cyber-world as well as performing the path planning and localization functions effectively.

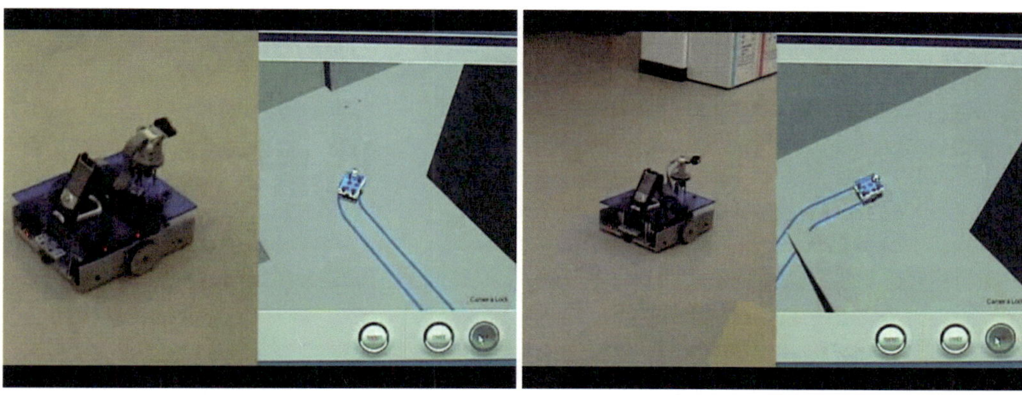

(a) Turning (b) Going straight

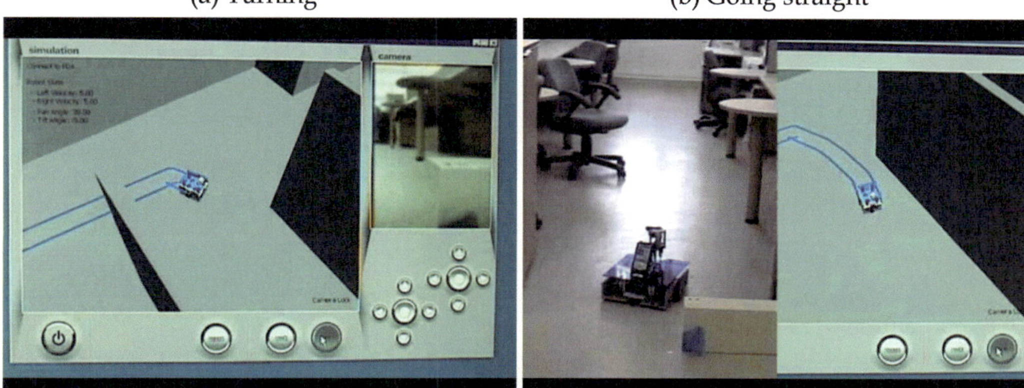

(c) Showing camera image (d) Getting goal position

Fig. 19. The experimental results of path planning and localization.

4.3 Obstacle avoidance and map building

When new obstacles are detected, the robot should make a new path to avoid the obstacle and update the map. As the PBMoRo system uses the PDA main system, we use the simplified algorithm for obstacle avoidance and map building based on the HAuPIRS architecture. Fig. 20 shows the experimental results of obstacle avoidance and map building. In this experiment, the PBMoRo system detected a new obstacle, made a new path, and updated the map. The bottom right figure in 20 shows the result of this map building.

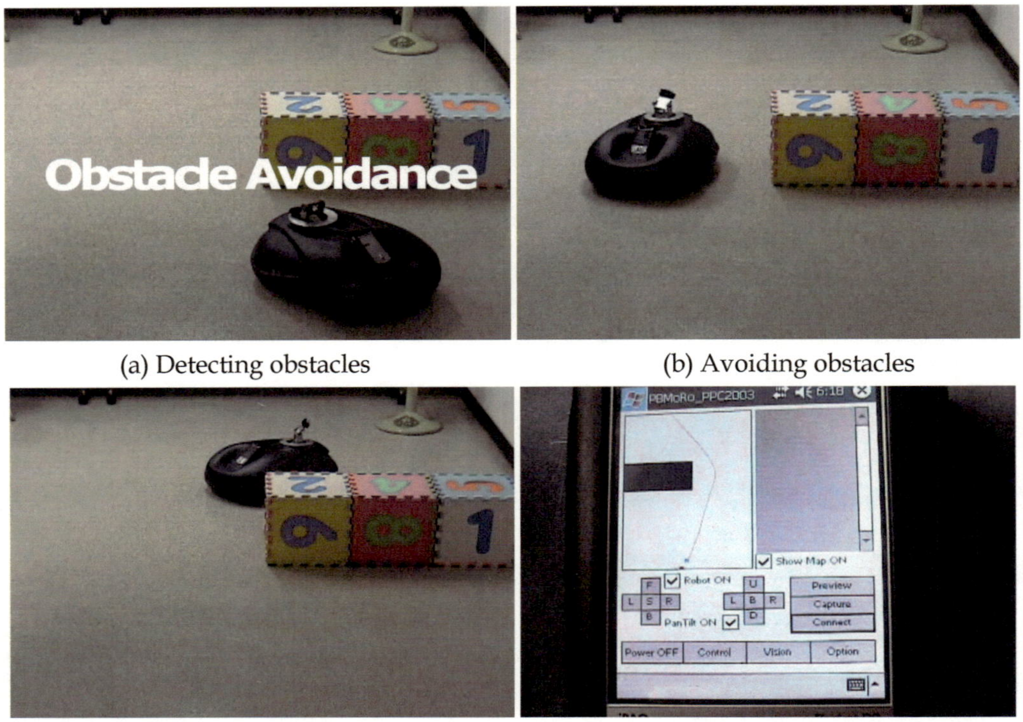

(a) Detecting obstacles (b) Avoiding obstacles

(c) Getting goal position (d) Showing the results on PDA

Fig. 20. The experimental results of obstacle avoidance and map building.

4.4 Home appliances controlling

This experiment is for controlling home appliances automatically. When a user orders the robot to turn home appliances on or off, the robot moves to the appliances like human beings if the distance is too far to use Bluetooth or an IR sensor. If the robot is near the appliance, the robot controls those using Bluetooth or IR sensors without moving. Fig. 21 shows the experimental results of home appliances control. We tested three experiments: turning on/off a fan, lamp, and television. The PBMoRo system moved near each appliance and controlled them. Using this function, a robot can manage a home instead of human beings and a user can monitor by camera the process as well as the results. This remote controlling is most useful when a robot controls dangerous appliances such as a gas range.

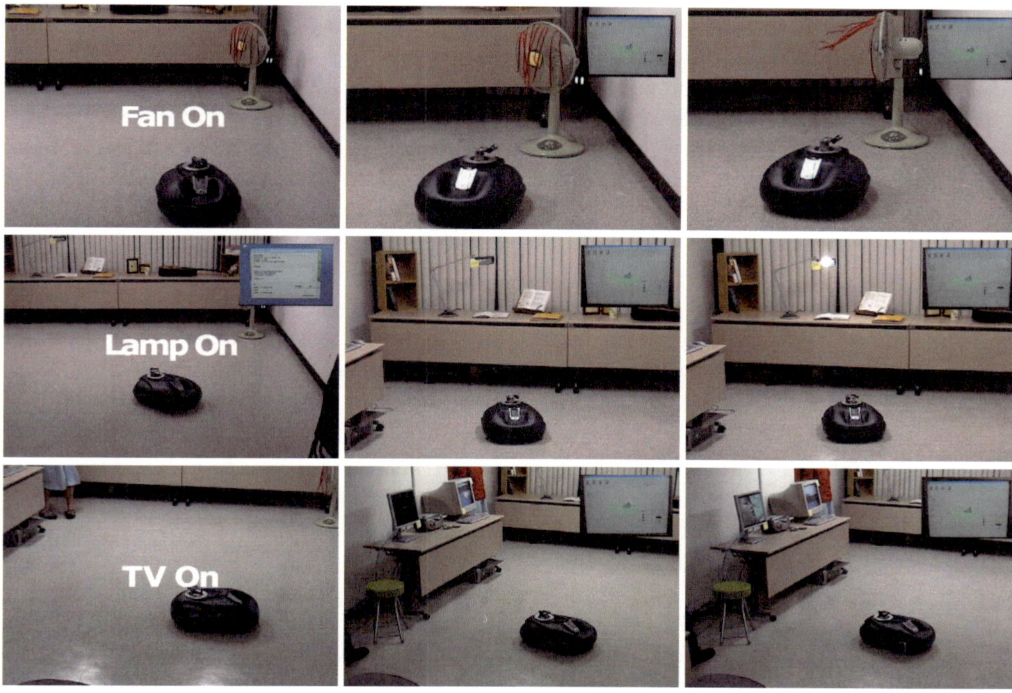

Fig. 21. The experimental results of home appliances controlling.

5. Conclusion

We have proposed the HAuPIRS architecture for organizing a more efficient and convenient home automation system which overcomes the limitations of conventional systems by using an intelligent service robot system. Intelligent service robot systems for the home environment should be designed to be human-friendly and should not draw unwelcome attention. The robot should also be light and small size for saving power and being conducive to an in home environment.

The HAuPIRS architecture solves the limitations of conventional home automation systems and intelligent robot systems for home environment using a PDA. Although a PDA has less performance than a PC which is used for conventional intelligent service robot systems, it is smaller and lighter while having long hours of operation. The robot system moves automatically as well as manually and users can control the robot system outside of home using a 3D monitoring system. The robot system also has a web camera and sends the streaming image to a 3D monitoring system.

Because the PBMoRo system uses a PDA, it is difficult to use the algorithms such as path planning and map building for conventional robot systems. To cure this problem, we simplified the algorithms and reduced the size of the streaming image. Service robots need many external ports for connecting hardware systems, but PDAs only have one external port and we used CAN communication.

We have tested many experiments: tracking, path planning, localization, obstacle avoidance, map building, and home appliance controlling. From these experiments, we verified that the proposed robot system can be one of the solutions for a home automation system.
In the future, we need to develop more efficient and robust algorithms with lower specification systems. The PBMoRo system can be a manager of any house and adapted to an apartment environment as well. It can also be more useful to control dangerous appliances using fire or water.

6. References

JongWhan, Kim (2002). Robot soccer technology, KAIST PRESS, ISBN, Place of Publication

Kuk-Jin Yoon, In-So Kweon. (2001). Landmark Design and real-time landmark tracking for mobile robot localization, SPIE2001, Sep, 2001, Korea

Kyung-Sang Bukdo (2004). Korea Intelligent Robot Contest 2004.

Ho Seok Ahn(2008), Advances in Service Robotics, InTech Education and Publishing.

Seung-Min Baek(2001), Intelligent hybrid control of mobile robotic system, The Graduate School of Sung Kyun Kwan University.

Arkin R. C. (1998), Behavior Based Robotics, The MIT Press.

Arkin R. C. (1989), Motor Schema Based Mobile Robot Navigation, International Journal of Robotics Research, vol. 8, no. 4, pp. 92-112.

Arkin R. C. (1987), Motor Schema Based Navigation for a Mobile Robot: An Approach to Programming by Behavior, Proceedings of the IEEE Conference on Robotics and Automation, pp. 264-71.

Beom H. R. and Cho H. S. (2000), Sonar-based Navigation Experiments on a Mobile Robot in Indoor Environments, Proceedings of the 15th IEEE International Symposium on Intelligent Control, July 2000, Greece.

Brooks R.(1986), A Robust Layered Control System for a Mobile Robot, IEEE Journal of Robotics and Automation, vol. RA-2, no. 1, pp. 14-23, 1986.

MostItech, http:// www.mostitech.com

NEC personal robot PAPERO, http://www.nec.co.jp/robot/english/robotcenter_e.html

Palm Pilot Robot kit from CMU, http://www.cs.cmu.edu/~pprk/

Vacuum cleaner robot from LG Electronics, http://www.lge.com/products/model/detail/v-r4000.jhtml

Sebastian Thrun (1999), et. Al., MINERVA: A second generation mobile tour-guide robot, Proc. of the IEEE International Conference on Robotics and Automation (ICRA'99), 1999.

Roland Siegwart (2007). Simultaneous localization and odometry self calibration for mobile robot, Autonomous Robots. Vol. 22, pp. 75–85.

Koide Y., Kanda T., Sumi Y., Kogure K. and Ishiguro H. (2004). An approach to integrating aninteractive guide robot with ubiquitous sensors, In Proceedings of the 2004 IEEE/RSJ International Conference on Intelligent Robots and Systems (IROS 2004), Vol. 3, pp. 2500-2505.

19

Onboard Mission Management for a VTOL UAV Using Sequence and Supervisory Control

Florian Adolf and Franz Andert
Institute of Flight Systems, Unmanned Aircraft Dept.,
German Aerospace Center (DLR)
Germany

1. Introduction

This chapter addresses the challenges of onboard mission management for small, low flying unmanned aerial vehicles (UAVs) in order to reduce their dependency on reliable remote control. The system presented and tested onboard an UAV provides different levels of autonomy, switchable at runtime either manually by the operator or automatically due to absence of a data link. This way, it is a feasible approach towards autonomous flight guidance within the low-altitude domain (e.g. urban areas) where unpredictable events are likely to require onboard decision-making.
In the following sections the problems of onboard mission management, embedded high level architectures and their implementation issues are discussed. The design of a onboard Mission Management System for a test platform with vertical take-off and landing (VTOL) capabilities is presented, followed by discussions of the implemented system and a research outlook.

2. Autonomy Management Problem

For many UAVs, an operator at a remote control station performs joystick control and plans the mission. The operator often commands the UAV using joystick remote control (e.g. rate or velocity commands) or sets a target location for a position command. With an onboard world model and path planning capabilities, more autonomy is on board the system such that an operator might issue higher-level commands, e.g. directing the vehicle to search a collision free path automatically and fly back to base.
This implies different abstraction levels within the onboard system such that each level of system autonomy is clearly represented. The level of autonomy at which an operator commands the UAV might vary during a mission. For example, while the UAV performs waypoint navigation the operator interrupts the flight in order to manually direct the UAV towards an object of interest that just appeared in a live video feed from the onboard camera.
The design and implementation of different levels of control necessitates provisions for operational safety and certain user requirements. In particular, the operator must remain in

the loop at all levels of autonomy whenever the data link is available. Also, the operational environment is characterized by events that can occur in an unknown order and at sporadic time instances. It must implement input checks for syntactical plausibility and even semantic correctness, wherever possible.
Beside this autonomy management problem, the organization and abstraction of the system into a suitable architecture is a challenge. Thus, in the next section existing architectural concepts are discussed.

3. Related High-Level Control Architectures

More self-reliance and decision-making autonomy poses questions regarding a suitable architecture according to which the management system is designed.
Knowledge-based systems establishing the concept of a cognitive process as decision-making entity were presented in the UAV domain (Hill 1997, Putzer 2003). Concepts exist that are based on the behavior-based paradigm (Weiss 2005), where a set of elementary behaviors (so-called skills, such as movement primitives) is combined in such a way that a new emergent behavior is created. Furthermore, layered architectures (Freed 2005) have been proposed that comprise distinct system modes also known as hybrid control (Egerstedt 1999).
Using knowledge-based systems, classical artificial intelligence spent over five decades trying to model human-like intelligence. Inspired by these systems, several research projects seek to produce a human-like thinking process (also known as cognition) in order to achieve high-level control in decision-making systems (Hill 1997, Putzer 2003). A commonly used cognitive architecture is implemented in SOAR (Laird 1987). Since real-time properties are one crucial design aspect for a UAV decision-making system, a real-time derivate of SOAR, Hero-SOAR exists.
However, there are major implementation issues related to cognitive production systems (Musliner 1995). First, "chunking", a pattern matching technique, might be hard to confine with respect to execution time and memory usage. Second, real-time reflexive actions (a direct connection of a sensor to an actuator) invoke a high-variance of unpredictable system events. Furthermore, problems were experienced when trying to effectively coordinate and mediate reflexive behaviors with the overall deliberative behavior of the system. If the reflexive actions can bypass the normal deliberation mechanisms, it may be difficult or impossible for the deliberation processing to reason about and affect the real-time reaction.
Hence, the architecture for any UAV decision making system should particularly focus on "embedding real-time in artificial intelligence" rather than "embedding artificial intelligence in real-time" (Musliner 1995). Moreover, a principle shortcoming of the cognitive approach is the emphasis on representation at a high, symbolic level. This yields to control strategies that may make conceptual sense to a human designer but the intelligence in such systems belongs to the designer. Additionally, it is questionable whether humans deploy a complex thinking process for every intended behavior rather than think in a more reactive way (Agre 1995).
These disadvantages are addressed by the behavior-based control with the Subsumption Architecture (Toal 1996, Brooks 1990), which does not necessarily seek to produce cognition. It rather uses a hierarchy of fast reactive loops where each loop is capable of executing a distinct behavior. Moreover, higher reactive loops modify the behavior of lower ones. The

concept of arbitration allows to automatically select among behaviors, and the so-called action-oriented perception frames the perceptual input according to the task. Some approaches interconnect elementary behaviors and superposition them, which results in a new, emergent system behavior. The ultimate goal in many behavior-based approaches is to enable robot-learning techniques such that a system can automatically deduce which behaviors must be compiled together in order to achieve a goal. Admittedly, one of the side effects is that they produce complex system topologies if behaviors are interconnected. It then is almost impossible to explain the system behavior. Moreover it is hard to achieve a notion of optimality (Pirjanian 1999).

UAVs are supposed to be semi-autonomous, remotely guided, assistant systems rather than anthropoid, autonomous systems. One of the key requirements of having several levels of system autonomy cannot be achieved with solely deliberate nor reactive architectures. Deliberate architectures relate "autonomy" to human-like intelligence and rational acting, whereas reactive architectures consider it as a system's "ability to act independently in the real world environment" (Makowski 2004).

Thus, it is desireble to combine advantages of knowledge-based and behavior-based architectures. Current robotic development created architectures combining both ideas into one system. Inspired by the Subsumption Architecture and empirical observations, the 3T architecture (Bonasso 1997) separates intelligent control into three interacting layers (or tiers). The first layer comprises a set of so-called reactive skills. These behaviors represent control laws tightly coupled with the environment through sensor readings and actuators. Skills make so-called simple-world assumptions such as, the sensor input is always valid and the desired goal can be achieved.

In order to accomplish a specific task, the sequencer on the second layer assembles an appropriate task network of skills by activating and deactivating respective skills. When more than one skill is active, they form a so-called task network.

The third layer is the deliberative layer, which comprises a planner that reasons about goals, resources and timing constraints with well-known rational techniques.

4. Mission Management System

In the following the overall system design is presented with respect to particular design decisions. The effective architecture of the onboard mission management system is based on ideas discussed in the previous section and yields two main system components: The Sequence Control System and the Supervisory Control System.

4.1 Design Decisions

The major requirements with respect to real-time execution, predictable system behavior and the need for different levels of autonomy at runtime yield the following principal design decisions:

- The embedded system architecture must be separated into interacting layers, to enable the implementation of deliberate and reactive approaches. This leaves room for a behavior-based reactive layer and allows several kinds of artificial intelligence techniques in the deliberative layer(s).

- The layered architecture chosen for this hybrid control problem is the 3T architecture. It offers a flexible way of modularization, centralizes the execution of actions and does not rely on interacting skills.
- The behavior-based paradigm, as a bottom-up strategy for intelligent systems, is worth being considered for the reactive layer, since it enables real-time execution and relatively simple behavior development. This paradigm allows a way to compile elementary problem solutions (e.g. moving to a position) into a library of behaviors.
- Known shortcomings of the behavior-based approach with respect to online learning, behavior interaction and arbitration techniques, are eliminated intentionally.
- When behavior interaction and abstract behaviors are not available in the reactive layer, the discussed disadvantages of the 3T approach can be neglected.

As a result, this design concept for onboard mission management combines the 3T architecture with ideas from the behavior-based paradigm.
In the following, the overall system is described from three points of view. The illustrations in Figure 1 outline how the principles of 3T's high level control decomposition are represented in the system. In order to highlight a system wide context, Figure 2 describes the component organization from an implementation point of view.

4.2 High-Level Control Architecture

The high-level control architecture onboard the UAV is based on the 3T architecture for a hierarchical decomposition of system autonomy (Figure 1). Moreover, the behavior-based paradigm yields distinct behaviors that can be combined sequentially across each layer. The behaviors are either of a basic movement primitive type (e.g. flying a linear trajectory) or of deliberate nature (e.g. searching and tracking an object on ground).

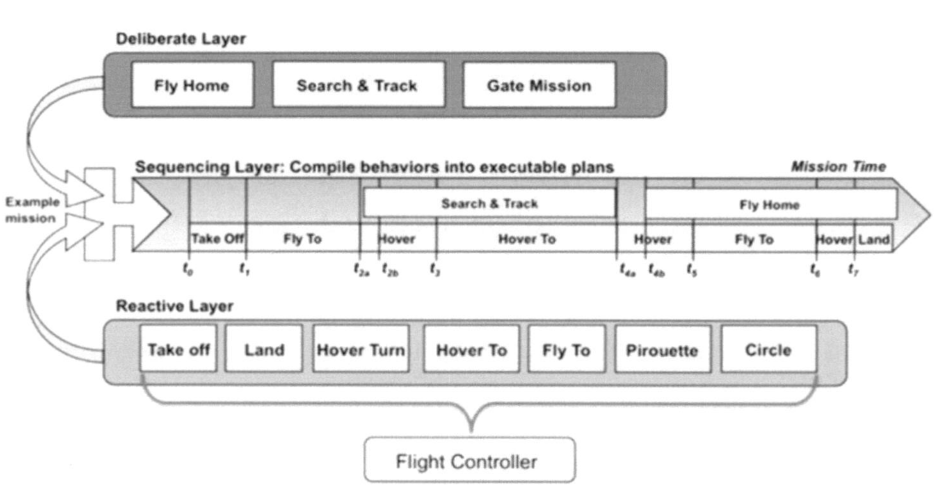

Fig. 1. Onboard high-level control based on the 3T architecture

Two basic prerequisites of the proposed mission management architecture in Figure 1 are implemented by two systems, sequentially executed at each instant of time. The first system implements deliberate behaviors and a set of operational safety features.
The second component generates flight control commands at every instant of time. It contains a library of basic movement behaviors from the reactive layer. Figure 1 illustrates which behaviors are located at the skill layer. These behaviors generate instantaneous trajectory-based control commands that are fed into the flight controller.
The deliberate layer shows examples of complex behaviors. These alter existing missions or create new missions. In this context, mission planning will output the list of sequential behavior commands shown in Figure 1.

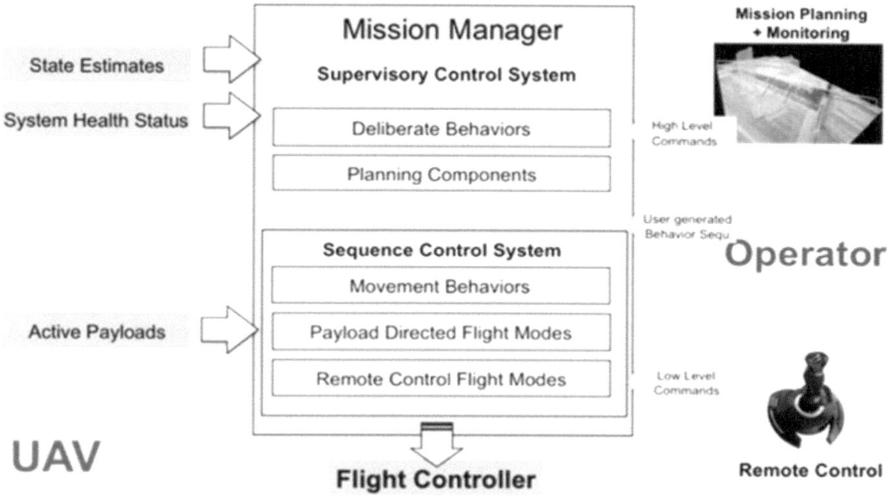

Fig. 2. The Mission Manager allows different levels of autonomy and comprises a supervisor and sequence controller implementing the 3T architecture

The set of basic behaviors from the reactive layer need to be represented as a system. It needs to coordinate and execute the reactive layer's basic movement behaviors that interface with the flight control system. Furthermore, a sequence of such behaviors needs to be executed automatically while handling unforeseen events like a sudden interruption from the remote operator. This is done in the Sequence Control System as it implements the executive component of the sequencing layer of the 3T architecture.
Since neither the deliberative layer nor the skills alone can handle all situations optimally, the Supervisory and Sequence Control System provide additional glue logic to store procedural knowledge that neither belongs clearly to the deliberative layer nor to the skill layer. For example, during flight testing a safety pilot may need to switch between manual or computer control, and thus the system must stop producing actuator commands and set its onboard components into a defined stand-by state.

4.3 The Sequence Control System

The Sequence Control System is exposed to a number of potentially concurrent events. Hence, the specification and implementation of the system is modelled as an event-based system. The majority of event-based systems are modelled using the Unified Mark-up Language (UML), an industry-wide standard notation. It supports the object oriented design pattern and provides dynamic modelling techniques such as state charts, sequence diagrams and activity diagram. State charts have been extensively studied such that abstract testing techniques allow verifying a model (semi-) automatically. Also, there exists good software tool support, which eases the development process significantly. Moreover, there are tools that provide code generators such that the implemented code is directly derived from the state chart-based specification. Otherwise, it is likely that specification and implementation begin to diverge over time.

Thus, the Sequence Control System is modelled as UML 1.2 State Charts. Basically, State Chart diagrams (also known as State Machines in UML 2) are finite automatons with a finite set of states where exactly one state is active at a time. They depict the dynamic behavior based on its response to events, showing how the model reacts to various events depending on its current state. Events can trigger a transition into another state, where so-called guards are the condition that must become true in order to traverse along the transition. The guards on similar transitions leaving a state must be consistent (deterministic) with one another.

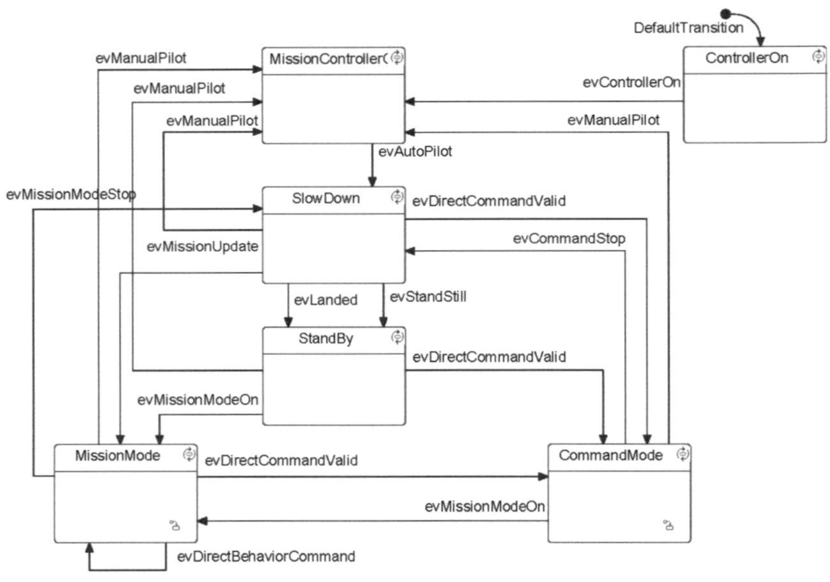

Fig. 3. The UML State Chart Model specifying the top level of the Sequence Control System

The UML model of the Sequence Control System is shown in Figure 3. It has two hierarchical levels where the top level models the procedural flow for a safe operation. The two composite states, "Mission Mode" and "Command Mode", model mission plan processing and direct command execution respectively. Every state of the top level has a transition to the "Mission Controller Off" to handle a manual control event, such that the Sequence Control System stands by in an idle state. If the system is in computer-based flight

mode (auto mode), another idle state "Stand By" lets the UAV hover at its current position when the state was entered; including a position on the ground. The state "Slow Down" is necessary to assure a smooth changeover into "Stand By" regardless of the flight maneuver being executed. In case the operator commands the UAV to stop, a transition from every auto mode state assures that the command is executed. Among events certain priorities exist. For example, an event switching to manual mode is more important than a stop command and requires processing. The order in which the event check is performed accounts for this obligation.
The state mission mode contains the actual library of behaviors. There are no transitions among behaviors assuring that only one can be active at a time. This is required in order to overcome emergent system behavior caused by overlapping and interactions. For each behavior there exists a termination condition, which transits into the command parser "Parse Command". Basically this state grabs behavior commands from an existing mission plan (Figure 4). It issues an event for traversing into the appropriate state. When the mission plan is processed, the "Mission Mode" composite state is exited. For payload directed flight, the composite state "Command Mode" can be entered from every state inside "Mission Mode".

4.4 The Supervisory Control System

The Supervisory Control System is responsible for taking high level decisions based on internal and external events. It is responsible for managing requests from the UAV operator (e.g. allow to load a mission or executing a complex command), as well as reacting to a loss of the data link. It is executed before the Sequence Control System at every instant of time. This allows the Supervisory Control System to modify a mission when conditions are recognized to imply a necessity of modification. Moreover, as long as there is no mission update incoming from the operator, it can act as a substitute commanding entity for the operator. It can command the Sequence Control System via the same type of commands that a remote operator can send to the Sequence Control System (e.g. start and stop the execution of a mission currently loaded). It is the entity in charge of managing the execution of deliberate behaviors. Therefore, the Supervisor retains planning capabilities, and recognizes associated high-level mission objectives (e.g. "Fly Home").

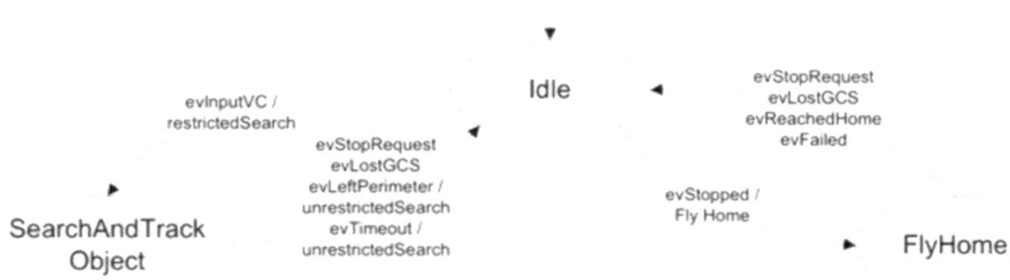

Fig. 4. Supervisory Control System as State Chart model managing two deliberate behaviors

Similarily to the Sequence Control System an UML model is be defined for the Supervisory Control System shown in Figure 4. In this example, it implements two high-level behaviors, "fly home" and "search and track object". The first lets the vehicle find a way to fly back to a start position, whereas the second lets the vehicle search and track an objects moving on the ground. The Fly Home behavior provides the vehicle with the capability of returning autonomously to the starting point of a given mission. It implements the replanning process shown in Figure 5 which is based on the properties of each basic movement primitives of the reactive layer. The Search and Track behavior can be used to find and track a moving object on the ground (Figure 6). Once spotted (e.g. using payload directed object detection) it is desirable to track it. Similar to the Fly Home behavior, it seizes information in the a-priori mission plan to implement tactical means, avoiding suboptimal search execution and tracking performance. Further details on the internal ongoings of these behaviors are presented in (Adolf 2009).

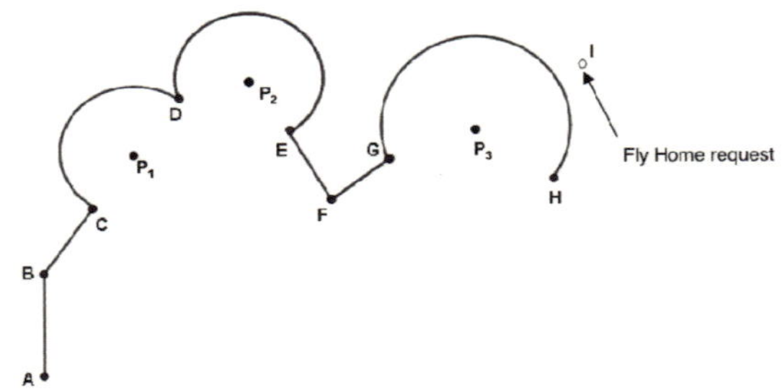

Origianl plan	Invert sequence order	Filter tasks	Recompute behavior parameters	Drop behaviors in Proximity
Take Off	Circle XY (P_3, θ_{P3})	Cirlce XY (P_3, θ_{P3})	Cirlce XY (P_3, - θ)	Cirlce XY (P_3, - θ)
Hover To (A, θ_A)	Hover To (G, θ_{G2})	Hover To (G, θ_{G2})	Hover To (G, $\theta_{G,F}$)	Hover To (G, $\theta_{G,F}$)
Pirouette Flight (B)	Hover To (G, θ_G)	Hover To (G, θ_G)	Hover To (F, $\theta_{G,F}$)	Hover To (F, $\theta_{G,F}$)
Waiting For T	Hover Turn (φ)	x Hover Turn (φ)	Hover To (F, $\theta_{F,E}$)	Hover To (F, $\theta_{F,E}$)
Hover To (C, θ_C)	Hover To (F, θ_F)	Hover To (F, θ_F)	Hover To (E, $\theta_{F,E}$)	Hover To (E, $\theta_{F,E}$)
Hover To (C, θ_{C2})	Circle XY (P_2, θ_{P2})	Cirlce XY (P_2, θ_{P2})	Hover To (E, $\theta_{2,E}$)	Hover To (E, $\theta_{2,E}$)
Circle XY (P_1, θ_{P1})	Circle XY (P_1, θ_{P1})	Cirlce XY (P_1, θ_{P1})	Circle XY (P2, - θ_{P2})	Cirlce XY (P2, - θ_{P2})
Cirlce XY (P_2, θ_{P2})	Hover To (C, θ_{C2})	Hover To (C, θ_{C2})	Hover To (D, $\theta_{1,D}$)	Hover To (D, $\theta_{\cdot,D}$)
Hover To (F, θ_F)	Hover To (C, θ_C)	Hover To (C, θ_C)	Circle XY (P1, - θ_{P1})	Cirlce XY (P1, - θ_{P1})
Hover Turn (φ)	Waiting For T	x Waiting For T	Hover To (C, θ_{C2})	Hover To (C, $\theta_{C,B}$)
Hover To (G, θ_G)	Pirouette Flight (B)	Pirouette Flight (B)	Hover To (C, θ_C)	Hover To (B, $\theta_{C,B}$)
Hover To (G, θ_{G2})	Hover To (A, θ_A)	Hover To (A, θ_A)	Pirouette Flight (B)	Hover To (B, $\theta_{B,A}$)
Circle XY (P_3, θ_{P3})	Take Off	x Take Off	Hover To (A, θ_A)	Hover To (A, θ_A)

Fig. 5. Planning process of the Fly Home deliberate behavior based on the basic behaviors

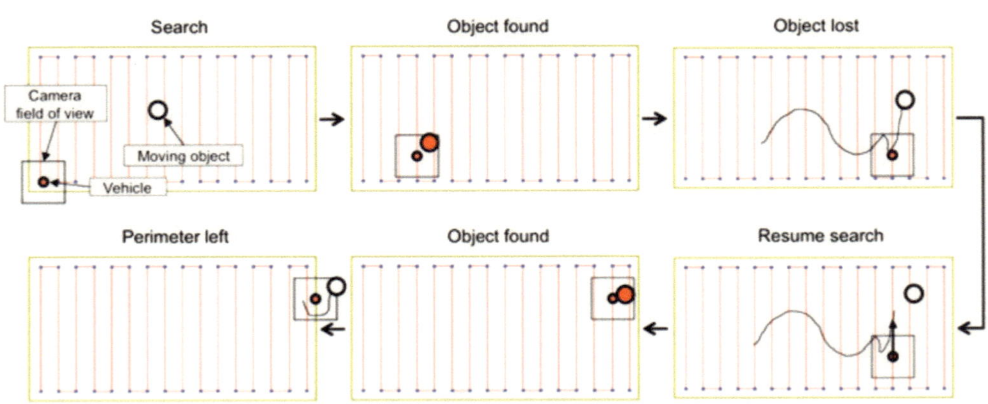

Fig. 6. The Search and Track deliberate behavior shown as it finds and track a ground object with a defined search area

4.5 Abstract System Testing

The overall complexity of the Supervisory and Sequence Control System imposes a thorough test strategy. Abstract tests relate to Model-based testing and as such assure that the model is free of principle design mistakes.

In the modelling stage a set of errors can occur. Some relate to potentially isolated or unreachable states, as well as, missing or erroneous triggers and guards. An even more fundamental problem is the theoretically infinite set of sequences that have to be tested in an abstract manner. That is, regardless of the meaning and function behind events, states and transitions, the tests must traverse through the model even if a certain test would make no sense from a practical point of view.

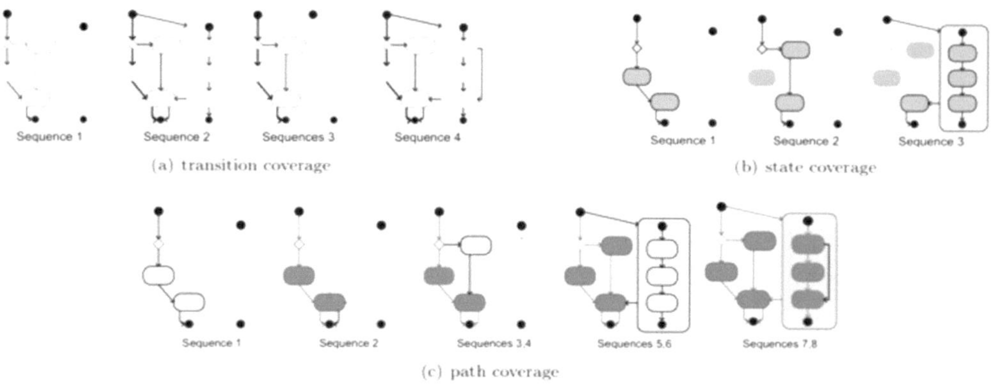

Fig. 7. Abstract testing strategies applied to the State Chart models

UML state charts represent a constructive model such that tests can make use of its internal structure, also known as white-box testing. These abstract tests can be divided into three groups, namely the path coverage, transition coverage and state coverage tests as highlighted in Figure 7. For each of these groups, a set of event chains is generated and executed. Once a full coverage of all possible combinations is reached, the tests are completed. However, there is no defined final state and states can have loops such that infinitely long test chains for path coverage tests occur. Therefore, a relaxation for the path coverage criteria is implemented such that loops must be passed exactly once. This is a strong and feasible criterion (Rumpe 2005). The system with greatest deepness is the Sequence Control System where at maximum six events yield from the start state to the deepest level. This implies 34^6 test sequences for the Sequence Control System.

Moreover, the McCabe metric (McCabe 1976) is used to finally estmate the cyclomatic complexity:

$$V(G)=edges+nodes+2 \tag{1}$$

The State Chart with most transitions and states is the Sequence Control System. The system has 43 edges and 23 nodes, thus a McCabe complexity of 22. This is commonly considered as a complex system but still with reasonable risk. A V(G)-value of more than 50 is considered a system not testable. This would be the case, if for example the system wasn't decomposed into a Supervisory and a Sequence Control System. According to this metric, the more deliberate behaviors the Supervisor contains, the more imminent becomes a further decomposition the Supervisory Control System. Since the deliberate behaviors are modelled using individual State Charts this problem is practically circumvented.

4.6 Plausibility Checks at Runtime

It is a relatively complex decision to determine whether payload directed flight should be permitted or not. First, compared to behaviors in "Mission Mode", every direct command may not have a termination condition. For example, one instantaneous velocity command from the operator cannot reach a timely bounded target state. For such cases, a quasi infinite behavior is bounded in execution time, such as the time passed since a velocity command was last received. Second, it depends on which data is coming on what input channels. These types of problems are addressed using a static truth table (Figure 8).

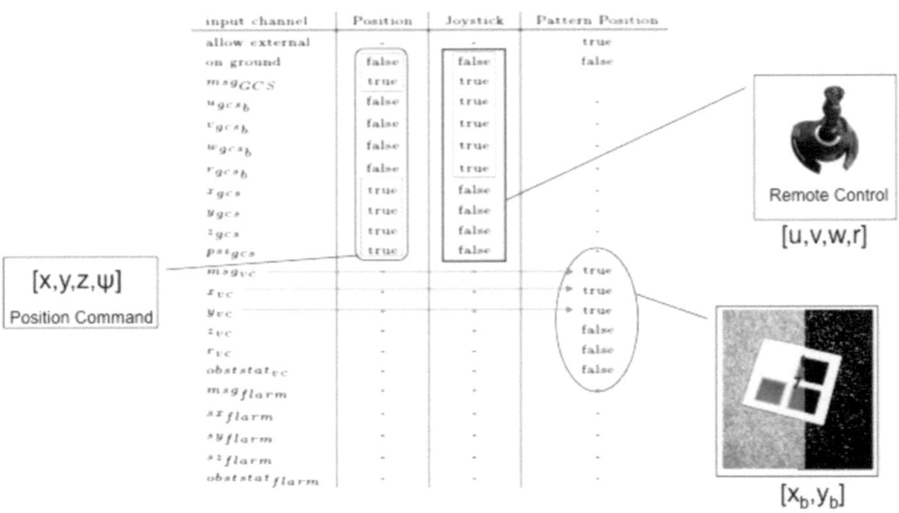

Fig. 8. Truth table assessing priority plausibility of payload or operator low-level commands

It checks valid combinations for payload directed flight and manual interruption of missions by the operator (e.g. joystick or position commands). The table is conflict-free, prioritses all combinations of relevant signals and allows plausibility checks of every incoming signal.

Likewise the command checks for low-level commands, missions defined as sequence of behavior commands need to be considered as one single complex command. Although it is hard to reason about the "sense" of a mission plan, it is possible to implement a plausibility check using a language grammar. This way, a behavior sequence is treated as a programming language that satisfies type 2 of the Chomsky hierarchy of languages. It can be expressed via a context-free language and thus it can be defined in an Extended Backus-Naur Form (EBNF, ISO-14977 2001). Furthermore, using attribute features (e.g. semantic checks of a height parameter), the attributed EBNF shown in Figure 8 has been implemented which fulfills the following general requirements:

- There's a need for a unique identifier for each mission by which the ground control station can recognize that a mission plan has been loaded by the onboard system successfully.
- Optionally, each mission can contain a coordinate transform header in order to transform certain behavior parameters from the ground control station's reference system into the onboard system's reference system.
- Enable delay for a mission start, e.g. by waiting on ground for a given period of time.

- Always command a take-off before any other locomotion behavior is performed, to assure any subsequent behavior command does not move the vehicle while still on ground.
- Only one mission repetition command is allowed and only at the end of the mission.
- Only at the end of a mission, aborting or pausing the execution of a mission plan.
- A single land behavior command without a consecutive take off is safe only when commanded at the end of a mission.

In general, EBNF descriptions specify syntax not semantics. Thus, in this notation each EBNF factor is optionally followed by a semantic action (here: parameters to nonterminals). If an EBNF expression is expected at the place the sequence stands, then a semantic action must either be an expression or be omitted. In the first case, semantic actions are particularly important tools to check floating point values against their "meaning" (=semantics), e.g. expected range for degree values. The implemented attributes show the extended plausibility check capabilities. Checks syntactically hard to perform can be handled. For example, it checks against an allowed maximum velocity of a movement behavior or maximum flight height restrictions. Moreover, behavior parameters are validated against a spatial discrepancy between the end and start position of movement behaviors. A start position of a behavior must always match the expected end position of a previous behavior.

mission → *uID* [*coordt*] [*wait*] {*bseq*} [*redo*] [*land*] [*off*]
coordt → coordinates: local_cartesian
datum: wge
#_ellipsoid: we
origin_latitude: *angle90*
origin_longitude: *angle180*
origin_height: *ureal*
orientation: *angle180*
bseq → *takeoff*
| *land* [*wait*] [*off*] *takeoff*
| *hoverto*
| *hoverturn*
| *waittime*
| *flyto* *flyto* *flyto* {*flyto*}
| *circle*
| *piroutte*
| *gatemission*
| *task*
| *off*
takeoff → TO $\langle h \rangle$ $\{ \forall h \in \mathbb{R}_{-} : h_{max} \leq h \}$
land → LD
hoverto → HV *position3D* *position3D* *angleAll* $\langle v \rangle$ $\{ \forall v \in \mathbb{R}_{+} : v \leq v_{max} \}$
hoverturn → HT *angleAll* $\langle \dot{v} \rangle$ $\forall \dot{v} \in \mathbb{R} : 0 < \dot{v}$
wait → WT $\langle t \rangle$ $\{ \forall t \in \mathbb{R}_{+} : 0 < t \}$
flyto → FT *position3D* $\langle t \rangle$ $\langle x \rangle$ $\langle y \rangle$ $\langle z \rangle$ $\langle v \rangle$ $\{ \forall t \in \mathbb{R}_{+}, \forall x, y, z \in \mathbb{R}, \forall v \in \mathbb{R}_{+} : v \leq v_{max} \}$
circle → CI *position3D* *position2D* $\langle v \rangle$ *angleAll* $\{ \forall v \in \mathbb{R}_{+} : v \leq v_{max} \}$
piroutte → PI *position3D* *position3D* $\langle \dot{v} \rangle$ $\{ \forall \dot{v} \in \mathbb{R} : 0 < \dot{v} \}$
gatemission → GM *position2D* *angleAll*
task → payloadcmd-on {*bseq*} payloadcmd-off
redo → REDO
off → *WO*
position2D → $\langle x \rangle$ $\langle y \rangle$ $\{ \forall x, y \in \mathbb{R} \}$
position3D → *position2D* $\langle z \rangle$ $\{ \forall z \in \mathbb{R} : h_{max} \leq h \}$
uID → ID $\langle i \rangle$ $\{ \forall i \in \mathbb{N}_{+} \}$
angleAll → $\langle \alpha \rangle$ $\{ \forall \alpha \in \mathbb{R} : -180° \leq \alpha \leq 360° \}$
angle360 → $\langle \alpha \rangle$ $\{ \forall \alpha \in \mathbb{R}_{+} : 0° \leq \alpha \leq 360° \}$
angle180 → $\langle \alpha \rangle$ $\{ \forall \alpha \in \mathbb{R} : -180° \leq \alpha \leq 180° \}$
angle90 → $\langle \alpha \rangle$ $\{ \forall \alpha \in \mathbb{R} : -90° \leq \alpha \leq 90° \}$

Fig. 8. Attributed EBNF for plausibility checks of sequence of behavior commands.

5. Integration and Flight Testing

The Mission Management system is integrated onboard the Autonomous Rotorcraft Testbed for Intelligent System (ARTIS) helicopters (Figure 9).

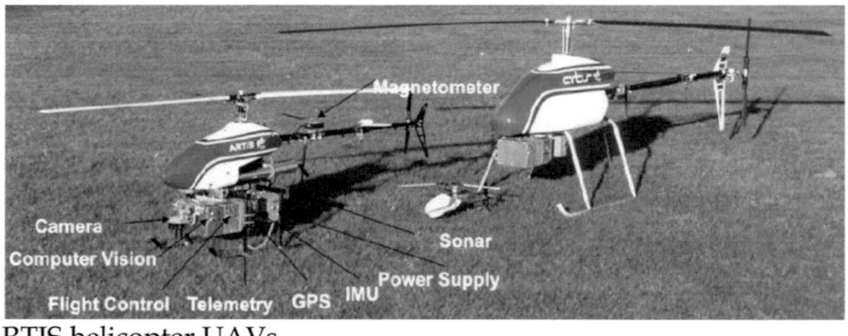

Fig. 9. The ARTIS helicopter UAVs.

The Mission Manager is integrated onboard the flight control computer as a component commanding directly and every cycle to the flight controller. The vehicle state estimates (e.g. position, velocities, acceleration) and further sensor states (e.g. ground distance sensor) are the main input to the system. Furthermore, the ground control station can send instructions on different levels of autonomy as described in the previous section.

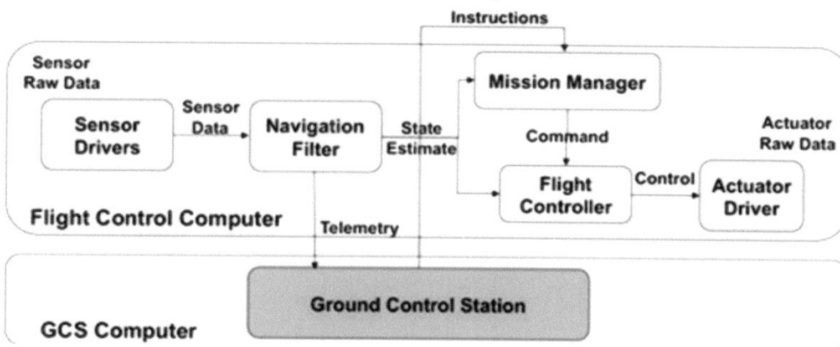

Fig. 10. Integration scheme of the Mission Manager onboard the ARTIS helicopter UAVs.

Before every flight test, the system is tested using the abstract testing method presented in the previous section. Once these tests are passed successfully, the integrated mission management software is tested Hardware-in-the-Loop (HITL) using a set of standard scenarios. Finally, the Mission Manager is flight tested to verify the generation of expected State Chart events (e.g. detection whether the vehicle is on ground or not), state transitions and feasibility of the state abstraction in the design.

Fig. 11. Flight testing onboard the ARTIS UAV (A) using the integrated Supervisory and Sequence Control System (B).

The example scenario in flight test of Figure 11 comprises an automatic take-off, navigation to a set of waypoints and an automatic landing. The behavior sequence is generated once before the test using an automated mission planning system (Figure 12). Once the mission has been accepted by the EBNF and loaded by the Supervisory Control System, it is automatically executed by the ARTIS helicopter UAV.

Fig. 12. Example for an automated planning is used to generate the sequence of behaviors.

The flight test shown in Figure 12 lets the UAV traverse along predefined waypoints, thus focusing on the waypoint navigation capabilities of the Sequence Control System. High-level behaviors of the Supervisory Control System were successfully tested as well. The flight test as shown in Figure 13 addresses the Search and Track behavior. The behavior successfully recognizes convex cells of a search area such that the vehicle does not exceed search cell perimeters while tracking. As a result, the Supervisory Control System autonomously commanded the Sequence Control System using a high-level.

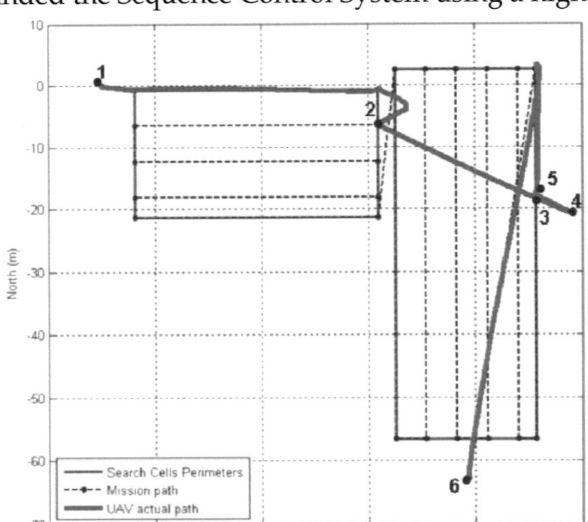

Fig. 13. The Search and Track deliberate behavior shown as it finds and track a ground object with a defined search area

The integrated mission management components system, the Supervisory and Sequence Control System, were first flight tested in September 2006 and since then continuously extended by new features. In particular the implementation of new high-level behaviors is an ongoing activity. While the control architecture remained untouched since its first deployment, the basic movement capabilities are also continuously extended in a "plug-

and-fly" fashion. This way, the architecture and the design consideration of both high-level control components showed to be a feasible solution to the software intensive task of onboard mission management.

6. Summary

This chapter present a 3T architecture combined with a behavior-based approach to integrate different levels of system autonomy onboard of UAVs. The presented approach comprises two State Chart modelled components, the Sequence and a Supervisory Control System, both implementing the sequencing layer of the 3T architecture. The Sequence Control System contains the reactive layer's movement primitives and the Supervisory Control System monitors and manages deliberate behaviors. Both systems are tested in an abstract way and flight-tested successfully.

By design both system components keep the operator in the loop as long as a data link is available. He has the possibility to intercept any running behavior by different kinds of commands, ranging from direct velocity commands (e.g. for remote joystick control), direct position commands, a single behavior (e.g. land) or a complex behavior command sequence. For each level of system autonomy, the system implements plausibility checks performed at runtime. In particular the attributed EBNF grammar enhances operational safety as it rejects a malformed behavior sequences harming the vehicle or exceeding allowed mission parameters (e.g. the maximum height).

As a result, the ARTIS UAV test platforms are controlled by a robust system that can handle unforeseen events deterministically. New behaviors, both deliberate and reactive, are added to the behavior library such that functional extensibility is facilitated.

Future work will address control architecture extensions as soon as the centralized sequencing layer needs to implement concurrency (e.g. multiple deliberate behaviors need to be active). Moreover, the current system does not allow concurrency between all behaviors and modules in the system. Moreover, a stronger formalism than State Chart models could address limitations in support for temporal design aspects.

7. References

Adolf, F. et.al. (2009). Behavior-based High-Level Control of a VTOL UAV, *Proceedings of AIAA Infotech@Aerospace Conference*, Seattle, WA, April 2009

Agre, P. et.al. (1995). Pengi : An Implementation of a theory of activity, *Computatioanl Intelligence: Collected Readings*, pages 635-644, American Association for Artificial Intelligence, Menlo Park, CA, USA, 1995, ISBN 0-262-62101-0

Bonasso, R. et.al. (1997). Experiences with an Architecture for Intelligent, Reactive Agents, *Journal of Experimental and Theoretical Artificial Intelligence*, Vol. 9, No. 2/3, pages 237-256, April 1997

Brooks, R. (1990). A robust layered control system for a mobile robot, *Readings in uncertain reasoning*, pages 204-213, Morgan Kaufmann Publishers Inc., San Francisco, CA, USA, 1990, ISBN 1-55860-125-2

Egerstedt, M. et.al. (1999). A Hybrid Control Approach to Action Coordination for Mobile Robots, *Proceedings of IFAC 99 14th World Congress*, Bejing, China, July 1999

Freed, M. et.al. (2005). An Architecture for Intelligent Management of Aerial Observation Missions, *Proceedings of AIAA Infotech@Aerospace Conference*, Arlington, VA, September 2005

Hill, R. et.al. (1997). *Intelligent agents for the synthetic battlefield: A company of rotary wing aircraft, The Ninth Innovative Applications of Artificial Intelligence Conference on Artificial Intelligence (IAAI-97)*, Providence, Rhode Island, July 1997

ISO-14977 (2001). Information Technology ó Syntactic Metalanguage ó Extended BNF, *International Organization for Standardization, ISO/IEC 14977*, 2001

Laird, J. et.al. (1987). SOAR: An Architecture for General Intelligence, *Journal of Artificial Intelligence*, Vol. 33, No. 1, pages 1-64, 1987

Makovski, P. et.al. (2004). Survey on Architecures and Frameworks for Autonomous Robots, November 2004

McCabe, T. (1976). A complexity measure, *IEEE Transactions On Software Engineering*, Vol. Se-2, No.4, December 1976

Musliner, D. et.al. (1995). The Challenges of Real-Time AI, *IEEE Computer*, Vol 28, No.1, January 1995

Pirjanian, P. (1999). The Notion of Optimality in Behavior-Based Robotics, *Journal of Robotics and Autonomous Systems*, 1999

Putzer, H. et.al. (2003). COSA *A generic cognitive system architecture based on a cognitive model of human behavior, Journal of Cognition, Technology and Work*, Vol. 5, 2003

Rumpe, B. (2005). Agile Modellierung mit UML, *Xpert.press*, Springer Verlag, 2005

Taol, D. et.al. (1996). Subsumption Architecture for the Control of Robots, *IMC-13*, Limerick, 1996

Weiss, L.-G. (2005). Intelligent Collaborative Control For UAVs, *Proceedings of AIAA Infotech@Aerospace Conference*, Arlington, VA, September 2005

20

Emotion Recognition through Physiological Signals for Human-Machine Communication

Choubeila MAAOUI and Alain PRUSKI
Laboratoire d'Automatique humaine et de Sciences Comportementales
University of Metz
7 Rue Marconi, 57070 Metz, France
choubeila.maaoui@univ-metz.fr

Abstract

The ability to recognize emotion is one of the hallmarks of emotion intelligence. This paper proposed to recognize emotion using physiological signals obtained from multiple subjects. IAPS (International Affective Picture System) images were used to elicit target emotions. Five physiological signals: Blood volume pulse (BVP), Electromyography (EMG), Skin Conductance (SC), Skin Temperature (SKT) and Respiration (RESP) were selected to extract 30 features for recognition. Two pattern classification methods, Fisher discriminant and SVM method are used and compared for emotional state classification. The experimental results indicate that the proposed method provides very stable and successful emotional classification performance as 92% over six emotional states.

1 Introduction

One of the interesting challenges in the community of human-computer interaction today is how to make computers be more human-like for intelligent user interfaces. In several experiments of Reeves and Nass [1], they show that humans impose their interpersonal behavioral patterns onto their computers. Thus, the design of recent human-computer interfaces should reflect this observation in order to facilitate more natural and more human-like interaction. Emotion, one of the user affect, has been recognized as one of the most important ways of people to communicate with each other. Given the importance and potential of the emotions, affective interfaces using the emotion of the human user are gradually more desirable in intelligent user interfaces such as human-robot interactions [2]. In order for such an affective user interface to make use of user emotions, the emotional state of the human user should be recognized or sensed in many ways from diverse modality such as facial expression, speech, and gesture. Thus, this paper investigates the automatic recognition of emotions in human-machine interaction using the combination of several feature sets from physiological signals. Research efforts in human-computer interaction are focused on the means to empower computers (robots and other machines) to understand human intention, e.g. speech recognition and gesture recognition systems [3]. In spite of considerable achievements in this area during the past several decades, there are still a lot of problems, and many researchers are trying to solve them. Besides, there is another important but ignored mode of communication that may be important for more natural interaction: emotion plays an important role in contextual understanding of messages from others in speech or visual forms.

There are numerous areas in human-computer interaction that could effectively use the capability to understand emotion [3]. For example, it is accepted that emotional ability is an essential factor for the next-generation personal robot, such as the Sony AIBO [4]. It can also play a significant role in 'intelligent room' [5] and 'affective computer tutor' [6]. The remainder of this paper is organized as follows: First, we describe related works to recognize the emotions of human user. The experience protocol and used equipment for emotion recognition are presented in section 3. In section 4, we specify the feature extraction method to classify our emotional categories from physiological signals. In section 5, we present the obtained experimental results. Finally, conclusion and future works are presented in section 6.

2 Related work

2.1 Modeling of discrete emotions

As people display the emotional expressions of others to their various degrees individually, it is not an easy task to judge or to model human emotions. The researchers often use two different methods to model emotions. One approach is to label the emotions in discrete categories, i.e. human judges have to choose from a prescribed list of word labels, e.g. joy, sadness, surprise, anger, love, fear, etc. One problem with this method is that the stimuli may contain blended emotions that cannot adequately be expressed in words since the choice of words may be too restrictive and culturally dependent. Another way is to have multiple dimension or scales to categorize emotions. Instead of choosing discrete labels or words, observers can indicate their impression of each stimulus on several continuous scales, for example, pleasant-unpleasant, attention-rejection, simple-complicated, etc. Two common scales are valence and arousal. Valence represents the pleasantness of stimuli, with positive (or pleasant) on the end, and negative (or unpleasant) on the other. Another dimension is arousal (activation level). The different emotional labels could be plotted at various positions on a two-dimensional plane spanned by these two axes to construct a 2D emotion model [7].

Recently, the low consistency of physiological configurations supported the hypothesis that the autonomic nervous system ANS activation during emotions indicates the demands of a specific action tendency and action disposition, instead of reflecting emotions [8].

The relation between physiological signals and arousal/valence is established in psychophysiology that argues that the activation of the autonomic nervous system (ANS) changes while emotions are elicited [9].

2.2 Automatic emotion recognition using physiological signals

There is a vast body of literature on the automatic recognition of emotions. With labelled data collected from different modalities, most studies rely on supervised pattern classification approaches to automatic emotion recognition.

Relatively little attention has been paid so far to physiological signals for emotion recognition compared to other channels of expression. A significant series of work has been conducted by Picard and colleagues at MIT Lab. For example, they showed that certain affective states may be recognized by using physiological measures including heart rate, skin conductivity, temperature, muscle activity, and respiration velocity [10]. Eight emotions deliberately elicited from a subject in multiple weeks were classified with an overall accuracy of 81%. Nasoz and al. [11] used movie clips to elicit target emotions from 29 subjects and achieved the best recognition accuracy (83%) by applying the Marquardt Backpropagation algorithm. More recently, Wagner and al [12] presented an approach to the recognition of emotions elicited by music using 4-channel biosignals which were recorded while the subject was listening to music songs, and reached an overall recognition accuracy of 92% for a 4-class problem.

3. Experimental Data Acquisition

3.1 Emotion induction protocol

A prevalent method to induce emotional processes consists of asking an actor to feel or express a particular mood. This strategy has been widely used for emotion assessment from facial expressions and to some extent from physiological signals [13]. However, even if actors are known to deeply feel the emotion they try to express, it is difficult to insure physiological responses that are consistent and reproducible by nonactors. Furthermore, emotions from actor-play databases are often far from real emotions found in everyday life.
The alternate approach for inducing emotions is to present particular stimuli to an ordinary participant. Various stimuli can be used such as images, sounds, videos [14] or video games. This approach presents the advantages that there is no need for a professional actor and that responses should be closer to the ones observed in real life.
It was essential to obtain a database of physiological signals representing specific emotional statuses. To acquire a database of physiological signals in which the influence of emotional status was faithfully reflected, we developed a set of elaborate protocols for emotion induction. We use the international affective picture system (IAPS) developed by LANG et al [15], and adopted for many psychophysiological studies involving emotion induction.
A preliminary test of the protocols was performed for 10 healthy subjects (7 males, 3 females) aged from 23 to 30 years. We have used five physiological signals (Blood Volume Pulse (BVP), Electromyography (EMG), Electrodermal activity (SC), Skin temperature (SKT) and Respiration (Resp)) (Figure 1). The EMG was measured from frontalis muscle. BVP and SKT were measured from little finger and the ring finger of the left hand, respectively. SC was measured from the index and middle fingers of the right hand. Resp was measured from abdomen subject's. We used a combination of these signals, to derive a set of features that can be used to train a classification algorithm.
For each subject, we presented six basic emotions: Amusement, Contentment, Disgust, Fear, No emotion (Neutrality) and Sadness. For each emotion, ten images are presented during 50 seconds.

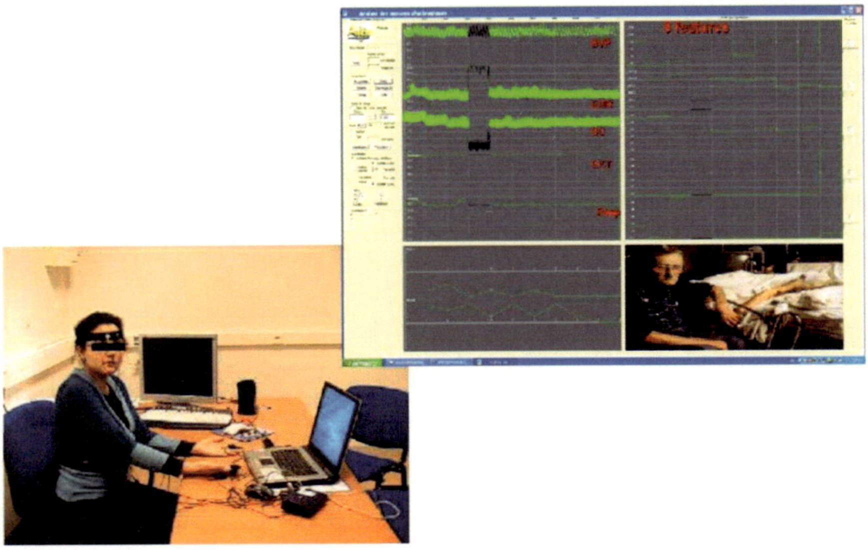

Fig. 1. Physiological signals acquisition system

3.2 Acquisition of physiological signals

The physiological signals were acquired using the PROCOMP Infiniti system [16]. The sampling rate was fixed at 256 samples per second for all the channels. Appropriate amplification and bandpass filtering were performed. One session of experiments took approximately 5 min. The subjects were requested to be as relaxed as possible during this period. Subsequently, emotional stimulus was applied, and physiological signals were recorded.

The participant was asked to self assess the valence and the arousal of his/her emotion using a Self Assessment Manikin (SAM [17]), with 9 possible numerical judgments for each dimension (arousal and valence), which will be used in future works. The used sensors are described in the following.

3.2.1 Blood Volume Pulse (BVP)

The Blood Volume pulse sensor uses photoplethysmography to detect the blood pressure in the extremities. Photoplethysmography is a process of applying a light source and measuring the light reflected by the skin. At each contraction of the heart, blood is forced through the peripheral vessels, producing engorgement of the vessels under the light source-thereby modifying the amount of light to the photosensor. The resulting pressure waveform is recorded.

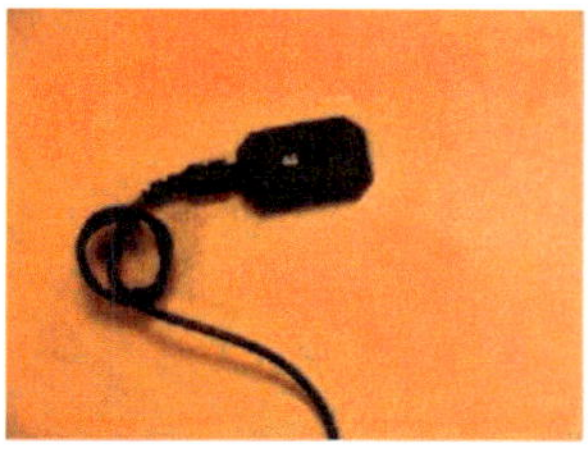

Fig. 2. BVP sensor

3.2.2 Electromyography (EMG)

The electromyographic sensors measure the electromyographic activity of the muscle (the electrical activity produced by a muscle when it is being contracted), amplify the signal and send it to the encoder. In the encoder, a band pass filter is applied to the signal. For all our experiments, the sensor has used the 0-400 microvolt range and the 20-500 Hz filter, which is the most commonly used position. (Figure 3)

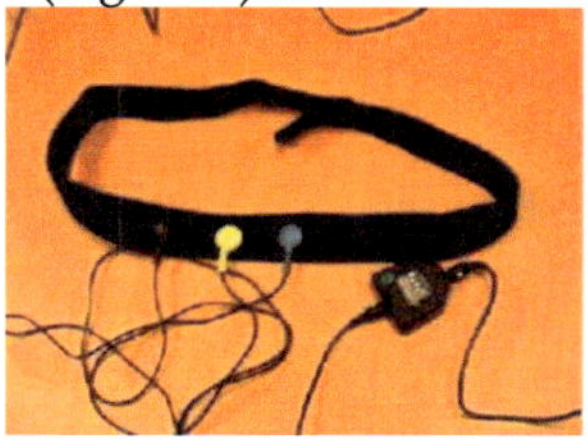

Fig. 3. EMG sensor

3.2.3 Electrodermal activity (EDA)

Electrodermal activity (EDA) is another signal that can easily be measured from the body surface and represents the activity of the autonomic nervous system. It is also called galvanic skin response [18]. It characterizes changes in the electrical properties of the skin due to the activity of sweat glands and is physically interpreted as conductance. Sweat glands distributed on the skin receive input from the sympathetic nervous system only, and thus this is a good indicator of arousal level due to external sensory and cognitive stimuli.

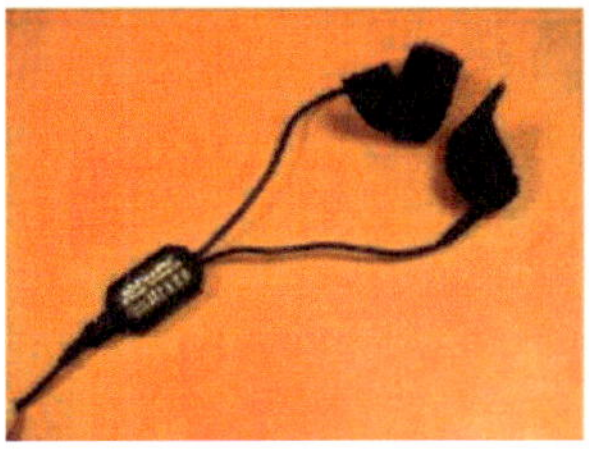

Fig. 4. Skin Conductivity sensor

3.2.4 Skin Temperature (SKT)

Variations in the skin temperature (SKT) mainly come from localized changes in blood flow caused by vascular resistance or arterial blood pressure. Local vascular resistance is modulated by smooth muscle tone, which is mediated by the sympathetic nervous system. The mechanism of arterial blood pressure variation can be described by a complicated model of cardiovascular regulation by the auto-nomic nervous system. Thus it is evident that the SKT variation reflects autonomic nervous system activity and is another effective indicator of emotional status.

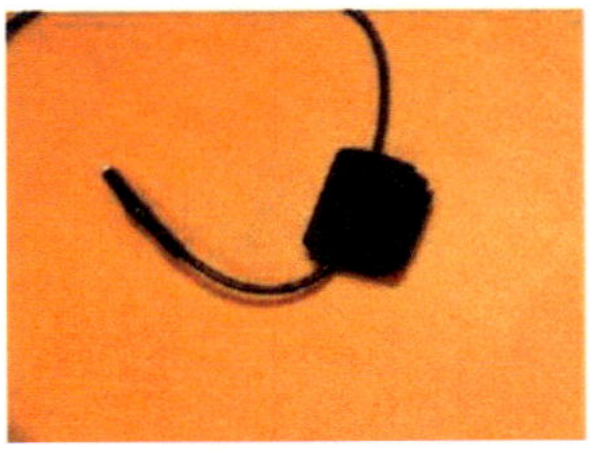

Fig. 5. Skin Temperature sensor

3.2.5 Respiration (Resp)

The respiration sensor can be placed either over the sternum for thoracic monitoring or over the diaphram for diaphragmatic monitoring (Figure 6). The sensor consists mainly of a large velcro belt which extends around the chest cavity and a small elastic which stretches as the subject's chest cavity expands. The amount of stretch in the elastic is measured as a voltage change and recorded. From the waveform, the depth the subject's breath and the subject's rate of respiration can be learned.

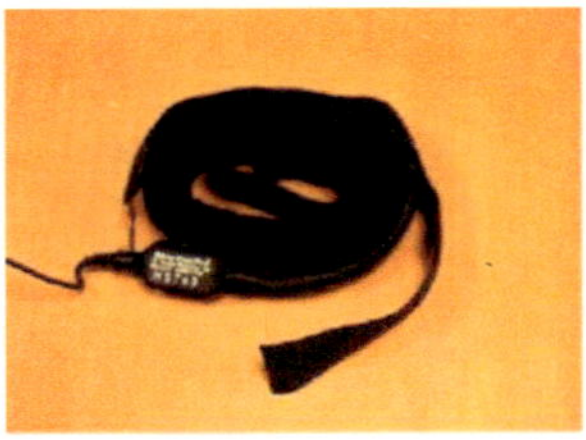

Fig. 6. Respiration sensor

4. Features extraction

Having established a set of signals which may be used for recognizing emotion, it is then necessary to define a methodology in order to enable the system to translate the signals coming from these sensors into specific emotions. The first necessary step was the extraction of useful information bearing features for pattern classification.

For emotion recognition training or testing, the features of each bio-potential data must be extracted. In this study, for each record, we compute the six parameters proposed by Picard [10] on the N values (5 seconds at 256 samples per second gives N=1280): the mean of the raw signals (Eq.1), the standard deviation of the raw signals (Eq.2), the mean of the absolute values of the first differences of the raw signals (Eq.3), the mean of the absolute values of the first differences of the normalized signals (Eq.4), the mean of the absolute values of the second differences of the raw signals (Eq.5) and the mean of the absolute values of the second differences of the normalized signals (Eq.6).

$$\mu_x = \frac{1}{T}\sum_{t=1}^{T} X(t) = \bar{X}(t) \quad (4.1)$$

$$\sigma_x = \sqrt{\frac{1}{T}\sum_{t=1}^{T}(X(t)-\mu_x)^2} \quad (4.2)$$

$$\delta_x = \frac{1}{T-1}\sum_{t=1}^{T-1}|X(t+1)-X(t)| \quad (4.3)$$

$$\bar{\delta}_x = \frac{1}{T-1}\sum_{t=1}^{T-1}|\bar{X}(t+1)-\bar{X}(t)| = \frac{\delta_x}{\sigma_x} \quad (4.4)$$

$$\gamma_x = \frac{1}{T-2}\sum_{t=1}^{T-2}|X(t+2)-X(t)| \quad (4.5)$$

$$\bar{\gamma}_x = \frac{1}{T-2}\sum_{t=1}^{T-2}|\bar{X}(t+2)-\bar{X}(t)| = \frac{\gamma_x}{\sigma_x} \quad (4.6)$$

where t is the sampling number and T is the total number of sample.

5. Emotion recognition method

5.1 Classification methods

After having extracted the features as described in the previous section, we then trained a statistical classifier, with the goal of learning the corresponding emotion for a set of features with which it is presented. A number of different classification algorithms have been proposed in the literature, Fernandez [19] for example used HMMs, while Healey [20] used Fisher linear discriminant projection. This paper will focus its attention on two of them: SVM algorithm and the Fisher linear discriminant. We chose to test and compare both methods. Feature vectors extracted from multiple subjects under the same emotional stimulus form a distribution in high-dimensional space.

- In SVM method and without dimensionality reduction, our system directly gives extracted feature vectors to the support vector machine (SVM) classifier.

-In the 2nd method, we reduce the dimension of the feature vector by Fisher projection and we use subsequent quadratic classifier for recognition.

Both methods will be now briefly described.

5.2 Support vector machine

Machine learning algorithms receive input data during a training phase, build a model of the input and output a hypothesis function that can be used to predict future data. Given a set of labeled training examples

$$S = ((x_1, y_1), \ldots, (x_l, y_l)), y_i \in \{-1,1\} \quad (5.1)$$

learning systems typically try to find a decision function of the form

$$h(x) = \mathrm{sgn}(\langle w.x \rangle + b)) \quad (5.2)$$

that yields a label $e\{-1, 1\}$ (for the basic case of binary classification) for a previously unseen example x. Support Vector Machines are based on results from statistical learning

theory, pioneered by Vapnik [21], instead of heuristics or analogies with natural learning systems.
SVM algorithms separate the training data in feature space by a hyperplane defined by the type of kernel function used. They find the hyperplane of maximal margin, defined as the sum of the distances of the hyperplane from the nearest data point of each of the two classes. The size of the margin bounds the complexity of the hyperplane function and hence determines its generalization performance on unseen data. The SVM methodology learns nonlinear functions of the form:

$$f(x) = \mathrm{sgn}\left(\sum_{i=1}^{l} (\alpha_i y_1 K(x_i x) + b) \right) \tag{5.3}$$

where the α_i are Lagrange multipliers of a dual optimization problem. Once a decision function is obtained, classification of an unseen example *x* amounts to checking on what side of the hyperplane the example lies.

5.3 Fisher linear discriminant

The Fisher's discriminant is a technique used to reduce a high dimensional feature set, x, to a lower dimensional feature set *y*, such that the classes can be more easily separated in the lower dimensional space. The Fisher discriminant seeks to find the projection matrix *w* such that when the original features are projected onto the new space according to

$$y = w^t x, \tag{5.4}$$

the means of the projected classes are maximally separated and the scatter within each class is minimized. This matrix *w* is the linear function for which the criterion function:

$$J(w) = \frac{w^t S_B w}{w^t S_W w} \tag{5.5}$$

is maximized. In this equation, S_B and S_W represent the between class scatter and within class scatter, respectively. This expression is well known in mathematical physics as the generalized Rayleigh quotient. This equation can be most intuitively understood in the two class case where is reduces to:

$$J(w) = \frac{\tilde{m}_1 - \tilde{m}_2}{\tilde{s}_1^2 + \tilde{s}_2^2} \tag{5.6}$$

where $\tilde{m}_1$ and $\tilde{m}_2$ are the projected means of the two classes and $\tilde{s}_1$ and $\tilde{s}_2$ are the projected scatter of the two classes. This function is maximized when the distance between the means of the classes is maximized in the projected space and the scatter within each class is minimized [22].

6. Experimental results

Figure 7 shows an example of the five physiological signals recorded during the induction of six emotions (Amusement, Contentment, Disgust, Fear, Neutrality and Sadness) for subject1 (male) and subject2 (female), respectively. It can be seen that each physiological signal, varies widely across emotion and also across subjects.

For emotion recognition, we have implemented the SVM method with a linear kernel and Fisher's discriminant classifier. A set of six examples for each basic emotion was used for training, followed by classification of 4 unseen examples per emotion.

Table 1 gives the percentage of correctly classified examples for ten subjects using SVM method and Fisher's discriminant. Using a linear classifier, we are able to correctly classify 6 emotions of 10 subjects.As it can be observed, Fisher and SVM classifiers give a good results (92%, 90% respectively) for emotion recognition.

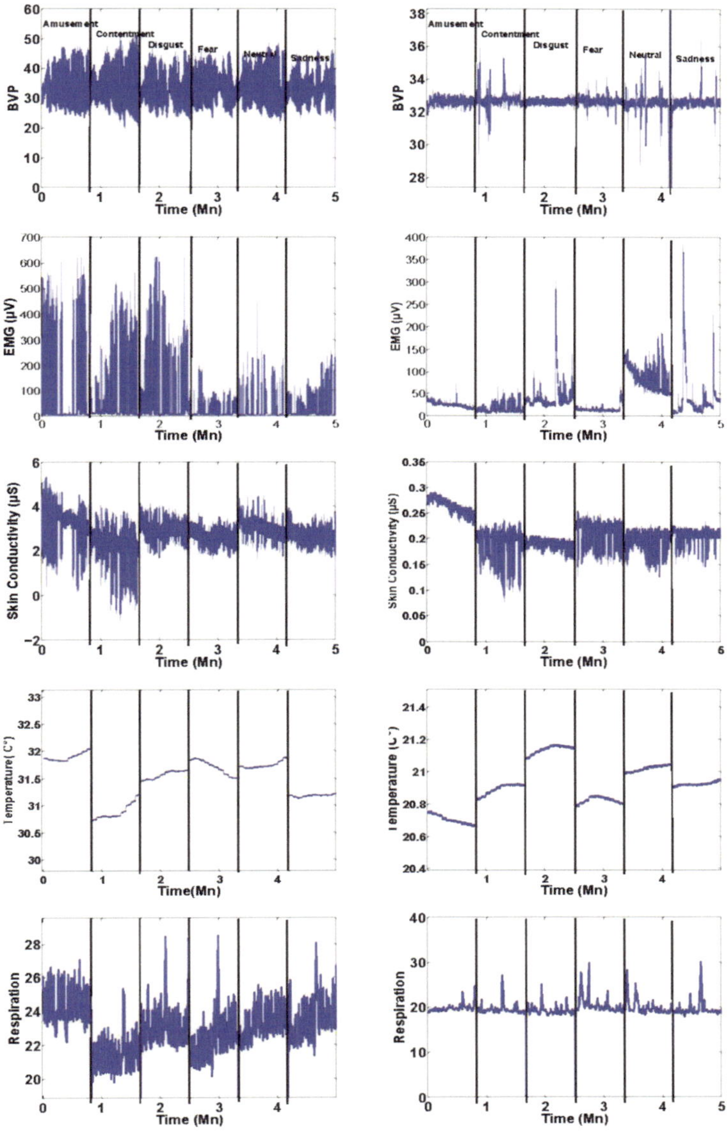

Fig. 7. An example of five physiological signals (BVP, EMG, SC, SKT and Resp) acquired during the induction of the six emotions (left: Subjectl, right: Subject 2)

Subject	Fisher Discriminant	SVM (linear kernel)
Subject1	90%	100%
Subject2	100%	90%
Subject3	100%	90%
Subject4	100%	90%
Subject5	100%	90%
Subject6	80%	90%
Subject7	80%	90%
Subject8	90%	80%
Subject9	90%	90%
Subject10	90%	90%
All Subjects	0%	45%

Table 1. Recognition accuracy of Fisher Linear Discriminant and SVM classification for 10 subjects

Knowing that SVM kernel choice is among the most important customizations that can be made when adjusting an SVM classifier to a particular application domain. It is interesting to test other kernel as: polynomial, sigmoid or gaussian radial basis function (RBF) and choose the best kernel in order to improve SVM recognition results.
Figures 8 and 9 present the results of the features signals separation. These features were projected down to a two dimensional space (Fisher features). Fisher transformation is often used to get a good representation of multidimensional class data in a two dimensional space. As expected, there were significant variations in the positions of data points for each emotion. The data are separated into well-defined clusters. Obviously, merging the features of all subjects does not refine the information related to target emotions (Figure 10). We can see that the data are unseparated, which explains the obtained Fisher rate as 0% for all subjects case (each emotion, varies widely across subjects).
Tables 2,3,4,5,6 and 7 give the confusion matrixes for the original training set of subject1, subject2 and all subjects, respectively, using Fisher discriminant and SVM method. It can be see, that the higher recognition ratio is always obtained for the corresponding and correct emotion for both methods, also, when we mixed the features signals. On the other hand, as it can be observed, the best results classification achievement was gained by the SVM method, especially, for all subjects case. These results are very promising in the sense that the application of Fisher linear discriminant analysis or SVM method to the task of emotion classification seems to provide very good results.

7. Conclusions and future works

in this paper we presented an approach to emotion recognition based on the processing of physiological signals. Physiological data was acquired in six different affective states and two pattern recognition methods have been tested: SVM method and Fisher linear discriminant. Recognition rates of about 90% were achieved for both classifiers. However, SVM classifier gives best results than Fisher discriminant using mixed features signals of different subjects. This study has shown that specific emotion pattern can be automatically recognized by a computer using physiological features.

Future work on arousal, valence assessment will be used in order to identify the emotion in the valence / arousal space. We intend to use wireless sensor in order to ensure a natural and without constraints interaction between human and machine. There is also much scope to improve our system to incorporate other means of emotion recognition. Currently we are working on a facial expression system which can be integrated with physiological signal features.

8. References

[1] B. Reeves and C. Nass, The Media Equation. Cambridge University Press, 1996.

[2] H.R. Kim, K.W. Lee, and D.S. Kwon, Emotional Interaction Model for a Service Robot, in Proceedings of the IEEE International Workshop on Robots and Human Interactive Communication, 2005, pp. 672-678.

[3] R. Cowie, E. Douglas-Cowie, N. Tsapatsoulis, G. Votsis, S.Kollias, W. Fellenz and J. G. Taylor, Emotion recognition in human-computer interaction, IEEE Signal Process. Magazine, Vol. 18, 2001, pp. 32-80.

[4] R. C. Arkin, M. Fujita, T. Takagi and R. Hasegawa, Ethological modeling and architecture for *An entertainment robot*, IEEE Int. Conf. Robotics & Automation, 2001, pp. 453-458.

[5] H. Hirsh, M. H. Coen, M. C. Mozer, R. Hasha and J. L. Flanagan, Room service, AI-style, IEEE Intelligent Systems and their application, Vol. 14, 1999, pp. 8-19.

[6] R. W. Picard, Affective computing, MIT Press, Cambridge, 1995.

[7] P. Lang, The emotion probe: Studies of motivation and attention, American Psychologist, vol. 50(5), 1995, pp. 372-385.

[8] R. J. Davidson, Parsing affective space: Perspectives from neuropsychology and psychophysiology, Neuropsychology, Vol. 7, no. 4, 1993, pp. 464—475.

[9] R. W. Levenson, Emotion and the autonomic nervous system: A prospectus for research on autonomic specificity, In H. L. Wagner, editor, Social Psychophysiology and Emotion: Theory and Clinical Applications, 1988, pp. 17-42.

[10] R. W. Picard, E. Vyzas, and J. Healey, Toward machine emotional intelligence: analysis of affective physiological state, IEEE Trans. Pattern Anal. Mach. Intell., Vol. 23, 2001, pp. 1175-1191.

[11] F. Nasoz, K. Alvarez, C. Lisetti,& N. Finkelstein, Emotion recognition from physiological signals for presence technologies, International Journal of Cognition, Technology, and Work - Special Issue on Presence, Vol. 6(1),2003.

[12] J. WagnerJ. Kim & E. Andre, From physiological signals to emotions: Implementing and comparing selected methods for feature extraction and classification, IEEE Internation Conference on Multimedia and Expo, Amsterdam, 2005, pp. 940-943.

[13] B. Herbelin, P. Benzaki, F. Riquier, O. Renault, and D. Thalmann, Using physiological measures for emotional assessment: a computer-aided tool for cognitive and behavioural therapy, in 5th Int. Conf on Disability, Virtual Reality, 2004, pp. 307-314.

[14] C.L. Lisetti and F. Nasoz, Using Noninvasive Wearable Computers to Recognize Human Emotions from Physiological Signals, Journal on applied Signal Processing, Hindawi Publishing Corporation, 2004, pp. 1672-1687.

[15] P. Lang, M. Bradley, B. Cuthbert, International affective picture system (IAPS): Digitized photographs, instruction manual and affective ratings. Technical report A-6. University of Florida (2005).
[16] www.thoughttechnology.com/proinf.htm
[17] J.D. Morris, SAM: The Self-Assessment Manikin, An Efficient Cross-Cultural Measurement of Emotional Response, Journal of Advertising Research, 1995.
[18] W. Boucsein, Electrodermal activity, New York: Plenum Press, 1992.
[19] R. Fernandez, R. Picard, Signal Processing for Recognition of Human Frustration; IEEE Int. Conf. on Acoustics, Speech and Signal Processing, Vol. 6, 1998, pp. 3773-3776.
[20] J. Healey and R. Picard, Digital processing of Affective Signals, ICASSP, 1988.
[21] V. N. Vapnik, An overview of statistical learning theory, IEEE Trans. Neural Network., Vol. 10, 1999, pp. 988-999.
[22] R. Duda, P. Hart, Pattern Classification and Scene Analysis. Bayes Decision Theory, John Wiley & Sons, 1973.

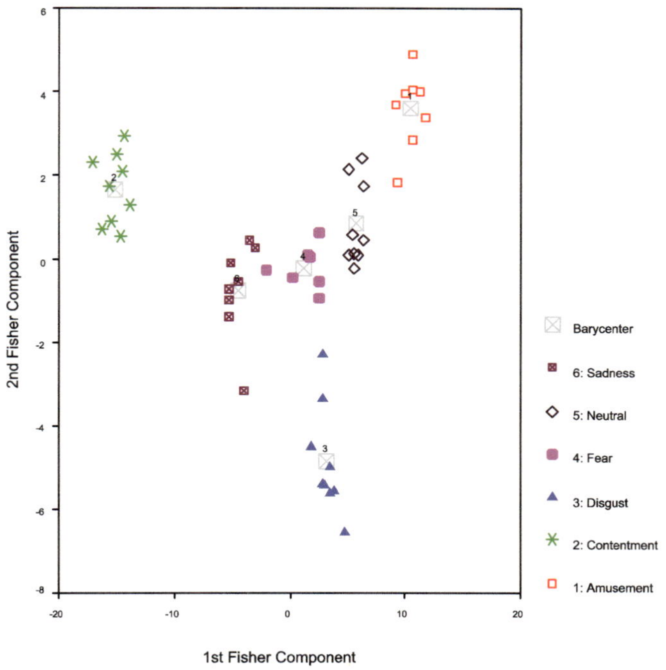

Fig. 8. Points representing emotion episodes are projected onto the first two Fisher features for subject!

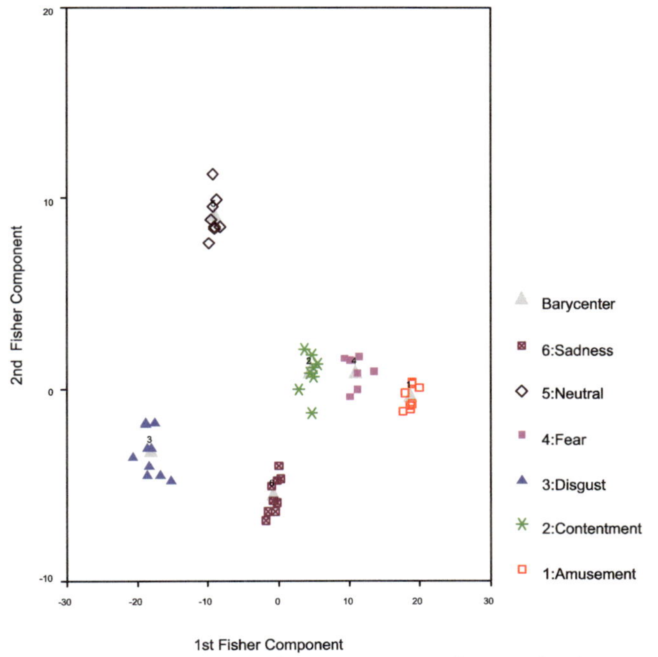

Fig. 9. Points representing emotion episodes are projected onto the first two Fisher features for subject2

Emotion	1	2	3	4	5	6
1	**65%**	0%	0%	10%	25%	0%
2	0%	**100%**	0%	0%	0%	0%
3	0%	0%	**100%**	0%	0%	0%
4	0%	0%	0%	**57.1%**	28.6%	14.3%
5	0%	0%	0%	0%	**100%**	0%
6	0%	0%	0%	12.5%	0%	**87.5%**

Table 2. Confusion matrix for emotion recognition using Fisher discriminant (subject1); 1: Amusement, 2: Contentment, 3: Disgust, 4: Fear, 5: Neutral, 6: Sadness

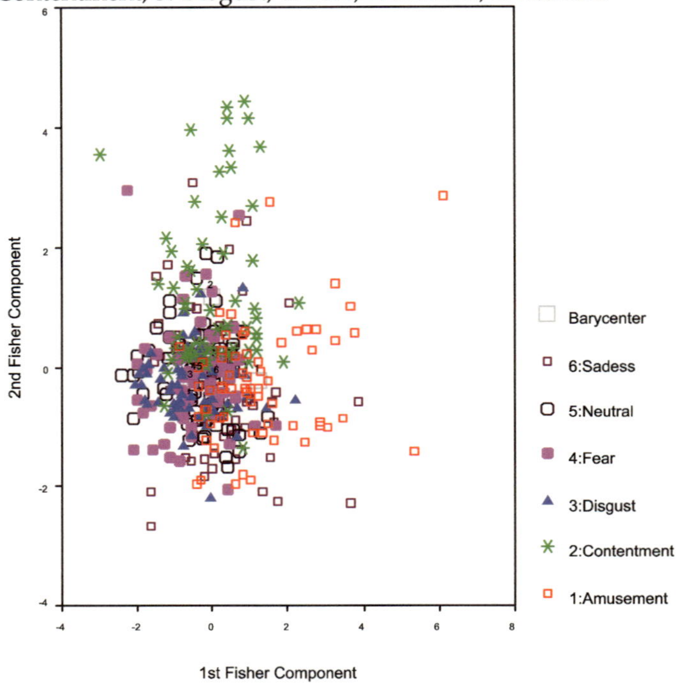

Fig. 10. Points representing emotion episodes are projected onto the first two Fisher features for all subjects

Emotion	1	2	3	4	5	6
1	**80%**	0%	0%	10%	10%	0%
2	0%	**90%**	0%	0%	0%	10%
3	0%	0%	**80%**	0%	10%	10%
4	0%	0%	0%	**80%**	10%	10%
5	0%	0%	0%	0%	**100%**	0%
6	0%	0%	0%	10%	40%	**50%**

Table 3. Confusion matrix for emotion recognition using SVM method (sub-ject1); 1: Amusement, 2: Contentment, 3: Disgust, 4: Fear, 5: Neutral, 6: Sadness

Emotion	1	2	3	4	5	6
1	**100%**	0%	0%	0%	0%	0%
2	0%	**87.5%**	0%	0%	0%	12.5%
3	0%	0%	**100%**	0%	0%	0%
4	14.3%	0%	0%	**85.7%**	0%	0%
5	0%	0%	0%	0%	**100%**	0%
6	0%	0%	0%	0%	0%	**100%**

Table 4. Confusion matrix for emotion recognition using Fisher discriminant (subject2); 1: Amusement, 2: Contentment, 3: Disgust, 4: Fear, 5: Neutral, 6: Sadness

Emotion	1	2	3	4	5	6
1	**90%**	0%	0%	0%	0%	10%
2	0%	**90%**	0%	0%	0%	10%
3	0%	0%	**100%**	0%	0%	0%
4	0%	0%	0%	**100%**	0%	0%
5	0%	0%	0%	20%	**80%**	0%
6	0%	0%	0%	20%	0%	**80%**

Table 5. Confusion matrix for emotion recognition using SVM (subject2); 1: Amusement, 2: Contentment, 3: Disgust, 4: Fear, 5: Neutral, 6: Sadness

Emotion	1	2	3	4	5	6
1	**35%**	5%	9%	4%	8%	5%
2	10%	**27%**	2%	9%	17%	1%
3	8%	3%	**35%**	6%	12%	0%
4	2%	5%	19%	**22%**	14%	5%
4	4%	4%	12%	12%	**29%**	6%
6	11%	10%	3%	7%	14%	**25%**

Table 6. Confusion matrix for emotion recognition using Fisher discriminant (all subjects); 1: Amusement, 2: Contentment, 3: Disgust, 4: Fear, 5: Neutral, 6: Sadness

Emotion	1	2	3	4	5	6
1	**46%**	4%	15%	1%	26%	8%
2	12%	**37%**	14%	4%	29%	4%
3	2%	6%	**59%**	5%	26%	2%
4	6%	7%	17%	**34%**	32%	4%
5	5%	10%	9%	3%	**69%**	4%
6	5%	11%	12%	6%	32%	**34%**

Table 7. Confusion matrix for emotion recognition using SVM (all subjects); 1: Amusement, 2: Contentment, 3: Disgust, 4: Fear, 5: Neutral, 6: Sadness

21

Robot Assisted Smile Recovery

Dushyantha Jayatilake, Anna Gruebler, and Kenji Suzuki
University of Tsukuba
Japan

1. Introduction

1.1 Facial Expressions

Facial expressions play a significant role in social information exchange and the physical and psychological makeup of a person because they are an essential method for non-verbal communication. Through facial expressions, human beings can show emotions, moods and information about their character.

Happiness, sadness, fear, surprise, disgust, and anger, are typically identified by psychologists as basic emotions with their corresponding characteristic facial expressions (Wang and Ahuja; 2003). Further, (Batty and Taylor; 2003) reported that humans have a very fast processing speed when it comes to identifying these six expressions and noted that positive expressions (e.g. happiness, surprise) are identified faster than the negative expressions (e.g. sadness, disgust).

Human beings share common characteristics in the way they express emotions through facial expressions which are independent from nationality, ethnicity, age or sex. It has been recorded that the ability to recognize the corresponding emotion in a facial expression is innate and is present very early, possibly form birth (Mandler et al.; 1997). However there is also evidence that universal expressions might be modified in social situations to create the impression of culture-specific facial expression of emotions. For example, (Ekman; 1992) noted that when an authority figure was present, the Japanese masked negative expressions with the semblance of a smile more than the Americans. If, because of accident or illness, a person looses the ability to make facial expressions this makes the face seem emotionless and leads to physical and psychological hardships.

1.2 Facial Paralysis

The medical term "Paralysis" is defined as the complete loss of muscle function of one or more muscle groups. Facial paralysis is the total loss of voluntary muscle movement of one or both sides of the face. It can happen to anyone of any age. Paralysis often includes loss of feeling in the affected area, and is mostly caused by damage to the nervous system or brain. The paralysis can be short term which is between a few minutes and a few hours, long term which is usually at most about 6-7 months but sometimes registering even 4-5 years, or permanent (Byrne; 2004; Garanhani et al.; 2007). The House classification system describes 7 grades of facial paralysis (6 according to (Beck and Hall; 2001)) form Normal to Total, where the latter is the complete facial paralysis with no tone (Quinn and Jr.; 1996).

Facial nerve paralysis is a fairly common issue that involves the paralysis of any parts stimulated by the facial nerve. Due to the lengthy and relatively convoluted nature of the pathway

Bell's Palsy	0.5/1000 per year
recurrence rate	7.0%
lifetime prevalence	0.64-2.0%

Table 1. Bell's Palsy statistics

of the facial nerve, several situations can result in facial nerve paralysis. The most common is Bell's palsy. As table 1 shows, the recorded cases of Bell's Palsy are 0.5 per 1000 persons per year with a recurrence rate of 7% and a lifetime prevalence of 6.4 to 20 per 1000. The occurrences show an equal male to female ratio however it is 3.3 times greater among pregnant females (Quinn and Jr.; 1996). About 84% of the Bell's Palsy patients said to have had a spontaneous recovery which is likely to occur within the first three weeks. However the remaining 16% show only moderate to poor recovery. Facial paralysis due to a brain tumor generally develops slowly and causes headaches, seizures, or hearing loss. Facial paralysis, although not life threatening, can cause severe distraction in many ways. Physically, it makes the patient less responsive as well as having difficulties in eating, drinking and talking due to the inability to purse the lips.

Apart from the loss of volitional facial muscle motion, facial paralysis has another major consequence, which is the loss of baseline muscle tone. This causes the changes in facial appearance, such as drooping of the ipsilateral face and and deviation of the nose to the contralateral side (Pensak; 2001). The face is such a salient feature of a person that such a facial disfigurement can result in severe social and vocational handicap. A paralysis or weakness of even only one side of the face can be an alarming and depressing event in the patient's life.

1.3 Current Medical Treatments and Rehabilitation for facial paralysis

The current treatment methods for facial paralysis can be divided into three main categories: physical therapy, medical therapy, and surgery. Physiotherapy, facial exercises and massage are used to treat all types of facial paralysis. Physical therapy uses facial neuromuscular retraining, which is based on selective motor training techniques to optimize the motor control of the facial muscles.

There have been reports on the use of facial electromyography (EMG) to analyze emotions through facial expressions. It has been stated that because of its high temporal resolution, facial EMG is well suited for measuring emotions, which have rapid onset and short durations (Harrigan et al.; 2005).Along those lines, visual and bio-feedback through specific mirror exercises and surface electromyography (sEMG) based augmented sensory information is used to enhance neural adaptation and learning. It has been stated that sEMG based treatments allow the therapist to quantify the muscle activity patterns, resulting in more effective and faster treatment. Based on the study on training of nasal muscles, (Vaiman et al.; 2005) stated that Electromyography-recorded amplitude of muscle tension of the nasal muscles significantly increased in all the patients considered in the study. (Sugimoto1 et al.; 2007) in their article on autogenic training explained how sEMG can be used and data can be processed to improve treatments. Paralysis due to a compromised immune system, for instance, bacterial external ear infection, is often treated by using intravenous antibiotics. Bell's palsy treatments may involve the use of steroids and anti-viral drugs. The prime target of rehabilitation would be to recover the normal facial tone and function.

In the event of an identified break in the facial nerve, repairing and grafting are performed to reestablish the connection. The hypoglossal to facial nerve transfer reestablishes neuronal im-

pulses to the facial muscles and supplies a baseline resting tone. However patients regaining the ability to show any spontaneous or emotive expression are said to be very rare with this method (Pensak; 2001).
Cross facial nerve grafting is the method of connecting the facial nerve of the paralyzed with the facial nerve of the healthy side by using a nerve obtained from the patient himself. This method said to produce symmetrical faces and since the grafting nerve is obtained from the same patient it eliminates the problems associated with rejection by the body. It is also worthwhile noting that to achieve the best results with facial reanimation techniques it is usually required to perform nerve repair or a combination of techniques that will lead to the strongest facial movement as soon after the injury as possible (May and Schaitkin; 2003).

1.4 Smile Recovery

In recent years, many researches paid attention to assistive technology. A number of robotic systems have been reported that support and assist human beings such as exoskeletons and prosthetic limbs. Most researches focus on supporting the human physical functions in terms of rehabilitation or healthcare. However, the human cognitive functions typified by facial or body expressions are as important as the physical functions.
By smiling a person can show affection, humor, and put others at ease, which are essential traits in natural human communication. The main objective of the "Smile Recovery" project is the design of a supportive device to recreate facial expressions and to put the "smile" back on the face. In general, the main noninvasive cure for facial paralysis is the use of physical therapy, although it is not able to support permanently paralyzed patients. Other standard treatment methods work together with invasive techniques to support permanently paralyzed patients. Although there is a possibility of using some other non conventional methods such as Functional Electrical Stimulus (FES), the technique is still in the experimental stage, and the effect also is only momentary (Dingguo and Kuanyi; 2004). On the other hand, *Robot Assisted Smile Recovery* investigates the support of facial expressiveness though the use of a robotic technology based wearable supportive device called *The Robot Mask* (Fig. 1). The four key features of this proposed design are:

1. Silent Actuation: a major characteristic of facial expressions is their silent nature of occurrence. Due to the mechanical noise present in traditional actuators such as electrical motor based actuators, it will be improper to use them for facial expression generation. In the Robot Mask this problem is solved by the use of specially designed Shape Memory Alloy (SMA) based actuators.
2. Use of bioelectrical signals from contralateral face to generate skin displacements in the ipsilateral face: This would help to reduce the facial disfigurement due to facial asymmetry. The use of bioelectrical signals of the mask wearer will facilitate the interpersonal timing of facial expressions associated with each individual.
3. Noninvasive: this will eliminate complications due to medical surgeries, reduce maintenance difficulties, and increase the use among both temporary and permanent facial paralysis.
4. Natural looking smile: based on a rigorous analysis of the facial morphology, artificial expressions that closely resemble natural expressions are recreated. Because of the importance and universality of facial expressions it is necessary for artificially reconstructed expressions to be as close to the natural ones as possible.

2. Analysis of Facial Morphology

Even though in case of illness or accident, muscle movement can be lost, the functions of natural muscles can be recreated artificially. In the case of muscles on the face, artificial muscles can be attached externally to the surface of the face to generate artificial facial expressions. This section introduces an analysis of both amount and direction of the facial surface skin displacement by using markers on the face and 3-dimensional imaging. It will be shown how artificially generated smiles can be generated and made to look similar to natural ones.

2.1 Preliminary experiments

In order to be able to recreate a smile adequately, it is necessary to perform a systematic analysis of the facial skin displacement due to natural muscle contraction. At the moment the focus is in the displacement between the neutral and the smiling face.

In a preliminary experiment, it became obvious that some areas of the face are subject to greater displacement than others during smiling. Markers were placed on the face in order to be able to track the displacement more accurately. Especially the area around the mouth shows greater displacement, as can be seen in figure 2, which shows a healthy subject smiling on command.

In order to recreate the "smile" it was attempted to displace that area of greatest natural displacement artificially. Figure 3 shows an artificially generated facial expression. It was possible to verify that artificial facial expressions can be achieved by displacing the surface of the skin using point based actuation. Because of the skin elasticity the displacement achieved by the actuators is propagated, displacing the area surrounding it. The surface skin is displaced artificially but it is not yet a good facsimile of the "smile". Therefore, a more in-depth analysis of facial displacement is necessary in order to recreate the smile in a convincing manner.

2.2 Systematic facial displacement analysis

For the systematic facial displacement analysis and accurate measurement of the skin displacement, markers were attached to the subject's face as in the previous experiment. However new patterns of markers were introduced. The first pattern used was in the form of a grid and it used the natural landmarks of the face, such as nose, eyes and chin as reference points. The second pattern used was based on the muscles activated while smiling: a series of markers was attached over the lines of the muscles *zygomaticus major*, *zygomaticus minor* and *risorius* . The location of the muscles on the face can be seen in figure 4.

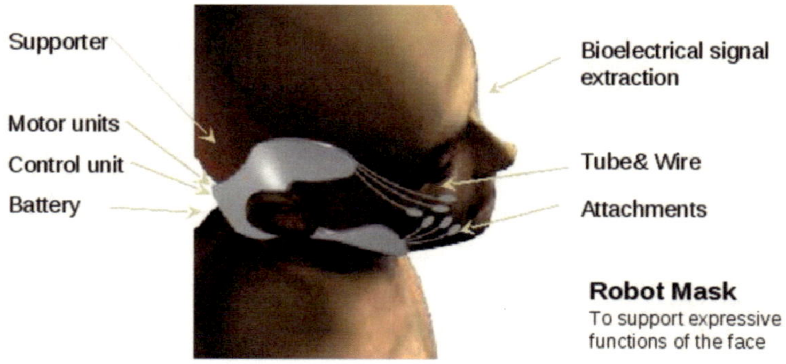

Fig. 1. The Non Invasive Robot Mask

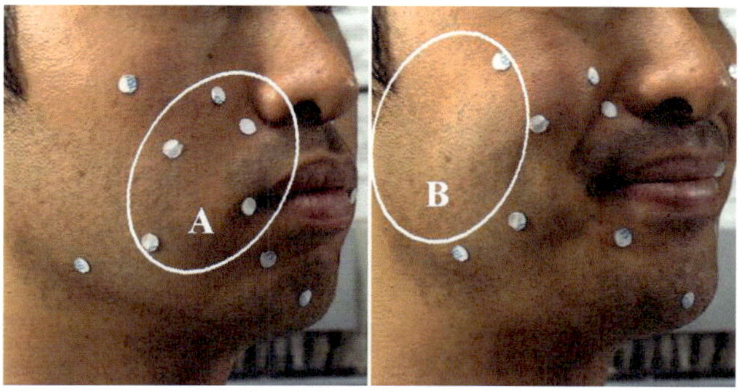

Fig. 2. Facial skin displacement in a healthy subject. The area designated with A shows the greatest displacement while smiling while B shows little displacement.

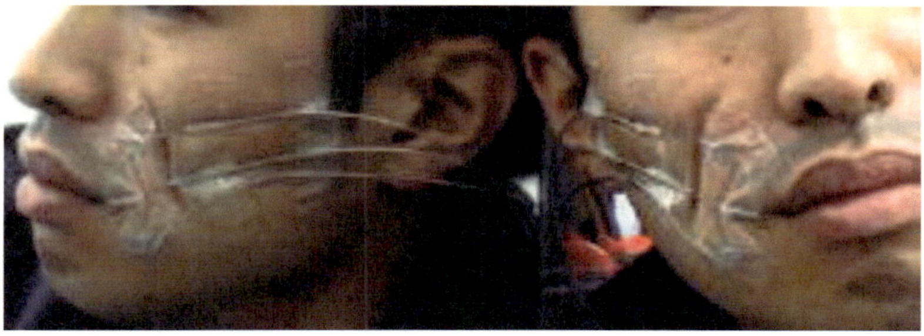

Fig. 3. Artificial facial skin displacement using 3 artificial muscles

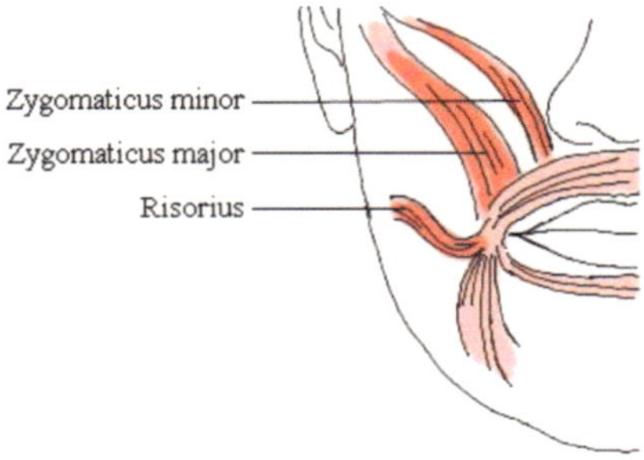

Fig. 4. Muscles on the face

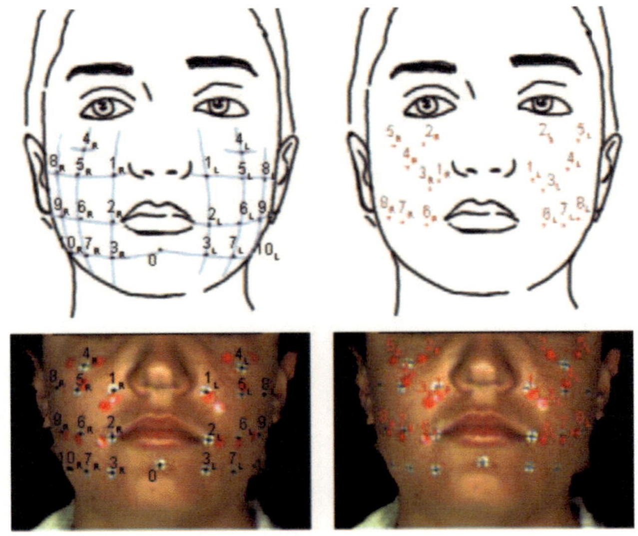

Fig. 5. Position of Markers on the face. Left: Grid-line markers. Right: Muscle markers

Figure 5 shows the position of both the grid-line and muscle markers. The grid-line markers are white with a back cross while the muscle markers are red. There are 10 grid-line markers on each side and one on the chin and 8 muscle markers, designated with the subindexes "L" and "R" for the left and right hand side respectively.
The subjects are photographed using a 3-dimensional imaging device. This serves the dual purpose of guaranteeing that the frontal and lateral views correspond to the same moment in time and makes rotation in 3 dimensions possible for calibration. After capture the images are rotated and scaled so that the eyes, ears and nose of both the neutral and smiling face match. Then, the frontal and lateral views were separated for analysis. These views can be seen in figure 6.
In order to calculate the total 3-dimensional displacement the horizontal and lateral displacement was measured and the Pythagoras equation was used. The total displacement of the dot (D_T) is the square root of the sum of the squares of the frontal displacement (D_F) and the lateral displacement (D_L) of the dot (Equation 1).

$$D_T = \left(D_L^2 + D_F^2\right)^{1/2} \tag{1}$$

The frontal displacements is calculated in a similar manner from the square root of the squares of the vertical displacement (h) and the horizontal displacement (l) of the dot in the frontal view (Equation 2).

$$D_F = \left(h^2 + l^2\right)^{1/2} \tag{2}$$

In the first part of the experiment the subject was asked to show a neutral expression and then to smile.
Figure 7 shows the 3-dimensional linear displacement of both grid-line and muscle markers between the neutral face and a natural smiling face. In accordance with previous studies that state that facial asymmetry is substantially left-sided regardless of emotional valence (Borod

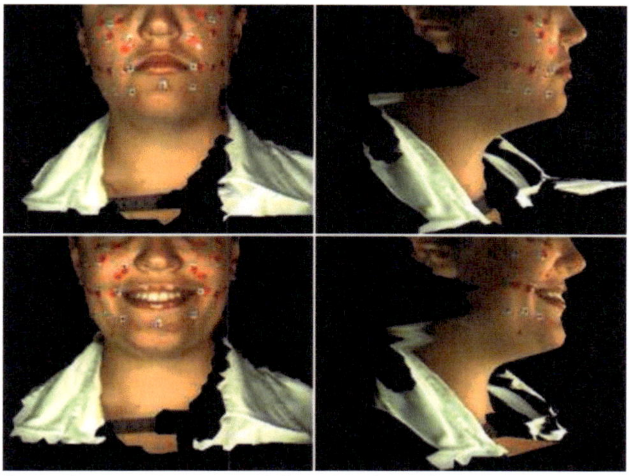

Fig. 6. Frontal and lateral views for neutral face (top) and smiling face (bottom)

et al.; 1998), a more pronounced displacement in the left side of the subject's face was observed.

By systematically analyzing the facial displacement between the neutral and the smiling face on a healthy subject it could be determined that the displacement on skin markers never exceeds 20 mm for the smile. It could also be noted that the greatest displacement correspond to grid-line points 2 and 6 which follow the horizontal line of the smile (Fig. 7), which is not surprising when the shape of the smile is considered.

In the second part of the experiment it was attempted to replicate the skin displacement during smiling using cables attached to the end points of the muscles. Cable 1 pulled from underneath the ear, simulating the muscle *risorius*, cable two pulled from above the ear simulating the muscle *zygomaticus major*. Three artificial expressions were created: artificial displacement by pulling cable 1 and cable 2 individually in turn and the facial displacement achieved by pulling both cables at once. 3-dimensional images were taken of the artificial facial expressions.

Figure 8 shows the displacement for artificially generated facial expressions for the artificial *risorius* simulator, artificial *zygomaticus major* simulator, and both simultaneously as well as the displacement during the natural smile for comparison. It should be noted that the artificial smiles were able to achieve their greatest displacement on the same points of greatest displacement as the natural smile (Fig. 8).

Also of interest when analyzing facial expressions, it is not only the amount of displacement that is important but also the direction towards which the displacement occurs. Studies have shown that angular deviations are more significant to the detection of facial symmetry than linear displacements (Rhodes et al.; 2005). The angle of displacement of both grid-line and muscle markers was analyzed and compared to the displacement during a natural smile. Figures 9 and 10 show the histograms for the difference in the displacement angles when compared with the natural smile for grid-line and muscle markers respectively. The values are given in degrees.

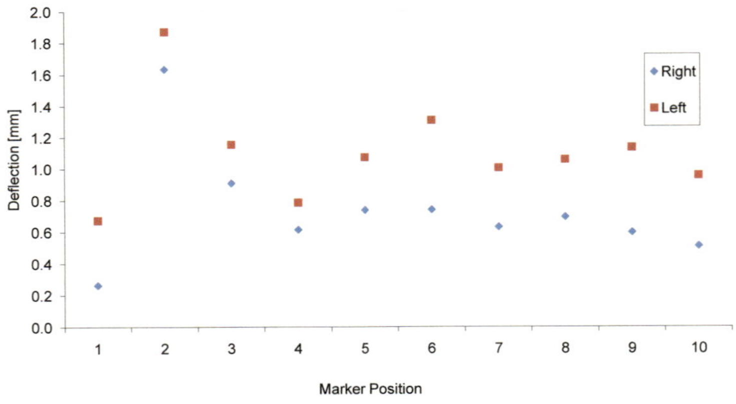

(a)

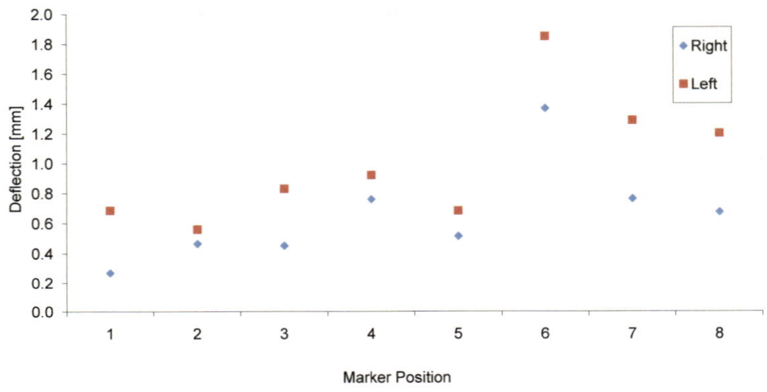

(b)

Fig. 7. 3-dimensional displacement [in cm] of (a)Grid-line markers and (b)Muscle markers between the neutral and the naturally smiling face.

An example of the artificial facial displacement, using the artificial *risorius* simulator and artificial *zygomaticus major* simulator can be seen in figure 11. The symmetrical image is constructed using a hemifacial actuation image and its mirror image.
From the experiments in artificial displacement of the facial skin it could be seen that pulling just from underneath the ear led to an amount of displacement similar to that of the natural smile both in grid-line and muscle markers, the angle of displacement was however not similar enough as the mouth was pulled downwards and not enough up. When pulling above the ear a smaller displacement than that of the natural smile could be observed, the angle of displacement however was very similar. Especially around the points of greatest displacement (grid-line points 2 and 6), the difference in displacement angle was under 5.5° which can be considered a good approximation of the smile. However the best approximation of the smile was achieved when using both artificial muscles simultaneously (Figures 9 and 10) which in accordance with the fact that in the healthy human being multiple muscles are used to create the smile.

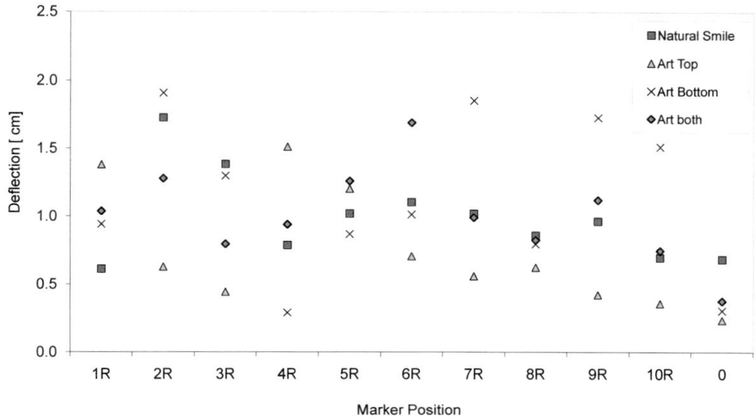

(a)

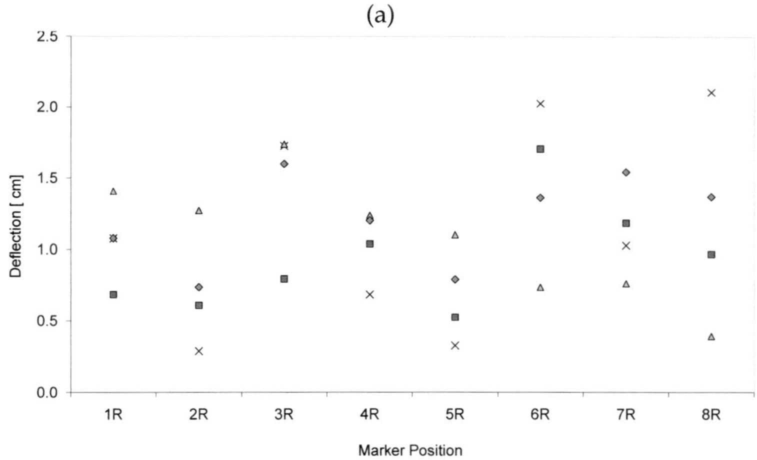

(b)

Fig. 8. 3-dimensional displacement [in cm] of (a)Grid-line markers and (b)Muscle markers between the neutral and natural smile, artificial *risorius* simulator ("Art Bottom "), artificial *zygomaticus major* simulator ("Art Top "), and both simultaneously.

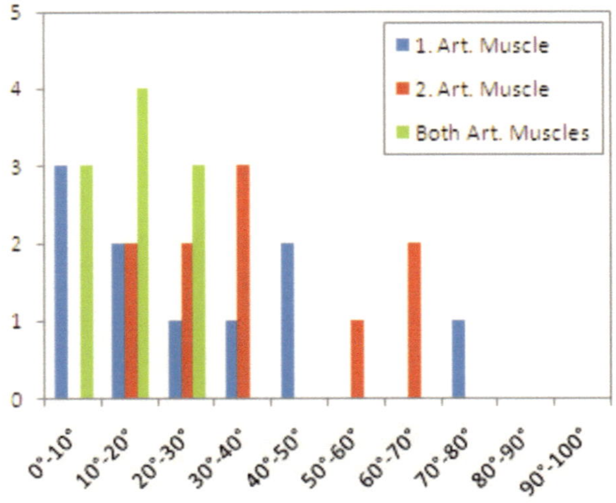

Fig. 9. Histogram of the error between the deviation angles of artificial and natural expressions for the artificial *risorius* simulator (1. Art. Muscle), artificial *zygomaticus major* simulator(2. Art. Muscle), and both simultaneously (Both Art. Muscles) of grid-line markers.

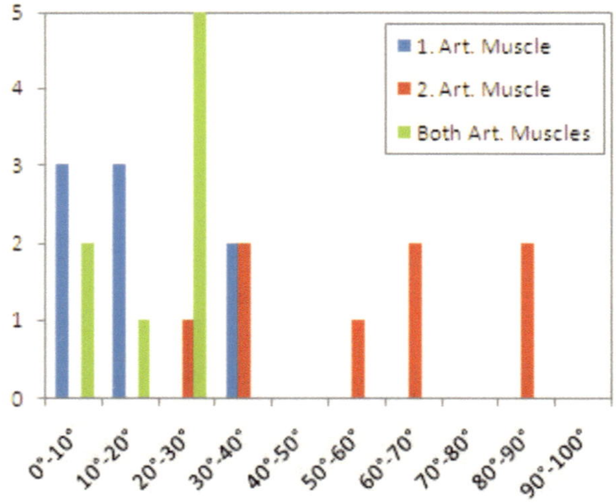

Fig. 10. Histogram of the error between the deviation angles of artificial and natural expressions for the artificial *risorius* simulator (1. Art. Muscle), artificial *zygomaticus major* simulator(2. Art. Muscle), and both simultaneously (Both Art. Muscles) for Muscle markers.

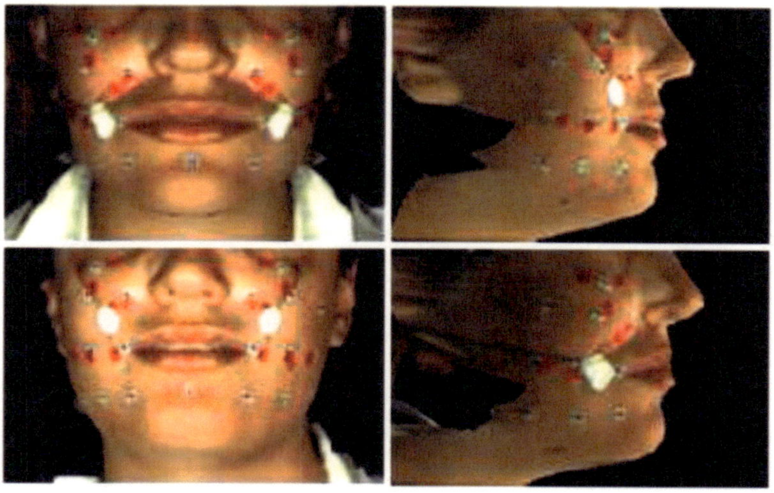

Fig. 11. Mirrored reconstruction of a hemifacial image where the smile is simulated using artificial *risorius* simulator (top), artificial *zygomaticus major* simulator (bottom)

Figures 7, 8 and 11 show that it is possible to simulate a smile artificially by using artificial muscles.

The method for the systematic analysis of facial displacement that was introduced in this chapter using grid-line and muscle markers simultaneously and using 3-dimensional image capturing is well suited for future calibration of the Robot Mask for individual users. Further, in the case of subjects suffering from only hemi-facial paralysis, it can be used to acquire the displacement amount and angles from the healthy side as a template for the artificial displacement on the paralyzed side.

In order to further analyze the 3-dimensional facial surface skin displacement, the artificial smile and natural smile's frontal protrusions were compared to the neutral face. The results can be seen in figure 12. The right side of the subject's face was displaced using two artificial muscles simultaneously. It can be seen that the same areas are displaced as in the natural smile: around the mouth and cheek. Notably, there is little displacement occurring in both the natural and the artificially generated smile in the area around the eyes. This shows that the efforts for smile recovery should concentrate on the area around the mouth.

Taking into account these results, the Robot Mask was designed in order to artificially displace the facial skin.

3. Robot Mask

The Robot Mask, which consists of a head supporter, motor unit and a pulling wire arrangement, attaches onto the face externally and pulls facial skin through cables attached to the skin (Fig. 13). The pulling movement is derived by a specially designed SMA based actuator called "Silent Actuation Unit" or SIAC unit in short. The basic principle is the contraction of SMA coil elements when undergoing Joule heating.

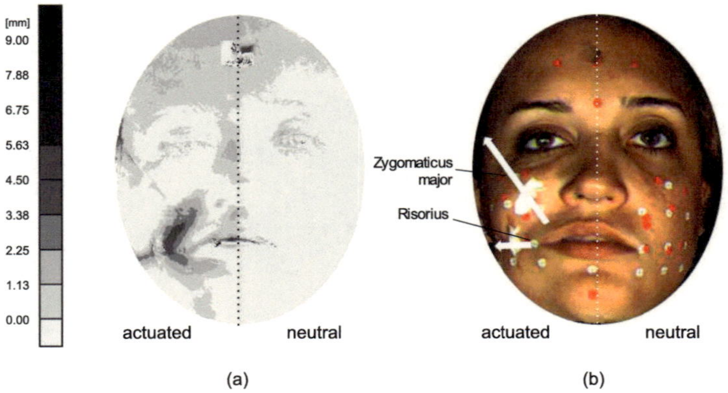

Fig. 12. Artificial facial displacement through actuation on the subject's right side. (a) Compared 3-dimensionally to the neutral face. (b)Actuation direction and corresponding muscles.

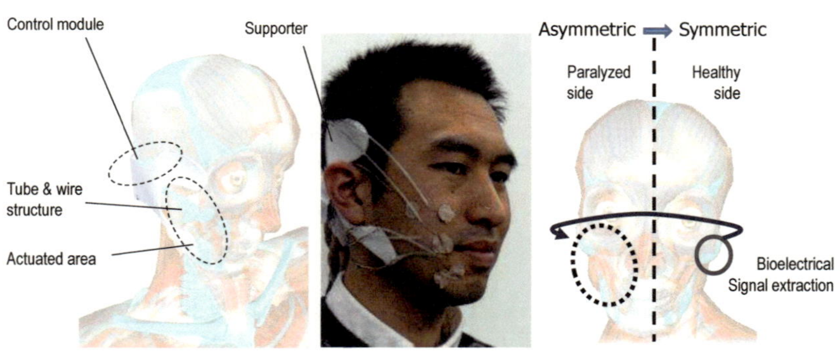

Fig. 13. The basic concept of Robot Mask

3.1 Shape Memory Alloys and Actuators

The Shape Memory Alloy group, which was first investigated in 1951 through a gold-cadmium alloy (Higgins; 1993) poses two distinct properties: the shape memory effect and the pseudo-elasticity. Pseudo-elastic materials show a phase transformation between Martensite and Austenite under mechanical loading and a complete reversal during subsequent reheating. Bellow a certain temperature (the critical temperature), a martensitic type of structure is formed with the alloy carrying a memory of the high and low temperature shapes. Furthermore, the superelastic characteristics of shape memory alloys (SMA) will make them capable of producing large recoverable strains (ASM; 1990).

There have been several reports, especially in recent times, about SMA based actuators. (Mascaro and Asada; 2003) reported on a wet shape memory alloy actuator which can be used for a robotic flesh. They kept the actuator immersed inside water and an electrical current was used to heat the SMA. They also proposed a hybrid electric/thermofluidic control method which uses water for cooling and either electric current or water for heating. Asada et al. (2006) presented a SMA actuator system for a 12 axis tendon actuator of a five fingered robot hand. (Yang and Gu; 2008) introduced the concept of a planar bending embedded SMA actuator, which shows better characteristics in both heating and cooling. Rediniotis et al. (2002) proposed an SMA actuator system which uses Fuel as the energy source, with convection heating and cooling using water. (Yambe et al.; 2001) in Japan have proposed a series of biomedical applications of SMA actuators. In 2001 they reported on SMA for internal artificial organs. They include the circulatory assistance for a heart failure and an artificial sphincter. Another biomedical application of Ni-Ti SMA is to create compressive stresses that promote knitting of broken bones. Shape-memory Ni-Ti alloys also have been employed experimentally to dilate blood vessels, thus increasing the flow of blood to vital organs (ASM; 1990).

3.1.1 Mechanical Properties of Actuation elements

Since it is possible to generate heat in an SMA coil though Joule heating by applying an electrical potential across it, several experiments were conducted to investigate its extension/contraction characteristics against different controlled actuation conditions. Figure 14 shows the performance of SMA coils under various such controlled conditions. These experiments were done using artificial muscles of initial length 20mm and initial expanded length 45mm. Figure 14(a) shows the wire deflection with time for low actuation voltages. Three tests were done using actuation voltages 1.5, 1.75 and 2.0 and the characteristics were compared with the 4.0V actuation. As can be seen from the graphs, the maximum actuation, as well as the actuation rate, show a directly proportional relationship to the actuation voltage. Figure 14(b) shows the time-displacement characteristics in heating and cooling. The graph shows a significant difference in performance during heating compared to cooling. Although deflection during heating is completed in 1.1 seconds, as many as 32 seconds were necessary for the cooling process to finish completely. Both the tests in figures 14(a) and 14(b) were carried out under no load conditions with natural drafting.

3.2 Supporting Facial Expressiveness

As indicated by Fig. 15, the Robot Mask has been designed to emulate a set of five muscles of the oral group of facial muscles, namely (1) *Zygomaticus minor*, (2) *Zygomaticus major*, (3) *Risorius*, (4) *Platysma*, and (5) *Depressor Anguli Oris*. Two actuators are been used to generate the movement of *Zygomaticus major*.

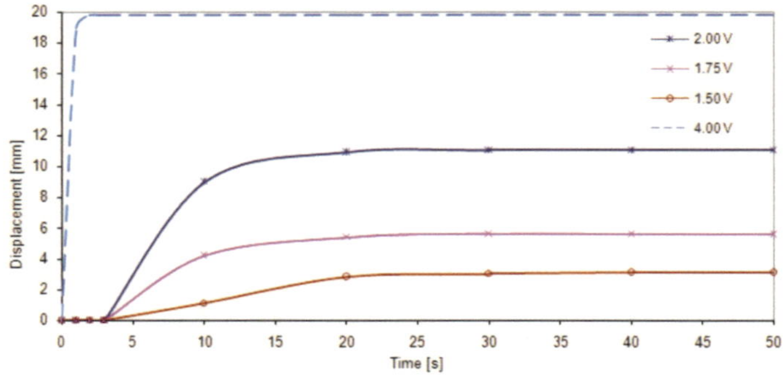

(a)

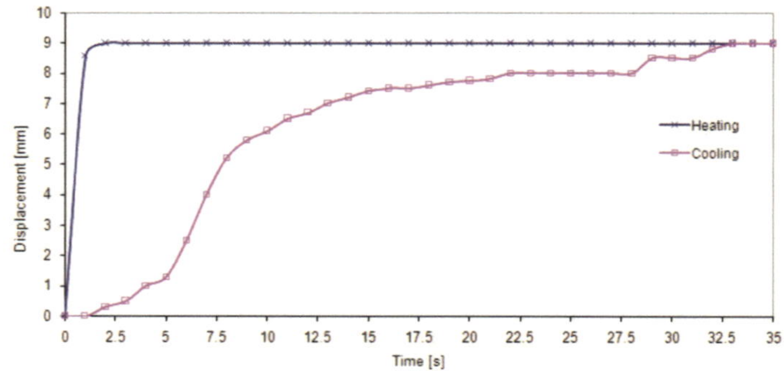

(b)

Fig. 14. Performance of SMA coils type actuation elements under various controlled conditions. (a) Low voltage actuation characteristics, and (b) Time-displacement characteristics during heating and cooling

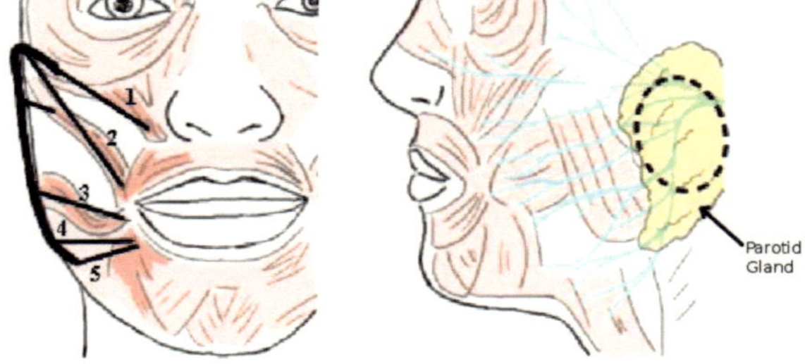

Fig. 15. Part of the oral group of facial muscles under consideration, actuator positioning, and the location of Parotid Gland with the emergence of the facial nerve highlighted with the dotted ellipse

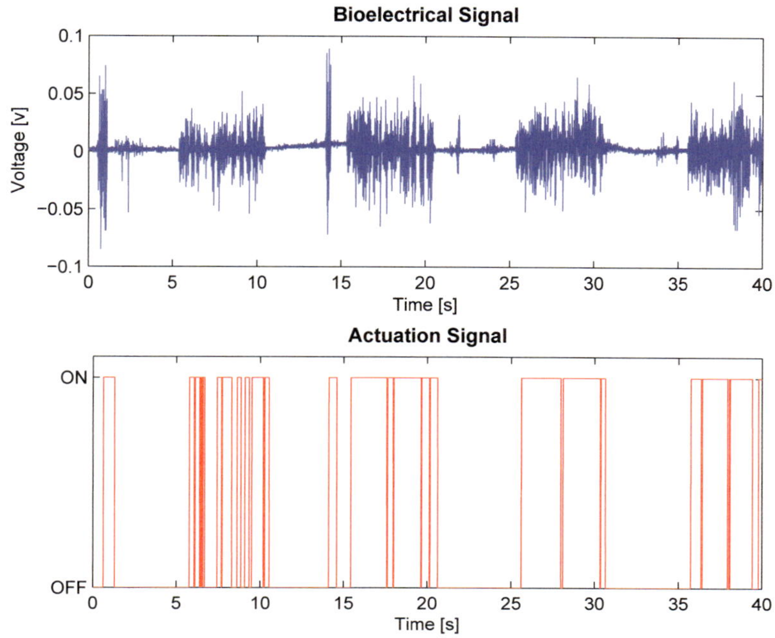

Fig. 16. Bioelectrical signal and the corresponding actuation signal. On top the raw signal and at the bottom the firing pattern

The voluntary responses of facial muscles are initiated from the appropriate regions of the brain, and the facial nerve helps to transmit these discharges to the facial muscles by using five terminal groups of branches, that emerge from the upper, anterior, and lower borders of the parotid gland, which is located below the lower half of the ear.

3.2.1 Bioelectric signal based control

The bioelectrical signals were extracted using Ag/AgCl flat type active electrodes which were placed near the parotid gland (Fig. 15) with the electrodes being 20mm apart. The third ground electrode was placed on the neck. Since the gel between the skin and the electrode takes some time to settle down, 15 min were left before taking any data. The signals were acquired through the 10bit ADC port of the master microcontroller at a sampling rate of 1kHz. The sampled data was sent to a 50 size FIFO buffer and then the squared values of buffer data were averaged and compared against a predefined threshold value. In this way it was possible to identify the presence of any bioelectrical signal that is different from the average neutral condition, however no attempt was made to identify the individual muscles associated with the signal.
Figure 16 shows a time sequence (represented in terms of it's sample number) of bioelectrical signal captured at a rate of 200Hz and the corresponding actuation logic derived by the master controller. The data corresponds to a time period of 40s.

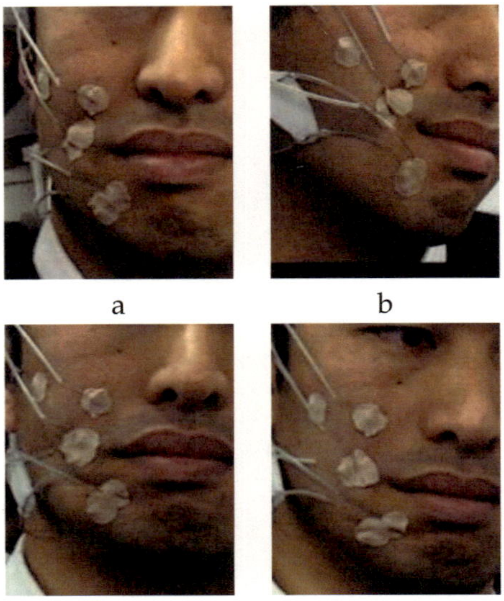

Fig. 17. Formation of artificial expressions through the Robot Mask.

3.3 Evaluation of Robot Mask Performances

The figure 17 shows three different sets of artificially generated facial actions. The tests were carried out using 6 artificial muscles in the arrangement explained in figure 15. Figures 17 (a) and (b) shows the front and side views of the effect of actuator 2 which is somewhat equivalent to the expression produced by the Zygomaticus major muscle. This resulted in a tiny wrinkle and a smile like pulling of a side of the mouth. Note that (a) and (b) are two different time instances and not the front and side views of same expression. 17(c) is the combined effect of actuators 4, 5 and 6 and Fig. 17(d) is the combined effect of all 6 actuators. In both those instances it is possible to see the circular type wrinkle around the mouth near the cheek. In (c) the wrinkle is more to the lower half of the face and in (d) it spreads fairly evenly around the mouth region. The pulling of wire 6 makes the back area of the facial skin move together with the area close to the mouth, making a larger deflection in skin with a less salient wrinkle around the mouth.

4. Conclusion

The two measures that influence the recreation of the natural smile identified by this research are the linear amount and the angle of facial skin displacement. As shown in this chapter, both were adequately emulated by the Robot Mask.
In this chapter it was shown that it is possible to generate artificial facial expressions that bear a close similarity to natural facial expressions by attaching cables to the surface skin of the face over the endpoints of the muscles *zygomaticus major* and *risorius*. In order to achieve an adequate artificial expression both in displacement amount and direction, it is necessary to use at least two cables.
The amount of facial displacement in a healthy subject, did not exceed the displacement range of the actuators, making the actuator based facial mask feasible.

Because the individual artificial muscles used in the Robot Mask are capable of contracting up to 20 mm the feasibility of the Robot Mask has been further established.
The chapter also showed how SMA based soft actuators could be used effectively to pull facial skin and generate facial expressions artificially.

5. Future works

Enhancements to the *Robot Assisted Smile Recovery* project will be carried out along three directions:

1. Improvements to actuators: this will include point to point actuation control and improving actuation rate by employing forced SMA coolers
2. Extending to other areas of the face: will investigate on how the mask can be used to cover the whole face and emphasis will be given to Orbicularis muscles around eye and mouth
3. Aesthetic improvements: investigating on methods to attach wires to the face in a less visible manner

6. References

H. Wang and N. Ahuja. Facial expression decomposition. *Proc. of Ninth IEEE International Conference on Computer Vision (ICCV)*, Volume 2:pp. 958–965, Oct. 2003.
M. Batty and M.J. Taylor. Early processing of the six basic facial emotional expressions. *Cognitive Brain Research*, 17(3):pp. 613–620, Oct. 2003.
G. Mandler, J. A. Russell, and J. M. Fernandez-Dols. The psychology of facial expression (studies in emotion and social interaction. Technical report, Cambridge University Press, 1997.
P. Ekman. Facial expressions of emotion: an old controversy and new findings. *Philosophical Transactions of the Royal Society of London*, 335:pp. 63–69, 1992.
P.J. Byrne. "importance of facial expression in facial nerve rehabilitation". *Current Opinion in Otolaryngology and Head and Neck Surgery*, Volume 12:332–335, 2004.
M. R. Garanhani, J.R. Cardoso, A. de Mello, G. Capelli, and M. C. Ribeiro. "physical therapy in peripheral facial paralysis: retrospective study". *Brazilian Journal of Otorhinolaryngology*, Volume 73(1):106–109, Jan/Feb 2007.
D.L. Beck and J.W. Hall. The hearing journal: Electroneuronography. Volume 54(3), March 2001.
F. B. Quinn and M.D. Jr. "facial nerve paralysis". [Online] Available: `http://www.utmb.edu/otoref/grnds/face961.htm`, March 1996.
Myles L. Pensak. *Controversies in Otolaryngology*. Thieme, first edition edition, 2001.
J. Harrigan, R. Rosenthal, and K. Scherer. The new handbook of methods in nonverbal behavior research. *Oxford University Press*, 2005.
M. Vaiman, N. Shlamkovich, A. Kessler, E. Eviatar, and S. Segal. Biofeedback training of nasal muscles using internal and external surface electromyography of the nose. *American Journal of Otolaryngology*, Volume 26(Issue 5):pp. 302–307, Sep. 2005.
K. Sugimoto1, T. C. Theoharides, D. Kempuraj, and P. Conti. Response of spinal myoclonus to a combination therapy of autogenic training and biofeedback. *BioPsychoSocial Medicine*, 1(18):18, Oct. 2007. [Online] Available: `http://www.bpsmedicine.com/content/1/1/18`.

Mark May and Barry M. Schaitkin. *Facial Paralysis: Rehabilitation Techniques*. Thieme, 2003.

Z. Dingguo and Z. Kuanyi. "neural network control for leg rhythmic movements via functional electrical stimulation". *Proc. of IEEE International Joint Conference on Neural Networks*, Volume 2:1245–1248, 2004.

J.C. Borod, E. Koff, S. Yecker, C. Santschi, and J.M. Schmidt. Facial asymmetry during emotional expression: Gender, valence, and measurement technique. *Neuropsychologia, Elsevier Science Ltd.*, Volume 36(Issue 11):1209–1215, November 1998.

G. Rhodes, M. Peters, K. Lee, M.C. Morrone, and D. Burr. Higher-level mechanisms detect facial symmetry. *Proc. of Biol. Sci.*, 272(1570):1379–1384, July 2005.

R.A. Higgins. *Engineering Metallurgy*, volume Part 1: Applied Physical Metallurgy. Butterworth-Heinemann, 6 edition edition, 1993. pp. 402-403.

ASM Handbook: Properties and Selection: Nonferrous Alloys and Special-Purpose Materials, volume 2. ASM International, 10th edition edition, Nov. 1990.

S. A. Mascaro and H. H. Asada. Wet shape memory alloy actuators for active vasculated robotic flesh. *Proc. of ICRA '03, IEEE International Conference on Robotics and Automation*, Volume 1:pp. 282–287, Sep. 2003.

K. Yang and C.L. Gu. Modelling, simulation and experiments of novel planar bendign embedded sma actuators. *Elsevier*, Mechatronics(18):pp. 323–329, 2008.

T. Yambe, S. Amae, S. Maruyama, Y. Luo, H. Takagi, S. Nanka, A. Tanaka, N. Kamiyama, R. Ohi, K. Tabayashi, H. Takeda, M. Yamada, and S. Nitta. Application of a shape memory alloy for internal artificial organs. *J Artif organs*, 4(2):88–91, June 2001. The Japanese Society for Artificial Organs.

Augmenting Sparse Laser Scans with Virtual Scans to Improve the Performance of Alignment Algorithms

Rolf Lakaemper
Department of Computer and Information Science Temple University, Philadelphia, USA
lakamper@temple.edu

Abstract

We present a system to increase the performance of feature correspondence based alignment algorithms for laser scan data. Alignment approaches for robot mapping, like ICP or FFS, perform successfully only under the condition of sufficient feature overlap between single scans. This condition is often not met, e.g. in sparsely scanned environments or disaster areas for search and rescue robot tasks. Assuming mid level world knowledge (in the presented case the weak presence of noisy, roughly linear or rectangular-like objects) our system augments the sensor data with hypotheses ('Virtual Scans') about ideal models of these objects, based on analysis of a current estimated map of the underlying iterative alignment algorithm. Feedback between the data alignment and the data analysis confirms, modifies, or discards the Virtual Scan data in each iteration. Experiments with a simulated scenario and real world data from a rescue robot scenario show the applicability and advantages of the approach.

1 Introduction

Robot mapping based on laser range scans is a major field of research in robotics in the recent years. The basic task of mapping is to combine spatial data usually gained from laser range devices, called 'scans', to a single data set, the 'global map'. The global map represents the environment scanned from different locations, even possibly scanned by different robots ('multi robot mapping'), usually without knowledge of their pose (= position and heading). One class of approaches to tackle this problem, i.e. to align single scans, is based on feature correspondences between the single scans to find optimal correspondence configurations. Techniques like ICP (Iterative Closest Point, e.g. [2, 24] and [22]) or FFS (Force Field Simulation based alignment, [20]) belong to this class. They show impressive results, but are naturally restricted: first since they are feature correspondence based, they require the presence of a sufficient amount of common, overlapping features in scans belonging together. Second, since the feature correspondence function is based on a state describing the relation of the single scans (e.g. the robots' poses), these algorithms are depending on sufficiently good state initialization to avoid local minima. In this paper, we suggest a solution to the first problem: correct alignment in the absence of sufficient feature correspondences. This problem can e.g. arise in search and rescue environments (these environments typically show a little number of landmarks only) or when multiple robots team to build a joint global map. In this situation, single scans, acquired from different views, do not necessarily reveal the entire structure of the scanned object. The motivation to our approach is that even if the optimal relation between single scans is not known, it is possible to infer hypotheses of underlying structures from the non-optimal combination of single scans based on the assumption of certain real world knowledge. Figure 1 illustrates the basic

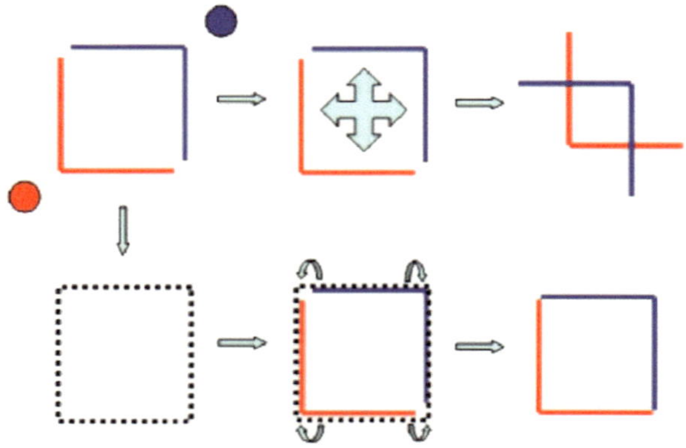

Fig. 1. Motivation of Virtual Scan approach (a-f in reading order): a) rectangular object scanned from two positions (red/blue robots). b) correspondence between single scans (red/blue) does not reveal the scanned structure c) misalignment due to wrong correspondences d) analysis of estimated global map detects structure e) structure is added as Virtual Scan f) correct alignment achieved due to correspondences between real world scans and Virtual Scans

idea. It shows a situation where the relation between features of single scans can not reveal the real world structure, and therefore leads to misalignment. Analysis from a global view estimates the underlying structure. This hypothesis then augments the real world data set, to achieve a correct result.

The motivational example shows the ideal case; it doesn't assume any error in the global map estimation (the relative pose between red and blue scan), hence it is trivial to detect the correct structure. Our system also handles the non ideal situation including pose errors. It utilizes a feedback structure between hypothesis generation and real data alignment response. The feedback iteratively adjusts the hypotheses to the real data (and vice versa). This will be discussed in more detail below. We first want to explain our approach in a more general framework.

Feature correspondence algorithms, e.g. in ICP or FFS, can be seen as low level spatial cognition processes (LLSC), since they operate based on low level geometric information. The feature analysis of the global map, which is suggested in this paper, can be described as mid level spatial cognition process (MLSC), since we aim at analysis of features like lines, rectangles, etc. Augmenting real world data with ideal models of expected data can be seen as an example of integration of LLSC and MLSC processes to improve the performance of spatial recognition tasks in robotics. We are using the area of robot perception for mobile rescue robots, specifically alignment of 2D laser scans, as a showcase to demonstrate the advantages of these processes.

In robot cognition, MLSC processes infer the presence of mid level features from low level data based on regional properties of the data. In our case, we detect the presence of simple mid level objects, i.e. line segments and rectangles. The MLSC processes model world knowledge, or assumptions about the environment. In our setting for search and rescue environments, we assume the presence of (collapsed) walls and other man made structures. If possible wall-like elements or elements somewhat resembling rectangular structures are detected, our system generates the most likely ideal model as a hypothesis, called 'Virtual Scan'. Virtual Scans are generated from the ideal, expected model in the same data format as the raw sensor data, hence Virtual Scans are added to the original scan data indistinguishably for the low level alignment process; the alignment is then performed on the augmented data set.

In robot cognition, LLSC processes usually describe feature extraction based on local properties like spatial proximity, e.g. based on metric inferences on data points, like edges in images or laser reflection points. In our system laser scans (virtual or real) are aligned to a global map using mainly features of local proximity using the LLSC core process of 'Force Field Simulation' (FFS). FFS was recently introduced to robotics [20]. In FFS, each data point can be assigned a weight, or value of certainty. It also does not make a hard, but soft decision about the data correspondences as a basis for the alignment. Both features make FFS a natural choice over its main competitor, ICP [2, 24], for the combination with Virtual Scans. The weight parameter can be utilized to indicate the strength of hypotheses, represented by the weight of virtual data.

FFS is an iterative alignment algorithm. The two levels (LLSC: data alignment by FFS, MLSC: data augmentation) are connected by a feedback structure, which is repeated in each iteration:

- The FFS-low-level-instances pre-process the data. They find correspondences based on low level features. The low level processing builds a current version of the global map, which assists the mid-level feature detection
- The mid level cognition module analyzes the current global map, detects possible mid level objects and models ideal hypothetical sources possibly

  being present in the real world. These can be seen as suggestions, fed back into the low level system by Virtual Scans. The low level system in turn adjusts its processing for re-evaluation by the mid level systems.

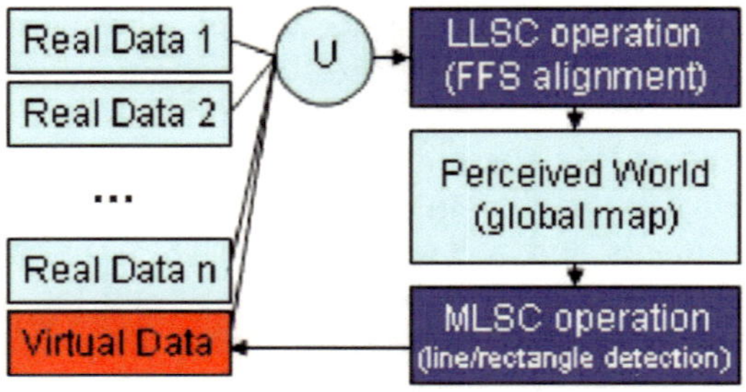

Fig. 2. LLSC/MLSC feedback. The LLSC module works on the union of real scans and the Virtual Scan. The MLSC module in turn re-creates a new Virtual Scan based on the result of the LLSC module.

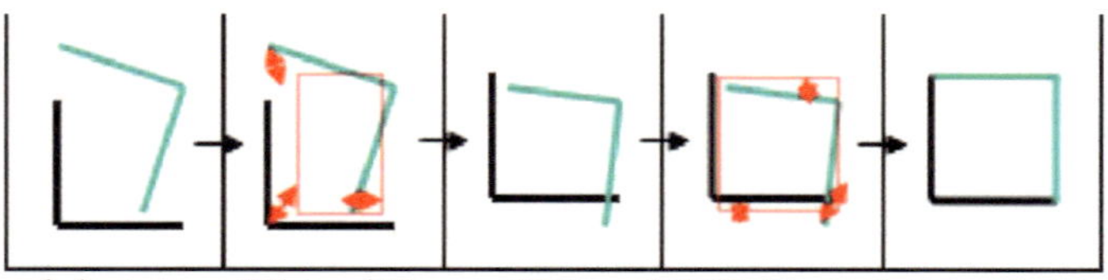

Fig. 3. Feedback between Virtual Scans (VS) and FFS. From left to right: a) Initial state of real data. b) Real data augmented by VS (red). c) After one iteration using real and virtual scans. d) new hypothesis (red) based on (c). e) next iteration. Since this results resembles an ideal rectangle, adding a VS would not relocate the scans. The system converged.

The following example will illustrate the feedback: Figure 3 assumes two scans, e.g. taken from robots in two different positions (compare to fig.1). An MLSC process detects a rectangular structure (the asumed world knowledge) and adds an optimal generating model to the data set. The LLSC module aligns the augmented data. The hypothesis now directs the scans to a better location. In each iteration, the relocated real scans are analyzed to adjust the MLSC hypothesis: LLSC and MLSC assist each other in a feedback loop.

2. Related Work in Spatial Cognition and Robot Mapping

The potential of MLSC has been largely unexplored in robotics, since recent research mainly addressed LLSC systems. They show an astonishing performance: especially advances in statistical inferences [5, 10, 13] in connection with geometric modeling of human perception [6, 9, 25] and the usage of laser range scanners contributed to a breakthrough in robot applications, with the most spectacular results achieved in the 2005 DARPA Grand Challenge where several autonomous vehicles were able to successfully complete the race [26]. But although the work on sophisticated statistical and geometrical models like extended Kalman Filters (EKF),e.g. [12], Particle Filters [10] and ICP (Iterative Closest Point) [2, 24] utilized in mapping approaches show impressive results, their limits are clearly visible, e.g. in the aforementioned rescue scenarios. These systems are still based on low level cognitive features, since they construct metric maps using correspondences between sensor data points. However, having these well-engineered low level systems at hand, it is natural to connect them to MLSC processes to mutually assist each other.

The knowledge in the area of MLSC in humans, in particular in spatial intelligence and learning, is advancing rapidly [7, 14, 27]. Research in AI models such results to generate generic representations of space for mobile robots using both symbolic, e.g. [16], and non symbolic, e.g. [8], approaches. Each is trying to identify various aspects of the cognitive mapping process. Naturally, SLAM (Simultaneous Localization and Mapping [4] is often used as an application example [23]. In [28], a spatial cognition based map is generated based on High Level Objects. Representation of space is mostly based on the notion of a hierarchical representation of space. Kuipers [16] suggests a general framework for a Spatial Semantic Hierarchy (SSH), which organizes spatial knowledge representations into levels according to ontology from sensory to metrical information. SSH is an attempt to understand and conceptualize the cognitive map [15], the way we believe humans understand space. More recently, Yeap and Jefferies [29] trace the theories of early cognitive mapping. They classify representations as being space-based and object-based. Comparing to our framework, these classifications could be described being related to LLSC and High Level Spatial Cognition (HLSC), hence the supposed LLSC/MLSC system would relate closer to space-based systems.

In [1], the importance of 'Mental Imagery' in (Spatial) Cognition is emphasized and basic requirements of modeling are stated. Mental Images invent or recreate experiences to resemble actually perceived events or objects. This is closely related to the "Virtual Scans" described in this proposal. Recently, Chang et al. [3] presented a predictive mapping approach (P-SLAM), which analyzes the environment for repetitive structures on the LLSC level (lines and corners) to generate a "virtual map". This map is either used as a hypothesis in unexplored regions to speed up the mapping process or as an initialization help for the utilized particle filters when a region is first explored. In the second case the approach has principles similar to the presented Virtual Scans. The impressive results of P-SLAM can also be seen as proof of concept of integrating prediction into robot perception.

The problem of geometric robot mapping is based on aligning a set of scans. On the LLSC level the problem of simultaneous aligning of scans has been treated as estimating sets of poses [22]. The underlying framework for such a technique is to optimize a constraint-graph, in which nodes are features, poses and edges are constraints built using various observations and measurements.

There are numerous image registration techniques, the most famous being
Iterative Closest Point (ICP)[2], and its numerous variants to improve speed and converge basins. Basically all these techniques do search in transformation space trying to find the set of pair-wise transformations of scans by optimizing some function defined on transformation space. The techniques vary in defining the optimization functions that range from being error metrics like "sum of least square distances" to quality metrics like "image distance". 'Force Field Simulation' (FFS), [20], minimizes a potential derived from forces between corresponding data points. The Virtual Scan technique presented in this paper will interact with FFS as underlying alignment technique.

3. Scan Alignment using Force Field Simulation

The understanding of FFS is crucial to the understanding of the presented extension of the FFS alignment using Virtual Scans. We will give an overview here. FFS aligns single scans S_i obtained by robots, typically from different positions. We assume the scans to be roughly pre-aligned (see fig.11), e.g. by odometry or shape based pre-alignment. This is in accord with the performance comparison between FFS and ICP described in [19]. FFS alignment, in

detail described in [20] is able to iteratively refine such an alignment based on the scan data only. In FFS, each single scan is seen as a non-deformable entity, a 'rigid body'. In each iteration, a translation and rotation is computed for each single scan simultaneously. This process minimizes a target function, the 'point potential', which is defined on the set of all data points (real and Virtual Scans: FFS cannot distinguish). FFS solves the alignment problem as optimization problem utilizing a gradient descent approach motivated by simulation of dynamics of rigid bodies (the scans) in gravitational fields, but " *replaces laws of physics with constraints derived from human perception*" [20]. The gravitational field is based on a correspondence function between all pairs of data points, the 'force' function. FFS minimizes the overlaying potential function induced by the force and converges towards a local minimum of the potential, representing a locally optimal transformation of scans. The force function is designed in a manner that a low potential corresponds to a visually good appearance of the global map. As scans are moved according to the laws of motion of rigid bodies in a force field, single scans are not deformed.
Fig. 4 shows the basic principle: forces (red arrows) are computed between 4 single scans (the 4 corners). FFS simultaneously transforms all scans until a stable configuration is gained.

$$V(v_1, v_2) = M(v_1, v_2)\frac{v_i - u_j}{\|v_i - u_j\|} \tag{1}$$

Its magnitude $\backslash\backslash M(v_i, Uj)\backslash\backslash = C(v_i, Uj)$ is defined as:

$$M(v_i, u_j) = \frac{1}{\sigma_t\sqrt{2\pi}} e^{\left(-\frac{\|v_i, u_j\|^2}{2\sigma_t^2}\right)} w_i w_j \cos(\angle(v_i, u_j)) \tag{2}$$

With Si, S_2 being two different scans, the force between two single data points v_i G S_1 and uj G S_2 is defined as a vector with parameters a_t, w_i, Wj, $Z(\mathbf{v_i}, \mathbf{u_j}\text{-})$ defined as follows: $Z(\mathbf{v_i}, \mathbf{u_j}\text{-})$ denotes the angle between the directions of points, which is defined as the angle between directions of assumed underlying locally linear structures. See fig. 5, left, for an example, which especially shows the influence of the cosine-term in eq.2: forces are strong between parallel structures only. In eq.2, the forces are strongly depending on a_t, which is a parameter steering the radius of influence. With a_t decreasing during the iterative process, FFS changes the influence of each data point from global to local. In addition, the weight $\mathbf{w_i}, \mathbf{w_j}$- (or mass) determines the influence of points $\mathbf{v_i}$, $\mathbf{Uj}$. The weight is a parameter which can e.g. express the certainty about a point, or it can model the feature importance. We utilize this feature of FFS to model the strength of hypothesis in the Virtual Scans. Hence in eq.2 the interfacing between LLSC and MLSC can be seen directly: distance and cosine term refer to LLSC, while the weights are derived from MLSC (in case of the Virtual Scans).
To compute the resulting movement from the forces of all point pairs between different scans, FFS re-assigns a constant mass to all data points and applies Newton's law of movement of rigid bodies in force fields. Constant mass causes data points participating in stronger force relations to influence the transformation stronger than those responding to weaker forces. For a single transformation step see fig. 5, right.
The step width A_t of the gradient descent step in FFS is determinded by a 'cooling process'. A_t It is monotonically decreasing, allowing the system in early iterations to jump out of local minima, yet to be attracted by local features in later steps. The interplay between a_t and A_t is an important feature of FFS. See figures 11 and 12 for an example of the performance of FFS on a laser range data set.

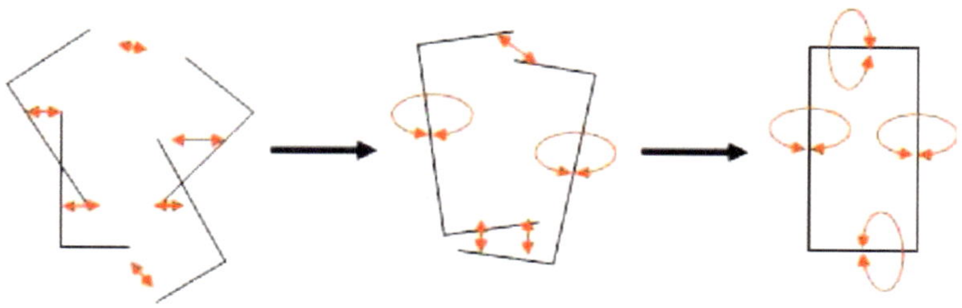

Fig. 4. Basic principle of FFS. Forces are computed between 4 single scans. Red arrows illustrate the principle of forces. The scans are iteratively (here: two iterations) transformed by translation and rotation until a stable configuration is achieved.

FFS is closely related to simultaneous ICP. A performance evaluation of both algorithms [19] showed similar results. In general, FFS can be seen as more robust with respect to global convergence with non near optimal initialization, since the point relations are not built in a hard (nearest neighbor) but soft(sum of forces) way. Also the inclusion of weight parameters makes it a natural decision for our purposes of extension using Virtual Scans.

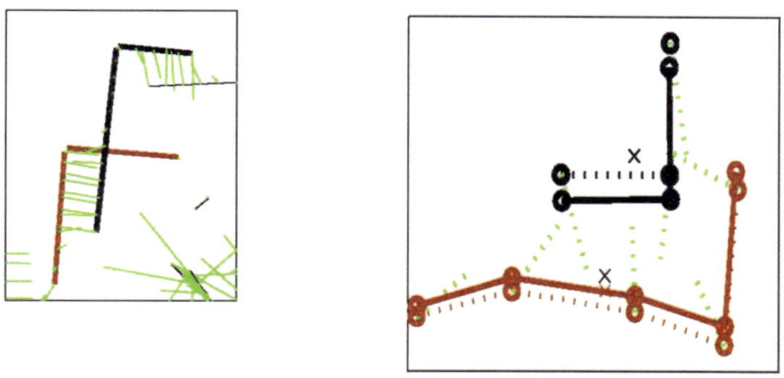

Fig. 5. FFS example. Left: Forces (green) between two rigid structures (brown, black). The black and brown lines connect the actual data set for display reasons only. The figure shows a magnification of the upper left corner of fig. 11, right. Right: example of force and movement. Dotted lines show 2 scans (black, brown) and their forces (green) in iteration t. Solid lines show the resulting transformed scans at iteration t + 1.

4. Creating Virtual Scans: Mid Level Analysis

The point set used in our system is not the original raw data, but a re-sampled version; two pre-precessing steps are performed before the algorithm is applied. First, underlying linear structures (line segments) are detected in each single scan. Since line segments rely on local linearity of the underlying data points, classic global approaches like Hough line detection are not feasible. A recently published technique [21], using a statistical approach, 'Extended Expectation Maximization', is specifically tailored to model laser scan data with line segments. Second, having the line segments, new data points are generated in an equidistant way along these segments. The original data is discarded in favor of the newly generated points. This solves certain problems of FFS with unequally distributed sample point densities. It also reduces the number of points drastically. See fig.6 for an example.

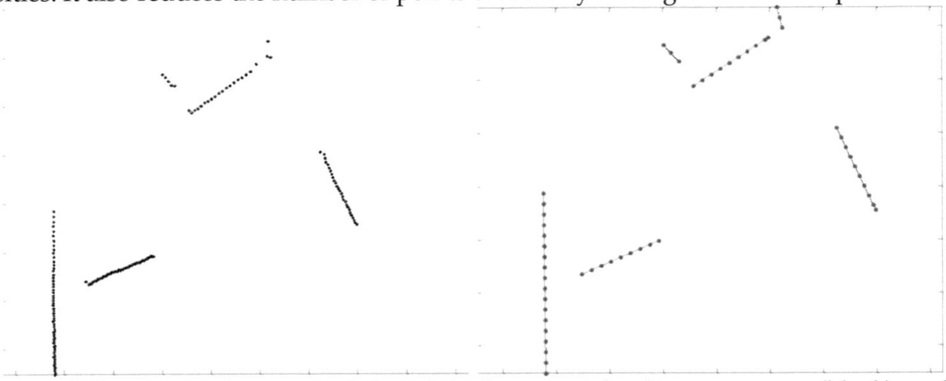

Fig. 6. FFS Pre-processing steps: left: original scan, right: line segments (black) and re-sampled point data (red)

The line and rectangle analysis is performed on the current, non augmented global map, i.e. the Virtual Scan of the previous iteration is discarded.

4.1. Lines

The usage of lines for our Virtual Scan approach is motivated by the world knowledge assumption of scanning a man made environment (e.g. a collapsed house): although these environments often locally don't show major linear elements any longer, a global view still often reveals an underlying global linear scheme, which we try to capture using a global line detection. Here (in contrast to our local line *segment* detection in the pre-processing step) we use the classic line detection approach of Hough transform [11], since it detects *globally* present linear structures. Hough transform does not only show location and direction of a line, but also the number of participating data points. We use this value to compute a certainty-of-presence measure, i.e. the strength of the line hypothesis. We only use lines above a certain threshold of certainty. We will specify below how the detected lines are utilized to create the Virtual Scan. The Hough detection is performed on the entire set of (re-sampled) data points. We do not use the local linear information of the line segment representation of the data here, since we aim at global linearity. This is more robustly detected by Hough transform.

4.2. Rectangles

The rectangle detection operates on the entire set of line segments of all scans, gained in pre-processing step 1. We use a rectangle detection approach described in [17]: each line segment (of each single scan) is translated into 'S,L,D space' (Slope,Length,Distance), which simplifies the detection of appropriate (rectangular like) configurations of four near parallel and near perpendicular segments. For details see [17]. Superimposing all line segments at an early stage of FFS leads to an additional problem, due to still imprecise pose estimation. Single lines in the environment, present in multiple single scans but not aligned perfectly yet, are represented by clusters of many segments, rather than the required single segment. We therefore merge similar lines in a cluster to a single prototype using a line merge approach described in [18], see figure 7. The rectangle detection module then predicts location, dimension and certainty-of-presence of hypothetical, ideal rectangles present in the data set of merged line segments. The certainty, or strength of the hypothesis is derived from properties (segment length, perpendicularity) of participating rectangle-generating line segments. This value is used to create the weight of the rectangle in the Virtual Scan.

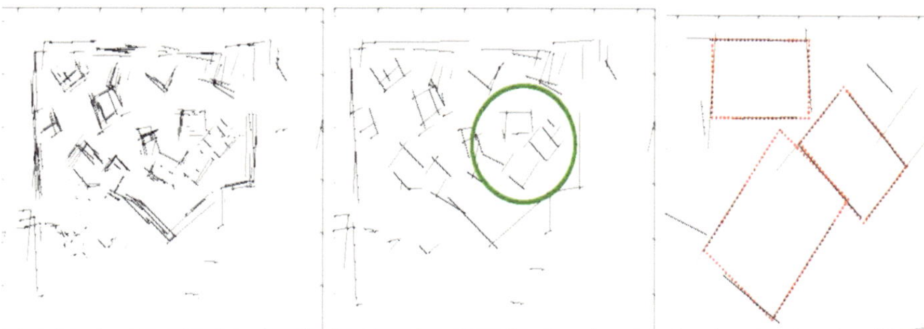

Fig. 7. Rectangle detection. a) Global map built by line segments of all single scans. b) result of global line merging c) red: detected rectangles (magnification of area encircled in b)

4.3. Creating a Virtual Scan

A Virtual Scan is a set of virtual laser scan points, superimposed over the entire area of the global map. The detected line segments and rectangles are 'plotted' into the Virtual Scans, i.e. they are represented by point sets as if they would be detected by a laser scanner. We assume a virtual laser scanner that represents each line and rectangle by a set of points, sub sampled equidistantly according to the point density of the underlying point data in the original data set. All detected elements (lines, rectangles) are plotted into a single Virtual Scan.

An important feature of the Virtual Scan is that each scan point is assigned a weight, representing the strength of hypothesis of the generating virtual structure. Utilizing this feature, we benefit from the weights that steer the FFS alignment. As defined in eq.2, the weight *Wi, Wj* directly influences the alignment process; stronger points, i.e. points with higher value w_i, have a stronger attraction. Hence, a strong hypothesis translates into a locally strongly attractive structure. The hypothesis value reflects the belief into the hypothesis relative to the real data; all data points of the real data are assigned a 'normal' weight of 1.

5. Alignment using Virtual Scans: Algorithm

The algorithm describes the interplay between LLSC (FFS) and the MLSC analysis. S_i, i = 1..n, denotes the real scan data, consisting of n scans. $V^{[t]}$ is the Virtual Scan in iteration t.
Init: t =1, V^{t° = 0, create set of line segments L_i for each scan S_i

1) Perform FFS on $(J_{i=1\ n}\ Si\ U\ V^{[t-1]}\backslash$ resulting in transformations (translation, rotation) $Ti^{[t]}$ for each scan $Si_{=i..n}$

2) Form global map G of points and GL of line segments, superimposing the transformed scans and their line segment representation: G = $(J_n$ T"J$^{t]}$(Si), GL = $(J_n$ T"J$^{t]}$(Li)

3) Detect set of lines L in G, set of rectangles R in GL

4) Create Virtual Scan VM. VM contains scan points representing the elements of L and R.

5) Compute parameters *at* and At for the FFS process

6) Loop: goto 1, or end if FFS converged (stable global map).

Fig. 8. Virtual Scans in an early stage of FFS. a) global map b) the Virtual Scan consisting of points representing detected lines and rectangles c) superimposition of real data and Virtual Scan. This is the data used in the next FFS iteration.

6. Experiments and Results

6.1. Sparse Scanning (Simulated Data)

This experiment shows the effect of Virtual Scans in a sparsely scanned environment. It features a simple environment to highlight the principle of Virtual Scans and to show the improvement in the alignment process. Please compare also to the motivational example in the introduction, as well as to figures 1 and 3. A simulated arena, consisting of 4 rectangular rooms, is scanned from 5 different positions. Each single scan is translated and rotated to simulate pose errors, and pre-processed see fig.9,a,b). We first try to align this data set using FFS without Virtual Scans. The performance of FFS depends on the initial value of at (see eq.2), at=0. at changes the radius of influence of neighboring points. We tried multiple initial

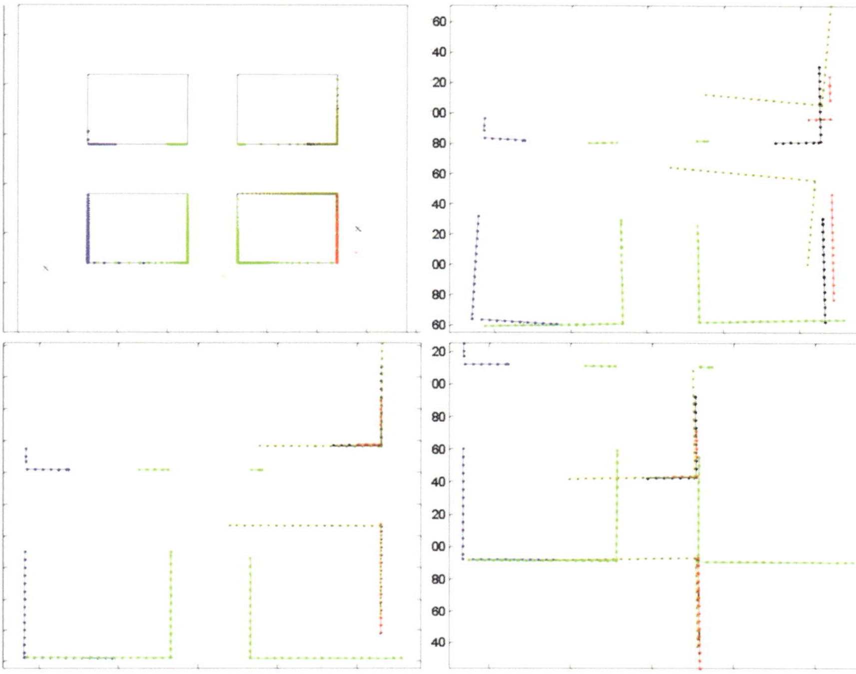

Fig. 9. (in reading order): a)simulated arena, scanned from 5 positions (crosses). Points of same color belong to same scan. b) After adding pose error to the data of (a) and pre-processing: underlying segments of (a) and re-sampled point data. c/d) result of FFS without Virtual Scans, intialized with configuration in (b). c) <7t=o = 30 d) <7t=o = 80

values, results of at=0 = 30 and at=0 = 80 are shown (a in units of the data set: the width of the simulated arena is 400 units), see fig.9c),d). In c), with a low at=0, local structures are captured and aligned correctly, but global correspondences can not be detected (the 'hallway' between the rooms shows an incorrect offset). Increasing a_t=$_0$ and therewith strengthening the influence of global structures in d) leads to wrong results since local correspondences become relatively unimportant: FFS optimizes correspondences of major structures (although they are distant from each other in the initial map). The disability of balancing the influence of local and global structures is is an inherent drawback of alignment processes which are based on point correspondences (e.g. ICP, FFS), and not a special flaw of FFS only (other values of *a* did not improve the alignment).

Fig.10 shows the improvement using Virtual Scans. This experiment uses the same setting as experiment fig.9,c) (a_t=$_0$ = 30)). FFS is able to detect correct local structures, and the global structures are captured through augmentation by Virtual Scans. Also the effect of the hypothesis adjustment by feedback is clearly visible: Fig.10a) shows an early hypothesis, which contains a wrong rectangle and misplaced lines. This early hypothesis is corrected by the feedback process between FFS and the rectangle/line detector. b)shows a later iteration, the line position is adjusted (though not perfect yet), 2 rectangle hypotheses compete (lower right corner). The final result is shown in c) and d). The detected lines adjusted expected global structures (the walls of the 'hallway') correctly, the winning rectangle hypothesis 'glued together' the corners of the bottom right room. Please notice that this room is a structure that is not entirely present in any single scan, but only detectable in the global map. Hence only the Virtual Scan enhanced FFS could perform correctly.

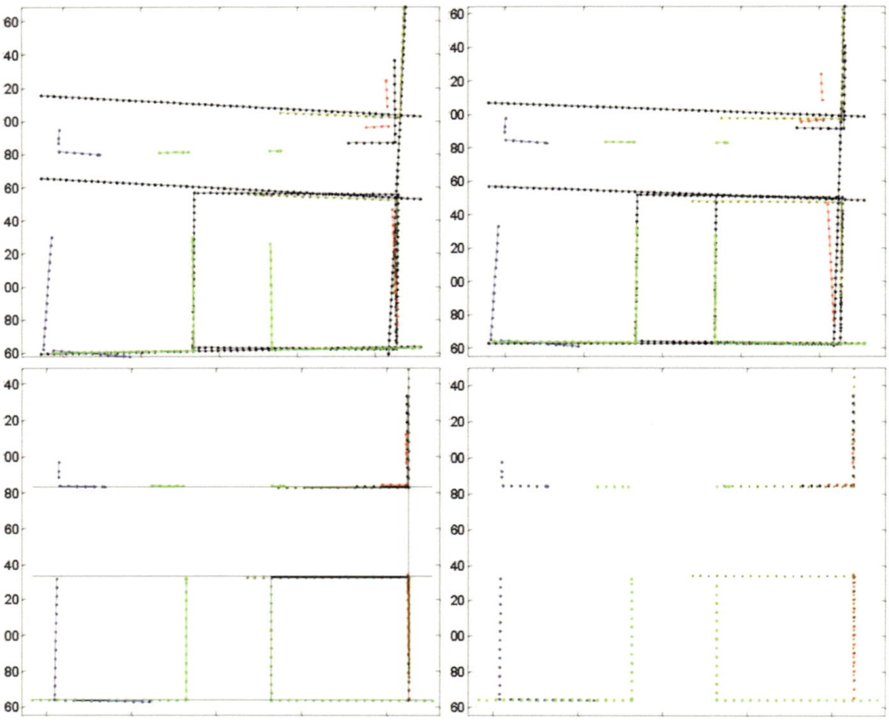

Fig. 10. Same experiment as in fig. 9, but using FFS (<7t=o = 30) and Virtual Scans. Elements of Virtual Scans (lines and rectangles) are shown in black. In reading order: a) iteration 10, b) iteration 20 c) iteration 30 (final result) d) same as (c), but virtual scan not shown.

6.2. NIST Disaster Area

This experiment shows the improved performance of the alignment process on a real world data set. The data set consists of 60 single laser scans, taken from 15 different positions in 4 directions (N,W,S,E) with 20° overlap. It can be interpreted as a scene scanned by 15 robots, 4 scans each. No order of scans is given. The scans resemble the situation of an indoors disaster scenery, scanned by multiple robots. The scans have little overlap and no distinct landmarks. The initial global map was computed using a shape based approach described in [19]. See fig. 11 for example scans and the initial map. We used the initial global map for two different runs of FFS, one using Virtual Scans, one without Virtual Scans. The increase in performance was evaluated by visual inspection, since for this data set no ground truth data is available. Comparing the final global maps of both runs, the utilization of Virtual Scans leads to distinct improvement in overall appearance and mapping details, see fig.12.

Overall, the map is more 'straight' (compare e.g. the top wall), since the detection of globally present linear structures (top and left wall in fig.12) adjusts all participating single segments

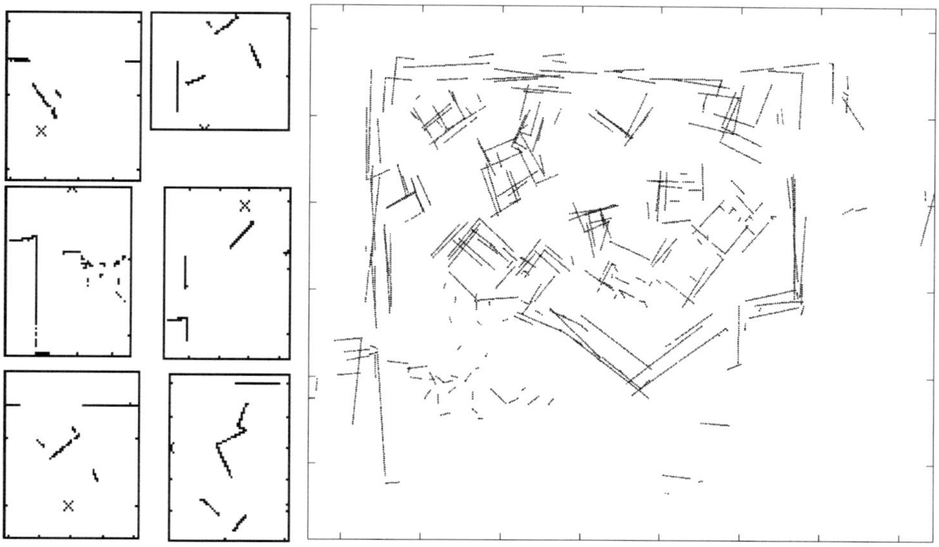

Fig. 11. The NIST disaster area data set. Left: 6 example scans (from a total of 60). Crosses show each robot's position. Right: 60 scans superimposed using a rough pre-estimation. This is the initial global map for the experiment in fig.12

to be collinear. These corrections advance into the entire structure.
The improvements can even better be seen in certain details, the most distinct ones encircled in fig.12, d),f). Especially the rectangle in the center of the global map is an excellent example for a situation where correct alignment is not achievable with low level knowledge only. Only the suggested rectangle from the Virtual Scan (see fig.12,c), can force the low level process to transform the scan correctly. Without the assumed rectangle the low level optimization process necessarily tried to superimpose 2 parallel sides of the rectangle to falsely appear as one (magnification of both situations in 12e).

7. Conclusion and Outlook

The presented implementation of an extension to the FFS alignment process using Virtual Scans containing hypothetical mid level real world structures could significantly improve the results for the alignment task. The implementation proves the applicability of the presented concept for the combination of LLSC and MLSC processes. The detection of simple elements (lines, rectangles) based on weak real world assumptions could improve the performance. We are aware that adding domain knowledge certainly enhances the risk of wrong inferences. The proposed system handles errors caused by premature belief in mid level features by implementing the feedback principle, which evaluates a single hypothesis. It is known that single hypothesis systems introducing higher knowledge tend to be not

robust. Under certain circumstances this behavior could also be observed in experiments with our system, which needed manual parameter adjustment to steer the influence of the hypotheses. It can be embedded into a multiple hypotheses framework, e.g. particle filters, which will be part of future work.
Additional future work also has to determine the (geometric) level of elements which are meaningful enough to improve the alignment process, yet not too dominant. The current elements were chosen to model assumptions of indoor disaster areas. Future research about assumed real world elements will adjust to outdoor disaster settings.

8. References

[1] S. Bertel. Thomas barkowsky, dominik engel, christian freksa. computational modeling of reasoning with mental images: basic requirements. D. Fum, F. del Missier, A. Stocco (Eds.), Proceedings of the 7th International Conference on Cognitive Modeling ICCM06, 2006.

[2] P. Besl. N. mckay. a method for registration of 3.d shapes. IEEE PAMI, 14(2), 1992. [3] H. J. Chang. C.s.george lee and y.h. lu and y.c.hu. p-slam: Simultaneous localization and mapping with environmental structure prediction. IEEE Transactions on Robotics, Vol.23, No.2, April 2007.

[4] G. Dissanayake. H. durrant-whyte and t. bailey. a computationally efficient solution to the simultaneous localization and map building (slam) problem. ICRA2000 Workshop on Mobile Robot Navigation and Mapping, 2000.

[5] A. Doucet. N. defreitas, n. gordon. sequential monte carlo methods in practice. Springer, 2000.

[6] D. J. Field. A. hayes, r.f. hess. contour integration by the human visual system: evidence for a local association field. Vision Research, 33, pages 173—193, 1993.

[7] C. Freksa. M. knauff, b. krieg-brueckner, b. nebel, t. barkowsky. spatial cognition iv: Reasoning, action, interaction. Springer, 2004.

[8] T. Ghiselli-Crippa. Hirtle s.c., munro, p. connectionist models in spatial cognition. The Construction of Cognitive Maps, Kluwer Academic Publishers. 1-51, pages 87—104, 1996.

[9] W. Grimson. Object recognition by computer: The role of geometric constraints. Boston, MA: MIT Press, 1991.

[10] G. Grisetti. C. stachniss, burgard w. improving grid-based slam with rao-blackwellized particle filters by adaptive proposals and selective resampling. ICRA, 2005.

[11] P. V. C. Hough. Methods and means for recognizing complex patterns. US patent 3,069,654, 1962.

[12] S. Huang. G. dissanayake. convergence analysis for extended kalman filter based slam. IEEE International Conference on Robotics and Automation, 2006.

[13] D. C. Knill. W. richards. perception as bayesian inference. New York: Cambridge, 1996.

[14] G. J. M. Kruijff. H. zender, p. jensfelt, h. i. christensen. situated dialogue and spatial organization: What, where and why? Int. J. of Advanced Robotic Systems, 2007.

[15] B. J. Kuipers. The cognitive map: 'could it have been any other way?'. Spatial Orientation: Theory, Research, and Application. New York: Plenum Press, pages 345—359, 1983.

[16] B. J. Kuipers. The spatial semantic hierarchy. Artificial Intelligence 119, pages 191 – 233, 2000.

[17] D. Lagunovsky. Ablameyko s. fast line and rectangle detection by clustering and grouping. Proc. of CAIP'97, Kiel, Germany, 1997.

[18] R. Lakaemper. L. j. latecki, d. wolter. geometric robot mapping. Int. Conf. on Discrete Geometry for Computer Imagery (DGCI), 2005.

[19] R. Lakaemper. A.nuechter, n.adluru. performance of 6d lum and ffs slam. Workshop on Performance Metrics and Intelligent Systems (PerMIS), Gaithersburg, MD, August 2007.

[20] R. Lakaemper. N.adluru, l.j.latecki, r.madhavan. multi robot mapping using force field simulation. Journal of Field Robotics, Special Issue on Quantitative Performance Evaluation of Robotic and Intelligent Systems, 2007.

[21] L. J. Latecki. R. lakaemper. polygonal approximation of laser range data based on perceptual grouping and em. IEEE Int. Conf. on Robotics and Automation (ICRA), Orlando, Florida, May 2006.

[22] F. Lu. Milios, e. globally consistent range scan alignment for environment mapping. Auton. Robots, 4(4), pages 333-349, 1997.

[23] A. Martinelli. A. tapus, k.o. arras, r. siegwart. multi-resolution slam for real world navigation. Proceedings of the 11th International Symposium of Robotics Research, Siena, Italy, 2003.

[24] A. Nuechter. K. lingemann, j. hertzberg, h. surmann, k. pervoelz, m. hennig, k. r. tiruchinapalli, r. worst, t. christaller. mapping of rescue environments with kurt3d. Proceedings of the International Workshop on Safety, Security and Rescue Robotics (SSRR '05),Kobe, Japan, June 2005.

[25] A. Pentland. Perceptual organization and the representation of natural form. AI Journal, Vol. 28, No. 2, pages 1-38, 1986.

[26] e. a. Thrun. Stanley: The robot that won the darpa grand challenge. Journal of Field Robotics 23(9), pages 661-692, 2006.

[27] D. Uttal. Seeing the big picture: Map use and the development of spatial cognition. Developmental Science, 3, pages 247 – 286, 2000.

[28] S. Vasudevan. V. nguyen, r. siegwart. towards a cognitive probabilistic representation of space for mobile robots. Proceedings of the IEEE International Conference on Information Acquisition (ICIA), Shandong, China, pages 20 – 23, August 2006.

[29] W. K. Yeap. M.e. jefferies. on early cognitive mapping. Spatial Cognition and Computation 2(2), pages 85-116, 2001.

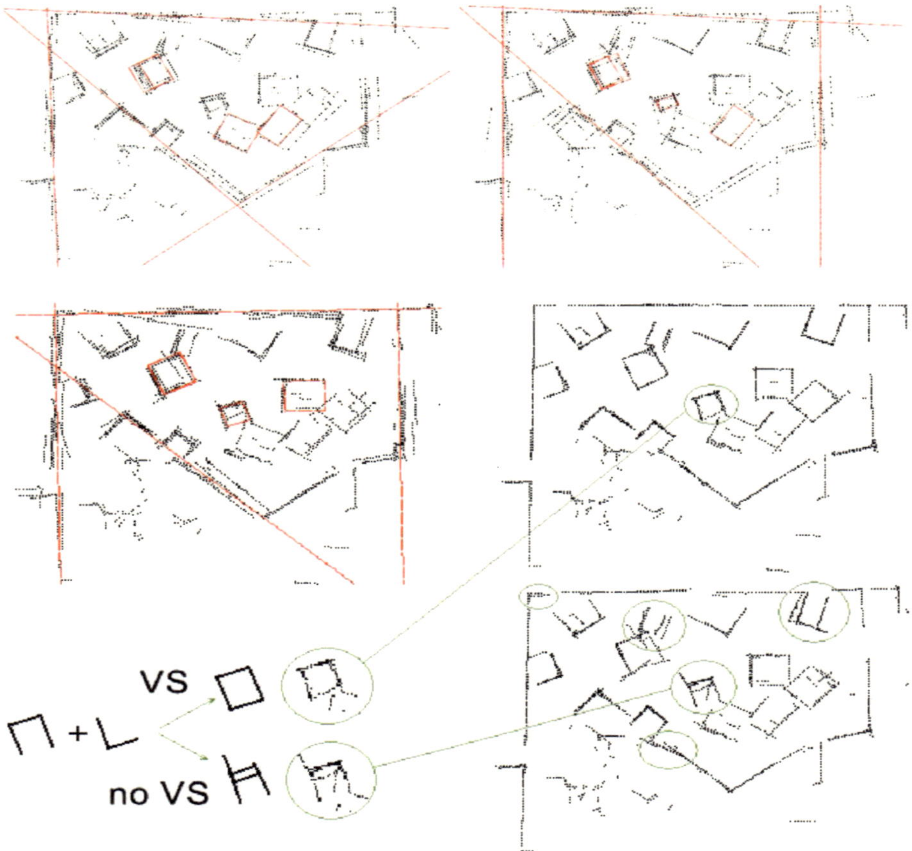

Fig. 12. Alignment of NIST data set (initial alignment see fig. 11). (a-f in reading order) a) iteration 4. b) iteration 8. c) iteration 10. In (a-c) red lines show the data of the Virtual Scan (VS). Please notice the mutual adjustment of hypotheses and real data. d) final result using Virtual Scans, after 100 iterations. The VS is not shown for clarity of display. Compare to f): final result of alignment without VS. Encircled areas show examples of improvement between (d) and (f). e) The center rectangle could only be aligned correctly using VS information. Please compare (e) to motivational examples fig. 1 and 3.

23

Sensor network for structuring people and environmental information

S. Nishio, N. Hagita, T. Miyashita, T. Kanda, N. Mitsunaga and M. Shiomi
ATR Intelligent Robotics and Communication Laboratories
Japan

T. Yamazaki
National Institute of Information and Communications Technology
Japan

1. Introduction

Recently numbers of studies have been performed to make robots behave and interact with people more friendly and effectively. For example, nonverbal human-like behaviors such as eye-contact or maintaining suitable distance are known to be important in forming social relationships with people (Yoshikawa et al., 2006; Huettenrauch et al., 2006; Walters et al., 2005). To provide services effectively and behave humanlike, robots needs to obtain detailed information on where people exist or what they have been doing. Moreover, robots can provide services to people in two different ways: people find and approaches robots, or robots approach person that needs assistance. When we look for services in the current world, shops, information kiosks or wall maps in public spaces are typical examples of the first type. In order to receive the service, people need to find out where the service is provided and have to approach the place. On the contrary, there had been few examples for the second form. Lifeguards at the swimming pool may be a rare example. When someone drawn is found, s/he dives in to help the person. Although the latter form is much more convenient, why can they be rarely seen in our everyday life? This is because services of this form need monitoring and recognition of people requirements which is still a difficult task. However, with the progress in ubiquitous sensing systems, this convenient form of service is becoming real.

Several research projects have been running under this intent. One typical example is the *aware home* (Kidd et al., 1999)0, where sensors are installed in the residence and making the home a smart, intelligent space for supporting life. However, due to the high cost required in building such system on individual houses, it is hard to imagine every home having these facilities. Instead, public space such as shopping malls, stations or airports are more likely to be equipped with these sensing systems. However, public spaces show typical natures that cannot be seen in individual homes such as: 1. they are typically crowded with people, 2. the area is wide, and 3. most of the places are designed for certain purpose. The first two shows that sensing is much more challenging in these places. But by the last feature, there is a possibility that robots may perform better utilizing the background knowledge specific to each environment.

Several robotic projects in public spaces have been performed, that provides guidance or shop recommendations. For example, in the *Robotics* project (Jensen et al., 2005), numbers of mobile robots were placed in the national expo for 5 months. Here the robots cooperated with each other, in order to locate themselves or find persons that need help, using the predefined knowledge on several areas in the expo pavilion. However, as you can experience by walking in a crowded station, it is not so easy to recognize what's happening in a place not nearby. With the help from the sensors equipped in the environment, this can be made easier. Also, when considering robotic services in real, knowledge on spatial area shall be better acquired mostly automatically, in order to follow the changes in the environment.

Also from the aspect of developing robotic services, raw position information such as coordinates may not be sufficient. For example, when making a navigation services that find a person in lost, what is required most is the status or behavior of the person, not the coordinate s/he resides. This is also useful for providing services appropriate in each context or situation the target person is in. The same is also true for spatial knowledge as described previously. The abstract, symbolic information is better suited for easily describing scenarios for human-robot interaction or robotic services (Fig. 1). Several research efforts have begun to obtain this abstract information from sensing results (Kanda, et al, 2007; Gerkey et al., 2003).

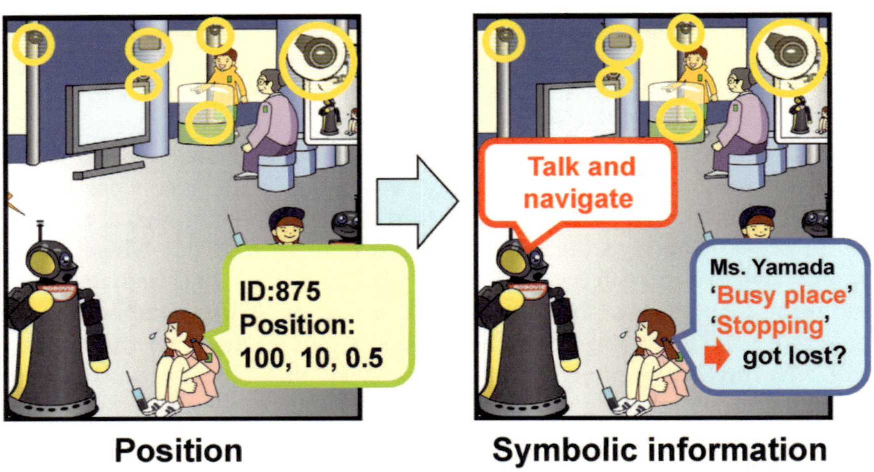

Fig. 1. Human-robot interaction with and without symbolic information.

Another issue related to the development of robotic services lies on their high cost in preparing a robotic system. Currently, developers need to prepare mostly every element by themselves that are required in composing a robot, such as localization, human identification or speech recognition. What is required is a common framework or platform system that can be used for building robots easily. Several projects have been conducted to prepare such common frameworks (Kranz et al., 2007; Thompson et al., 2007), but there seems to be nothing standard yet.

Our intention is to prepare an infrastructural system that senses human behavior and spatial nature in the environment. This system will provide robots information obtained from the environment in a standard manner. In past, factories have been designed so that industrial robots can perform effectively, albeit of their limited sensing ability. However, it is not likely that public spaces are going to be built in the same way suitable for robots. Thus, instead of structuring the environment as in factories, by utilizing the advanced sensing and pattern recognition techniques, we intent to make the information extracted from the environment to be structured and to be easily processed in robotic systems. Our server system is based on four-layer architecture, the *Structuring model of Environmental Information* (Figure 2). Measurements from sensors are first processed in the *sensor layer* and are then integrated to produce uniform positioning results in the *segment layer*. At the same time, positions are accumulated and processed to produce symbolic information on human behavior and space in the *primitive layer*. Both the coordinates and symbolic information produced are provided to robots, enabling robotic services to be designed and run based on both numerical and symbolic descriptions. In past, there have been several attempts to combine robotic services with environmental sensing systems (Thompson et al., 2007). There were also studies on sensing systems that produce abstract symbolic information (Jensen et al., 2003). But as far as we know, no such integrated system exists which can be used in the real world and which provides standard access methods for robotic services in general.

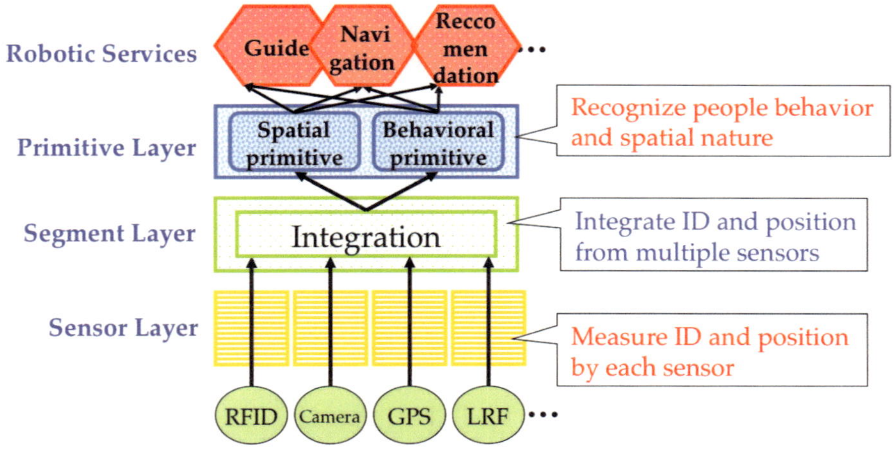

Fig. 2. Four-layer architecture of structured environmental information

In the following, we will first describe the first three layers in our server platform: robust sensing methods for human positioning in real environments, integration of positioning results by individual sensors and methods for recognizing human behaviors and spatial nature. Based on the proposed architecture and these technologies, we have built two trial fields in public spaces and have been providing these to researchers for testing purpose. In the last section, we describe these two trial fields and briefly show some robotic service experiments held using the proposed server features.

2. The Sensor Layer: Measuring Human Position

Human positioning is one major field of research. Various methods and sensing devices have been actively developed (Hightower & Borriello, 2001). One typical and frequently used device is the Global Positioning System (GPS). Many cellular phones nowadays are equipped with GPS receivers, and are widely used for location-based services such as navigation or restaurant recommendation. Although these systems are sufficient for rough navigation, the accuracy of positioning is still not enough for their use in crowded public spaces. Moreover, for robotic use, much accuracy is required for approaching a person or for interacting movements such as making gestures or keeping eye contacts.
Finding people by on-board sensors installed on robots are another field that is actively studied (Gockley et al., 2007; Fransen et al., 2007). However, this is effective only after the robots have approached close enough to the person in target. Thus, the issue still remains on how to find the person who needs help in the public environment. On the contrary, by installing and utilizing sensors in the environment, we can not only obtain a more accurate observation, but also be able to degrade the cost required in robot development. By accessing the positioning results from the environment, developers can get rid of positioning part and can concentrate on service development.
Environmental sensing has been long studied in the field of computer vision or pattern recognition. Unfortunately, most of these methods do not perform robustly in the real environment, where people behavior or conditions such as lightening is not under control. In order to implement a robust real-time human positioning, we decided to obtain and integrate results from four sensors of different modalities.

2.1 LRF tracking of human position

Laser range finder (LRF) is one of the most commonly used sensors in the field of robotics. It can quickly and precisely scan the environment. However, in a public environment where many obstacles or person exists, the sensing is often obscured by occlusions. Our solution was to perform tracking using a network of multiple LRFs. Each LRF was mounted at average human torso height, approximately 85 cm high from the ground. A particle filter is used to track the position and velocity of each people, by using a model of a torso and two arms (Dylan et al., 2007)0. Fig. 3 shows the raw sensor data during the walking motion and how the model is fitted, and the results of tracking in real field.

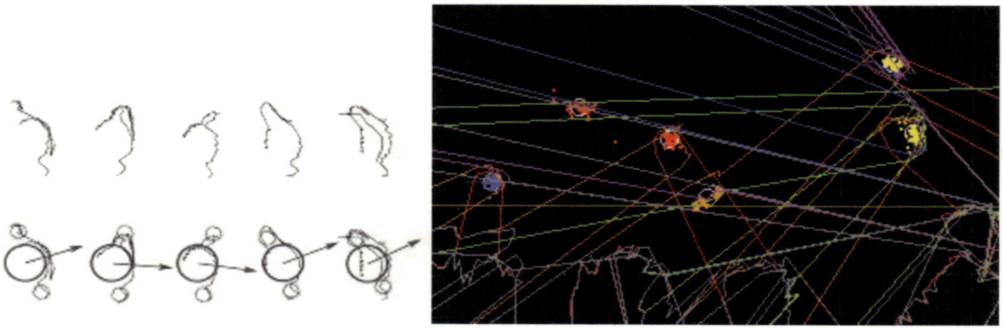

Fig. 3. Estimating pedestrian arm and torso movement from LRF output.

2.2 Multi-camera and LRF based human tracking

One of our goals for human positioning was to obtain the position of the human head. The head information is required for robots to effectively communicate with people, such as having conversation or maintaining eye contacts. This is done by integrating a head tracking system using multi-viewpoint images captured by a set of distributed cameras and the LRF based tracking system described above. Based on the 2D torso position estimation by the LRF tracker, head position is estimated by applying a human head model to edge images obtained (Matsumoto et al., 2008). As described previously, our purpose is to settle up this positioning system in a semi-outdoor public space, where there exists numbers of obstacles and the lightening condition frequently changes. This is a crucial condition for applying image processing techniques. Our solution was to combine both vision-based method and non-vision sensing method, in order to obtain detailed positioning estimation of people in the real environment. Fig. 4 shows an example of the tracking results.

Fig. 4. Tracking results by the combination of multi-camera and LRF. The grey circles indicate head tracking result and the yellow dot clouds show hypotheses. The green dots on the bodies indicate base LRF tracking results.

2.3 Human tracking by RFID tag and GPS

Radio-frequency identification (RFID) tags and Global Positioning System (GPS) receivers carried by people are also used for obtaining their IDs and positions. As for the RFID tags, position estimation are done by a particle filter using the measured signal strength of active RFID tags (Shiomi et al., 2005). Although the positioning result obtained by RFID tags and GPS receivers are low both in accuracy and precision, these estimations are useful for associating identities (ID) to precise positioning results obtained from LRF or camera trackers. Moreover, GPS receivers can be used to track people in the wide outdoor environments where other sensors are not suitable. As people often move between indoor and outdoor in public spaces, continuous and seamless positioning in both fields are useful for keeping track of each person, especially for capturing their intention.

3. The Segment Layer: Integration of Position Estimations

In the second layer, the segment layer, estimation results from individual trackers in the first layer are integrated to produce a final estimation of each person in the environment. That is, the 2D / 3D positions and IDs are integrated to compose a final 3D position of a human head accompanied with ID information (
Fig. 5). We currently use a simple nearest-neighbor method to integrate the resulting position and ID measurements, making concerns for error distribution of each of the initial estimations. Table 1 shows the positioning accuracy in the two trial fields and Fig. 6 shows the result of the integrated position estimation. We have found that our system is able to continuously track more than 20 people in real time. This was tested in real environments where various conditions such as number of people or illumination changes frequently.

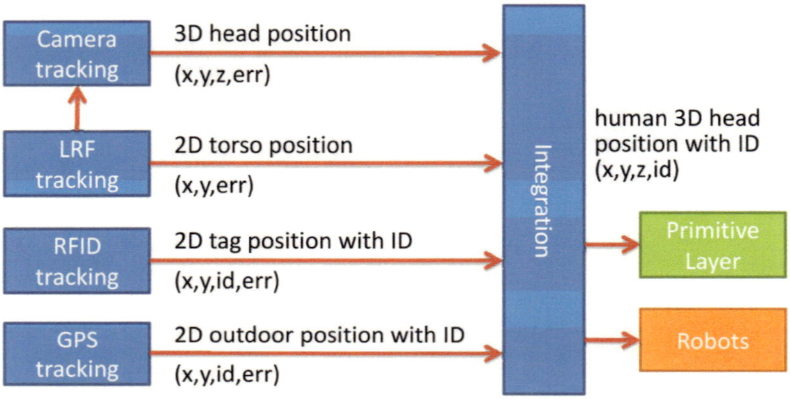

Fig. 5. Integration of individual tracking results in the segment layer. By combining estimation results from the four trackers, the segment layer outputs human head 3D positions with IDs.

	Keihanna environment		UCW environment	
Sensor type	sensor #	RMS	sensor #	RMS
LRF (x,y)	6	49 mm	6	58 mm
Camera (x,y,z)	16	13 cm	16	12cm
RFID (x,y)	17	287 cm	9	249 cm
GPS (x,y)	-	10 m	-	10-100 m

Table 1. Accuracy of position estimation in the two trial fields.

Fig. 6. Example of the integrated position estimation. The red dots show the head estimation results and the green dots show the torso estimation where head was not detected or when person was out of camera tracking area.

4. The Primitive Layer: Recognition of Human Behaviors and Spatial Natures

In order to efficiently design robotic services suitable for each environment, developers often need symbolic descriptions on how each person is behaving and what kind of attribute each area in the environment holds. Also, by knowing the typical behavioral patterns of people in the environment, robots will be able to provide context-aware services suitable for each person. These recognitions can be done by accumulating and learning people trajectories in the environment.

By obtaining symbolic information based on each environment, developers can easily adapt their own robotic services to various places. We call these types of symbolic information *primitives*. Using the measured human position history, the platform server calculates and provides the primitives to the robots. In the following, the three types of primitives are described: *local behavioral primitive*, *global behavioral primitive* and *spatial primitive*.

4.1 Local behavioral primitive

One sort of information useful in designing robotic services is the short-time trend in people movement, such as walking or running. This kind of symbolic information is called *local behavioral primitive*. Recognition of local behavioral primitives is performed by a simple classifier based on support vector machine (SVM) using the obtained people trajectories. Features such as the velocity or degree of curvature of the trajectory are used for training the classifier.

In the current implementation, two types of local behavioral primitives are computed, each with its own classifier. The first is for the *style* of walking (straight, turning right/left, wandering, u-turn). The durations of trajectories used were 5.1 seconds. This means that observation of at least 5.1 seconds is required to decide the style of walking. We tested the classification accuracy by leave-one cross validation test which resulted in 82.5 % accuracy. Another is about the *type* of walking (running, busy/idle walking, stopping, waiting). We trained the classifier with 166 labeled data. The durations of trajectories used are 4.9 seconds. Leave-one cross validation test resulted in 87.8 % accuracy. Fig. 7 shows recognition results in a real environment.

Fig. 7. Primitive recognition results. The box above a person indicates the detected local behavioral primitive (style of walking) and the one below shows the spatial primitive.

4.2 Global behavioral primitive

Global behavioral primitive shows what kind of typical visiting pattern the person is following. The difference in calculating between global and local behavioral primitives mainly exists on the length of the trajectory required. However, as the length of the trajectory increases, we can assume that the pattern holds more abstract, semantic information on the intent of each person.

In an experiment was held in a science museum, where visitor trajectories were collected through RFID tag tracking (Kanda et al, 2007). By analyzing more than 8,000 moving patterns acquired, 5 typical visiting patterns were extracted. As we can think that these patterns, *i.e.* global behavioral primitives, indicate topic of interest for each visitor, robots can easily provide services or information adapted to each person.

4.3 Spatial primitive

A *spatial primitive* indicates the nature of an area. How people have been typically behaving in the area, or what kinds of events have frequently occurred there in the past. This can be manually specified by system designers. Fig. 7 shows an example of manually specified spatial primitive. Here, the 'front_shop' indicator shows that the person in target is in the front of a shop. Otherwise, spatial primitives can be calculated based on visitor trajectory histories as global behavioral primitive (Kanda et al, 2007).

5. The Kansai Environment: Trial Field for Robotic Services

Using these technologies, we have built two trial fields for robotic services: the *Kansai Environment*. One is located at the entrance of an office building (NICT Keihanna building) and another in a shopping mall (Universal City Walk Osaka, UCW for short), both in the Kansai district of Japan. These two locations were chosen due their difference in nature of the place and in the people behavior. The first location is mainly aimed for business or exhibition purpose and the second is for shopping. As the underlying systems in both fields are common, here we will mainly describe about the UCW field. The measured positioning accuracy for both fields is shown in Table 1. The only difference was, due to high buildings in the surroundings, we were not able to use GPS in the UCW field.
The shopping mall where the UCW field is settled is located between a train station and an amusement park (Universal Studio Japan). Every day thousands of people walk through the mall for shopping, dining or just for the sake of passing for their way to the park. Fig. 8 shows sensor alignments in UCW trial field.

Fig. 8. Sensor alignments in the Kansai Environment (UCW).

These trial fields were intended to be a common environment for testing various robotic service implementations, assisted by the environmental server functionalities. Thus, a common, standard interface was required for the server system. For this purpose, we have used the Robotic Localization Service standard (Nishio, et al., 2009). The access interface, including the outputs from the segment and primitive layer, was implemented to follow this standard. Also, a repository server was prepared to exchange the various metadata definition such as 3D coordinate reference systems or primitive symbolic coordinate systems used for the output.

5.1 Basic robotic service in the Kansai Environment

In order to see the effectiveness of the platform server system, we have conducted a robotic service experiment at the UCW field. A humanoid robot (Robovie-IIF) was used. This robot is equipped with several sensors, such as a LRF, an omni-directional camera and ultrasound distance sensors. However, none of them were used in this experiment. Instead, the robot completely relied on the information provided from the platform server and never used its own sensing system for interacting with people.

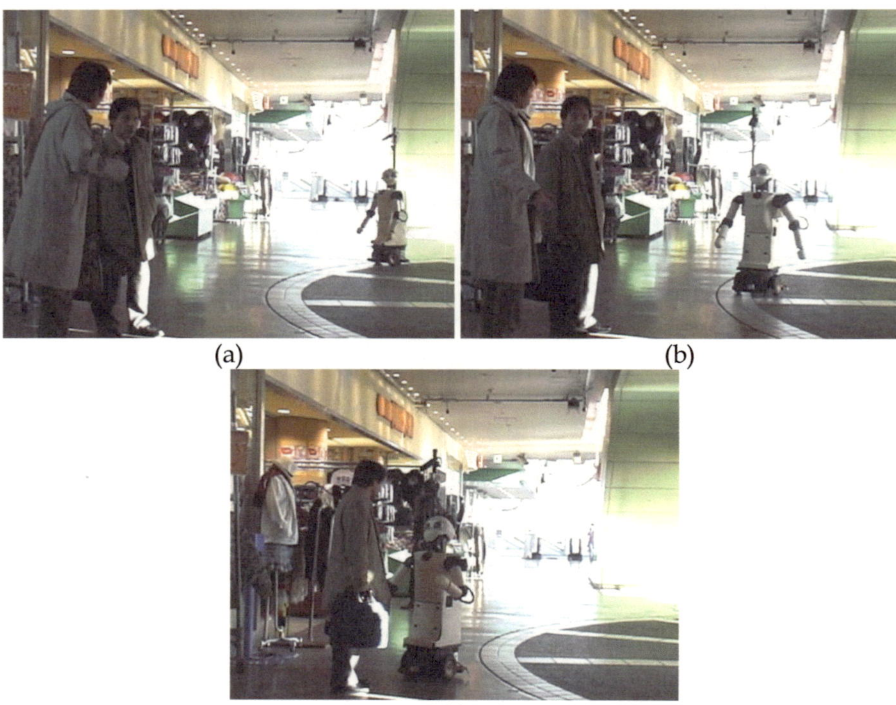

(a) (b) (c)

Fig. 9. Scenes from the robotic service experiment using the Kansai Environment (UCW). (a) Server system detects a person standing in front of a shop. (b) Robot starts approaching the people based on the positioning results from the server. (c) The robot starts shop guidance service, while keeping eye contact with the target person. The robot's eye contact motion is also performed based on the head position received from the server.

Fig. 9 shows a sample sequence from the experiment. When the platform server detects a person with a local behavioral primitive 'stopping' and with a spatial primitive 'shop front' or 'map', Robovie starts approaching the person using the estimated position from the platform server. After the robot reaches nearby the person, it begins providing route

guidance or shop information, depending on the spatial primitive it has received. While providing information, Robovie also makes gestures and keeps eye contact with the person, using the head position estimation given from the environmental server.
Although the robotic service designed is quite a simple one, the fact that it worked out quite stably and effectively seems to indicate the validity of our idea in decreasing the robot development cost. At the same time, this also shows the possibility of developing robotic applications independent to the sensing ability of each robot.

5.2 Using the Kansai Environment

From June 2008 to March 2009, we made the Kansai Environment available to public. Interface specification was released on the web, and robot researchers were invited to use the environment for trying the environmental server functionality with their robots free of charge. Since its release, eight research organizations or projects used the field.

6. Acknowledgement

The authors would like to thank Sumisho Urban Kaihatsu Co. Ltd. and Robot Laboratory for their corporation in constructing the Universal City Walk Osaka trial field. This work was supported in part by the Special Coordination Funds for Promoting Science and Technology of the Ministry of Education, Culture, Sports, Science and Technology, the Japanese Government.

7. References

Fransen, B.; Morariu, V.; Martinson, E.; Blisard, S.; Marge, M.; Thomas, S.; Schultz, A. & Perzanowski, D. (2007). Using Vision, Acoustics, and Natural Language for Disambiguation, *Proceedings of ACM/IEEE Int. Conf. Human-robot interaction,* pp. 17-24

Glas, D.; Miyashita, T.; Ishiguro, H. & Hagita, N. (2007). Laser Tracking of Human Body Motion Using Adaptive Shape Modeling, *Proceedings of IEEE/RSJ Int. Conf. Intelligent Robots and System,* pp. 602-608

Gockley, R.; Forlizzi, J. & Simmons, R. (2007). Natural Person Following Behavior for Social Robots, *Proceedings of ACM/IEEE Int. Conf. Human-robot interaction,* pp. 17-24, 2007.

Hightower J. & Borriello, G. (2001). Location Systems for Ubiquitous Computing, *IEEE Computer*, vol. 34, no. 8, pp. 57–66

Huettenrauch, H.; Eklundh, K.; Green, A. & Topp, E. (2006). Investigating Spatial Relationships in Human-Robot Interaction, *Proceedings of IEEE/RSJ Int. Conf. Intelligent Robots and Systems*, pp. 5052 - 5059

Gerkey, B.; Vaughan, R. T. & Howard, A. (2003). The Player/Stage Project: Tools for Multi-Robot and Distributed Sensor Systems, *Proceedings of Int. Conf. Advanced Robotics*, pp. 317-323

Jensen, B.; Tomatis, N.; Mayor, L.; Drygajlo, A. & Siegwart, R. (2005). Robots Meet Humans - Interaction in Public Spaces, *IEEE Trans. Industrial Electronics*, Vol. 52, No. 6, pp. 1530-1546

Jensen, B.; Philippsen, R. & Siegwart, R. (2003). Narrative Situation Assessment for Human-Robot Interaction, *Proceedings of IEEE Int. Conf. Robotics and Automation*, pp.1503-1508

Kanda, T. ; Shiomi, M.; Perrin, L.; Nomura, T.; Ishiguro, H. & Hagita, N. (2007). Analysis of People Trajectories with Ubiquitous Sensors in a Science Museum, *Proceedings of IEEE Int. Conf. Robotics and Automation*, pp.4846-4853

Kidd, C.; Orr, R.; Abowd, G.; Atkeson, C.; Essa, I.; MacIntyre, B.; Mynatt, E.; Starner, T. & Newstetter, W. (1999). The Aware Home: A Living Laboratory for Ubiquitous Computing Research, *Proceedings of Second Int. Workshop on Cooperative Buildings, Integrating Information, Organization, and Architecture*, pp.191-198

Kranz, M.; Maldonado, A.; Rusu, R. B.; Hoernler, B.; Rigoll, G.; Beetz, M. & Schmidt, A. (2007). Sensing Technologies and the Player-Middleware for Context-Awareness in Kitchen Environments, *Proceedings of Fourth Int. Conf. Networked Sensing Systems*, pp. 179-186

Matsumoto, Y.; Wada, T.; Nishio, S.; Miyashita, T. & Hagita, N. (2008). Scalable Multi-People Head Tracking for Robotic Services Combining Multiple Sensors, *Proceedings of 5th International Conference on Ubiquitous Robots and Ambient Intelligence (URAI2008)*, pp. 581-586

Nishio, S.; Lee, J.; Yu, W.; Sakamoto, T.; Noda, I.; Tsubouchi, T. & Doi, M. (2009). A New Standard for Robotic Localization Service, In: *Cutting Edge Robotics 2009*, Kordic, V., Lazinica, A. and Merdan, M. (Eds.), xxx-xxx, IN-TECH, ISBN978-3-902613-46-2, Vienna

Shiomi, M.; Kanda, T.; Kogure, K.; Ishiguro, H. & Hagita, N. (2005). Position estimation from multiple RFID tag readers, *Proceedings of 2nd Int. Conf. Ubiquitous Robots and Ambient Intelligence*

Thompson, S.; Kagami, S.; Sasaki, Y.; Nishida, Y.; Enomoto, T. & Mizoguchi, H. (2007). Combining Ubiquitous and On-Board Audio Sensing for Human-Robot Interaction, *Proceedings of ACM/IEEE Int. Conf. Human-robot interaction,* pp. 209-216

Walters, M.; Dautenhahn, K.; Boekhorst, R.; Koay, K.; Kaouri, C.; Woods, S.; Nehaniv, C.; Lee, D. & Werry, I. (2005). The influence of subjects' personality traits on personal spatial zones in a human-robot interaction experiment, *Proceedings of Int. Workshop Robot and Human Interactive Communication*, pp.347-352.

Yoshikawa, Y.; Shinozawa, K.; Ishiguro, H.; Hagita, N. & Miyamoto T. (2006). The effects of responsive eye movement and blinking behavior in a communication robot, *Proceedings of IEEE/RSJ Int. Conf. Intelligent Robots and Systems*, pp. 4564 – 4569.

Minimally invasive force sensing for tendon-driven robots

Alberto Cavallo, Giuseppe De Maria, Ciro Natale and Salvatore Pirozzi
*Dipartimento di Ingegneria dell'Informazione, Seconda Università degli Studi di Napoli
Aversa (CE), 81031, Italy*

1. Introduction

The purpose of this chapter is to present a sensor solution for measurement of tendon tension or joint torque in force control applications of tendon-driven robots. In the robotics community, it is well-known that the use of torque sensors is useful not only when joint friction is significant (Pfeffer et al., 1989), (Vischer & Khatib, 1995), but also when motion or impedance control of flexible joints robots is in order (Ott et al., 2003), (Ott et al., 2004). This is the case, for example, of lightweight robots used for space applications (Ferretti et al., 2005). But probably the most important field of application of torque sensors is the control of robotic hands, in fact such complex systems are specifically designed to allow the robot to interact with the environment, usually unstructured and so generic that a safe interaction can be ensured only if the mechanism possesses a compliant behaviour. When integrated robotic hands are considered (Butterfaß et al., 2001), such compliance has to be provided by torque control (Liu et al., 1999), and therefore finger joint torque sensors appear mandatory. The same objective exists for tendon-driven artificial hands which are the most used solutions for prosthetic applications (Carrozza et al., 2005) and when the torque control could be useful not only for overcoming friction and other disturbances but specifically also for reproducing nonlinear characteristic of human-like tendons (Gialias & Matsuoka, 2004). Since the first prototypes of robotic hands (Salisbury & Craig, 1982), it was clear how important were the requirement of limited invasiveness for every kind of sensor to be integrated into the device. From the first solutions provided for tendon-driven fingers (Kaneko et al., 1990) only limited progress has been made and the sensor solutions adopted appear still cumbersome and quite invasive (Jung et al., 2007). These are usually based on strain gauges, which are cumbersome and require additional mechanical components and interfaces in order to provide the necessary measurement (Biagiotti et al., 2005).
The alternative solution described in this chapter is based on a Fibre Bragg Grating (FBG) used as strain sensor. Owing to the minimally invasive nature typical of optical fibres, it can be easily integrated in any type of tendon-driven robot. One of the most promising applications can be the integration of the sensor into anthropomorphic robotic hands for accurate impedance or compliance control. The sensing element of the proposed torque sensor can be easily bonded directly to the tendon following its natural routing with a significantly reduced invasiveness with respect to conventional sensors. In fact, the Bragg

sensor is embedded into the optical fibre whose typical diameter is about 125µm which allows its integration into the tendon itself. Furthermore, the greatest advantage of a solution based on the FBG, bonded directly to the tendon, is its immunity to electromagnetic disturbances typical of optical fibres, and thus it can be used without any concern in conjunction with the electrical motors, usually adopted for robotic actuation, even in close vicinity as it happens for robotic hands.

The experimental results presented here have been obtained on a simple test-bench realized by using off-the-shelf and cheap components conceived to demonstrate the potentiality of the sensor and its effectiveness in an actual compliance control scheme of a single-joint mechanism. In such type of control schemes, it is of major importance to ensure that the actuator exerts on the tendon a prescribed force or a prescribed torque when pulleys are used for mechanism actuation. The main limiting factor in achieving such a goal is the presence of dry friction at motor side, especially in the gear trains usually adopted to optimize torque transfer from the motor to the tendon. As mentioned before, such limitation can be easily overcome by means of torque or force feedback. In the experimental setup used here, the fibre has been bonded to a steel tendon with a diameter of 420 µm used to drive the servomechanism built to emulate a single articulation of a tendon-driven robot. The optical sensor measures the strain variation that an external torque generates on the tendon. Two different demodulations schemes are compared, i.e. a conventional narrowband (Zhao & Liao, 2004) and a new modified broadband interrogation circuit are implemented to convert the optical signal into an electrical one. For each demodulation scheme, the sensor has been calibrated using load cells and the differences between the two schemes have been highlighted in terms of sensitivity and dynamic range.

Taking into account the requirements of the implemented compliance control system, one of the two demodulation schemes has been implemented in the real feedback control application to evaluate the quality and the effectiveness of the proposed sensor.

The chapter is organized as follows. Section 2 presents the working principle of the optoelectronic sensor together with the two demodulation schemes designed as conditioning electronics. In Section 3, the sensor calibration curves are measured for both demodulation schemes with a suitable experimental apparatus using precision load cells. Section 4 reports the experimental results of the exploitation of the proposed tension sensor in an actual feedback control law, i.e. a compliance control of a single-joint mechanism actuated through a steel tendon.

2. Optical tension sensor based on FBG

As already mentioned, the basic idea is to exploit the FBG as a strain sensor since such a strain is proportional to the tendon tension through the elasticity of the tendon itself. Therefore, to improve sensor sensitivity, loss of strain transfer from the tendon to optical fibre has to be avoided. To this aim, before bonding the FBG to the steel tendon, the portion of fibre jacket corresponding to the grating position was removed. A picture reporting a detail of the optical sensor is presented in Fig. 1, where it is evident how minimally invasive is the sensing element.

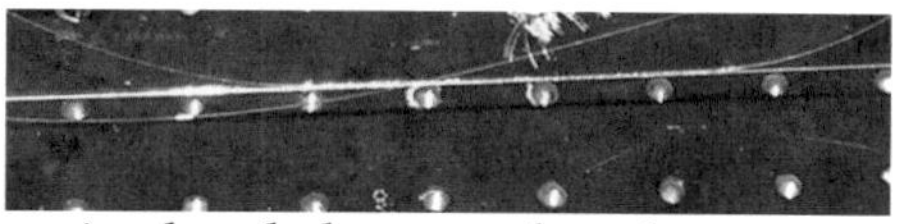

Fig. 1. Detail of the Bragg grating bonded to a steel tendon.

It is well-known that when the FBG is subject to a strain, its reflectivity spectrum shifts and thus a suitable conditioning electronics can be used to detect such a shift converting it into an electrical signal. The most common demodulation technique is based on the acquisition of the whole grating reflectivity spectrum through an optical spectrum analyzer detecting the shift of its peak (Zhao & Liao, 2004). However, such solutions are charaterized by relatively long response time and thus can be used to measure static and slowly varying dynamic strains. Also, the conditioning electronics used to implement such techniques is quite complex, cumbersome and expensive. In other applications, like in force measurement in robotics, the needed response time is significantly lower and, at the same time, simple, cheap and with limited weight and size electronics are desireble. Therefore, different demodulation techniques are mandatory, like those presented in the following.

2.1 Narrow band demodulation

The first technique presented to interrogate the proposed sensor is based on a narrow-band demodulation scheme. The output light wave from a single longitudinal mode Distributed FeedBack (DFB) diode laser was used to probe the wavelength shift of the FBG reflection curve imposed by the strain signal to be detected. If the laser frequency is within the linear range of the FBG reflection slope, the strain signals will change the reflected power, which can be simply measured using a photodetector (Zhao & Liao, 2004). It can be seen that this technique has several advantages, such as fast response, ease of use and high sensitivity depending on leading edge slope of the FBG reflection curve. The reflectivity spectrum of the FBG used in our experiments had a quasi-flat top from 1552 nm to 1556 nm, with a peak reflectivity ≥ 75% (see Fig 2). The leading edge of the FBG reflection curve extended from 1547.28 nm to 1550.95 nm (measured from 10% to 90% of maximum reflectivity), whereas the right portion extended from 1556.37 nm to 1557.24 nm. The trailing edge of the FBG reflection slope allowed for a 72GHz linear slope width, and it was chosen as the operating range due to the higher linearity exhibited by the FBG reflectivity in this spectral portion. In order to keep a linear relationship between the reflected optical power and the strain signal, the DFB laser emitting wavelength must lie within the trailing edge of the FBG reflectivity spectrum (see Fig. 2). Moreover, assuming a typical FBG curve shift of 1 nm for an applied strain of 1000 με (Kersey et al., 1997), the strain level must be kept lower than 870 με in order to avoid sensor output saturation.

The optical set-up employed for FBG interrogation is schematically illustrated in Fig. 3. Light emitted by a DFB laser was sent to a Y-coupler, which directed the light to the FBG. Light reflected from the FBG was then re-directed to a photodiode, whose bandwidth was on the order of hundreds of MHz, thus much higher than the bandwidth of the strain signals to be detected. The output signal from the photodiode was finally sent to the conditioning electronics, basically formed by a low-pass filter used to eliminate the high-frequency noise.

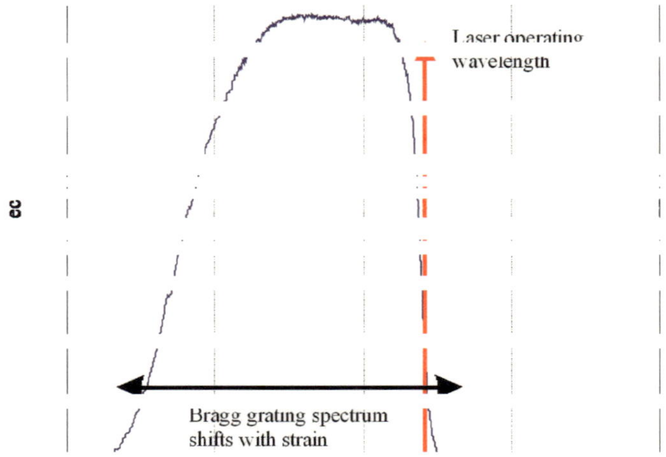

Fig. 2. Reflectivity spectrum of the grating.

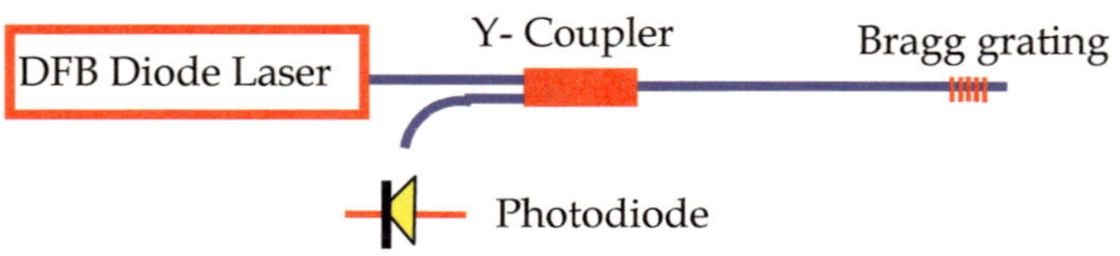

Fig. 3. FBG narrow band demodulation scheme.

2.2 Broadband demodulation

The second technique presented to interrogate the proposed sensor uses the Bragg grating together with a broadband SLED optical source. Differently from classical demodulation techniques using a broadband source, the technique proposed here is not based on a spectral analysis and thus allows to measure fast varying dynamic strain. The output light of SLED was used to illuminate the FBG, which presents a very narrow spectrum compared to that of the source. If the strain signal to be detected shifts the Bragg grating reflection curve within a monotonous range of the broadband source spectrum, the variations of the reflected power can be simply measured using a photodetector (see Fig. 4). With respect to the technique described in the previous subsection this one shows the same advantages in terms of fast response and ease of use. In this case, the sensitivity depends both on leading edge slope of the SLED spectrum and on Bragg grating reflection curve. Depending on the application requirements an optimal combination of these parameters can be selected. Moreover, the maximum strain level measurable with this technique is limited only by the maximum applicable deformation before the breaking of the grating (about 10000 με). In fact, a strain level of 10000 με corresponds to a FBG spectrum shift of about 10 nm which is well below the length of the broadband source leading edge that is about 50 nm. The reflectivity spectrum of the FBG used in our experiments for this demodulation technique had a peak reflectivity ≥ 90% and a full-width-half-maximum bandwidth of 0.168 nm with a centre wavelength of 1520.132 nm. The optical set-up employed for broadband interrogation is essentially similar to the narrow band case with the broadband source instead of DFB laser (see Fig. 5). Light emitted by broadband source was sent to a Y-coupler, which directed the light to the FBG. Light reflected from the FBG was then re-directed to the photodiode connected to the conditioning electronics. Moreover, the cost and the encumbrance of this solution are lower since the DFB diode laser is more expensive and more cumbersome with respect to the SLED source.

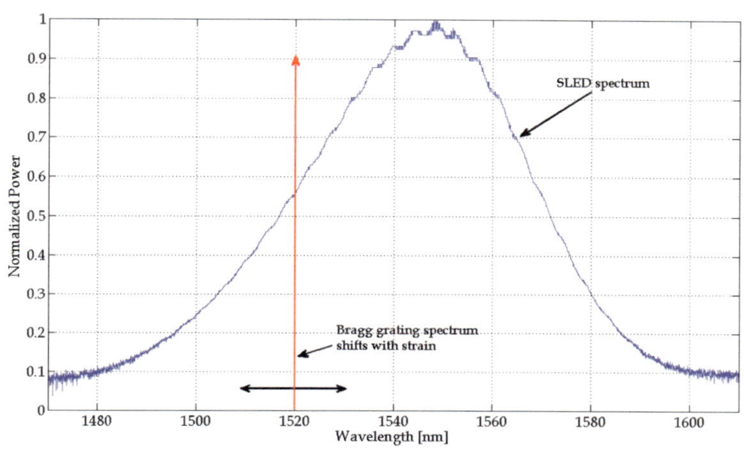

Fig. 4. Spectrum of broadband source with respect to the FBG.

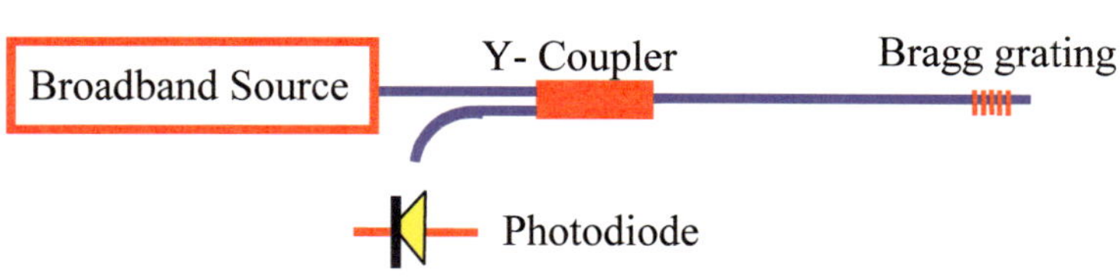

Fig. 5. FBG broadband demodulation scheme.

3. Sensor calibration

The force sensor has been calibrated using a setup constituted by a steel tendon with a diameter of 250 μm connected, at one end, to a micro-positioning stage and to a load cell used to calibrate the tension sensor, at the other end (Fig. 6). When the micro-positioning stage moves, a force causes a strain variation of the tendon proportional to the stiffness of the material used to realize the tendon. The FBG measures this strain and consequently the tension after a proper calibration procedure. The force f is related to the output voltage v_B obtained from the optical sensor signal after the demodulation and a suitable electronic conditioning as

$$f = k_s k_\varepsilon v_B = k_B v_B \tag{1}$$

where k_ε is the strain sensitivity, and k_s is the tendon stiffness sketched as a lumped spring in Fig. 7, being k_B the overall sensor sensitivity. During the calibration of the sensor, the force applied to the tendon, shown as f in Fig. 7, has been measured using a load cell and the corresponding output voltage v_B of the conditioning electronic circuit has been evaluated.

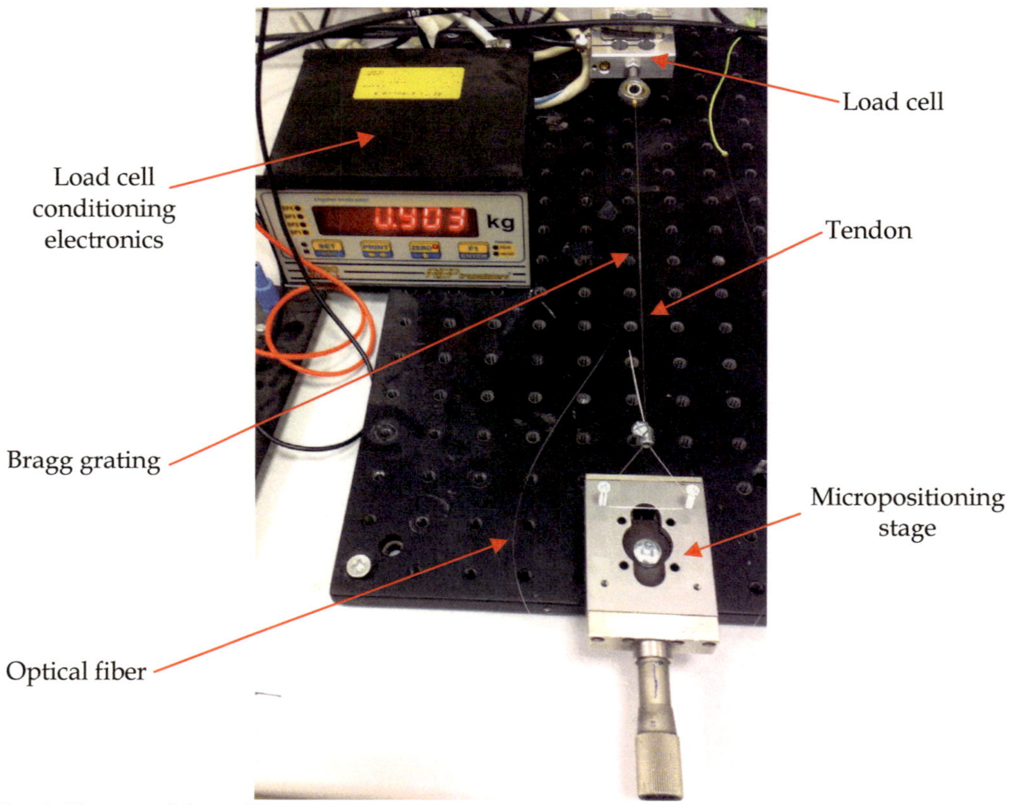

Fig. 6. Picture of the calibration setup

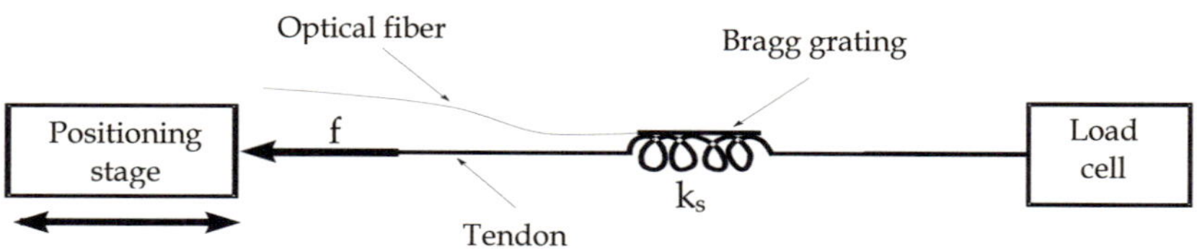

Fig. 7. Sketch of the calibration setup.

The same setup has been used to calibrate the sensor using the two demodulation techniques described above. Different set of experimental measurements has been carried out to calibrate the optical sensor in the two cases. The measurements have been fitted with a linear curve, whose parameters have been estimated with a least mean square algorithm. Fig. 8 shows the results of the sensor calibration with a narrow band demodulation technique reporting the experimental data and the fitting curve, whose equation is

$$f = 2.61 v_B + 0.11 \tag{2}$$

The calibration curve and the fitting curve of the sensor with a broadband demodulation technique is reported in Fig. 9. In this case the equation of fitting curve is

$$f = 46.78 v_B + 0.54 \tag{3}$$

The constant offset term in both cases, due to sensor electronics, and reported in (2),(3) can be easily eliminated via software whenever a digital control system is adopted to exploit the sensor measurement. In conclusion, the Bragg sensor used with a narrow band demodulation technique can be considered a force transducer with a calibration constant of 2.61 N/V, while with a broadband demodulation technique the calibration constant is 46.78 N/V.

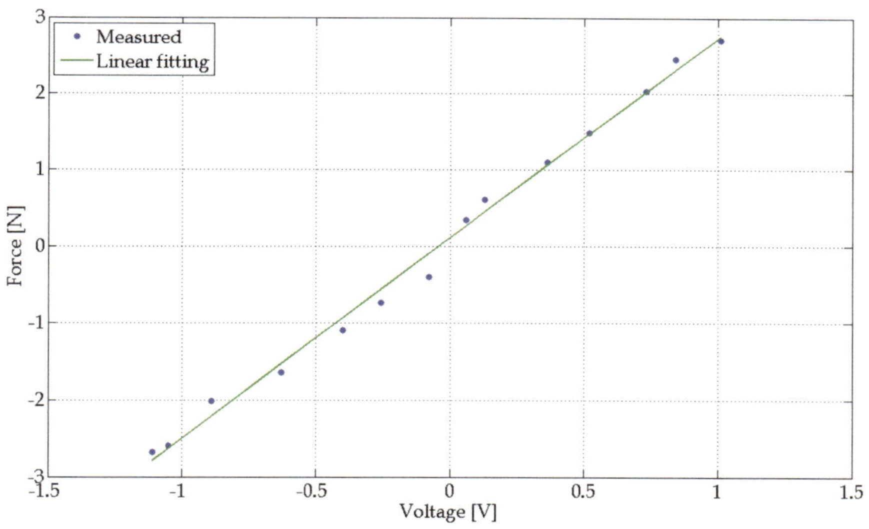

Fig. 8. Calibration curve with narrow band demodulation.

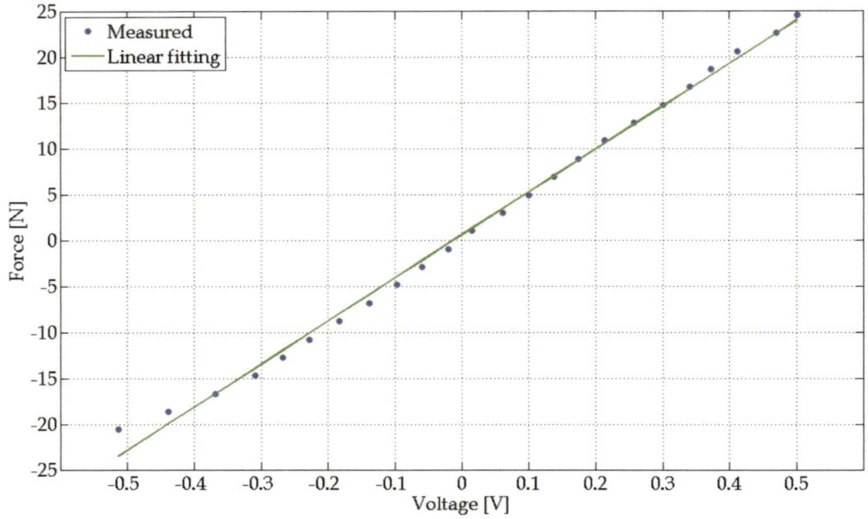

Fig. 9. Calibration curve with broadband demodulation.

4. Testing

The test-bench, whose picture is reported in Fig. 10, is constituted by a steel tendon stretched between two pulleys. The first one is connected via gear train with the motor shaft (left side of the picture), the other pulley is fixed to the inertial load (right side of the picture). The load shaft is connected to the supporting structure through ball bearings to reduce the load-side friction as much as possible. In fact, one of the expected results of the force feedback is to counteract the disturbances (mainly dry friction) acting only on the motor side, while no rejection of load-side disturbances is expected. A potentiometer is mounted to the motor side in order to measure angular position for implementation of compliance control. The Bragg sensor written into the optical fibre is bonded to the tendon and a conditioning electronics module, based on the narrow band technique (see Section 3), has been designed and produced to suitably demodulate the optical output signal. The implemented solution has been selected taking into account the range within the force amplitudes vary during the testing.

4.1 Test-bench Mathematical Modelling

The test-bench described so far can be schematically represented as in Fig. 12 where the tendon stiffness is sketched as a lumped spring. The mathematical model of the proposed test-bench is the classical two-mass system whose equations are

$$J_m \ddot{\vartheta}_m + \beta_m \dot{\vartheta}_m = k_r k_t i_m - \tau_l - \tau_d \tag{4}$$

$$J_l \ddot{\vartheta}_l + \beta_l \dot{\vartheta}_l = \tau_l + \tau_e \tag{5}$$

$$\tau_l = 2rk_s(\vartheta_m - \vartheta_l)r \tag{6}$$

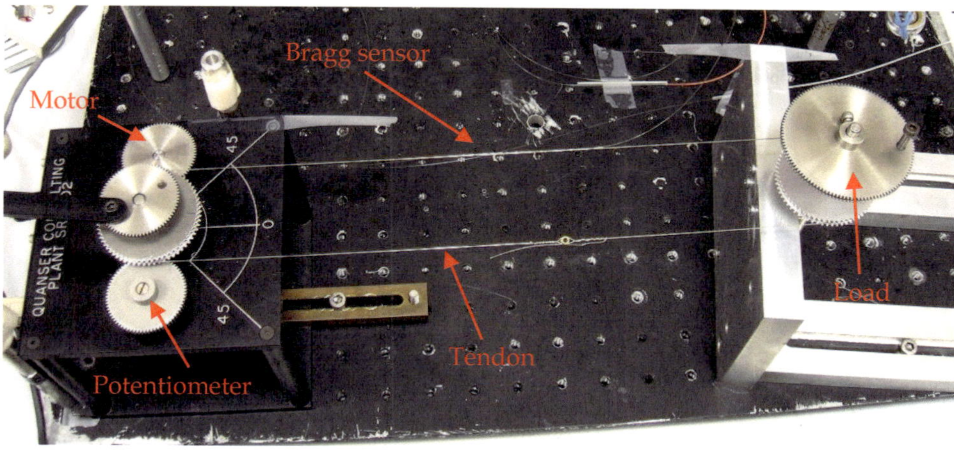

Fig. 10. Test bench setup for testing.

where J_m, β_m, ϑ_m and J_l, β_l, ϑ_l are the overall inertia, friction coefficient and angular position of the motor side and load side respectively. The motor is driven by the armature current i_m with torque constant k_t, and it is connected to the pulley via gear train with gear ratio k_r. Both pulleys have diameter 2r and the tendon stiffness is k_s. The load torque τ_l is the torque transferred to the load through the tendon, which has been suitably prestressed in order to prevent buckling thus allowing measurement of both positive and negative torques. The disturbance torque τ_d takes into account all the dynamic effects not explicitly modelled, e.g. dry friction and high frequency dynamics. Finally, τ_e represents the external torque applied to the load which models the interaction of the servomechanism with the environment. If a load torque is applied to the system the tendon exhibits a strain variation proportional to the stiffness of the material used to realize the tendon. Using the presented sensor, it is possible to measure this strain and consequently the load torque after a proper calibration procedure. The narrow band demodulation technique has been selected considering the maximum strain that the implemented setup exerts on tendon.

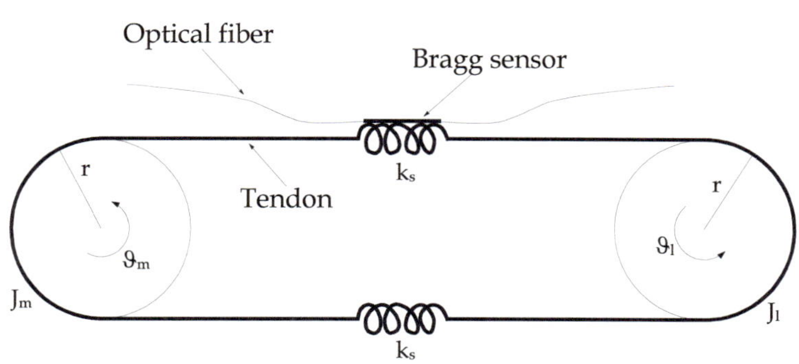

Fig. 11. Sketch of the test bench.

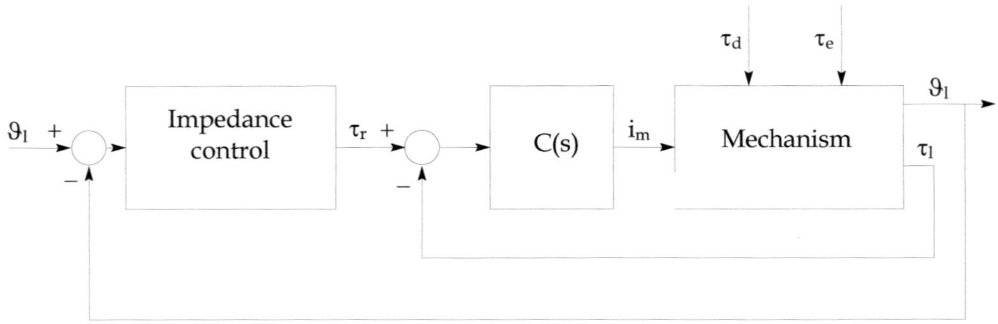

Fig. 12. Compliance control scheme.

4.2 Compliance Control

The compliance control system used to test the proposed torque sensor is based on a double control loop, a typical control scheme reported in the literature (see Fig. 12). The control input of the mechanical system is the motor driving current i_m, while the measured outputs are the angular position ϑ_l and the load torque τ_l. The unmanipulable inputs of the system are the disturbance torque τ_d and the external torque τ_e. The torque controller C(s) operates so as to reduce the error between the reference torque τ_r and the measured load torque. The reference torque is generated by the outer compliance controller evaluating the error between the reference angular position ϑ_r and the actual angular position ϑ_l. The performances of a compliance or an impedance control broadly depend on the disturbances at motor torque level and the objective of the torque inner controller is to reject the disturbance τ_d.

Taking into account the mathematical model of the mechanism, reported in (4),(5) and that the main source of torque disturbances is constituted by the Coulomb friction, i.e. a constant friction torque, the torque controller has to include an integral action on the torque error so as to completely reject the constant input disturbances. The proposed linear controller has transfer function

$$C(s) = \frac{k_c(1+sT_1)}{s(1+sT_2)} \frac{1+s2\zeta/\omega_n + s^2/\omega_n^2}{(1+sT_3)^2} \qquad (7)$$

where k_c is the controller gain and the time constants T_1 and T_2 are chosen such that $T_1 > T_2$ resulting into a phase-lead term aimed at ensuring the stability of the closed-loop system. In order to improve the performance in terms of bandwidth, and thus tracking accuracy, a couple of complex zeros is introduced to avoid that the closed-loop resonant mode related to the flexible connection between the motor and the load gets unstable. Finally, two high frequency poles with the same time constant T_3 have been introduced to keep a sufficient roll-off of the loop gain to filter measurement noise.

The compliance control law of outer loop computes τ_r as

$$\tau_r = k_p(\vartheta_r - \vartheta_l) - k_d\dot{\vartheta}_l + \beta_l\dot{\vartheta}_l \qquad (8)$$

and, under the assumption of perfect tracking of the inner loop, i.e. $\tau_l = \tau_r$, the closed-loop system behaviour is described by the equation

$$J_l\ddot{\vartheta}_l + k_d\dot{\vartheta}_l + k_p(\vartheta_l - \vartheta_r) = \tau_e \tag{9}$$

where the user can select the desired stiffness value k_p and the desired friction parameter k_d.

4.3 Experimental results

Two different sets of experiments have been carried out to compare both the static and the dynamic performance of the compliance control making use the proposed torque sensor with those of a compliance control implemented with the sole position feedback. Note that in both cases, the load angular position used in the outer loop controller has not been measured directly, but it has been considered equal to the motor angular position measured through the potentiometer as in Fig. 10, under the assumption of a reference position ϑ_r with significant spectral components within a frequency range much lower than the resonant mode of the system. In the case of pure position feedback, the torque inner loop has been replaced with the transfer function

$$C_F(s) = \frac{1}{k_r k_t (1 + sT)} = \frac{1}{14 \cdot 0.00767(1 + s/200)} \tag{10}$$

which transforms the reference torque τ_r, computed by the compliance controller (8), into the motor driving current i_m by taking into account both the gear ratio k_r and the torque constant k_t in (4) and by filtering high frequencies with the same roll-off of the torque loop gain.

A set of static measurements has been made to compare the actual stiffness to the desired one when the pure position feedback is implemented. In detail, with different values of k_p, several values of external torque τ_e have been applied to the system and the corresponding position ϑ_l has been measured so as to estimate the actual stiffness k_{pa} of the servomechanism via a least-mean square method. The desired and actual stiffness values are reported in Table 1 together with the relative error defined as

$$e_{\%} = 100\frac{k_{pa} - k_p}{k_p} \tag{11}$$

The table evidently shows how the performance of the pure position control approach is quite limited for low values of desired stiffness, while it improves for higher values of desired stiffness.

k_p [Nm/rad]	0.100	0.200	0.300	0.400	0.500
k_{pa} [Nm/rad]	0.131	0.212	0.328	0.484	0.502
$e_{\%}$	31.3	6.17	9.44	21.0	0.406

Table 1. Actual servomechanism stiffness without torque feedback

k_c [A/Nm]	T_1 [s]	T_2 [s]	ζ	ω_n [rad/s]	T_3 [s]
250	1/5	1/50	0.025	$2\pi 180$	1/2200

Table 2. Parameters of the controller C(s)

As expected, when the torque feedback is used, with the parameters selected for the torque controller C(s) in (7) as reported in Table 2, the error in (11) is zero for every value of desired stiffness, since the pole in the origin of C (s) ensures a null tracking error at steady-state.
A second set of experiments has been performed to evaluate the dynamic characteristics of the compliance control obtained with the proposed control system. Then, the actual dynamic behaviour is compared with both the desired one and with that achieved using the pure position feedback.
A square-wave reference position ϑ_r with a frequency of 0.5Hz has been imposed to the system and experiments have been carried out with different values of k_p and of k_d. First of all, the load inertia J_l has been estimated using the torque controller as described in the following. Several couples of values for k_d and k_p have been imposed so as to obtain lightly damped responses and then natural frequencies ω_n have been evaluated from the analysis of the transient responses in terms of load position. With reference to (9), the actual value of the stiffness is equal to the imposed one owing to the torque control action, thus, the inertia J_l can be computed as k_p/ω_n^2 for each experiment. Eventually, the mean value has been chosen as the estimated value of the load inertia, namely $J_l = 10^{-4}$ Nms^2/rad. Therefore, by knowing the impedance parameters J_l, k_d and k_p, it is possible to calculate the theoretical dynamic response of the load position and compare it with the actual response of the system under the compliance control.
The results when torque feedback control is implemented with impedance parameters $k_d=10^{-3}$ and $k_p = 0.1$ are reported in Fig. 13 displaying the reference angular displacement ϑ_r together with the angular displacement (desired ϑ_l) computed from the theoretical equation (9) and the actual one. It is evident that the natural frequency and the settling time of the experiment are very close to the theoretical ones, even though the two curves are not identical due to some residual nonlinearities, which can mainly be attributed to friction disturbances acting on the load shaft, against which the torque loop has no effect. Similar remarks can be made for the results shown in Fig. 14, where the same test is reported for different values of the impedance parameters $k_d = 8\cdot 10^{-3}$ and $k_p = 0.1$ leading to a damped response. In such a case, the non null steady-state error is due to the effect of dry friction on the load shaft. The same two experiments have been repeated when the compliance control, with the same impedance parameters k_p and k_d, using only position feedback is implemented. The results are shown in Fig. 15 and Fig. 16, and it is evident how different from the expected one is the poor dynamic performance obtained with this approach, in fact in both cases a highly damped response as well as a large steady-state error are obtained due to the effect of friction acting on the motor and gear shafts.

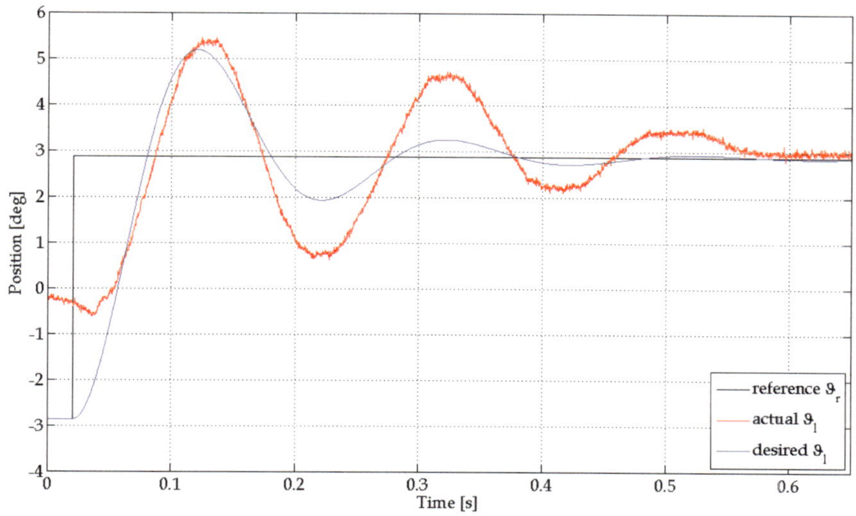

Fig. 13. Compliance control with position and torque feedback ($k_d=10^{-3}$ and $k_p=0.1$)

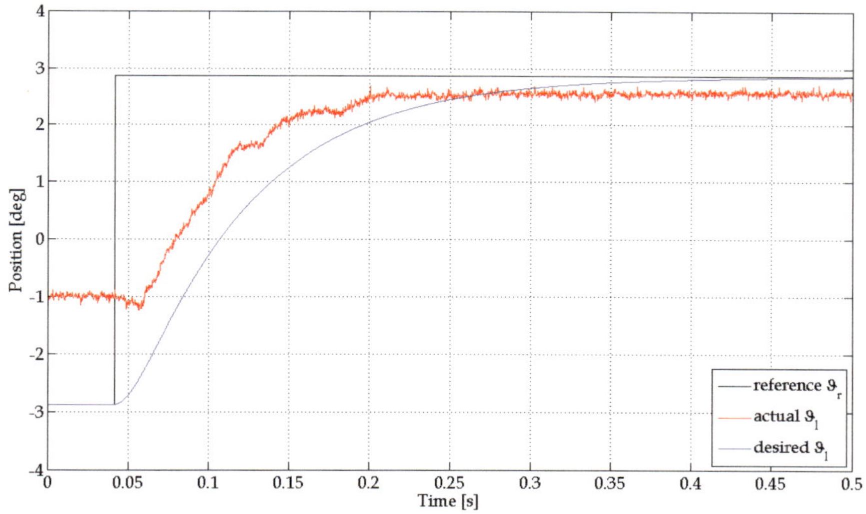

Fig. 14. Compliance control with position and torque feedback ($k_d=8\cdot10^{-3}$ and $k_p=0.1$)

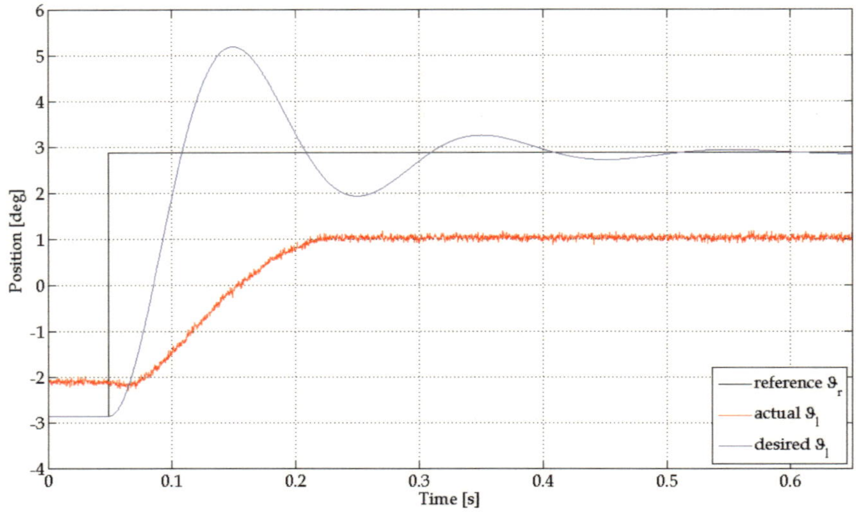

Fig. 15. Compliance control with only position feedback (k_d=10^{-3} and k_p=0.1)

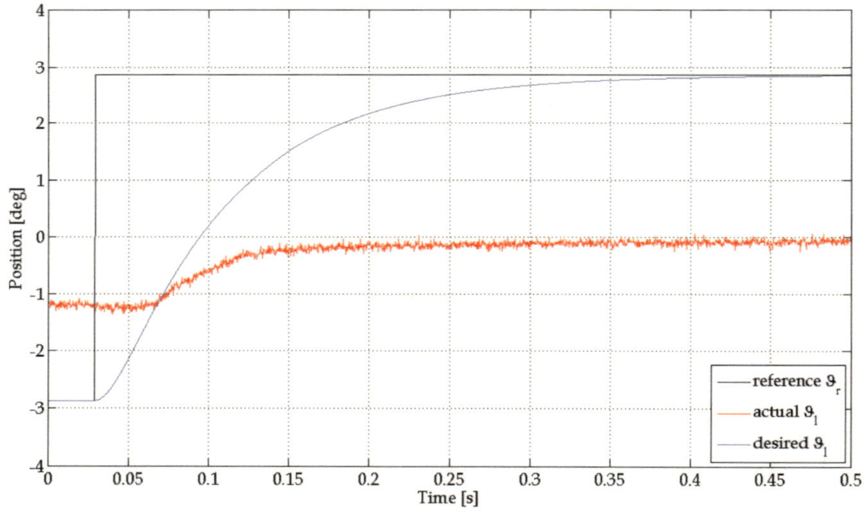

Fig. 16. Compliance control with only position feedback (k_d=$8\cdot10^{-3}$ and k_p=0.1)

5. Conclusions

The chapter tackles the problem of the design of a sensor to measure the tendon tension for tendon-driven robots. The requirements of this application are the minimally invasiveness of the sensing element itself and the limited complexity, weight, cost and size of the conditioning electronics. The proposed solution is based on optoelectronic technology, i.e. a

FBG used as force sensor. Such a choice permits to satisfy the minimally invasiveness requirement since the grating is integrated into a fibre of sole 125 μm and can be directly bonded to the tendon. Two different demodulation schemes have been proposed to be selected according to application requirements in terms of dynamic range and sensitivity. Both techniques allow to measure fast varying dynamic strain in contrast to classical demodulation schemes based on spectral analysis. Moreover, the second one can be implemented with cheaper optoelectronic components and with lighter conditioning electronics. The two schemes have been implemented for the realization of two tendon tension sensors with different calibration curves, which have been experimentally identified. Finally, the sensor interrogated using the narrow band demodulation technique has been exploited as a torque feedback sensor in a compliance control system of a single joint mechanism. The experimental results showed the advantage provided by torque feedback when a highly compliant behaviour is required to the servomechanism without losing position tracking performance.

6. Acknowledgement

The research leading to these results has received funding from the European Community's Seventh Framework Programme (FP7/2007-2013) under grant agreement no. 216239 (DEXMART project).

7. References

Biagiotti, L.; Lotti, F.; Palli, G.; Tiezzi, P.; Vassura, G. & Melchiorri, C. (2005). Development of UB Hand 3: Early Results, *Proc. of IEEE Int. Conference on Robotics and Automation*, pp. 4488-4493, Barcelona.

Butterfaß, J.; Grebenstein, M.; Liu, H. & Hirzinger, G. (2001). DLR-Hand II: Next Generation of a Dextrous Robot Hand, *Proc. of IEEE Int. Conf. on Robotics and Automation*, pp. 109-114, Seoul.

Carrozza, M. C.; Cappiello, G.; Stellin, G.; Zaccone, F.; Vecchi, F.; Micera, S. & Dario, P. (2005). On the Development of a Novel Adaptive Prosthetic Hand with Compliant Joints: Experimental Platform and EMG Control, *Proc. of IEEE/RSJ Int. Conforence Intelligent Robots and Systems*, pp. 1271-1276, Edmonton.

Ferretti, G.; Magnani, G.; Viganò, L. & Rusconi, A. (2005). On the use of torque sensors in a space robotics application, *Proc. of IEEE Int. Conference on Robotics and Automation*, pp. 1947–1952, Edmonton.

Gialias, N. & Matsuoka, Y. (2004). Muscle Actuator Design for the ACT Hand, *Proc. of IEEE Int. Conference on Robotics and Automation*, pp. 3380–3385, New Orleans.

Jung, S. Y.; Kang S. K.; Lee, M. J. & Moon, I. (2007). Design of Robotic Hand with Tendon-driven Three Fingers, *Proc. of IEEE International Conference on Control and Automation*, pp. 83-86, Seoul.

Kaneko, M.; Yokoi, K. & Tanie, K. (1990). On a New Torque Sensor for Tendon Driven Fingers, *IEEE Tranactions on Robotics and Automation, 6* (4), pp. 501-507.

Kersey, A. D.; Davis, M. A.; Patrik, H. J.; LeBlanc, M.; Poo, K. P.; Askins, A. G.; Putnam, M. A. & Friebele, E. J. (1997). Fiber grating sensors, *Journal of Lightw. Technol. , 15* (8), pp. 1442-1462.

Liu, H.; Butterfaß, J.; Knoch, S.; Meusel, P. & Hirzinger, G. (1999). A new control strategy for DLR's multisensory articulated hand, *Control Systems Magazine* , *19* (2), pp. 47-54.

Ott, C.; Albu-Schaffer, A.; Kugu, A. & Hirzinger, G. (2003). Decoupling based Cartesian impedance control of flexible joint robots, *Proc. of IEEE Int. Conference on Robotics and Automation*, pp. 3101–3107, Taipei.

Ott, C.; Albu-Schaffer, A.; Kugu, A.; Stramigioli, S. & Hirzinger, G. (2004). A passivity based Cartesian impedance controller for flexible joint robots Part I: torque feedback and gravity compensation, *Proc. of IEEE Int. Conference on Robotics and Automation*, pp. 2659–2665, New Orleans.

Pfeffer, L.; Khatib, O. & Hake, J. (1989). Joint Torque Sensory Feedback in the Control of a PUMA Manipulator, *IEEE Tran. on Robotics and Automation* , *5* (4), pp. 418-425.

Salisbury, J. & Craig, J. (1982). Articulated Hands: Force Control and Kinematic Issues, *Int. J. of Robotic Research* , *1* (1), pp. 4-17.

Vischer, D. & Khatib, O. (1995). Design and Development of High-Performance Torque-Controlled Joints, *IEEE Tran. on Robotics and Automation* , *11* (4), pp. 537-544.

Zhao, Y. & Liao, Y. (2004). Discrimination methods and demodulation techniques for fiber bragg grating sensors, *Opt. Lasers Eng.* , *41*, pp. 1-18.

Tweezers Type Tool Manipulation by a Multifingered Hand Using a High-speed Visual Servoing

Satoru Mizusawa[*11], Akio Namiki[*2],
Taku Senoo[*] and Masatoshi Ishikawa[*1]
[*1]*University of Tokyo,* [*2]*Chiba University*
Japan

1. Introduction

Recently, there has been increasing need for a dexterous robot hand beyond the capability of the human hand. There has been considerable work concerning grasping and manipulation by a robot hand [1], however most of the researches are focused on the stable grip. In order to achieve a skilful handling like a human hand, it is necessary that a robot hand has the capability to manipulate a target dexterously with fingers having multi degrees of freedom.
One of the dexterous tasks of a human hand is tool manipulation. A human hand can operate many types of tools much like a part of his body. The tool manipulation is more difficult than the typical manipulation. The hand has to control a target through the tool while the tool is gripped by the hand stably. The object also has to be controlled by the other grasped object. In order to achieve such a task, sensory feedback control is very important.
It is not a requirement that such a tool manipulation is directly applied to the actual application such as industrial use. However, tool manipulation is one of the most dexterous human skills, and it should help us to solve the problem of why a human hand is so dexterous. By analyzing the skill involved in tool manipulation, various types of information which may be useful in developing a robot hand more dexterous than a human hand may be acquired.
In this chapter we propose a tool manipulation system by a multifingered hand. As a handled tool, we adopt tweezers. The manipulation of a tweezers-type tool is analyzed and achieved using a visual servoing control strategy. In the control system, the contact point between a robot finger and a tool is regarded as a kind of passive joint which is controlled by external force and friction force. And by using high-speed vision, relative positions among a robot hand, a tool and a handled object are measured in realtime, and realtime sensory feedback is achieved.

2. Background

There have been several researches from a biological viewpoint concerning how to manipulate tools by the human hand. However, there have been few researches about tool manipulation by a robot hand.
Sugiuchi et al. studied manipulation of scissors by a dual multi-finger hand arm system [2]. Kawasaki et al. developed a haptic interface to manipulate plural tool devices for their multifingered hand [3]. By changing the hold part of the tools, the hand could easily manipulate the tools. Kemp et al. detected the tip of a tool by image processing, and a robot hand manipulated it [4]. Bernardin et. al. developed a system which extracted manipulation skills including tool manipulation from a human skill based on sensor fusion approach[5].
Some of these previous researches are heuristic. In some of them, the process of manipulation is the main research interest. There are few researches that focus on dynamic properties of tool manipulation and its sensory control feedback method.
On the other hand, recently a multifingered hand system which has the capability to move quicker than a human hand has been developed [6]. By using high-speed visual feedback at the rate of 1kHz, various types of manipulation has been achieved, for example, quick regrasping of an object [7], pen spinning [8] and rope knotting[9]. This precise and rapid motion is appropriate for as a test-bed for tool manipulation.
In this research, some essential elements for tool manipulation are extracted: the passive joint and visual servoing control method. And a new control strategy is proposed based on the elements. By using high-speed hand system, the control strategy is verified.

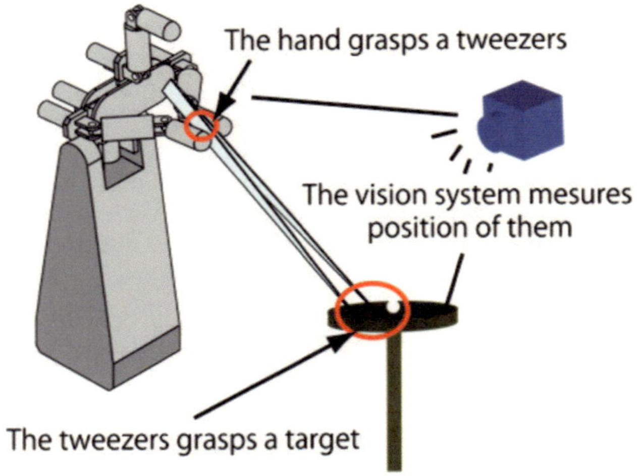

Fig. 1. Concept

3. Dynamic property of tool manipulation

Fig.1 shows the experimental setup. A multifingered hand grasps a tweezers with two fingers. The hand manipulates the tip of the tweezers so that the tweezers pick up an object placed on the table. The position of hand, tweezers and object are measured by the vision system. In this configuration, the hand and the tweezers are integrated, and they are regarded as one mechanism. The contact point between the finger and the tool is regarded as a kind of passive joint. This passive joint cannot be controlled actively because there is no actuator between the finger and the tool in general. However, the translational and rotational friction force at the contact point can be controlled by changing the grip force. Also, using an external force such as gravity, the condition at the contact point can be controlled (Fig.4). If the hand can make another contact between the tool and the other external object, the contact point also can be controlled. This means that this system has a multi-joint structure that includes a passive joint. There are following features in such a system.

3.1. Posture change of a passive joint structure during manipulation

In tool manipulation, it is possible that the condition at the contact point (passive joint) may changes during manipulation. First, the initial grasp condition of the tool may be changed at each trial, because of the error involved in placing the tool. This means that the condition for grasping the tool is changed for each trial. Next, if some external force is added to a part of the tool during manipulation, a rotational moment at the contact point occurs. When this rotational moment is larger than the static frictional force, the relative posture of the tool to the hand is also changed.

For these reason, it is necessary to measure the relative positions in realtime, and to control the hand so as to cancel the changes. A visual servoing control method is effective for this problem. If the rate of feedback is sufficiently fast, the control can be achieved more smoothly. In this case the camera's focal length is set at 28[mm] , because of the small size of the target. The calculated camera's horizontal angle of view is about 0.3[mm/pixel]. If target moves around by 1.5[m/s].

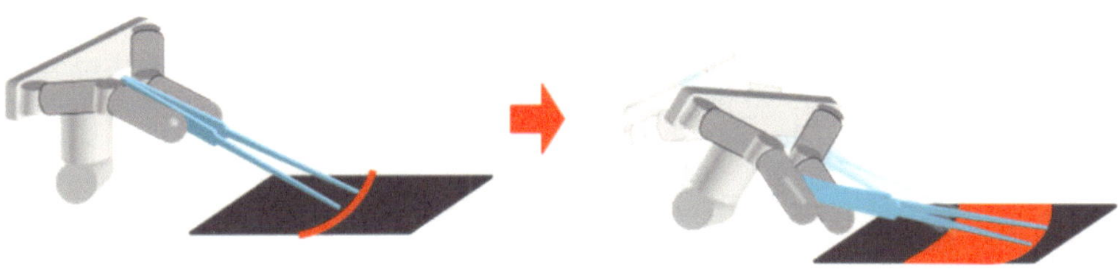

Fig. 2. Tool manipulation using a passive joint.

Fig. 3. The passive joint absorbs impact force.

For a camera with normal speed, the 30[fps] camera, the target moves about 5[mm/frame]. For 500[fps] camera, the target moves about 1[pixel/frame]. This means that the high-speed vision provides a low cost of search range and high accuracy of manipulation.

3.2 Manipulation using passive joint

When a human hand manipulates a tool, the contact point between hand and tool is not always fixed. Corresponding to the situation, the finger may slide on the tool by reducing grip force. As a result flexible manipulation can be achieved. There are two advantages of the passive joint.

The first advantage is that the passive joint extends the range of manipulation. An example is shown in Fig.2. In this case, the robot hand is fixed on the ground, and only the wrist joint can be rotated. If the tweezers is rigidly fixed on the finger, the tip of the tweezers can only reach to the circumference as shown in Fig.2 left. Using rotation of passive joint, one DOF (degrees of freedom) is added. As a result, the tip of the tweezers can reach over an area other than circumference as shown in Fig.2 right.

The second advantage is that unexpected impact force added to the tool can be absorbed by gripping the tool softly (Fig.3) . These advantages mean that not only fixed grip but also soft flexible grip are important in tool manipulation.

4. System Configurations

4.1 High-speed multifingered hand [8]

Fig.5 shows the mechanical design of the hand. It has three fingers, and we call the fingers "left thumb", "index finger" and "right thumb" by looking from the left side. The index finger has 2-DOF, and the other fingers have 3-DOF, and the hand has 8-DOF in total.

A newly developed small harmonic drive gear and a high power mini actuator are fitted in each finger link. The design of this actuator is based on the new concept that maximum power output, rather than rated power output, should be improved. The hand can close its joints at 180deg per 0.1s. The hand has two wrist joints. The wrist part has a differential rotation mechanism, and comprises two axes: the flexure-extension axis (WJ1 in Fig. 5) and the adductor axis (WJ2 in Fig.5). To minimize the inertia of the hand-wrist unit, actuators are equipped with the root of the elbow, and two belts are used to transmit power from the actuators to the wrist. Its maximum velocity is 300rpm, and the maximum output is 12N.

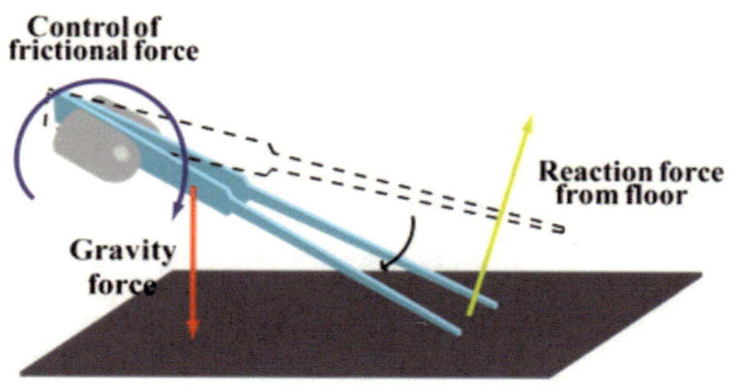

Fig. 4. The passive joint feature.

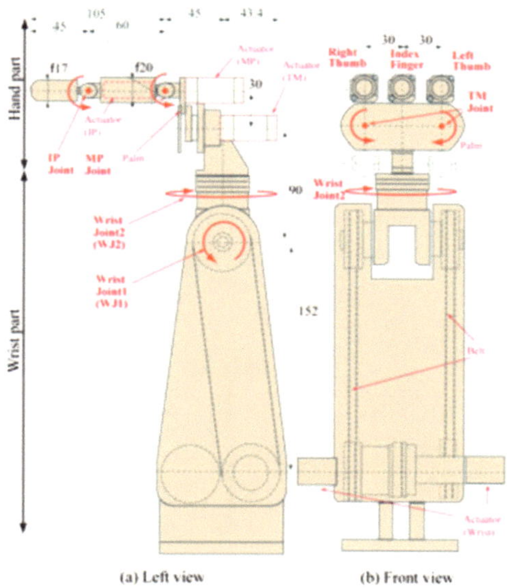

Fig. 5. High-speed multifingered hand[12]

4.2 High-speed vision system

The vision system uses high-speed camera called BaslerA504k. Its resolution is 1280× 1024. It is connected to the image processing PC (CPU: Pentium 4 (3.20GHz) Mem: 1.0GB) by cameralink. It takes 8bit gray-scale images and the maximum speed is 500frames/s by 1280× 1024. The image processing PC calculates feature value at almost 500frames/s and send the value to the control processing system (dSPACE) by CAN connection.

The tips of tweezers are marked and we can take moment feature of image and get gravity of image. In this system we assume that marker and target are detected. Those are checked by hand. Threshold value is constant as one chooses.

4.3 Others

The control processing system (dSPACE, CPU: PowerPC 750GX (1Ghz)) receives the value from the image processing PC and control the multifingerd hand at 1kHz. The tweezers is 20cm long and the material is SUS304. Fig.6 shows system configuration.

5. Control Algorithm

The hand is controlled through image based visual servoing. Image based visual servoing is a control method by feedback of position error in image plane. It is robust against calibration error, and the computation of the target 3D position is not needed.

The fingers are used to control the tweezers grip. In this case, two fingers which have four joints θ_1, θ_2, θ_3, and θ_4 are used. The 2-axis wrist joints $\boldsymbol{\theta}_w = (\theta_{w1}, \theta_{w2})$ are used to control the position of the tip of the tweezers.

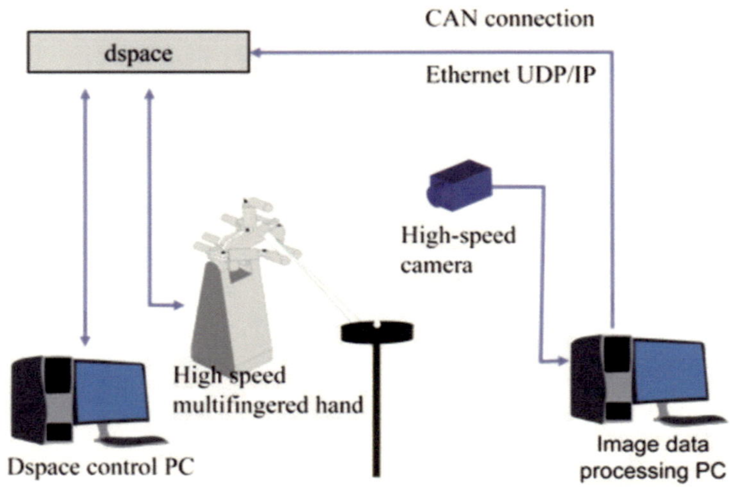

Fig. 6. System configuration.

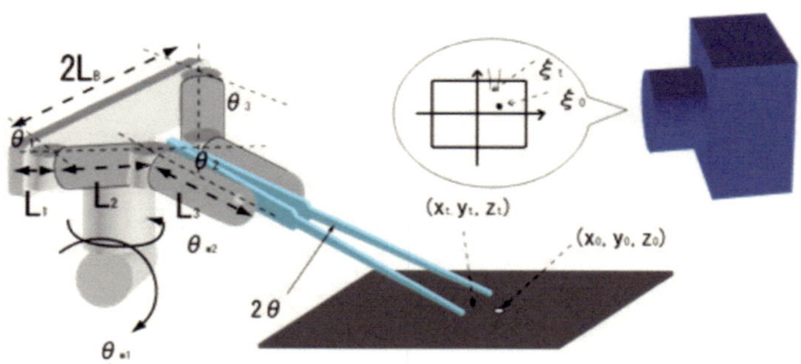

Fig. 7. Parameters of the system.

5.1 Target tracking of tweezers

The target position in the image plane is defined as $\xi = (\xi_x, \xi_y)$ and the 3D position of the target is defined as $\boldsymbol{x} = (x, y, z)$. Using the perspective projection function f, we can write $\xi = f(\boldsymbol{x})$. From the kinematics of the hand and tweezers structure, we can write the tip of tweezers point as $\boldsymbol{x} = g(\boldsymbol{\theta})$.
By differentiating this relation,

$$\dot{\xi} = \frac{\partial f}{\partial x}\frac{\partial g}{\partial \theta_w}\dot{\theta} \equiv J\dot{\theta}_w \tag{1}$$

with respect to

$$\boldsymbol{\theta}_d = k\boldsymbol{J}^{-1}(\xi_0 - \xi_t) + \boldsymbol{\theta} \tag{2}$$

where k is constant.

Using this equation, we can calculate the desired angle of the wrist joint. When the tip of the tweezers touches the table plane where the target is placed, by using the passive joint a target impact force is absorbed, and the tip of the tweezers can move along the surface of the table plane.
In this condition, the image Jacobian $\boldsymbol{J}$ is changed because of the change in kinematics of the grip. Although the speed of convergence is changed, tracking is achieved in most cases because of robustness of the visual servoing.

5.2 Control of passive joint: joint structure changes

The following expressions are introduced:

$$k = k + a\left(b - \left|\frac{\partial \xi_t}{\partial x}\right|\right) \tag{3}$$

where $\frac{\partial \xi_t}{\partial x}$ is the rate of change of image feature, and a, b are constant coefficients. It corresponds to the tracking speed of the tweezers, and the gain is automatically adjusted.

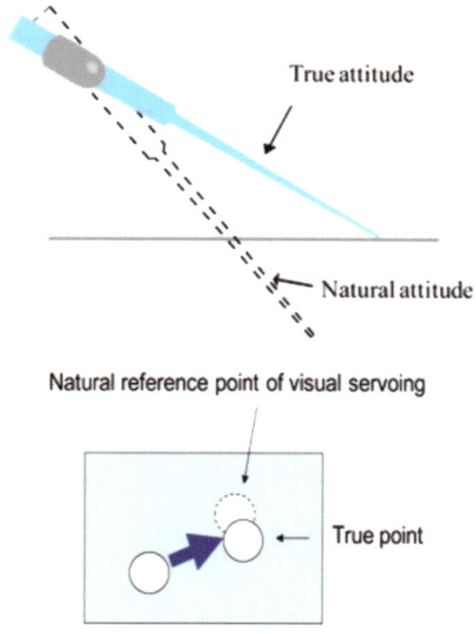

Fig. 8. Kinematic error and image error caused by the passive joint.

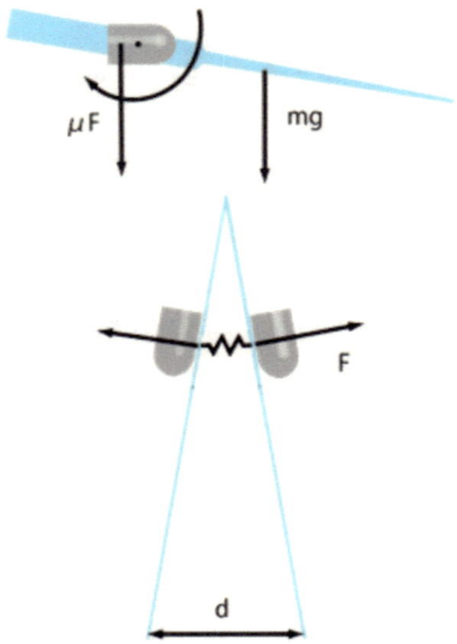

Fig. 9. Simple spring model of tweezers.

5.3 Control of passive joint: use of gravity

If we regard the tweezers as a simple spring model as in Fig.9, the following expressions can be led:

$$d_0 > d_b > d_0 - \frac{mg}{\mu k} \tag{4}$$

where d_0, d_b are respectively the initial width and the reference width of the tweezers tip, μ is the coefficient of static friction, m is the weight of tweezers, g is an acceleration of gravity, k is the spring constant.

5.4 Target grasp by tweezers

The tweezers is controlled with the visual servoing according to the following expressions:

$$\boldsymbol{\theta} \leftarrow \boldsymbol{\theta} + \boldsymbol{j}\,(u_d - u) \tag{5}$$

where $\boldsymbol{\theta}$ is finger joints angle, $\boldsymbol{j}$ is image Jacobian, u_d is reference of tips of tweezers width an d u is image feature.

5.5 Algorithm flow

Fig. 10 shows the algorithm flow in this system.

Tweezers grasping phase: The hand grasps the tweezers. Fig.11 shows how the tweezers are grasped in the case of "precision prismatic thumb-index finger" grasp [3]. The width of tweezers tips decreases if the fingers touch the tweezers and grasp it. This decrease is observed by the vision system. The operation moves through the following phase.

Target tracking phase: The hand manipulates tips of the tweezers to the target. The center of the target and the center of the tweezers tips are set to ξ_0, ξ_t in eqn.(2). When the tip of the tweezers touches the table surface plane where the target is placed, the tracking speed is controlled based on eqn.(3).

Target grasping phase: The hand grasps the target with the tweezers. Using eqn.(5), u is set to width of tips of the tweezers and u_d is set to the target width.

Target moving phase: The hand tracks the target to place it. Using eqn.(2), $\xi 0$ is set to place the target. ξt is set to the center of the tweezers tips.

Target releasing phase: The hand releases the target with tweezers. Using eqn.(5), u is the tweezers tips width and u_d is set to larger than the target width.

Tweezers moving phase: The hand moves the tweezers to place it. Using eqn.(2), ξ_{0x} is set to the initial value of the center of tweezers tip. Now, joint structure between the hand and the tweezers is changed. So by using eqn.(5), the width of the tweezers tip is set to the point contact situation and the kinematic structure of the passive joint is changed.

Tweezers releasing phase: Tweezers is released with pushing it.

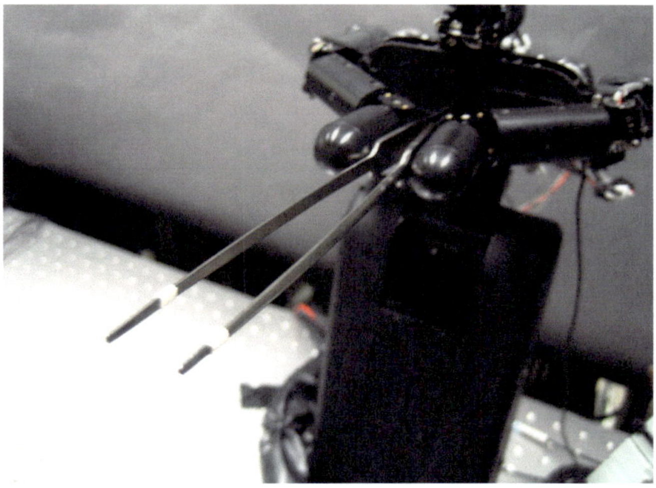

Fig. 11. Precision Grasp

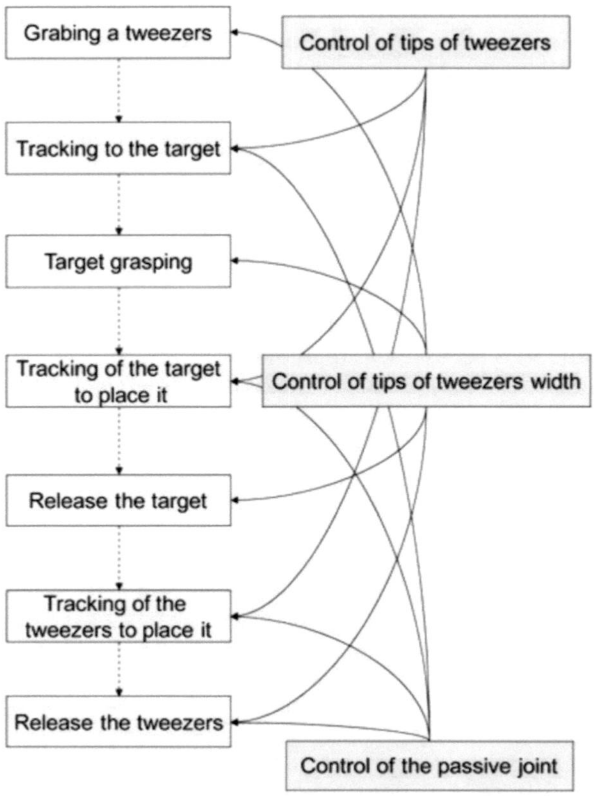

Fig. 10. Algorithm flow.

6. Experimental result

We used a single rice grain as a target; its long axis is roughly 5mm and its short axis is roughly 3mm. Another target is a screw; Its is 7.5mm long and 2mm across. Fig.12, Fig.13 and Fig.16 show the results.

Fig.12 shows experimental results under different initial conditions. We put the target at the different position for each trial. Regardless of difference in the initial conditions the tweezers grasped the target well.

Fig.13. shows experimental results by using the algorithms previously described. The hand grasps the tweezers, grasps the target by tweezers, moves the target for place point and places the tweezers. At 0.5 second, the hand grasps the tweezers. At 2 second, the grabbing state changed. At 2.7 second, the tweezers placed the target. At 3.5 second, the hand changed the grabbing state of tweezers by using gravity. Fig.14 shows the part of the tracking of tweezers to the target. When Y pixel of the center of tweezers tips was 745, the tip of tweezers touched the table plane. After that, the tweezers tips moved along the surface. Normally, if the passive joint structure changes, the tracking speed changes rapidly. But in this case the speed changed smoothly by changing the coefficient of the control method.

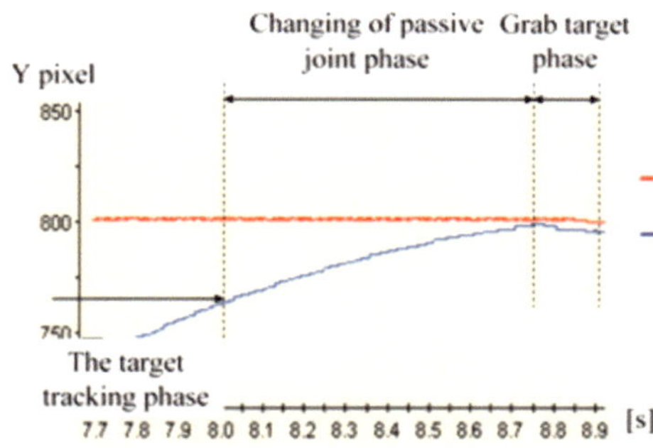

Fig. 14. Experimental result of target tracing phase.

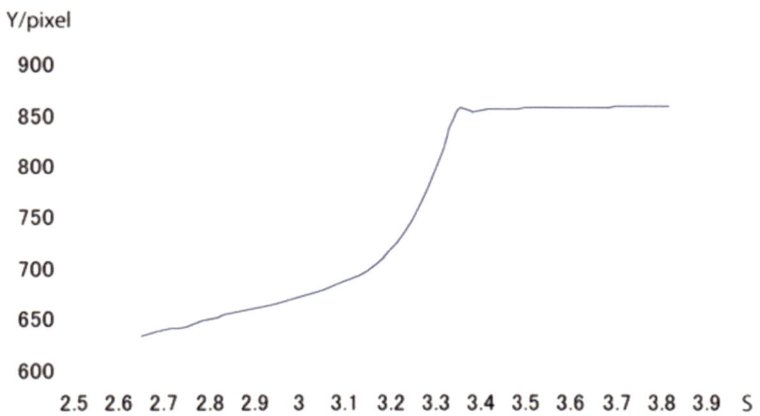

Fig. 15. The tweezers tips moving on image plane of Y pixel on gravity use phase.

Fig. 15. shows the tips of tweezers on the image plane of Y pixel. This figure shows the hand makes good use of the gravity.

Fig. 16. shows experimental results of moving target. The target is moving from opposite side of the hand to the hand, and the hand grasps the target. Regardless of motion of the target, the hand grasps well. Fig. 17 and 18 shows the tips of tweezers and the target on image plane of X pixel and Y pixel. In this case the time between vision recognition and grabbing is 0.1 second. The tips of tweezers follows the target well. This experiment concludes the high speed vision and high speed hand are effective for grabbing in this case.

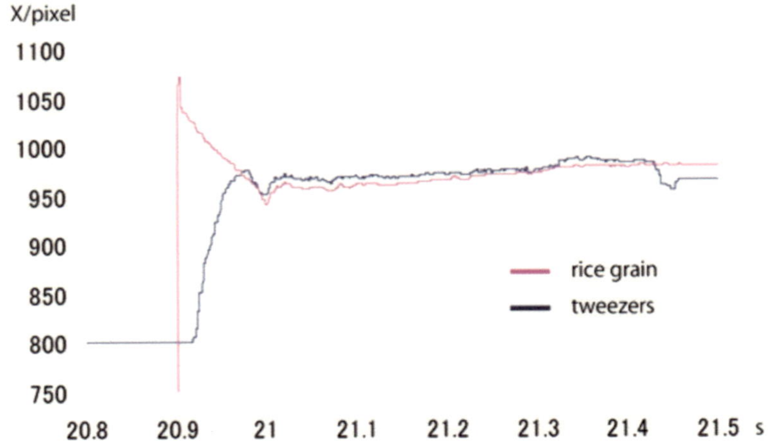

Fig. 17. The target and the tweezers tips moving on image plane of X pixel.

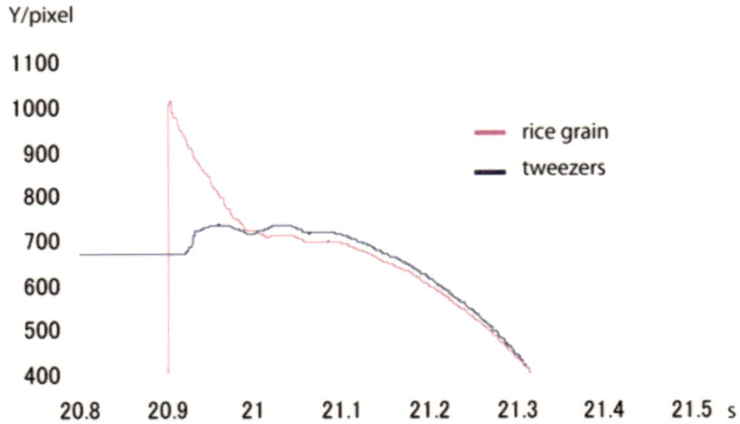

Fig. 18. The target and the tweezers tips moving on image plane of Y pixel.

7. Conclusion

The effectiveness of the visual servoing in the use of the tool was shown. It was shown that a more flexible control was possible by the use of the passive joint. In the future work we will check the applicability of this approach to other tools and its commonization to other this problem.

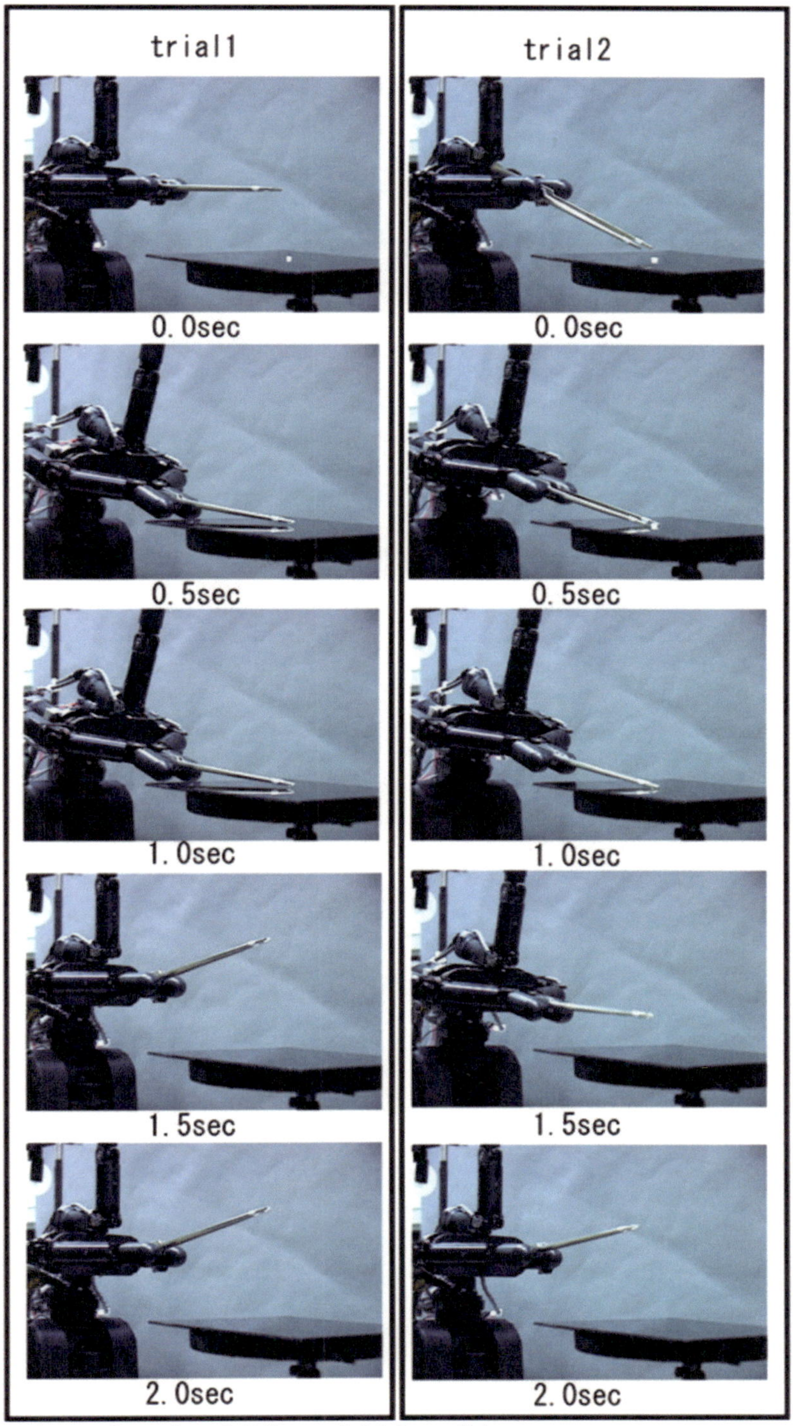

Fig. 12. Experimental results of different initial conditions.

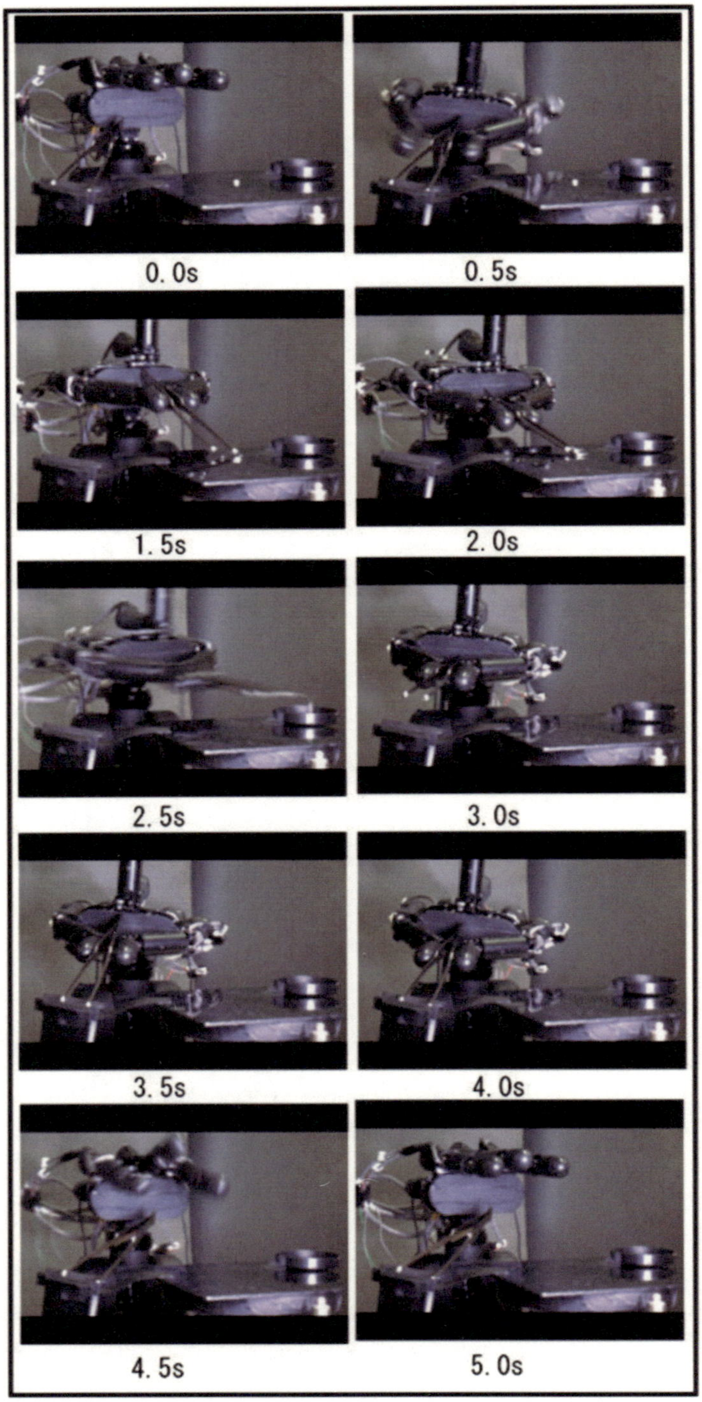

Fig. 13. Experimental results of rice pick up.

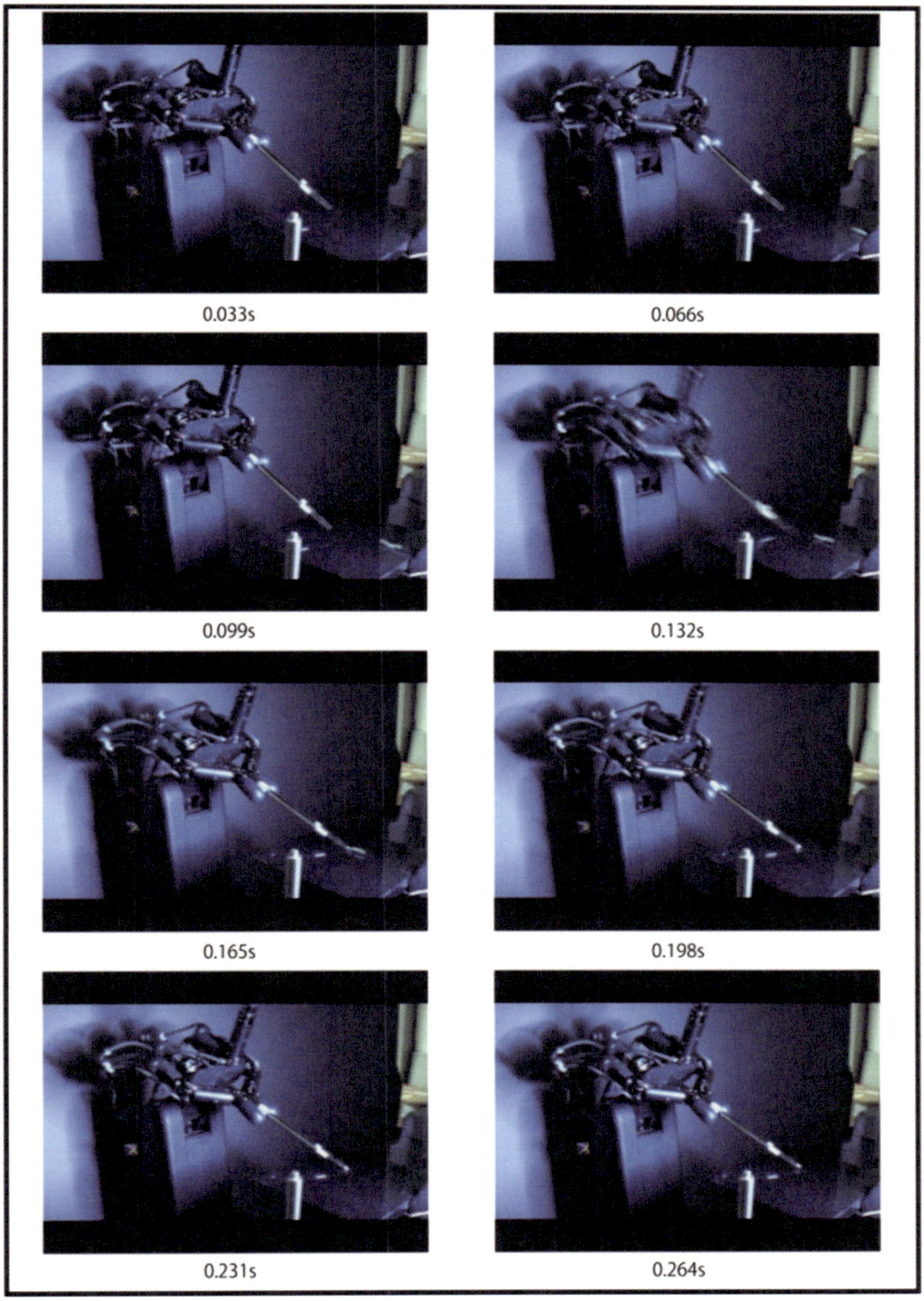

Fig. 16. Experimental results of moving target.

8. References

[1] A. Bicchi. Hands for dexterous manipulation and robust grasping: A difficult road toward simplicity. *IEEE Trans. Robot. and Automat.*, 16(6):652–662, 2000.

[2] S. Hutchinson, G.D. Hager, P.I. Corke, A tutorial on visual servo IEEE Transactions on Robotics and Automation, Vol. 12, No. 5, pp. 651-670, 1996

[3] M. R. Cutkosky. On grasp choice, grasp models, and the design of hand for manufacturing tasks. Proc. IEEE Int. Conf. Robot. and Automat., Vol. 5, No. 3, pp. 269-279, June 1989.

[4] H.Sugiuchi and T.Morino. A Pair of Scissors Handling Task with Dual Multi-Finger Hand Arm System : Add vision support functions JSME Conference on Robotics and Mechatronics, Vol. 2003 p. 66

[5] H.Kawasaki, T.Mouri, S.Ikenohata, Y.Ohtsuka and T.Endo. Multi-Fingered Haptic Interface Robot Handling Plural Tool Devices: Proceedings of the Second Joint EuroHaptics Conference and Symposium on Haptic Interfaces for Virtual Environment and Teleoperator Systems table of contents Pages 397-402

[6] Kemp, Charles C. and Edsinger, Aaron. Robot Manipulation of Human Tools: Autonomous Detection and Control of Task Relevant Features Proceedings of the Fifth Int Conf on Development and Learning, Special Session on Classifying Activities in Manual Tasks. 2006.

[7] K. Bernardin, K. Ogawara, K. Ikeuchi, R. Dillmann, A Sensor Fusion Approach for Recognizing Continuous Human Grasping Sequences Using Hidden Markov Models, IEEE Transactions on Robotics, Vol. 21, No. 1, pp 47-57, 2005.

[8] A.Namiki, Y.Imai, M.Ishikawa, and M.Kaneko. "Development of a High-speed Multifingered Hand System and Its Application to Catching," *Proc. Int. Conf. Intelligent Robots and Systems*, pp. 2666-2671, 2003.

[9] N. Furukawa, T. Senoo, A. Namiki, and M. Ishikawa, Dynamic regrasping using a high-speed multifingered hand and a high-speed vision system, *Proc. IEEE Int. Conf. on Robotics and Automation*, pp. 181–187, 2006.

[10] T. Ishihara, A. Namiki, M. Ishikawa, and M. Shimojo, Dynamic Pen Spinning Using a High-speed Multifingered Hand with High-speed Tactile Sensor, *Proc. IEEE RAS Int. Conf. on Humanoids Robots*, pp. 258–263, 2006.

[11] Y. Yamakawa, A. Namiki, M. Ishikawa and M. Shimojo, One-Handed Knotting of a Flexible Rope with a High-Speed Multifingered Hand having Tactile Sensors. Proc. of IEEE/RSJ 2007 Int. Conf. on Intelligent Robots and Systems, pp. 703–708, 2007.

[12] Akio Namiki, Yoshiro Imai, Makoto Kaneko, Masatoshi Ishikawa: Development of a High-speed Multifingered Hand System, International Conference on Intelligent

26

Vision-Based Haptic Feedback with Physically-Based Model for Telemanipulation

Jungsik Kim[1] and Jung Kim[1]
[1]*Korea Advanced Institute of Science and Technology (KAIST)*
South Korea

1. Introduction

Haptic feedback offers the potential to increase the quality and capability of human-machine interactions as well as the ability to skillfully manipulate objects by exploiting the sense of touch (Lin & Salisbury, 2004). Previous studies on haptic feedback systems typically dealt with virtual reality (VR)-based simulations, and telemanipulation systems. VR-based simulation systems used haptic information for various applications such as gaming (Morris, 2004), surgical simulations (Basdogan et al., 2004), or molecular simulations (Ferreira, 2006) in order to provide realistic virtual experiences along with sound and graphic rendering. In telemanipulation, haptic feedback has been studied in the fields of robotic guidance and obstacle avoidance (Hassanzadeh et al., 2005), robotic surgery (Mayer et al., 2007; Wagner et al., 2007) and micro/nano manipulation (Sitti & Hashimoto, 2003, Ammi et al., 2006). According to these studies, the feedback of haptic information to an operator can improve performance and provide telepresence. For example, in nano- or bio-manipulation applications, where the operator manipulates a micro-scale object with limited two-dimensional vision feedback through a microscope, haptic assistance can be used to provide the depth information, generate virtual fixtures or guides and thus improve the operator manipulation final quality (e.g., operation time and efficiency).

The goal of telemanipulation is to create a human operator interaction with a remote environment as closely as possible. Such a goal can be realized by (i) obtaining the available information of the slave site, such as the geometry, kinematic information, and material properties; (ii) applying this information to a user with high-fidelity master devices; and (iii) efficiently conveying the user response to the slave environment through actuating systems. Although many studies on the technical issues encountered in telemanipulation have been carried out, sensing the force information and its reflection to a user still constitutes a challenging issue because of problems associated with sensor design and force rendering.

Sensing the force information of a slave environment is a prerequisite in order to display a user force feedback during manipulation tasks. For example, the realization of a force feedback in telemanipulation has mainly been done thus far by integrating force sensors into a slave site to measure reaction forces between a slave robot and the environment. The

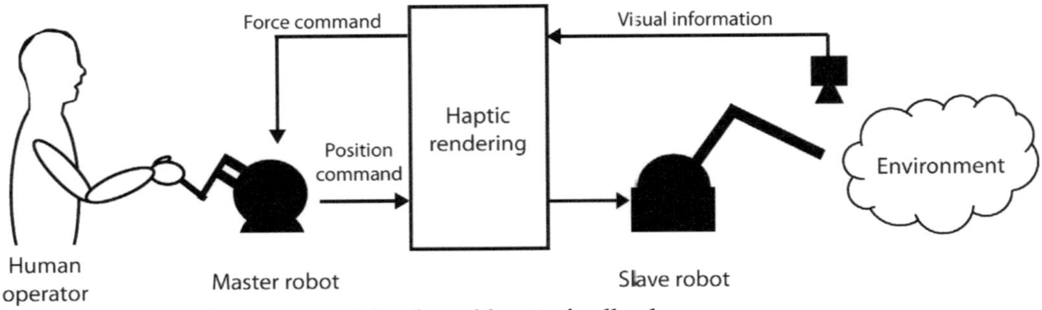

Fig. 1. Telemanipulation with vision-based haptic feedback

measured force signals are then filtered to guarantee the stability of the haptic device and offer an improved quality of the force feedback. The force sensor, however, has a low signal-to-noise ratio (SNR) for force feedback and can be damaged through physical contact with the environment or by exposure to biological and chemical materials. Although the use of a strain-gauge sensor or a commercial six-axes force/torque sensor in teleoperated robotic surgery has been examined (Mayer et al., 2007; Wagner et al., 2007), current commercial surgery robots hardly provide an adequate haptic feedback due to safety and effectiveness issues, partially associated with the reliability of the force sensor in a noisy environment. Very-small-scale force sensing for micromanipulation is more difficult because of the design of small force sensors that needs to meet challenging requirements for such applications, including micro-sensing for multiple degrees of freedom (DOF) with high resolution and accuracy while maintaining a high SNR. In addition, sufficient reliability and repeatability of the force sensor must be preserved. In particular, micro-scale measurements for biomanipulation are subject to severe disturbances due to the liquid surface tension (e.g., when cells are in a medium) and adhesion forces (Lu et al., 2006; Gauthier & Nourine, 2007). Therefore, new methods capable of avoiding the use of the force sensors have recently become very prevalent.

This chapter presents a new method for rendering the interaction forces of a slave environment based on visual information rather than on direct force measurements using a force sensor (Fig. 1). The visual information measured from optical devices is transformed into haptic information by modeling the slave environment. The interaction forces are rendered from this environment using a mechanical model representing the relationship between the object deformation and the applied forces. Therefore, it is not necessary to use force sensors. Originally, the term "haptic rendering" was defined as the process of computing and generating forces in response to a user interaction with virtual objects (Salisbury et al., 1995), including collision detection, force response, and control algorithms (Salisbury et al., 2004). The proposed algorithm also incorporates these components in order to compute and generate forces due to the user interaction with the visually modeled slave environment.

The interaction force prediction algorithm is investigated using image processing and physically-based modeling techniques. The geometry (boundary) information of a deformable object is obtained from images of the slave site in pre-process, and the kinematic information of a slave tool tip can be obtained using a fast image processing algorithm for the input of the physically-based model to estimate the interaction forces. In this Chapter, the boundary element method (BEM) is used as a physically-based modeling technique for the modeling while a priori knowledge of the material properties is assumed. During the interactions, the boundary conditions are updated using a real-time motion analysis of the slave environment. The interaction forces are then calculated based on the model, and are then conveyed to the user through a haptic device. The proposed algorithm only requires the material properties and the object edge information. Thus, this algorithm is robust to topological changes of the model network. In addition, measuring the deformation of an entire object body and applying it to the model as nodal displacements can be a very time-consuming work. Therefore the position update of a slave robot (tool tip) is used to recover the forces, similarly to the haptic interaction point (HIP) in VR applications (Massie & Salisbury, 1994). Moreover, the proposed system addresses the force sensing issues in both micro- and macro-scales so that a very small- or very large-scale slave environment can be rendered using the proposed algorithm.

This chapter is organized as follows: Section 2 presents the previous work related to vision-based force estimation methods. Section 3 provides an overview of the proposed haptic rendering algorithm, which is based on image processing and physically-based modeling techniques. In order to demonstrate the effectiveness of the proposed method, macro- and micro-scale telemanipulation systems were developed. In Section 4, the experimental results of the developed telemanipulation systems are presented. Finally, conclusions and suggestions with regard to future work are given in Section 5.

2. Previous Work

A large number of computer vision and image processing techniques have been investigated with regard to the object recognition and tracking (Ogawa et al., 2005), the characterization of material properties (Tsap et al., 2000; Liu et al., 2007a), the collision detection (Wang et al., 2007), and the modeling of deformable objects (Metaxas & Kakadiaris, 2002). In this context, the force estimation from visual information has also received much attention. Forces are usually computed based on the geometric information of an object (or a manipulator) for the known input displacements, for which the measured geometrical information is applied to a force estimation algorithm. For instance, Wang et al. (2001) computed the deformation gradients of elastic objects from images and estimated the external forces using the stress-strain relationships. Luo and Nelson (2001) presented a method fusing force and vision feedback for a deformable object manipulation, in which the measured deformation was applied to a finite element (FE) model to obtain the force estimates. Greminger and Nelson (2004) showed a force measurement through the boundary displacements of elastic objects using a Dirichlet-to-Neumann map. Nelson et al. (2005) measured the applied forces for biological cells with a point-load model for cell deformation. DiMaio and Salcudean (2003) measured the tissue phantom deformation to estimate the applied force distribution during the insertion of a needle. Anis et al. (2006) used the force-displacement relationship of a micro-gripper in a microassembly process. Liu et al. (2007b) measured the contact forces of a biological single-cell using the deflection of a polydimethylsiloxane (PDMS) post in a cell holding device.

A few researchers have studied the real-time force estimation algorithms for haptic rendering based on visual information. Owaki et al. (1999) introduced a concept in which the visual data of real objects were used as haptic data to simulate the virtual touching of an object, but not for telemanipulation tasks. They used a high-speed active-vision system allowing to obtain visual data at 200 Hz. Ammi et al. (2006) used microscopic images to provide haptic feedback in a cell injection system. A cell nonlinear mass-spring model was used to compute the interaction forces for haptic rendering. However, mass-spring models offer limited accuracy (Kerdok et al., 2003). Other significant disadvantages of their method include its weak connection to biomechanics. For example, there was no mechanically relevant relationship between the model parameters and the object material properties. Moreover, the parameters were calculated from off-line finite element method (FEM) simulations; this required extra FE modeling efforts and the results were influenced by the network topology. Kennedy and Desai (2005) proposed a vision-based haptic feedback system in the case of robot-assisted surgery. A rubber membrane was modeled using a FE model, and a grid located on the rubber membrane was visually tracked in order to measure its displacement. The FE model then reflected the interaction forces using the displacement values as boundary conditions. With this method, however, it was necessary to stamp a grid pattern on the object to generate the internal meshes and track each node for the FE model, which made this method inconvenient and impractical for biological- and micro-scale objects. In addition, real-time solution of FEM is usually not feasible (Delingette, 1998).

In conclusion, the mass-spring system and FEM model in the aforementioned studies present severe shortcomings, often requiring additional efforts. FEM models were not efficient enough to be used in real-time applications. Finally, in many of the previous systems, the FEM required a controlled slave environment to model the membrane. The mass-spring model was usually non-realistic and highly-sensitive to the tuning of the model, such as in the spring constant of the mesh, through additional experiments. To circumvent the issues related to the use of FEM and mass-spring models, the present paper uses BEM as an alternative approach to estimate the forces required for the haptic feedback. BEM is a numerical solution technique to solve the differential equations representing an object model that computes the unknowns on the model boundary instead of on its entire body. The proposed method uses the object edge information and known material properties, which make it highly adaptive to the network topology changes by reducing the amount of additional effort required in previous systems.

3. Vision-Based Haptic Interaction Method

3.1 Overview

Fig. 2 represents the coordinates of the developed system. A master interface has a master space with frame $\mathbf{\Phi}$ in which the position of the haptic stylus is given by the three-dimensional (3D) vector ${}^{\Phi}\mathbf{p}$. The physical interactions between a manipulator and a deformable object are introduced in the slave space $\boldsymbol{\varphi}$. The shape of an object can be expressed by ${}^{\varphi}\mathbf{q}$ and the position of the manipulator ${}^{\varphi}\mathbf{p}$ is related to ${}^{\Phi}\mathbf{p}$ by the transform $\mathbf{T}^{P}$. The interactions in the slave space are mapped to the image space $\mathbf{I}$ to measure the position ${}^{\varphi}\mathbf{p}$ and ${}^{\varphi}\mathbf{q}$ and to estimate the interaction force ${}^{\varphi}\mathbf{F}=f({}^{\varphi}\mathbf{q}, {}^{\varphi}\mathbf{p})$, where $f(\cdot)$ represents the continuum mechanics method. The interaction force ${}^{\varphi}\mathbf{F}$ is then transformed into ${}^{\Phi}\mathbf{F} = \mathbf{T}^{F} \cdot {}^{\varphi}\mathbf{F}$ using the transform $\mathbf{T}^{F}$. The transforms $\mathbf{T}^{P}$ and $\mathbf{T}^{F}$ contain scaling factors between the master and slave spaces. If a position scaling factor in $\mathbf{T}^{P}$ is set to scale down (or up), the forces are scaled up (or down) by a force scaling factor in $\mathbf{T}^{F}$.

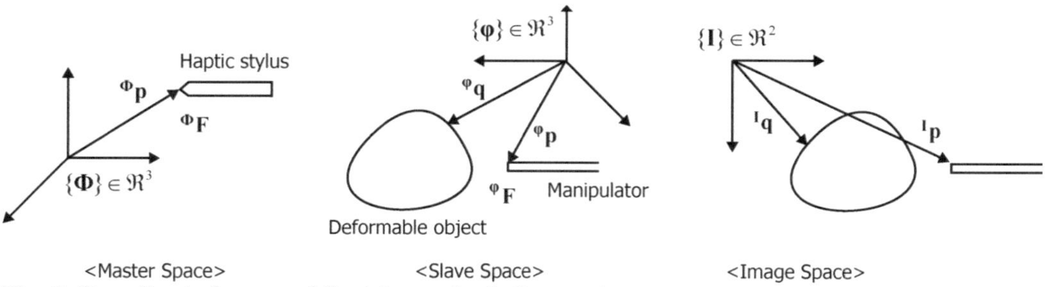

Fig. 2. Coordinate frames of the telemanipulation system

The algorithm consists of two parts (Fig. 3): the construction of a deformable object model (preprocess) and the interaction force update for each frame (run-time process). In the preprocess phase, the edge information of the object is obtained using image processing techniques, and a boundary mesh is constructed based on the edge information. The boundary element (BE) model is then created with the object mesh and known material properties. Using this model, the system of equations is built and pre-computed; it is used for a fast update of the system matrix in the run-time process.

In the run-time phase, collision detection and force computations are performed at a rate of 1 kHz. When a user interacts with a deformable object, the displacement at the contact point is applied to the model as a boundary condition. The boundary contact force is then computed using the BEM. If the displacement magnitude or the contact point changes, new force values can be obtained by updating the boundary conditions using real-time image processing and by applying them to the pre-computed system matrix in the preprocess phase.

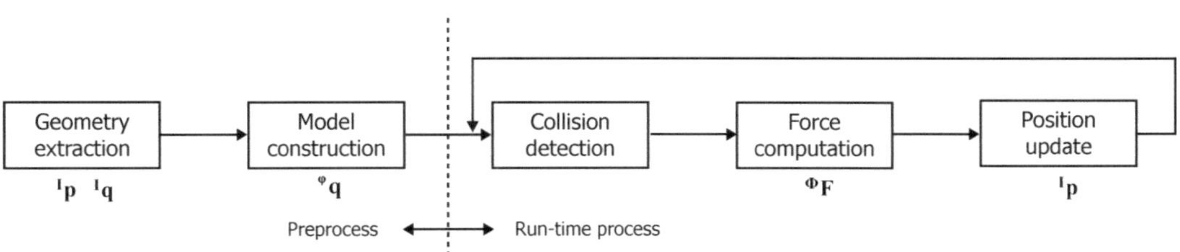

Fig. 3. The force prediction algorithm pipeline

The key parts of the algorithm consist of the geometry extraction from images, the object modeling and the real-time computation of the interaction forces. The remainder of this Section concretely explains each part of the algorithm.

3.2 Geometry Extraction

Fast and accurate motion tracking and edge detection techniques are important for modeling a deformable object. The edge (${}^{I}\mathbf{q}$) of the object along with the tool tip position (${}^{I}\mathbf{p}$) of a slave-manipulator is extracted and tracked using the following methods.

A template matching is used to track the tool tip position (${}^{I}\mathbf{p}$), which is a process that determines the location of a template by measuring the degree of similarity between an image and the template. Although there are several methods that can measure the degree of similarity, such as the summation of the squared difference (SSD), a normalized cross-correlation coefficient was implemented to reduce the degree of sensitivity to contrast changes in the template and in the video image (Aggarwal et al., 1981). The correlation between the pixel of the template (w × h) and every pixel in the entire image is given by

$$\tilde{C}({}^{I}x,{}^{I}y)=\frac{\sum_{{}^{I}y'}^{h-1}\sum_{{}^{I}x'}^{w-1}\tilde{T}({}^{I}x',{}^{I}y')\tilde{I}({}^{I}x+{}^{I}x',{}^{I}y+{}^{I}y')}{\left(\sum_{{}^{I}y'}^{h-1}\sum_{{}^{I}x'}^{w-1}\tilde{T}({}^{I}x',{}^{I}y')^{2}\sum_{{}^{I}y'}^{h-1}\sum_{{}^{I}x'}^{w-1}\tilde{I}({}^{I}x+{}^{I}x',{}^{I}y+{}^{I}y')^{2}\right)^{1/2}} \tag{1}$$

where $\tilde{I}({}^{I}x+{}^{I}x',{}^{I}y+{}^{I}y')=I({}^{I}x+{}^{I}x',{}^{I}y+{}^{I}y')-\bar{I}({}^{I}x,{}^{I}y)$, $\tilde{T}({}^{I}x',{}^{I}y')=T({}^{I}x',{}^{I}y')-\bar{T}$. $I({}^{I}x,{}^{I}y)$ and $T({}^{I}x,{}^{I}y)$ are the corresponding values at location $({}^{I}x,{}^{I}y)$ of the image and template pixels, respectively. $\bar{I}({}^{I}x,{}^{I}y)$ and $\bar{T}$ are the average pixel value in the template and the average pixel value in the image under the template window, respectively. In order to reduce the computational load of the pixel-by-pixel operation (Equation 1), a moving region-of-interest (ROI) is adopted. As the movement of the tool tip is very small in the sequential frames, the ROI is determined around the identified position via a template matching. The template matching is then performed in the ROI to obtain the new position.
To represent the geometry (${}^{\varphi}\mathbf{q}$) of a deformable object, the two-dimensional object boundary (${}^{I}\mathbf{q}$) is extracted using the active contour model (snake) developed by Kass et al. (1988). The contour with a set of control points is initially manually placed near the edge of interest. The energy function defined surrounding each control point is then computed, and the contour is drawn to the edge of the image where the energy has a local minimum. In this paper, a fast greedy algorithm (Williams & Shah, 1992) for energy minimization is used and the energy function E_{snake} is defined by

$$E_{snake}=\int(\alpha(s)\cdot E_{cont}+\beta(s)\cdot E_{curv}+\gamma(s)\cdot E_{image})ds \tag{2}$$

Here, s is the arc-length along the snakes contour taken as a parameter. The continuity energy E_{cont} minimizes the distance between control points and prevents all control points from moving toward the previous control point. E_{curv} represents the curvature energy and it is responsible for the curvature of the contour corner. The image energy E_{image} indicates the normalized edge strength. The values of α, β and γ determine the factors of each energy term. The edge of the object is finally represented by the positions of the control points which are used to mesh the boundary of the object for the BE model.

3.3 Continuum Mechanics Model

For realistic and plausible force estimation, the continuum mechanics modeling of a deformable object has been widely studied and developed in haptic applications (Meier et al., 2005). In continuum mechanics, differential equations for the stress- or strain-equilibrium have to be solved and numerical methods such as FEM and BEM are usually used with a discretization of the object into a number of elements.

The BEM directly uses mechanical parameters and handles various interactions between the tools and the objects. Due to its physically-based nature and computational advantages over the FEM, it has been used in computer animation and haptic applications. James and Pai (2003) successfully applied BEM to the simulation of a deformable object with haptic feedback. The reaction force and deformation were computed based on pre-computed reference boundary value problems known as Green's functions (GFs) and a capacitance matrix algorithm (CMA).

In this work, the BE model of a deformable object was built using the extracted object edge information using the control points of an active contour model and the related material properties (Young's modulus E and Poisson's ratio ν). The boundary of the object was discretized into N elements. The points representing the unknown values, tractions (forces per unit area) and displacements are defined as nodes. In the present study, we have selected constant elements for simplicity, namely the nodes are assumed to be in the middle of each element and the unknowns have a constant value over each element. The resulting system of equations is given by Equation 3 (Kim et al., 2009).

$$\mathbf{HP} = \mathbf{GV} \tag{3}$$

Here, the $\mathbf{H}(E, \nu, \mathbf{q})$ and $\mathbf{G}(E, \nu, \mathbf{q})$ matrices are 2N × 2N dense matrices in the case of 2D problems. $\mathbf{P}$ and $\mathbf{V}$ are the displacement and traction vectors, respectively. The boundary conditions, displacements or tractions, are applied at each node to solve these algebraic equations. When the displacement value is given on a node, the traction value can be obtained, and vice versa. Equation 3 can be rearranged as

$$\mathbf{AY} + \overline{\mathbf{A}}\overline{\mathbf{Y}} = \mathbf{0} \rightarrow \mathbf{Y} = \mathbf{A}^{-1}(-\overline{\mathbf{A}}\overline{\mathbf{Y}}), \tag{4}$$

where $\mathbf{Y}$ is the unknown vector consisting of unknown boundary nodal values, and $\overline{\mathbf{Y}}$ represents the known boundary conditions. $\mathbf{A}$ and $\overline{\mathbf{A}}$ consist of the columns of the $\mathbf{H}$ and $\mathbf{G}$ matrices according to the indices of $\mathbf{Y}$ and $\overline{\mathbf{Y}}$, respectively. $\mathbf{Y}$ can be obtained by solving Equation 4.

When an object is deformed, the boundary conditions at the collision nodes change. Therefore, Equations 3 and 4 must be rewritten to take the new boundary conditions into account and they must be solved in real-time.

3.4 Real-Time Force Computation

For a real-time and realistic haptic interaction, it is necessary to provide a haptic feedback with updating rates greater than 500 Hz (Chen & Marcus, 1998). In other words, the interaction forces must be computed within 2 msec. In order to solve the linear matrix system of Equation 4 in real-time, a CMA is used (James & Pai, 2003). If the S boundary conditions change for the linear elastic model, the $\mathbf{A}$ matrix for a new set of boundary conditions can be related to the pre-computed $\mathbf{A}_0$ matrix by swapping simple S block columns. Using the Sherman-Morrison-Woodbury formula, the relationship between $\mathbf{A}$ and $\mathbf{A}_0$ can be obtained as follows:

$$\mathbf{A}^{-1} = \mathbf{A}_0^{-1} - \mathbf{A}_0^{-1}(\bar{\mathbf{A}}_0 - \mathbf{A}_0)\mathbf{I}_S\mathbf{C}^{-1}\mathbf{I}_S^T\mathbf{Y}_0 \tag{5}$$

Equation 4 can be then represented by

$$\mathbf{Y} = \mathbf{A}^{-1}(-\bar{\mathbf{A}}\bar{\mathbf{Y}}) = \mathbf{Y}_0 + (\mathbf{I}_S + \Xi\mathbf{I}_S)\mathbf{C}^{-1}\mathbf{I}_S^T\mathbf{Y}_0$$

$$\begin{aligned} \mathbf{C} &= -\mathbf{I}_S^T\Xi\mathbf{I}_S \\ \Xi &= -\mathbf{A}^{-1}\bar{\mathbf{A}} \\ \mathbf{Y}_0 &= \left[\Xi(\mathbf{I} - \mathbf{I}_S\mathbf{I}_S^T) - \mathbf{I}_S\mathbf{I}_S^T\right]\bar{\mathbf{Y}} \end{aligned} \tag{6}$$

Here, $\mathbf{I}_S$ is an 2N × 2S submatrix of the identity matrix, $\mathbf{C}$ is known as the capacitance matrix (2S × 2S) and $\mathbf{Y}_0$ is computed using Equation 4. The GFs Ξ is computed for a predefined set of boundary conditions in the preprocess phase. Equation 6, known as the capacitance matrix formulae, can then be implemented to reduce the amount of re-computation. The solution $\mathbf{Y}$ for the tractions and displacements over the entire boundary can be obtained by computing the inverse of the smaller capacitance matrix. For example, in the case of a point contact, S =1, only a 2 × 2 matrix inversion is required.
It is not necessary to compute the global deformation because the visual feedback is provided through real-time video images rather than using computer-generated graphic images. Given the nonzero displacement boundary conditions at the contact S nodes, the resulting contact force can be computed by

$${}^{\Phi}\mathbf{F} = \alpha_E\mathbf{V}_S = \alpha_E\mathbf{I}_S^T\mathbf{Y} = -\alpha_E\mathbf{C}^{-1}\mathbf{I}_S^T\bar{\mathbf{Y}} = -\alpha_E\mathbf{C}^{-1}\bar{\mathbf{Y}}_S \tag{7}$$

Here, α_E is the effective area. It consists of the nodal area and a scaling factor for different-scale manipulation tasks in order to magnify (or reduce) the contact force while providing a haptic feedback to the user.
Although the contact forces are rapidly computed using locally updated boundary conditions, the forces are obtained at a visual update rate (of approximately 60 Hz) because of the boundary conditions that are updated from the images. It is insufficient to achieve a good fidelity haptic feedback. Therefore, a force interpolation method (Zhuang & Canny, 2000) is used to derive the forces at high rates (1 kHz).

3.5 Collision Detection

The collision detection is achieved utilizing hierarchical bounding boxes and a neighborhood watch algorithm (Ho et al., 1999). The BE model is hierarchically represented as oriented bounding box trees and stored in a preprocess phase. If a line segment between the previous and current tool tip positions is inside the bounding box, potential collisions are sequentially checked along the tree. When the last bounding box for the line element collides with the line segment, the ideal haptic interface point is constrained at the collision node. The distance between the tool tip and the collision node is used as the displacement boundary condition of the node. During interactions, the collision nodes are rapidly updated using a neighborhood watch algorithm, which is based on a predefined linkage between the nodes.

4. Case Studies and Results

The developed algorithm was evaluated for the manipulation of elastic materials with different scales. Two experiments were conducted to demonstrate the effectiveness of the algorithm in macro- and micro-telemanipulation tasks. In both systems, the deformation of the objects and the motion of a slave robot were captured by a CCD camera (SVS340MUCP, SVS-Vistek, Seefeld, Germany with 640 × 480 pixels resolution and maximum of 250 fps) and the images were transmitted to a computer (Pentium-IV 2.40 GHz). The 2D geometry information can be known through image processing techniques using OpenCV. A commercial haptic device (SensAble Technologies, PHANToM OmniTM, USA) was used for force feedback and a priori knowledge of the material properties was obtained through the experiment and from the literature. The behavior of the model during manipulation was compared with that from a real deformable object. The overall system block diagram is shown in Fig. 4.

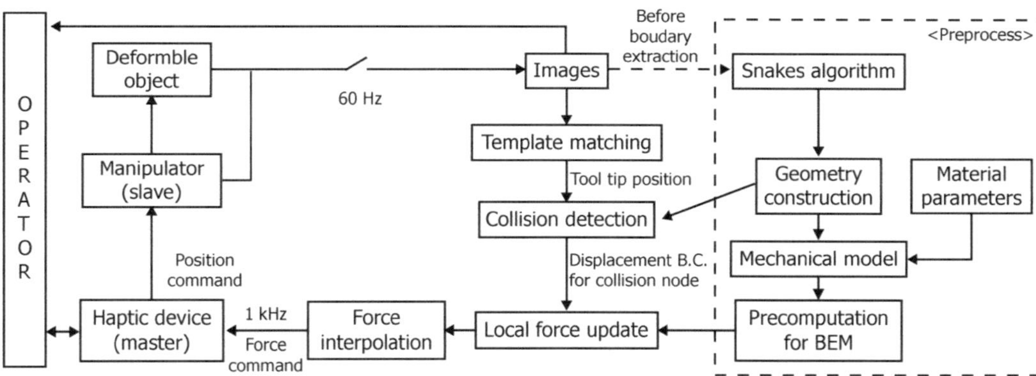

Fig. 4. Overall system block diagram

4.1 Experiment 1: Macro-Scale Telemanipulation System

The macro-scale manipulation system consists of an inanimate deformable object and a planar manipulator with an indenter tip as a slave robot. Fig. 5 shows the setup for the experimental platform. A 3 DOF planar manipulator (500 mm × 500 mm) performs indentation tasks on a rectangular-shaped object made from silicone gel (88 mm × 88 mm × 9 mm, GE, TSE3062, USA). The Young's modulus of the silicone block is 127 kPa (Kim et al., 2008; Kim et al., 2009). The images obtained using a CCD camera have a size of 640×480 pixels and a resolution of 0.35 mm/pixel. In addition, the indentation force is measured using a one-axis force sensor (Senstech, SUMMA-5K, Korea) with a resolution of 50 mN. The force sensor is used to validate the estimated force from visual information.

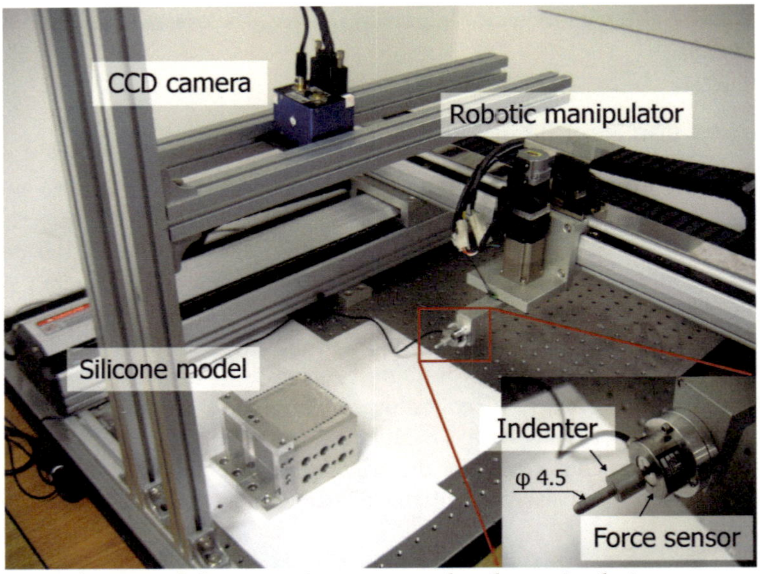

Fig. 5. Experimental setup of slave part in macro-scale telemanipulation system

The geometry of the rectangular-shaped block was represented using 60 control points along the active contour. Hence, the BE model consisted of 60 line elements with 60 nodes. As one side of the block was fixed to the platform, zero displacement boundary conditions were applied on this side. When the indenter deformed the block, the resulting contact force was computed based on the proposed method. Simultaneously, the actual contact force along the indenter insertion axis was measured by the force sensor.
The model prediction was compared with the block response. Fig. 6 shows a comparison between the actual block deformation and the global deformation of the BE model according to dissimilar indentation locations. The dotted line represents the nodes of the BE model; it is determined as a result of the input displacement at the contact point. Each nodal displacement of the BE model is in good agreement with the deformation of the object. The interaction forces at the contact point are shown in Fig. 7. The results show a reasonable match between the actual and estimated force values. While the local strain was raised, the difference between the values was increased due to the linear approximation of the silicone block nonlinearities. A measure of bias (0.0576 N) was also observed due to errors coming from the object buckling along the perpendicular direction to the plane and from measurement errors occurring in the image analysis (e.g., edge detection noise, minor illumination changes). The bias could be overcome using a scaling factor in the case of the micromanipulation system, where the scaled-up reaction force must be reflected to the user.

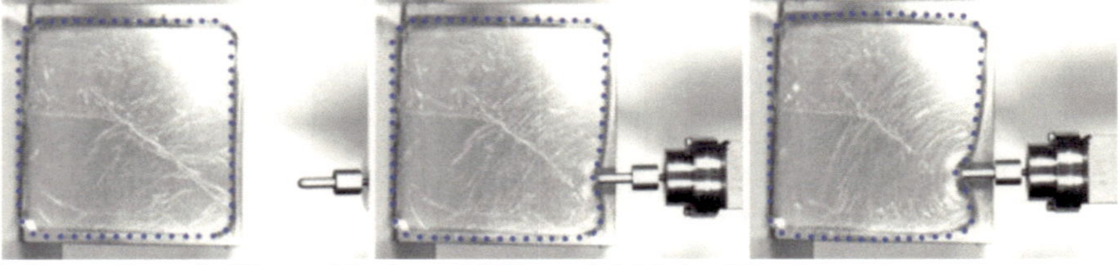

Fig. 6. Deformation of silicone block and BE model (dotted line)

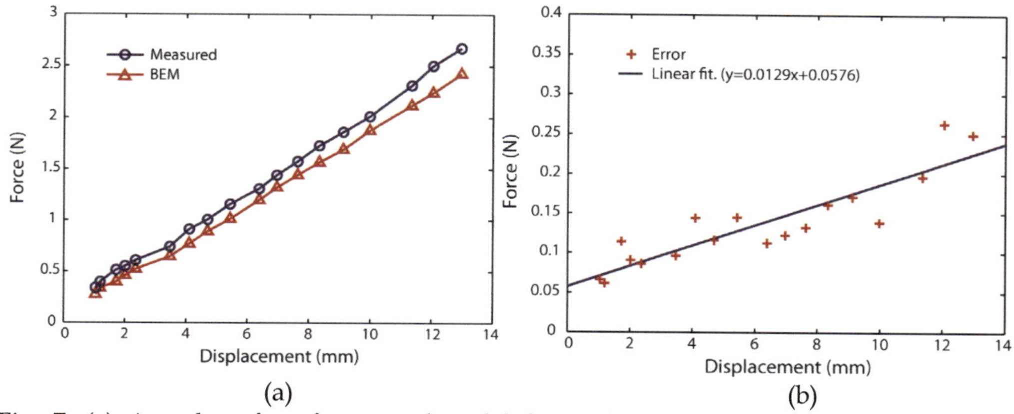

Fig. 7. (a) Actual surface forces and nodal forces from BEM, and (b) errors along the indentation axis

4.2 Experiment 2: Cellular Manipulation System

In this experiment, an application to cellular manipulation is presented. Cellular manipulations such as a microinjection are now increasingly used in transgenics and in biomedical and pharmaceutical research. Some examples include the creation of transgenic mice by injecting cloned deoxyribonucleic acid (DNA) into fertilized mouse eggs and intracytoplasmic sperm injections (ICSI) with a micropipette. However, most cellular manipulation systems have primarily focused to date on visual information in conjunction with a dial-based console system. The operator needs extensive training to perform these tasks, and even an experienced operator can have low success rates and a poor reproducibility due to the nature of the tasks (Kallio & Kuncova, 2003; Sun & Nelson, 2002).

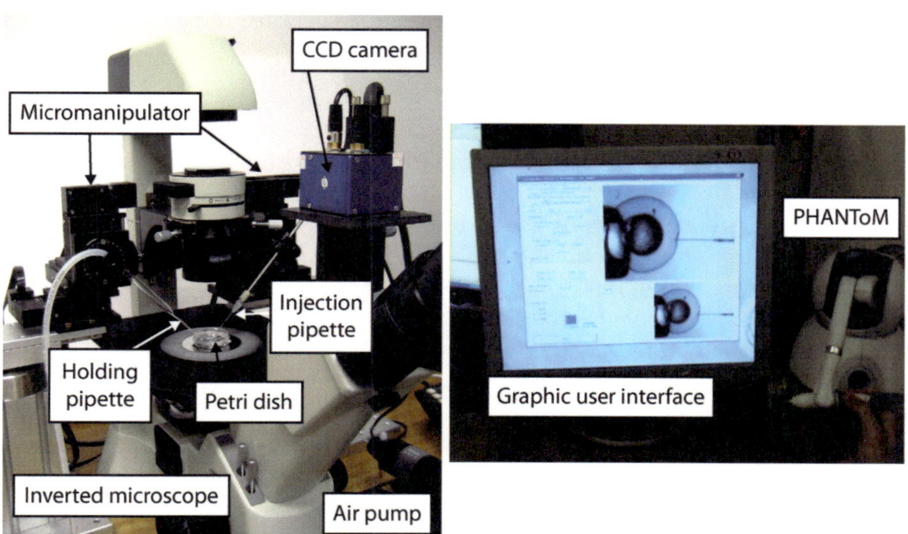

Fig. 8. Developed cellular manipulation system

The developed cell injection system is shown in Fig. 8. It consists of an inverted microscope (Motic, AE31, China) and two 3 DOF micromanipulators (Sutter, MP225, USA) to guide the cell holding and injection units. An injection micropipette (Humagen, MIC-9 μ m-45, USA) is connected to a micromanipulator, whereas a glass capillary with an air pump (Eppendorf, CellTram Air, Germany) is connected to another micromanipulator to hold the cell. Each micromanipulator has a resolution of 0.0625 µm along each axis and a travel distance of 25 mm. Images were captured at a 40× magnification. The obtained images have a size of 640 × 480 pixels and a resolution of 2 µm/pixel.
Zebrafish embryos were used as a deformable object in the experiments. Zebrafish have been widely used as a model in developmental genetic and embryological research due to their similarity to the human gene structure (Stainer, 2001). The embryos are considered as a linear elastic material for research in the small deformation linear theory. It has been reported that the Young's modulus of the chorion of the zebrafish embryo is approximately 1.51 MPa with a standard deviation of 0.07 MPa and that the Poisson's ratio is equal to 0.5 (Kim et al., 2006). These properties were used in the BE model of the cell.
Conventionally, the cell injection procedure involves (i) guiding the injection pipette, (ii) puncturing the membrane, (iii) and depositing the materials. In this work, the task was to puncture the chorion of a zebrafish embryo and to guide the injection pipette to a targeted position. The location of the targeted position was randomly chosen and changed for every test.

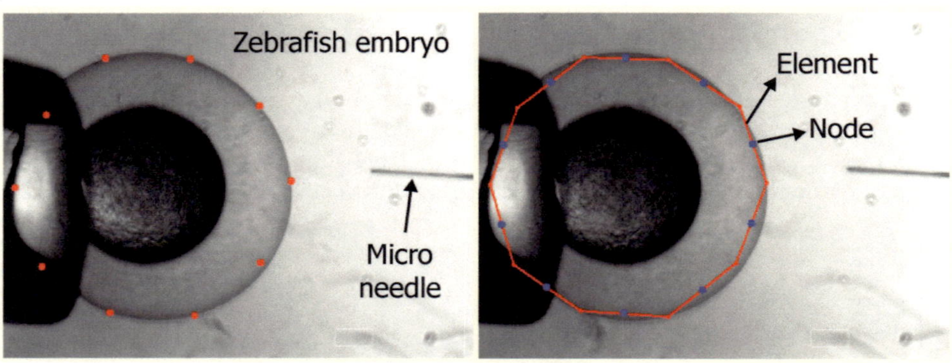

Fig. 9. Edge detection of a zebrafish embryo and BE model with 10 elements.

Fig. 9 shows the edge detection of the zebrafish embryo and the BE model with line elements. The nodes attached to the holding pipette (a glass capillary) have zero displacement boundary conditions..
Unlike macro-scale experiments for the silicone block, and as a result of excessive forces, the cell membrane was punctured in this case using an injection pipette. Therefore, it was necessary to provide the user with a puncturing cue. As the BEM cannot compute the membrane puncturing, the overshoot of the injection pipette after the breaking of the membrane was measured. Published work revealed that the penetration force significantly decreases after puncturing (Kim et al., 2006). Accordingly, when the position overshoot occurred, the magnitude of the reaction force was set to zero.
Fig. 10 shows the estimated force response for the deformation created by the injection pipette. The membrane was punctured when the deformation length ranged approximately between 50 µm and 200 µm. According to previously-published work (Kim et al., 2006), the

force-deformation relationship for a zebrafish embryo is characterized by a nonlinear behavior that can be approximated as linear for small deformations (up to 100 μm). This allows us to use the proposed linear elastic model for small deformations.

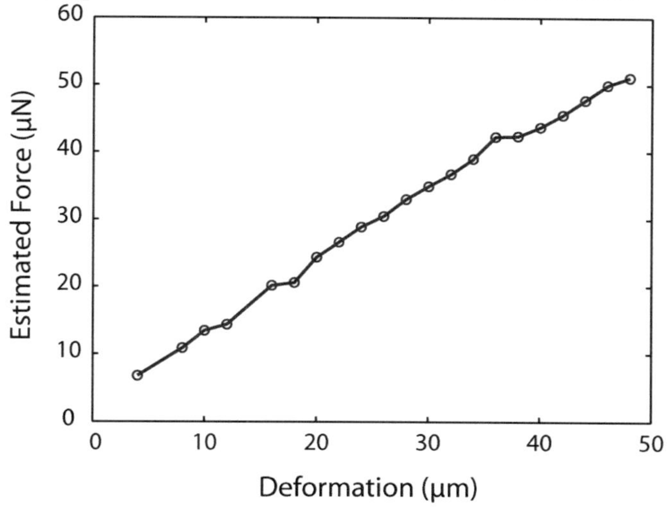

Fig. 10. Estimated force of a zebrafish embryo using vision-based haptic interaction method.

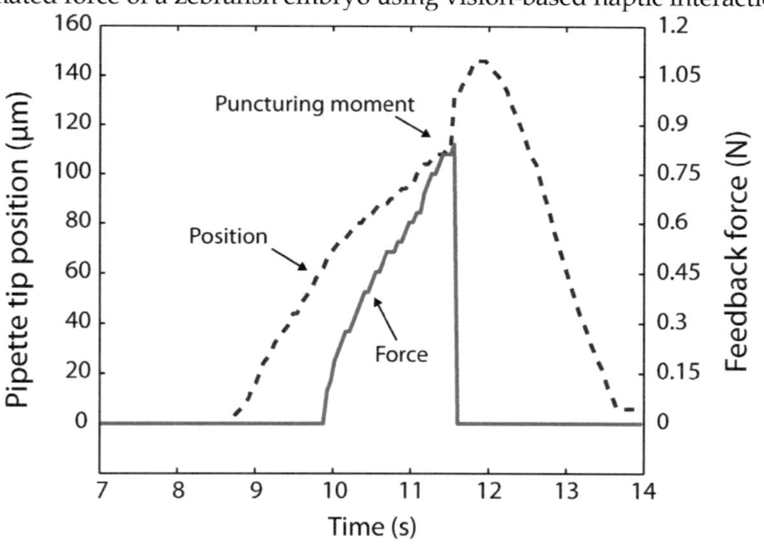

Fig. 11. Amplified cell injection and puncturing force computed using vision-based haptic interaction method

In order to display the force response to a user, the micro contact forces need to be magnified. Specifying and varying the appropriate force scaling factor has been an issue in micromanipulation (Lu et al., 2006; Menciassi et al., 2004). The scaling factor was experimentally chosen within the maximum applicable force of the haptic device (3.3 N). Fig. 11 shows the scaling forces over time for haptic rendering. The forces increase during the insertion of the micropipette, and drop to zero when puncturing occurs.

5. Conclusions and Discussions

In this paper, a haptic rendering algorithm of deformable objects was investigated while inferring the force information of a slave environment using visual information. This method is based on image processing techniques (active contour model and template matching) for the modeling of the slave environment and on a continuum mechanics model for the interactive haptic rendering. Experiments for different scales of telemanipulation systems were performed to demonstrate the effectiveness of the algorithm. The main result is that the developed method can be simply used to estimate the forces without a direct force measurement. The results of two different experiments also showed that the algorithm allows the users to feel reaction forces in real time during the indentation and injection tasks by means of haptic devices.

The advantages of the proposed method over direct force measurements using force sensors can be summarized as follows.

(i) The proposed system only requires a priori knowledge of the object material properties and edge information. These fewer requirements allow the algorithm to be robust to potential topological changes of the model network and do not imply a controlled slave environment.

(ii) The scale of the slave environment does not affect the rendering method. The same algorithm can not only be used in a micro- (or nano-) scale but also in a macro-scale environment. The cellular manipulation system of a zebrafish embryo and the macro-scale telemanipulation experiment of a silicone block showed the potential of the proposed method when applied at different scales. Therefore, it is expected that the developed rendering algorithm can be used in telemanipulation systems with various scales. Examples may include a cellular manipulator, a microassembly system or a telesurgery system. The proposed algorithm is particularly well suited for micromanipulation due to difficulties associated with reliable micro force sensing.

(iii) As the forces are inferred from the object model and the tracked tool tip position, it is not necessary to integrate a force sensor. As a non-contact (indirect) measurement, the developed algorithm will only be slightly affected by breakdowns caused by physical or biochemical interactions. In addition, the visual information of the slave environment is consistently available, as optical devices are installed in the manipulation system.

In the proposed method, the accurate modeling of the deformable objects is a key part for getting a high-fidelity haptic feedback. A number of assumptions and model parameters were required for the physically-based modeling. These could be determined by considering the characteristics of the objects, such as the material properties, geometry and contact conditions. This study assumed that a manipulated object was characterized by linear elastic responses having isotropic and homogeneous properties. However, in reality many deformable objects (e.g., biological cells, soft tissues) are inhomogeneous, anisotropic and made of nonlinear materials. If the aforementioned assumptions enable a rapid computation speed for a better stability of the haptic feedback, the unmodeled behavior might lead to registration problems (modeling error). For example, because the linear elasticity assumption will fail once the model deformation is sufficiently large, the model behavior diverges from that of a deformable object when a large deformation is produced during a manipulation. This modeling error can also be observed due to friction modeling. In our future work, the detrimental effects of modeling errors on the telemanipulation performance will be studied. If a manipulation task requires a large object deformation or deep interaction, the modeling error in the proposed algorithm might be overcome by adopting a nonlinear modeling approach (Wu et al., 2001) and even an inhomogeneous modeling technique (Jun et al., 2006). The added values will be accompanied by additional computational difficulties introduced by the techniques adopted. An analysis of the trade-off between the added values and the computational burden will also be required.

The BE model was characterized by a priori knowledge of the material properties and geometry obtained from images. The material parameters of many animate and inanimate objects have been measured and determined for various applications including motion analysis, flaw identification and haptic rendering. In this study, the unknown material properties of the deformable objects (the zebrafish embryo and the silicone block) were obtained from literature and using experiments. However, the parameters of other objects of interest may not be readily obtainable. Additional efforts would then be required to objectively determine the physical parameters. In the future work, the authors will strongly consider increasing the available information from the imaging sources. To achieve this goal, an image-based method for the identification of material parameters will be developed by applying an efficient and robust prediction algorithm. The parameters and the interaction forces will be estimated for the input displacements.

At two-dimensional modeling together with a mono image analysis was suitably established in the present experiments in the case of thin planar objects and planar manipulation tasks. An extension of this work to 3D models will be more helpful for many applications. Indeed, a 3D approach will provide additional cues for visual constraints such as those associated with depth information and occlusion.

The developed algorithm has only considered point-contacts between the object and the instrument. However, the measurement of the distribution forces present on the object or the instrument can be achieved using the proposed method without difficulty, while a direct measurement using conventional force sensors is often difficult and sometimes impossible. Another interesting extension will also include the integration of additional haptic feedback modalities, such as a torque feedback.

6. References

Aggarwal, J. K.; Davis, L. S. & Martin, W. N. (1981). Correspondence Processes in Dynamic Scene Analysis. *Proceedings of the IEEE*, Vol. 69, No. 5, 562–572.

Ammi, M.; Ladjal, H. & Ferreira, A. (2006). Evaluation of 3D pseudo-haptic rendering using vision for cell micromanipulation. *Proceedings of IEEE/RSJ International Conference on Intelligent Robots and Systems*, pp. 2115–2120, Beijing, China.

Anis, Y. H.; Mills, J. K. & Cleghorn, W. L. (2006). Vision-based measurement of microassembly forces. *Journal of Micromechanics and Microengineering*, Vol. 16, No. 8, 1639–1652.

Basdogan, C.; De, S., Kim, J., Muniyandi, M., Kim, H. & Srinivasan, M. (2004). Haptics in minimally invasive surgical simulation and training. *IEEE Computer Graphics and Applications*, Vol. 24, No. 2, 56–64.

Chen, E. & Marcus, B. (1998). Force feedback for surgical simulation. *Proceedings of the IEEE*, Vol. 86, No. 3, 524–530.

Delingette, H. (1998). Towards Realistic Soft Tissue Modeling in Medical Simulation. *Proceedings of IEEE: Special Issue on Surgery Simulation*, Vol. 86, No. 3, 512–523.

DiMaio, S. P. & Salcudean, S. E. (2003). Needle insertion modeling and simulation. *IEEE Transactions on Robotics and Automation*, Vol. 19, No. 5, 864–875.

Ferreira, A. & Mavroidis, C. (2006). Virtual reality and haptics for nano robotics: A review study. *IEEE Robotics and Automation Magazine*, Vol. 13, No. 3, 78–92.

Gauthier, M. & Nourine, M. (2007). Capillary Force Disturbances on a Partially Submerged Cylindrical Micromanipulator. *IEEE Transactions on Robotics*, Vol. 23, No. 3, 600–604.

Greminger, M. A. & Nelson, B. J. (2004). Vision-based force measurement. *IEEE Transactions on Pattern Analysis and Machine Intelligence*, Vol. 26, No. 3, 290–298.

Hassanzadeh, I.; Janabi-Sharifi, F., Akhavan, R. & Yang, X. (2005). Teleoperation of mobile robots by shared impedance control: a pilot study. *Proceedings of IEEE International Conference of Control Applications*, pp. 346–351, Toronto, Canada.

Ho, C. H.; Basdogan, C. & Srinivasan, M. A. (1999). Efficient point-based rendering techniques for haptic display of virtual objects. *Presence: Teleoperators and Virtual Environments*, Vol. 8, No. 5, 477–491.

James, D. L. & Pai, D. K. (2003). Multiresolution Green's function methods for interactive simulation of large-scale elastostatic objects. *ACM Transactions on Graphics*, Vol. 22, No. 1, 47–82.

Jun, S.; Choi, J. & Cho, M. (2006). Physics-based s-Adaptive Haptic Simulation for Deformable Object. *Proceedings of 14th Symposium on Haptic Interfaces for Virtual Environment and Teleoperator Systems*, pp. 477–483, Alington, USA.

Kallio, P. & Kuncova, J. (2003). Manipulation of living biological cells: Challenges in automation. *Workshop on microrobotics for biomanipulation in the IROS'03*, Las Vegas, USA.

Kass, M.; Witkin, A. & Terzopoulos, D. (1988). Snakes: active contour models. International *Journal of Computer Vision*, Vol. 1, No. 4, 321–331.

Kennedy, C. W. & Desai, J. P. (2005). A vision-based approach for estimating contact forces: Applications to robot-assisted surgery. *Applied Bionics and Biomechanics*, Vol. 2, No. 1, 53–60.

Kerdok, A. E.; Cotin, S. M., Ottensmeyer, M. P., Galea, A. M., Howe, R. D. & Dawson, S. L., (2003). Truth Cube: Establishing Physical Standards for Soft Tissue Simulation. *Medical Image Analysis*, Vol. 7, No. 3, 283–291.

Kim, D. H.; Hwang, C. N., Sun, Y., Lee, S. H., Kim, B. & Nelson, B. J. (2006). Mechanical analysis of chorion softening in prehatching stages of zebrafish embryos. *IEEE Transactions on Nanobioscience*, Vol. 5, No. 2, 89–94.

Kim, J. S.; Janabi-Sharifi, F. & Kim, J. (2008). A Physically-Based Haptic Rendering for Telemanipulation with Visual Information: Macro and Micro Applications. *Proceeding of the IEEE/RSJ Int. Conf. Intelligent Robots and Systems*, pp. 3489-3494, Nice, France.

Kim, J. S.; Janabi-Sharifi, F. & Kim, J. (2009). Haptic Interaction Method Using Visual Information and Physically-Based Modeling. *IEEE/ASME Trans. Mechatronics*, On review for publication.

Lin, M. & Salisbury, K. (2004). Haptic rendering - Beyond visual computing. *IEEE Computer Graphics and Applications*, Vol. 24, No. 2, 22–23.

Liu, X.; Wang, Y., & Sun, Y. (2007a). Real-time high-accuracy micropipette aspiration for characterizing mechanical properties of biological cells. *Proceedings of IEEE International Conference on Robotics and Automation*, pp. 1930–1935, Rome, Italy.

Liu, X.; Sun, Y., Wang, W., & Lansdorp, B. M. (2007b). Vision-based cellular force measurement using an elastic microfabricated device. *Journal of Micromechanics and Microengineering*, Vol. 17, No. 7, 1281–1288.

Lu, Z.; Chen, P. C. Y. & Lin, W. (2006). Force sensing and control in micromanipulation. *IEEE Transactions on Systems, Man, and Cybernetics, Part C: Applications and Reviews*, Vol. 36, No. 6, 713–724.

Luo, Y. & Nelson, B. J. (2001). Fusing force and vision feedback for manipulating deformable objects. *Journal of Robotic Systems*, Vol. 18, No. 3, 103–117.

Massie, T. H. & Salisbury, J. K. (1994). The PHANToM haptic interface: A device for probing virtual objects. *Proceedings of ASME Dynamic Sys. Control Div.*, pp. 295–301, Chicago, USA.

Mayer, H.; Nagy, I., Knoll, A., Braun, E., Bauernschmitt, R. & Lange, R. (2007). Haptic feedback in a telepresence system for endoscopic heart surgery. *Presence: Teleoperators and Virtual Environments*, Vol. 16, No. 5, 459–470.

Meier, U.; Lopez, O., Monserrat, C., Juan, M. C. & Alcaniz, M. (2005). Real-time deformable models for surgery simulation: a survey. *Computer Methods and Programs in Biomedicine*, Vol. 77, No. 3, 183–197.

Menciassi, A.; Eisinberg, A., Izzo, I. & Dario, P. (2004). From macro to micro manipulation: models and experiments. *IEEE/ASME Trans. Mechatronics*, Vol. 9, No. 2, 311–320.

Metaxas, D. N. & Kakadiaris, I. A. (2002). Elastically Adaptive Deformable Models. *IEEE Transactions on Pattern Analysis and Machined Intelligence*, Vol. 24, No. 10, 1310–1321.

Morris, D.; Neel, J. & Salisbury, K. (2004). Haptic battle pong: High-degree-of-freedom haptics in a multiplayer gaming environment. *Experimental Gameplay Workshop, GDC 2004*, San Jose, USA.

Nelson, B. J.; Sun, Y. & Greminger, M. A. (2005). Microrobotics for molecular biology: Manipulating deformable objects at the microscale. In: *Springer Tracts in Advanced Robotics*, Vol. 15, 115–124, Springer Berlin/Heidelberg.

Ogawa, N.; Oku, H., Hashimoto, K. & Ishikawa, M. (2005). Microrobotic visual control of motile cells using high-speed tracking system. *IEEE Transactions on Robotics*, Vol. 21, No. 4, 704–712.

Owaki, T.; Nakabo, Y., Namiki, A., Ishii, I. & Ishikawa, M. (1999). Real-time system for virtually touching objects in the real world using modality transformation from images to haptic information. *Systems and Computers in Japan*, Vol. 30, No. 9, 17–24.

Salisbury, K.; Brock, D., Massie, T., Swarup, N. & Zilles, C. (1995). Haptic rendering: programming touch interaction with virtual objects. *Proceedings of the 1995 symposium on Interactive 3D graphics*, pp. 123–130, Monterey, California, United States.

Salisbury, K.; Conti, F., & Barbagli, F. (2004). Haptic rendering: introductory concepts. *IEEE Computer Graphics and Applications*, Vol. 24, No. 2, 24–32.

Sitti, M. & Hashimoto, H. (2003). Teleoperated touch feedback from the surfaces at the nanoscale: modeling and experiments. *IEEE/ASME Trans. Mechatronics*, Vol. 8, No. 2, 287–298.

Stainier, D. Y. R. (2001). Zebrafish genetics and vertebrate heart formation. *Nature Reviews Genetics*, Vol. 2, No. 1, 39–48.

Sun, Y. & Nelson, B. J. (2002). Biological cell injection using an autonomous microrobotic system. *International Journal of Robotics Research*, Vol. 21, No. 10-11, 861–868.

Tsap, L. V.; Goldgof, D. B., Sarkar, S. & Powers, P. S. (2000). A method for increasing precision and reliability of elasticity analysis in complicated burn scar cases. *International Journal of Pattern Recognition and Artificial Intelligence*, Vol. 14, No. 2, 189–211.

Wagner, C. R.; Stylopoulos, N., Jackson, P. G. & Howe, R. D. (2007). The Benefit of Force Feedback in Surgery: Examination of Blunt Dissection. *Presence: Teleoperators and Virtual Environments*, Vol. 16, No. 3, 252–262.

Wang, W. H.; Liu, X. Y. & Sun, Y. (2007). Contact detection in microrobotic manipulation. *The International Journal of Robotics Research*, Vol. 26, No. 8, 821–828.

Wang, X.; Ananthasuresh, G. K. & Ostrowski, J. (2001). Vision-based sensing of forces in elastic objects. *Sensors and Actuators A*, Vol. 94, No. 3, 142–156.

Williams, D. J. & Shah, M. (1992). A fast algorithm for active contours and curvature estimation. *CVGIP: Image Understanding*, Vol. 55, No. 1, 14–26.

Wu, X. L.; Downes, M. S., Goktekin, T. & Tendick, F. (2001). Adaptive nonlinear finite elements for deformable body simulation using dynamic progressive meshes. *Computer Graphics Forum*, Vol. 20, No. 3, 349–358.

Zhuang, Y. & Canny, J. (2000). Haptic interaction with global deformations. *Proceedings of IEEE International Conference on Robotics and Automation*, pp. 2428–2433, San Francisco, USA.

Image Stabilization for *In Vivo* Microscopic Imaging

Sungon Lee
Korea Institute of Science and Technology
Republic of Korea

1. Introduction

Robotics enjoys its growing number of applications in various fields. In this chapter, a robotic system for bio-medical application will be introduced. By adding a robotic system to the conventional microscope, we have solved one of the challenging problems with *in vivo microscopic imaging*. *In vivo microscopic imaging* refers to the imaging technology that visualizes the function of biological process within intact living organism with microscopes. The technology is thought to be a very powerful tool in biological research by enabling biologists to observe what happens inside living organs in a live body, which was impossible before. This useful tool will also play a critical role in many bio-related industries as well. For example, it can greatly enhance the drug discovery process (Bullen, 2008).

However, observing inside a living body with great magnification is not easy. There are some challenges such as insufficient spatial resolution, physical access issues and so on. One of the challenges includes observation problem. Observation itself is significantly disturbed by the physiological motions such as breath, heartbeat, and peristalsis. Even though the animal under observation is usually put under the anesthesia, these motions keep occurring simply because the animal is alive. The motions shake the whole body. So, even very small trembling can happen at any organ. You may not feel it with your own eyes. However, a microscope, the magnifying device, enlarges this trembling as well as organs. As a result, the trembling of the organ sometimes distorts the images from scan-based microscopes such as confocal laser scanning microscopes, or sometimes makes the images totally black by causing out of focus in optical microscopes. All the times, *in vivo* motion brings about observational difficulty.

We tackle this problem. By employing motion canceling robotic technology, we have proposed twp image stabilization methods. After explaining on a fundamental difficulty with *in vivo* microscopy more detail in the next section, two image stabilization systems will be explained with experimental results.

2. *In Vivo* Microscopic Imaging and Its Problem

A fundamental difficulty of *in vivo* microscopic imaging lies in that the microscopy is highly sensitive to motion, which naturally and necessarily occurs at cells of living animals. The causes of this motion include breathing, heartbeat and peristalsis. Since these motions are parts of life processes, they occur even when the subject is put under anesthesia. Those motions significantly disturb the microscopic observation. At worst, they make the observation impossible by causing out-of-focus in the microscope view. Fig. 1 shows an example of unstable observation. Images are from a confocal microscope. Black parts in images are often observed due to out-of-focus by subject's motion. So, continuous observation is impossible being disturbed by the motion.

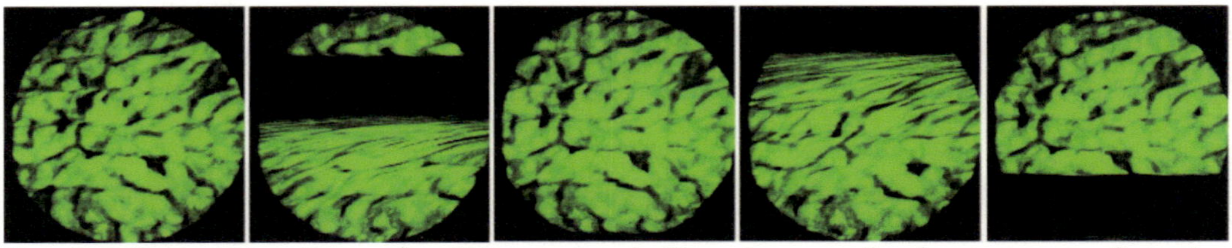

Fig. 1. Microscopic views of a living mouse's liver in a confocal microscope (invasive observation); images are unstable due to body trembling caused by physiological motions such as breathing, heartbeat, and peristalsis.

We have measured this motion. Fig. 2 shows the height of a live mouse liver measured by a laser-displacement sensor. The mouse was under anesthesia. In the graph, the big and periodic impulse-like motion turns out to be caused by breathing. Breathing vibrates the whole body once per one or two seconds. Between the respirations, heartbeat also trembles the body slightly with approximately 10 Hz, which is also periodic. Another low frequency motion, which moves the body slowly, is also observed. This motion is thought to be caused by peristalsis.

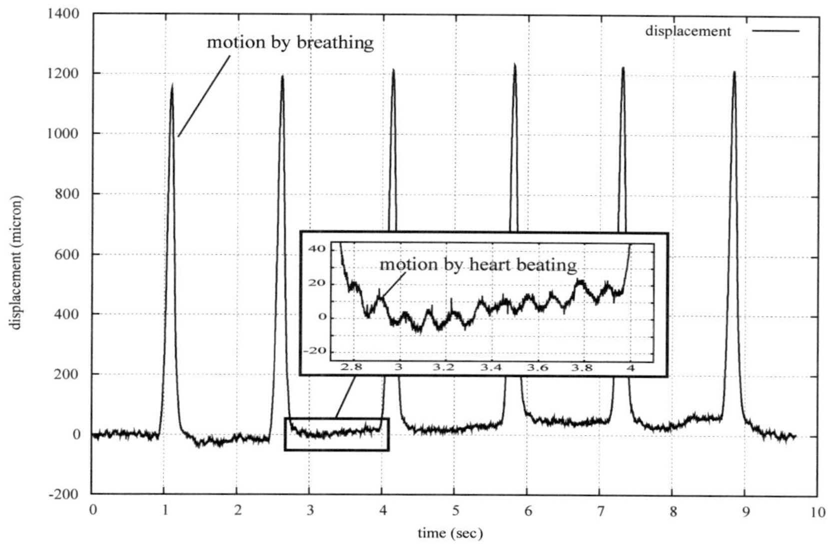

Fig. 2. Motions at a live mouse liver under anesthesia.

In the following sections, we introduce two robotic systems stabilizing observed images through motion synchronization. An objective lens will be controlled to synchronize itself with the subject's motion. This synchronization will virtually remove the relative motion between the lens and the subject, leading to stabilized images.

3. Motion Compensation by Visual Servoing

3.1 System

The first solution is a vision based compensation system (Lee et al. 2008a). We use a high-speed camera for detecting the *in vivo* motion, and move the objective lens to follow the detected motion. To implement this idea, a high-speed camera with 1000 fps is installed into one port of the microscope to measure motion on the image plane, and a robotic closed arm with enough accuracy and power was designed to move the objective lens. In robot technology terms, the system can be classified into an image-based visual servoing system (Hutchinson et al. 1996). In the image-based visual servoing, the motion signal f is defined in the image space. The image Jacobian J_{im} and the robot Jacobian J_{rob} map the motion signal f to the joint velocities q as follows:

$$\dot{f} = J_{im}\dot{r} = J_{im}J_{rob}\dot{q} \tag{1}$$

where $\dot{f}$ is the velocity of the motion, $\dot{r}$ is the end-effector velocity, and $\dot{q}$ is the joint velocity. From (1), we design a stable control law based on the resolved motion rate control.

$$\dot{q} = KJ^{-1}\Delta f \tag{2}$$

where the Jacobian matrix $J = J_{im}J_{rob}$, the error vector $\Delta f = f_d - f$ and K is a gain matrix. Then, Δf behaves as follows:

$$\Delta\dot{f} + K\Delta f = 0. \quad (3)$$

Before applying the visual feedback solution to the problem, we need planarize the in vivo motion because a single camera can only detect 2-D motion. If the motion moves the body in the direction of the light, the images becomes blurred by out-of-focus, making no image processing available. In order to prevent the subject from moving in that direction, we employ a simple pressing mechanical device (we call it *mechanical stabilizer*). The stabilizer presses the observed area with a small cover glass. Then, the motion was successfully restricted to the horizontal motion. And, since the translational motion is dominant compared to the rotational motion through the observation, our target motion to be stabilized is set as the 2-D translational motion.

We have developed a piezo-driven robotic closed arm with two DOFs to move the objective lens. It is a five-bar linkage with living hinges. Two accurate piezo-actuators push the mechanism, and then the enlarging mechanism amplifies the insufficient motion of the piezo-actuators. The living hinge, a thin section of the material, is widely used in the design of the MEMS due to its lack of any friction and very little wear.

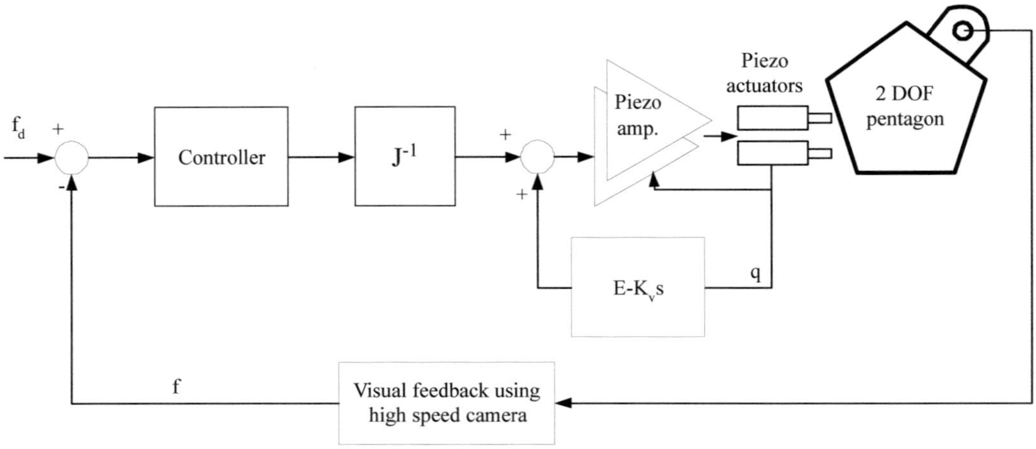

Fig. 3. Block diagram of visual feeback control for microscope image stabilization

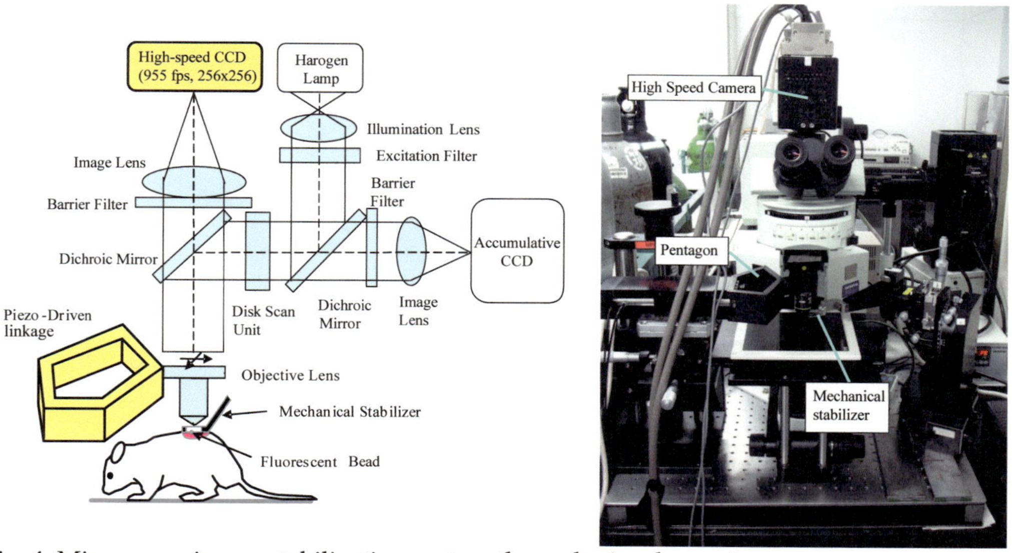

Fig. 4. Microscope image stabilization system through visual servoing

3.2 Result and Discussion

The *in vivo* experimental results show success of the visual servoing based compensation; motions were almost canceled, and as a result, we were able to get stationary image sequences.

Fig. 5 represents the compensated and the remaining motions. The solid line represents the residual motion while the compensated motion is also plotted with the dotted line in X axis (In Y axis, similar result was obtained). The residual motion was less than ±10 μm, while the maximum amplitude of the compensated motion was more than 150 μm. Thus, the image stabilization system removed more than 90% of the motion. The successful motion synchronization consequently generates stable image sequences, as shown in Fig. 7, which would be shown as in Figs. 6, without image stabilization. As we can compare with these image sequences, the vision-based image stabilization system greatly has improved *in vivo* image sequences. The stabilized image sequence is surely much easier to observe. Seamless and stable observation has become possible.

The experimental results have been very satisfactory, meeting our expectation. For improvement and broader applications, the following points should be considered in the next design.

1) *Coping with more complex motions*: Current design can only compensate 2-D translational rigid-body motion. Motion in the direction light axis can cause out-of-focus blurring in the images, and nonrigid-body motions or rotational motions still remain even though these are small compared to the 2-D translational motion.

2) *Observing a subject as intact as possible*: The pressure from the cover glass of the mechanical stabilizer may have unwanted effects on tissues or the living subjects.

3) *No artificial fiducials for image processing*: The fluorescent beads, the artificial fiducials, restrict the observation. It is difficult to locate them at a specific spot, and the beads themselves block the viewing below them.

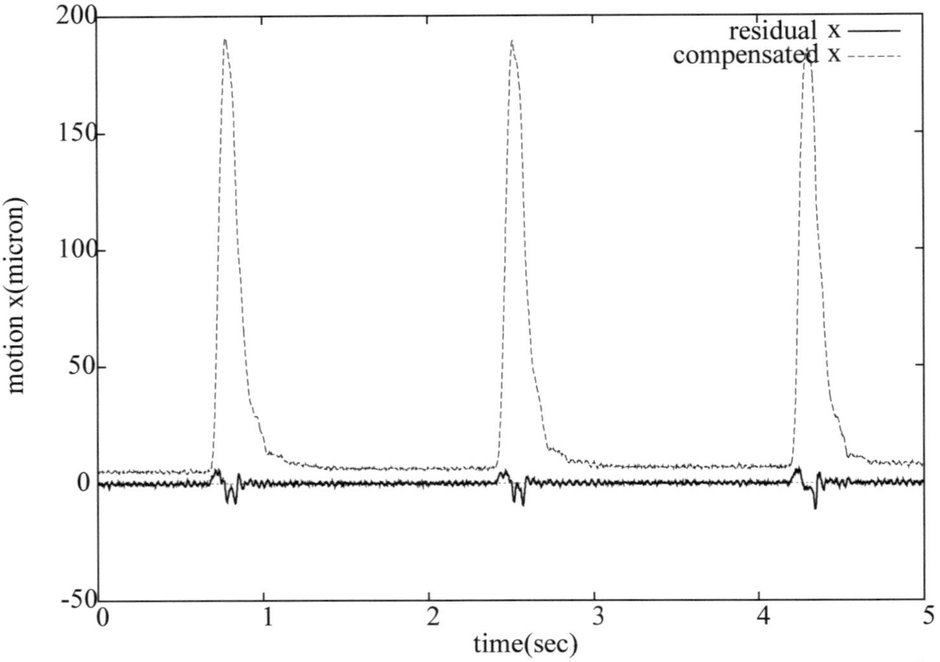

Fig. 5. Solid line: residual motion detected by a high-speed camera, and dotted line: compensated motion caculated from control inputs.

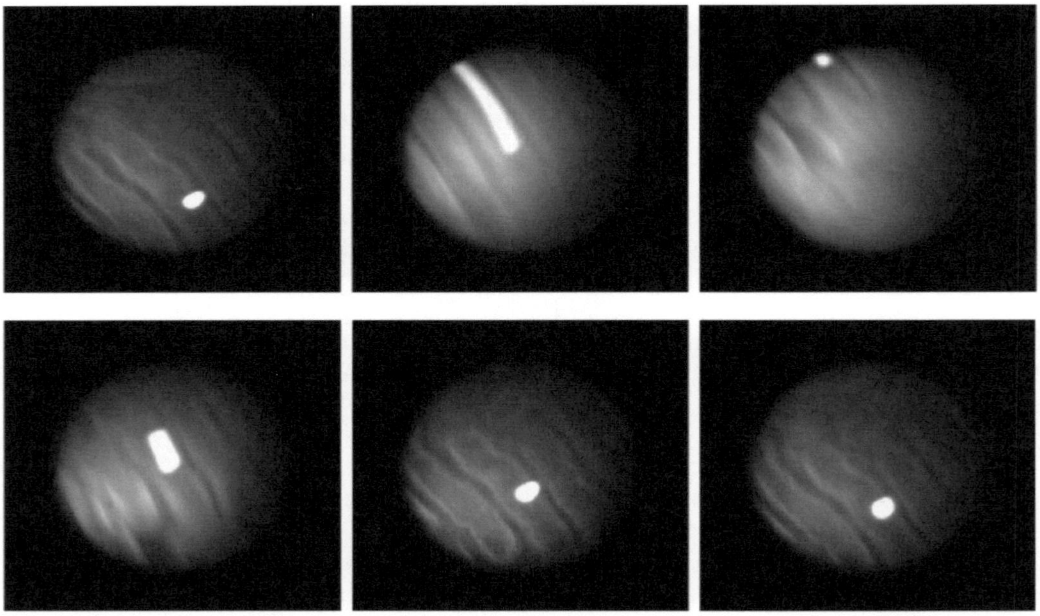

Fig. 6. Microscope image sequence of a mouse kidey (field of view is 200 μm, image sequence was captured by a cooled CCD camera with 37 fps).

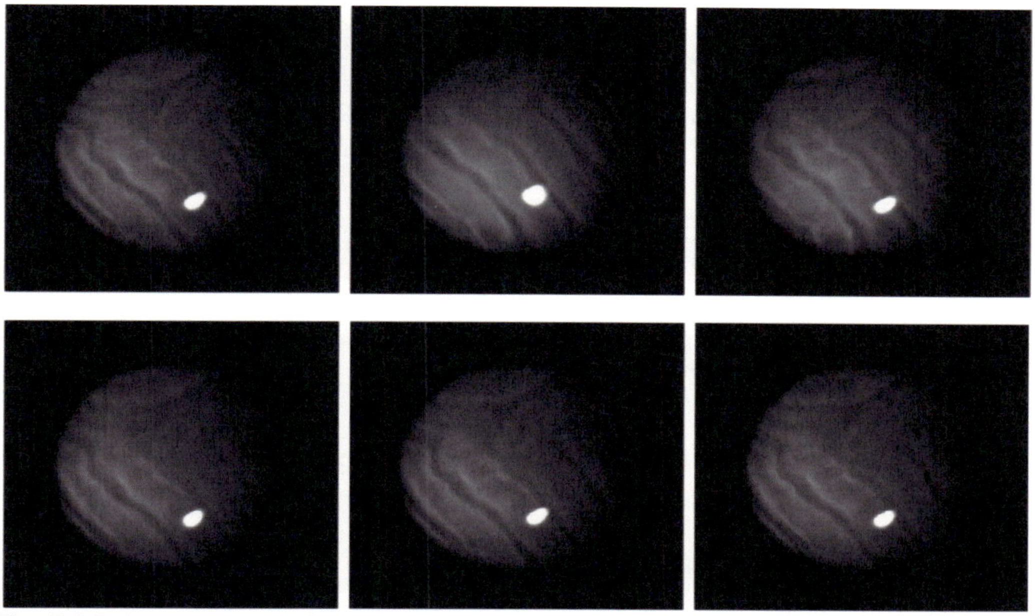

Fig. 7. Motion-compensated microscope image sequence of a mouse kidey (field of view is 200 μm, image sequence was captured by a cooled CCD camera with 37 fps).

4. Motion Compensation by Contact-sensing

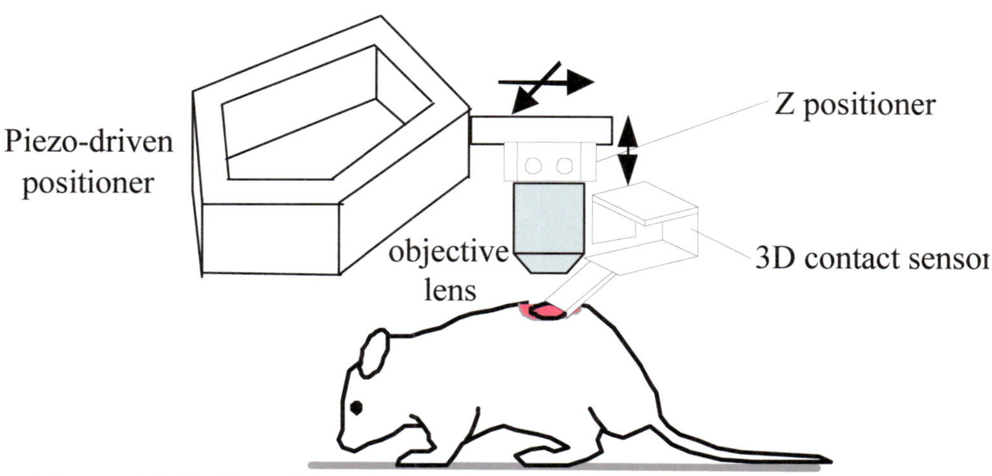

Fig. 8. Image stabilization with contact-sensing

Although the previous solution using a visual servoing system was very successful in removing 2-D motions, there are two weak points. One is that it can only compensate 2-D motion and the other weak point is that the high speed camera system and image processing is too a heavy and expensive solution. This section presents 3-D motion compensation using a developed simple contact-type sensor which is able to detect 3-D motion *in vivo* (Lee et al., 2008b).

4.1 System

The system consists of a developed contact-type sensor and a 3-D motion compensator. *In vivo* motion is estimated by the developed contact-type sensor, and this estimated motion becomes input to the 3-D motion compensator. The 3-D motion compensator moves the objective lens. Fig. 8 illustrates the overall system. The contact-type sensor consists of three thin beams. The end tip of the sensor is placed on a subject while the other end is fixed to the microscope. The body of the tangible sensor is designed to be elastically bent with small force. To keep the contact, the sensor is initially installed on the subject with a pretension. The tip of the sensor ideally moves together with the tissue under it, leading the bending of the sensor body. Strain gauges attached on the sensor body catch this bending to estimate the motion of the tip. It is a three dimensional cantilever. Cantilevered beams are now the most ubiquitous structures in the field of micro-electro-mechanical systems (MEMS) specifically as sensors. The signals from three strain gauges in the sensor body are used to estimate the displacement of its tip. We assume that the change of the signals and the displacement has linear relation as

$$\Delta r = C\Delta s \tag{4}$$

where r and s are a 3 × 1 displacement vector and a 3 × 1 signal vector, respectively. C, a 3 × 3 matrix, describes the relation of the two. It is determined experimentally.
For 3-D motion of the objective lens, we use the same actuator developed in visual servoing system in the previous section. To this mechanism, we have added one DOF actuator for vertical which is a commercial product for fast auto-focusing.

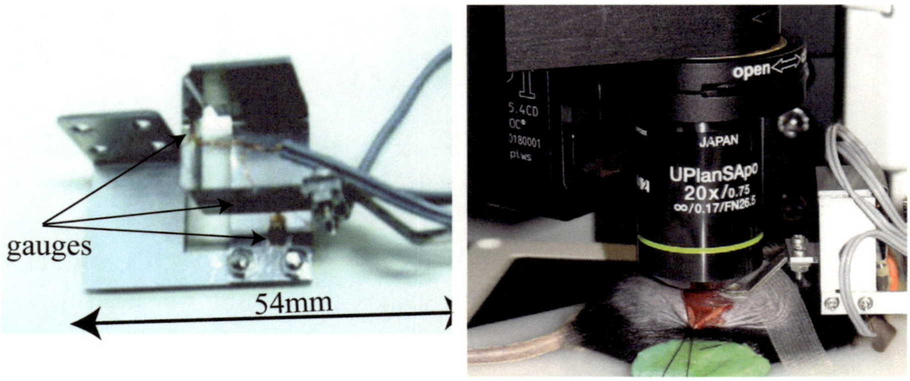

Fig. 9. The developed contact sensor (left) and a contact-sensing based *in vivo* motion compensation (right)

4.2 Result and Discussion

Tests were performed with respect to artificial motion. We used motorized micro-stages for artificial motion generation. Two stages produce horizontal and vertical motion. This motion is estimated by the developed contact-type sensor and compensated by the 3-D compensator. We put a sample tissue of a mouse liver on the micro-stage. Sine wave at the frequency of 1 Hz was generated. The amplitude of the wave is 100 µm both in vertical and horizontal directions. Fig. 10 plots the remaining motion without and with the motion compensation. Fig. 11 and 12 are the image sequences at those times, respectively.

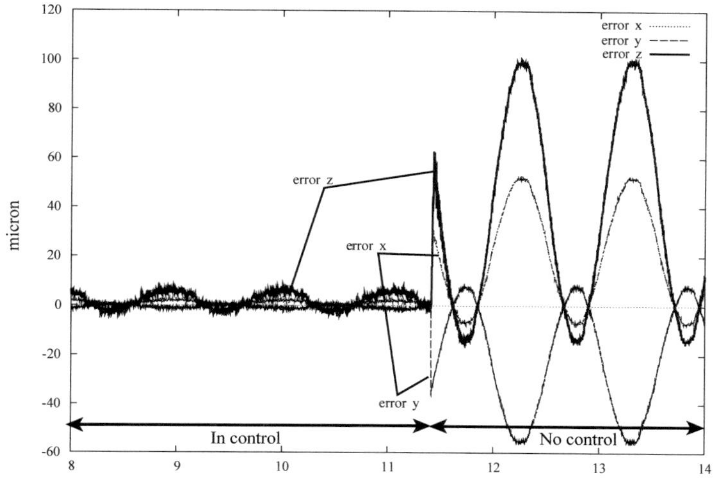

Fig. 10. The remaining motion without (front part) and with (latter part) motion compensation

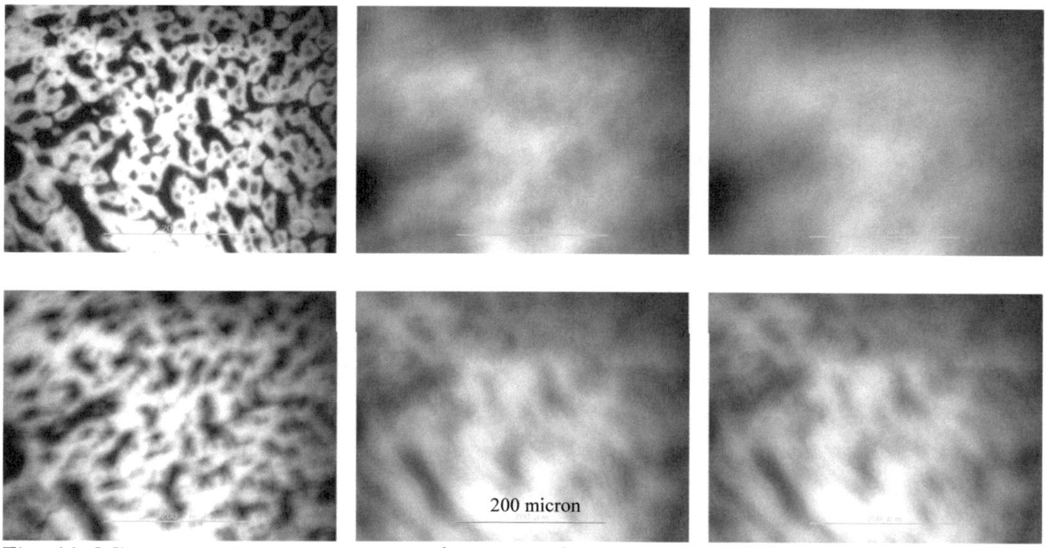

Fig. 11. Microscope image sequence of a mouse liver sample; Artificial sine motion with 1 Hz frequency was generated both in vertical and horizontal directions.

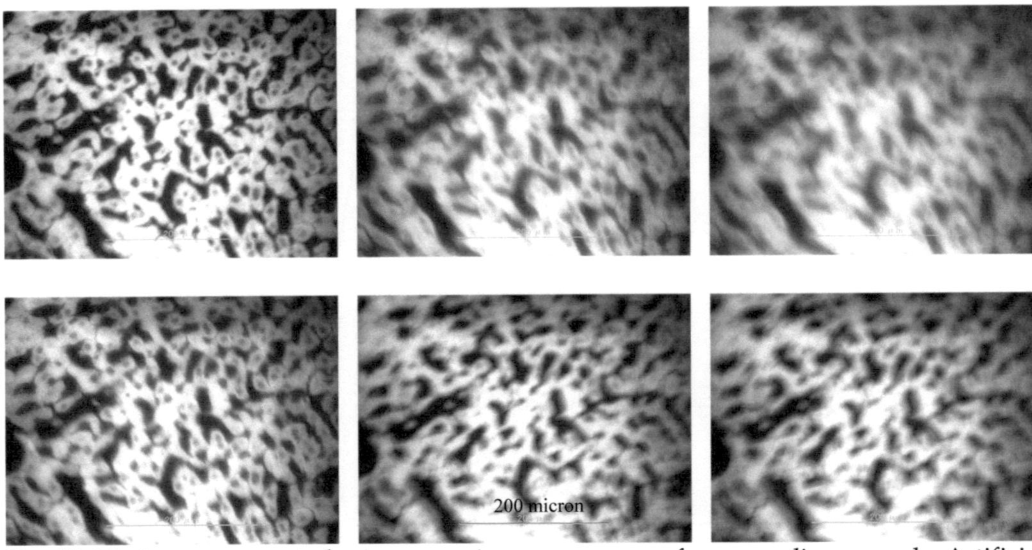

Fig. 12. Motion compensated microscope image sequence of a mouse liver sample; Artificial sine motion with 1 Hz frequency was generated both in vertical and horizontal directions

5. Conclusion

In this chapter, we have introduced microscope image stabilization methods. Two configurations were examined being based on the idea of synchronization. The developed systems mechanically compensate *in vivo* motion by moving objective lens to follow the subject's motion. By removing the relative motion between the objective lens and its subject, we could obtain more stabilized images than ones without compensation. Depending on the sensing methods, we examined two systems: a visual feedback system and a contact-sensing based system. The visual feedback system employs a high-speed camera for detecting fast *in vivo* motion, and the contact-sensing based system uses our developed contact-type sensor which can estimate 3-D motion. In both systems, experimental results show that the image stabilization markedly reduces the effect of the *in vivo* motion, stabilizing the image from the microscope.

The examples introduced in this chapter have shown that the robotic technology can make significant contribution to biological research which is thought to be not directly related to the robotics. Like this, the robotic technology will have great impact on the various fields in the future.

6. References

Bullen, A. (2008), Microscopic imaging techniques for drug discovery, *Nature Reviews Drug Discovery*, Vol. 7, 54-67.

Hutchinson, S. A., Hager, G. D. & Corke, P. I. (1996), A tutorial on visual servo control, *IEEE Transaction on Robotics and Automation*, Vol. 12, No. 5, 651–670.

Lee, S.; Nakamura, Y., Yamane, K., Toujo, T., Takahashi, S., Tanikawa, Y., & Takahashi, H. (2008a), Image stabilization for in vivo microscopy by high speed visual feedback control, *IEEE Transaction on Robotics*, Vol. 24, 45–54.

Lee, S.; Ozaki, T., & Nakamura, Y. (2008b), In vivo microscope image stabilization through 3-d motion compensation using a contact-type sensor, *Proceedings of IEEE Int. Conf. on Intelligent Robots and Systems,* pp. 1192-1197, 2008.

Nakamura, Y.; Kishi, K., & Kawakami, H. (2001), Heartbeat synchronization for robotic cardiac surgery, *Proceedings of IEEE Int. Conf. on Robotics and Automation,* pp. 2014–2019, 2001.